Molecular Systematics and Plant Evolution

The Systematics Association Special Volume Series

Series Editor

Alan Warren
Department of Zoology, The Natural History Museum,
Cromwell Road, London, SW7 5BD, UK.

The Systematics Association provides a forum for discussing systematic problems and integrating new information from genetics, ecology and other specific fields into taxonomic concepts and activities. It has achieved great success since the Association was founded in 1937 by promoting major meetings covering all areas of biology and palaeontology, supporting systematic research and training courses through the award of grants, production of a membership newsletter and publication of review volumes by its publishers Taylor & Francis. Its membership is open to both amateurs and professional scientists in all branches of biology who are entitled to purchase its volumes at a discounted price.

The first of the Systematics Association's publications, *The New Systematics*, edited by its then president Sir Julian Huxley, was a classic work. Over 50 volumes have now been published in the Association's 'Special Volume' series often in rapidly expanding areas of science where a modern synthesis is required. Its *modus operandi* is to encourage leading exponents to organise a symposium with a view to publishing a multi-authored volume in its series based upon the meeting. The Association also publishes volumes that are not linked to meetings in its 'Volume' series.

Anyone wishing to know more about the Systematics Association and its volume series are invited to contact the series editor.

Forthcoming titles in the series:

Homology in Systematics
Edited by R. W. Scotland and R. T. Pennington

Other Systematics Association publications are listed after the index for this volume.

The Systematics Association Special Volume Series 57

Molecular Systematics and Plant Evolution

Edited by Peter M. Hollingsworth,
Richard M. Bateman and
Richard J. Gornall

London and New York

First published 1999
by Taylor & Francis Limited
11 New Fetter Lane, London EC4P 4EE

Simultaneously published in the USA and Canada
by Routledge
29 West 35th Street, New York, NY 10001

Transferred to Digital Printing 2002

Taylor & Francis is an imprint of the Taylor & Francis Group

© 1999 Systematics Association

Every effort has been made to ensure that the advice and
information in this book is true and accurate at the time of
going to press. However, neither the publisher nor the authors
or editors can accept any legal responsibility or liability for
any errors or omissions that may be made. In the case of drug
administration, any medical procedure or the use of technical
equipment mentioned within this book, you are strongly
advised to consult the manufacturer's guidelines.

British Library Cataloguing in Publication Data
A catalogue record for this book is available from the British
Library

Library of Congress Cataloguing in Publication Data are available

ISBN 0–7484–0908–4 (pbk)
       0–7484–0907–6 (hbk)

Contents

Contributors

Abbott, R. J., School of Environmental and Evolutionary Biology, University of St Andrews, St Andrews, Fife, KY16 9TH, UK.

Albert, V. A., The Lewis B. and Dorothy Cullman Program for Molecular Systematics Studies, The New York Botanical Garden, Bronx, New York 10458–5126, USA.

Bailey, J. P., Department of Biology, University of Leicester, Leicester, LE1 7RH, UK.

Bakker, F. T., Plant Science Laboratories, University of Reading, Whiteknights, Reading, RG6 2AS, UK.

Barrett, J. A., Department of Genetics, University of Cambridge, Downing Street, Cambridge, CB2 3EH, UK.

Barrett, S. C. H., Department of Botany, University of Toronto, M5S3B2, Ontario, Canada.

Bateman, R. M., Royal Botanic Garden, 20A Inverleith Row, Edinburgh, EH3 5LR, UK.

Charleston, M. A., Department of Zoology, University of Oxford, South Parks Road, Oxford, OX1 3PS, UK.

Clokie, M., Department of Biology, University of Leicester, Leicester, LE1 7RH, UK.

Comes, H. P., Institut für Spezielle Botanik und Botanischer Garten, Johannes Gutenberg-Universität, Mainz, 55099 Mainz, Germany.

Cronk, Q. C. B., Institute of Cell and Molecular Biology, University of Edinburgh, King's Buildings, Mayfield Road, Edinburgh, EH9 3JH, UK.

Cubas, P., Institut Nacional de Investigaciones Agrarias, Departamento de Mejora Genética y Biotecnologia, Carretera de la Coruna, km 7.5, 28040, Madrid, Spain.

Culham, A., Plant Science Laboratories, University of Reading, Whiteknights, Reading, RG6 2AS, UK.

Davis, J. I., L. H. Bailey Hortorium, 462 Mann Library, Cornell University, Ithaca, NY 14853, USA.

Doyle, J. J., L. H. Bailey Hortorium, 462 Mann Library, Cornell University, Ithaca, NY 14853, USA.

Doyle, J. L., L. H. Bailey Hortorium, 462 Mann Library, Cornell University, Ithaca, NY 14853, USA.

Ennos, R. A., Institute of Ecology and Resource Management, University of Edinburgh, Darwin Building, King's Buildings, Mayfield Road, Edinburgh, EH9 3JU, UK.

Ferris, C., Department of Biology, University of Leicester, University Road, Leicester, LE1 7RH, UK.

Gibby, M., Plant Molecular Biology Laboratory, Department of Botany, The Natural History Museum, Cromwell Road, London, SW7 5BD, UK.

Goremykin, V., Hans-Knöll-Institut für Naturstoff-Forschnung, Postfach 100813, D-07708, Jena, Germany.

Gornall, R. J., Biology Department, University of Leicester, Leicester, LE1 7RH, UK.

Gustafsson, M. H. G., The Lewis B. and Dorothy Cullman Program for Molecular Systematics Studies, The New York Botanical Garden, Bronx, New York 10458–5126, USA.

Hahn, W. J., Center for Environmental Research and Conservation, Columbia University, 1200 Amsterdam Avenue, MC 5557, New York, NY 10027, USA.

Hansmann, S., Institut für Genetik, Technische Universität Braunschweig, Spielmannstr. 7, D-38023, Braunschweig, Germany.

Harris, S. A., Department of Plant Sciences, University of Oxford, South Parks Road, Oxford, OX1 3RB, UK.

Hershkovitz, M. A., Laboratory of Molecular Systematics, MSC MRC 534, Washington, DC 20560, USA.

Hewitt, G. M., School of Biological Sciences, University of East Anglia, Norwich, NR4 7JT, UK.

Hu, X-S., Institute of Ecology and Resource Management, University of Edinburgh, Darwin Building, King's Buildings, Mayfield Road, Edinburgh, EH9 3JU, UK.

King, R. A., Department of Biology, University of Leicester, University Road, Leicester, LE1 7RH, UK.

Kowallik, K. V., Botanisches Institut, Heinrich-Heine-Universität Düsseldorf, Universitätsstr. 1, D-40225, Düsseldorf, Germany.

Langdon, A., Institute of Ecology and Resource Management, University of Edinburgh, Darwin Building, King's Buildings, Mayfield Road, Edinburgh, EH9 3JU, UK.

Martin, W., Institut für Genetik, Technische Universität Braunschweig, Spielmannstr. 7, D-38023, Braunschweig, Germany.

McNicol, J. W., Biomathematics & Statistics Scotland, Scottish Crop Research Institute, Invergowrie, Dundee, DD2 5DA, UK.

Möller, M., Royal Botanic Garden, 20A Inverleith Row, Edinburgh, EH3 5LR, UK.

Morgan-Richards, M., Otago University, Department of Zoology, P.O. Box 56, Dunedin, New Zealand.

Morgante, M., Dipartimento di Produzione Vegetale e Tecnologie Agrarie, Universita di Udine, Via dell Scienze 208, 1–33100 Udine, Italy.

Page, R. D. M., Division of Environmental and Evolutionary Biology, Graham Kerr Building, Institute of Biomedical and Life Sciences, University of Glasgow, Glasgow, G12 8QQ, UK.

Pannell, J. R., Department of Plant Sciences, University of Oxford, South Parks Road, Oxford, OX1 3RB, UK.

Powell, W., DuPont Agricultural Biotech, Suite 200, 1 Innovation Way, Newark, DE 19711, USA.

Provan, J., Department of Cell and Molecular Genetics, Scottish Crop Research Institute, Invergowrie, Dundee, DD2 5DA, UK.

Rumsey, F. J., Plant Molecular Biology Laboratory, Department of Botany, The Natural History Museum, Cromwell Road, London, SW7 5BD, UK.

Sinclair, W. T., Department of Plant Biology, The Scottish Agricultural College, Auchincruive, AYR, UK.

Soranzo, N., Department of Cell and Molecular Genetics, Scottish Crop Research Institute, Invergowrie, Dundee, DD2 5DA, UK.

Stace, C. A., Department of Biology, University of Leicester, Leicester, LE1 7RH, UK.

Stoebe, B., Botanisches Institut, Heinrich-Heine-Universität Düsseldorf, Universitätsstr. 1, D-40225, Düsseldorf, Germany.

Vogel, J. C., Plant Molecular Biology Laboratory, Department of Botany, The Natural History Museum, Cromwell Road, London, SW7 5BD, UK.

Wilson, N. J., Department of Cell and Molecular Genetics, Scottish Crop Research Institute, Invergowrie, Dundee, DD2 5DA, UK.

Wolff, K., Department of Agricultural and Environmental Science, University of Newcastle upon Tyne, Newcastle, NE1 7RU, UK.

Zimmer, E. A., Laboratory of Molecular Systematics, MSC MRC 534, Washington, DC 20560, USA.

Preface

The estimation of relationships within the hierarchy of individuals, populations and species is of fundamental importance to the understanding of evolutionary patterns and processes. The amounts, dynamics and spatial partitioning of genetic variability largely determine the evolutionary potential of a species. At higher taxonomic levels understanding the relationships among species can provide a retrospective window on how they evolved.

The study of relationships among plants has been revolutionised in recent years by advances in molecular genetic techniques. Topics that have been addressed include the discrimination of the environmental and genetic components of variability, finding markers to characterise individuals, quantifying breeding systems, assessing levels and patterns of gene flow, studying hybridisation, establishing phylogenetic relationships, testing biogeographical hypotheses, examining the relationship between genotype and phenotype, and comparing morphological and molecular evolution.

The papers in this volume stem from the conference on *Advances in Plant Molecular Systematics* held at the University of Glasgow (UK) in August 1997. The aim of the meeting was to provide a forum for the exchange of ideas among plant evolutionary biologists and systematists. Presentations covered a broad spectrum of the taxonomic hierarchy and included discussions of population genetics, phylogenetics, developmental genetics and molecular evolution. While these disciplines have their own terminologies, controversies and foci, it is clear that for a synthetic understanding of systematics and evolutionary biology the incorporation of concepts and data from all of these fields is desirable. The aim of this volume is to bring together papers from the conference in order to contribute to this synthesis.

The book progresses upwards through the demographic–taxonomic hierarchy, drawing heavily on empirical data to address specific conceptual issues. The opening chapter by Richard Ennos *et al.* provides a review of the population genetic theory of organelle markers and discusses the potential applications and limitations of such markers in elucidating the history and ecology of plant populations. This is followed by a discussion by Colin Ferris *et al.* (Chapter 2) on the use of organelle markers to trace the plant migrations that occurred during post-glacial recolonisation. Difficulties in detecting suitable levels of variability for infra-specific studies are emphasised in these first two chapters and this problem is addressed by Jim Provan *et al.* (Chapter 3) who discuss the potential of simple sequence repeats in the chloroplast genome as a source of more variable markers. Organelle markers are often the molecule of choice for phylogenetic and phylogeographic studies, and can be of great value to population geneticists. However, organelle DNAs differ from their nuclear counterparts in modes of inheritance and tempos of evolution, leading to different expectations of population genetic structure. These three chapters provide an accessible summary of the causes of population structure in the context of organelle DNA and discuss the value (and difficulties) of measuring this structure. In addition to infra-specific studies, the issues raised here are of interest above the level of the species. Studies of plant hybridisation can benefit considerably from data from non-recombinant, uniparentally inherited markers and, in addition, an understanding of the behaviour of organelle markers at the infra-specific level can aid the interpretation of inter-specific phylogenetic data among closely related species.

Shifting from the organelle genomes to the nuclear genome, the use of arbitrary fingerprinting techniques for examining population structure and species differentiation is discussed by Kirsten Wolff and Mary Morgan-Richards (Chapter 4), focusing on case studies of *Alkanna orientalis* and *Plantago major s.l.* The *Alkanna* study highlights the value of RAPDs (randomly amplified polymorphic DNAs) for detecting fine-scale structure within a single population, whereas in *Plantago* RAPDs and inter-SSRs (simple sequence repeats) have been used to investigate population structure, hybridisation and species differentiation within the complex. Wolff and Morgan-Richards argue that despite the criticisms often levelled at RAPDs (e.g. Harris in Chapter 11 of this volume), if used with sufficient caution they can provide easily accessible and biologically meaningful data.

In Chapter 5 Spencer Barrett and John Pannell use a combination of empirical data and theoretical models to highlight the importance of metapopulation dynamics in the evolution of plant mating-systems. Barrett and Pannell argue that there has been an over-emphasis on the local population as the study unit for investigations into the evolution of plant reproductive systems. They suggest that an incorporation of landscape level approaches, which take into account recurrent local extinction (extirpation) and recolonisation, are more appropriate for studies of those plant species whose populations tend to be ephemeral. Barrett and Pannell also point out that a greater understanding of the role of metapopulation dynamics on mating-system evolution is important from an inter-specific perspective, as breeding system transitions can lead to reproductive isolation and speciation. They suggest that 'studies at the metapopulation level may act as a conceptual bridge in the genealogical hierarchy linking micro- and macro-evolutionary enquiry' (p. 93).

Johannes Vogel *et al.* (Chapter 6) investigate the genetic evidence for multiple origins of polyploid homosporous pteridophytes. They argue that some of the published evidence is inconclusive, contending that a detailed understanding of the cytological relationships and genetic composition of potential progenitor species is essential before distinguishing between single and multiple origins. In addition they draw attention to potential pitfalls when deducing fixed heterozygosity, particularly stressing problems associated with inadequate sample sizes and the confounding effects of alloploid-like behaviour in some autoploids.

In Chapter 7 Richard Gornall reviews the available molecular data on agamo-spermous species to summarise our current knowledge of their population genetic structure. He finds that even narrowly defined microspecies can be multiclonal and concludes that most of the variability can be attributed to the sexual ancestry of the taxa concerned, with only a limited contribution being made by asexual processes.

The following two chapters are aimed explicitly at the interface of population genetics and phylogenetics. In Chapter 8 Jerry Davis addresses the conceptual problem of extending Hennigian cladistics to the infra-specific level when attempting to apply the concept of monophyly to sexually reproducing organisms whose relationships are reticulate (tokogenetic). He argues that attempts to redefine monophyly in order to do this are misguided. Davis concludes that 'if a successful synthesis of phylogenetics and biosystematics is to emerge, practitioners will need to take a full account of the distinction between reticulate and hierarchic descent, and acknowledge the key role of population systems as basal phylogenetic elements' (p. 167).

In Chapter 9 Peter Comes and Richard Abbott present the use of both population genetic and phylogenetic approaches to study reticulate evolution in Mediterranean *Senecio* species. They demonstrate the importance of both 'bottom up' and 'top down' approaches in which population processes are investigated in the context of well characterised species relationships. They found that a phylogenetic approach (top down) permitted an estimate of the degree of reticulation in inter-specific relationships. Comes and Abbott show how introgressive hybridisation can lead to taxonomically disparate species being erroneously grouped together due to both chloroplast *and* ITS capture but in the absence of marked introgression of morphological features or RAPD markers. These processes were studied in more detail at the population level (bottom up) and revealed a striking example of the differential and pronounced introgression of cytoplasmic rather than nuclear markers in interactions between two species. This has led to a greater appreciation of the effects species may have on each other when they come into contact or when they are at an early stage of differentiation, and clearly demonstrates the value of using more than one gene/marker type in molecular systematics investigations.

In Chapter 10 Clive Stace and John Bailey present a review of the applications of genomic *in situ* hybridisation (GISH) as a tool in taxonomic and evolutionary studies. They summarise four main topics: chromosome disposition, genome identification, recognition of parts of genomes, and studies of meiosis. They discuss the difficulties of predicting the lowest taxonomic level at which GISH is likely to provide discrimination, and suggest that it is often at (but frequently higher than) the species level. They also highlight the difference in probe binding between species with large amounts of evenly dispersed repetitive DNA and those where the repetitive DNA is either less well dispersed or less abundant. Although a lack of well dispersed repetitive DNA can lead to only partial labelling of genomes when complete labelling is expected, they note that the results are nevertheless interpretable.

Chapter 11 is a review by Stephen Harris of the principles, applications and limitations of RAPDs in systematics. Since the invention of the technique in 1990 there has been an explosion of studies using RAPDs as a source of markers in biodiversity studies. In addition there are now several related PCR-based arbitrary fingerprinting techniques that also provide anonymous, dominant DNA markers. Harris discusses the commonly reported reproducibility problems with RAPDs and the difficulties of cross-laboratory comparisons, as well as issues related to appropriate methods of data analysis. While some of the more recent and sophisticated arbitrary fingerprinting approaches such as AFLPs (amplified fragment length polymorphisms) are reported to be less susceptible to experimental artifacts, many of the problems of data analysis remain. While acknowledging that useful results have been obtained, Harris concludes that the promise of RAPDs as a simple, efficient and reliable source of DNA data in systematics has yet to be fulfilled.

In Chapter 12 Jeff and Jane Doyle discuss the role of nuclear protein-coding genes in phylogeny reconstruction. Although great advances in our knowledge of the relationships among plants have been made using chloroplast DNA and nuclear ribosomal DNA, the nuclear genome as a whole has remained largely untapped as a source of characters. The large size of the nuclear genome, the fact that different regions and classes of DNA evolve at markedly different rates, and that recombination and independent assortment provide numerous independent loci for estimating historical relationships, all suggest that analyses of nuclear sequence data will become more and more common. However, the use of low copy number DNA sequences also presents technical challenges. Low copy number sequences can be difficult to amplify and there can be considerable problems in disentangling orthologous and paralogous copies of genes. We still have much to learn about the role of concerted evolution in inter-locus homogenisation and, of course, heterozygosity leads to problems not only in reading sequences, but also in placing an organism uniquely at the end of a single branch in a phylogenetic analysis.

Mike Charleston and Rod Page (Chapter 13) consider the problems of evaluating whether data is sufficiently 'tree-like' to be analysed using tree construction algorithms and suggest spectral analysis as a first step in the evaluation of phylogenetic signal. They provide an introduction to the methodology and a brief summary of its applications.

One of the most commonly sequenced regions in plant systematics, ribosomal DNA, is the focus of a detailed review by Mark Hershkovitz et al. (Chapter 14). Sequence data from ribosomal DNA have been applied to questions ranging from infra-specific variability to the evolution of green plants. The application of different regions of rDNA at different taxonomic levels is discussed and evaluated, and future prospects for research are outlined. Hershkovitz et al. argue that although differential levels of homogenisation and substitution biases can impede phylogenetic reconstruction, when they are properly accounted for, phylogenetic results from rDNA will be as robust as those from other classes of DNA. Furthermore, they stress that the evolutionary processes behind patterns of rDNA sequence variability represent fascinating evolutionary phenomena that merit investigation in their own right.

In Chapter 15 Bettina Stoebe et al. describe an investigation into chloroplast gene evolution among those plastid genomes that have been completely sequenced. They document the presence/absence of 254 protein-coding genes among the 12 fully sequenced plastid genomes. For the 46 proteins common to metaphytes and algae, a phylogeny has been reconstructed (rooted with homologous proteins from a cyanobacterial genome) using 11,521 amino acid positions per species. Using this phylogenetic tree, the fates of individual genes across lineages have been plotted. Interestingly, parallel loss of genes outnumber unique losses, with 68 genes having a single loss and 122 genes undergoing parallel losses in independent lineages. Furthermore, Stoebe *et al.* show that of those genes that have been lost on more than one occasion, 62 have undergone three parallel losses and 15 have been lost four times. In this context parallel events are five times commoner than unique events. In addition they document the presence of 47 functional nuclear genes that appear to have been transferred from the chloroplast to the nucleus, and present a 'road-map' showing the phylogenetic distribution of these events.

The following two chapters describe case studies using sequence data to investigate phylogenetic relationships and the evolutionary biology of species in the Geraniaceae and Gesneriaceae. In Chapter 16 Freek Bakker *et al.* present a case study of *Pelargonium* using chloroplast DNA sequences to reconstruct a phylogeny as a vehicle to interpret growth-form diversification and biogeography. They found two major clades, the larger of which has diversified in the South African Cape region in response to changes in the pattern of rainfall and various edaphic factors. The smaller clade shows much less variation in habit but has spectacular Africa/Australia disjunctions reflecting dispersal.

Michael Möller *et al.* (Chapter 17) use sequence data from rDNA (ITS) and *trn*L, along with *Gcyc*, the putative homologue from Gesneriaceae of the nuclear developmental gene *cycloidea* (which regulates floral symmetry in *Antirrhinum*), to examine phylogenetic relationships among species in the Gesneriaceae. The *trn*L intron and *trn*L-*trn*F spacer are relatively conserved and suitable for resolving phylogenetic relationships at the level of the genus. In contrast, ITS appears to be more suitable for resolving relationships among species and evolves five times faster than *trn*L. *Gcyc* shows intermediate substitution rates (three times greater than the *trn*L intron/spacer). However, Möller *et al.* note that at very low taxonomic levels sequence divergence in *Gcyc* is higher than in ITS. Their explanation for this apparent discrepancy is that detection of genetic changes in multicopy DNA is hindered by the time taken for them to spread through the rDNA repeat arrays to an abundance sufficient to be detected by PCR sequencing. In the putative single copy gene *Gcyc*, however, mutations can be detected as soon as they have occurred. At higher taxonomic levels, functional constraints on *Gcyc* act as an anchor on rates of divergence, whereas ITS continues to accumulate mutations and diverge. All three sequences give similar phylogenetic trees on parsimony analysis. Three cases of transitions from zygomorphy to actinomorphy were found, with one of them subsequently reverting to zygomorphy. *Gcyc* appears to be intact and functional even in actinomorphic flowers.

In Chapter 18 Mats Gustafsson and Victor Albert discuss how advances in molecular systematics and molecular developmental genetics can enhance our understanding of the evolution of plant form. Using an *rbc*L phylogeny of the angiosperms as a framework, they show that the evolution of inferior ovaries has occurred on many occasions. The problems of assigning the homology of the inferior ovary are discussed, and the role that patterns of gene expression may have in determining this are explored. Gustafsson and Albert conclude that a synthesis of phylogenetic, positional, morphological and developmental genetic data is required for a full understanding of complicated questions of homology.

Lastly, Richard Bateman (Chapter 19) attempts to synthesise a disparate range of concepts (and establish links to several of the later chapters of the volume) in order to reappraise approaches to the definition and interpretation of evolutionary radiations. He offers a pattern-based definition of radiations, rejecting *a priori* assertions of adaptations and key innovations, and notes the prevalence and significance of the punctuated equilibrium pattern of evolution. His essay concludes by suggesting a protocol for interpreting radiations; it requires the study of morphology, a nuclear gene and a plastid gene, the generation and careful comparison of cladograms from each matrix, further analysis of a combined matrix, and the subsequent mapping of non-discrete morphological characters and extrinsic (broadly ecological) characters across the resulting trees.

We conclude by thanking the authors for their contributions and the many reviewers who provided helpful comments on drafts of the chapters. We also thank the Systematics Association (the major sponsors of the Glasgow conference), and PMH acknowledges the support of the NERC Taxonomy Initiative at the University of Glasgow (grant GST/02/0833).

Pete Hollingsworth[1], Richard Bateman[1] and Richard Gornall[2]

[1] Royal Botanic Garden Edinburgh, UK
[2] Department of Biology, University of Leicester, UK

Chapter 1

Using organelle markers to elucidate the history, ecology and evolution of plant populations

R. A. Ennos, W. T. Sinclair, X-S. Hu and A. Langdon

ABSTRACT

Recent advances in DNA analysis have made it possible to identify a variety of forms of genetic variation in the chloroplast and mitochondrial genomes of plants and to analyze this variation at the population level. A review of population genetic theory for these uniparentally inherited non-recombining genomes indicates that they are likely to show low levels of intraspecific sequence variation in coding regions, and that targeting of alternative forms of variation (duplications, rearrangements) or non-coding regions of the genome with higher mutation rates may be necessary to detect useful genetic markers. The properties of the markers set limits to their utility in addressing population level and phylogenetic questions. By developing and applying appropriate population genetic theory it is possible to exploit the unique properties of organelle genomes to address previously intractable problems in plant ecology and evolution. The use of organelle markers to study postglacial history, seed and pollen flow, and hybridization in plants are reviewed. A major limitation of genetic parameters estimated from analysis of organelle genomes is that they are based on unreplicated measurements. Each organelle genome behaves as if it were a single gene and provides only one piece of data. Population genetic analysis of organelle markers is likely to be most fruitful when integrated with studies of traditional nuclear markers.

1.1 Introduction

Advances in molecular biology and their application at the population level have recently provided us with the opportunity to explore the extent and distribution of genetic variation at a wide range of loci within plant species. When these data are interpreted in the light of population genetic theory it is often possible to make inferences about the evolutionary and ecological processes which have moulded the observed genetic structure (Avise, 1994). Most of the existing data on plant population genetic structure comprise studies of nuclear markers, predominantly co-dominant isozyme variants with low mutation rates. The population genetic behaviour of these biparentally inherited genetic markers is well understood. This has enabled the markers to be used as tools to make quantitative measurements of such attributes as mating systems (Brown, 1989), clonal spread (Ellstrand and Roose, 1987), and migration rates within and among plant populations (Ellstrand

In *Molecular systematics and plant evolution* (1999) (eds P. M. Hollingsworth, R. M. Bateman and R. J. Gornall), Taylor & Francis, London, pp. 1–19.

and Marshall, 1985; Meagher, 1986). They have also been employed to explore aspects of the population history of species (Kinloch *et al.*, 1986; Parks *et al.*, 1994; Konnert and Bergmann, 1995).

Within the last few years more sophisticated techniques of DNA analysis have made it possible to extend our knowledge of population genetic structure not only by using novel nuclear DNA markers (RAPD, AFLP, microsatellite), but also by exploiting our ability to detect genetic variation in the two organelle genomes of plants, the chloroplast and the mitochondrial genomes (Palmer, 1992; Soltis *et al.*, 1992). These genomes are sufficiently different from their nuclear counterpart in mode of inheritance and evolution that it has been necessary to develop further population genetic theory to predict the levels of variation and behaviour of organelle variants in plant populations (Birky *et al.*, 1989; Petit *et al.*, 1993; Ennos, 1994; Hu and Ennos, 1997). Theory also highlights the limitations of using such markers as well as suggesting those areas in which data from organelle markers may yield novel information not available from traditional studies of nuclear genetic markers.

The objective of this chapter is to review the development and application of recent theory concerned with the population genetic behaviour of organelle markers in plants. We will investigate first the levels of organelle marker diversity anticipated in plant populations, and compare these with levels of variation for nuclear markers. We will then consider how the unique properties of organelle markers can be exploited to yield novel information on migration, population history and hybridization in plants. We hope to identify the circumstances under which the additional effort required to analyze intraspecific variation in organelle markers can be justified, and to indicate the areas in which these markers can give unique insights into population processes.

1.2 Selectively neutral genetic variation in organelle genomes

The presence of adequate levels of selectively neutral intraspecific genetic variation is a prerequisite for applying the approaches of molecular ecology within any plant species. In order to assess the potential for plant organelle DNA to yield useful genetic marker data it is therefore necessary to compare the levels of genetic variation expected to be maintained in the chloroplast and mitochondrial genomes of plants with those maintained in the nuclear genome. To facilitate this a summary of the important differences in the structure and behaviour of plant genomes is given (Table 1.1).

The nuclear genome of higher plants comprises a diploid complement of freely reassorting linear chromosomes that are biparentally inherited. Recombination occurs among genes on homologous chromosomes. In contrast both the chloroplast and mitochondrial genomes of plants are circular DNA molecules that are predominantly uniparentally inherited (Harris and Ingram, 1991; Birky, 1995). In the absence of significant heteroplasmy (presence of more than one organelle genotype per cell) the organelle genomes can be treated as effectively haploid (Birky *et al.*, 1989). Over the time scale relevant to population genetic theory there is little or no recombination among the genes on the organelle chromosomes, though this

Table 1.1 Important properties of the three plant genomes affecting their population genetic behaviour.

Attribute	Nuclear genome	Chloroplast genome	Mitochondrial genome
Structure	Linear chromosomes	Circular chromosome	Internally recombining circular chromosome
Ploidy level	At least diploid (higher plants)	Haploid	Haploid
Inheritance	Biparental	Uniparental Maternal (Angiosperms) Paternal (Gymnosperms)	Uniparental Maternal
Recombination	Present	Absent	Absent

does not preclude the occasional generation of new structural variants by intragenomic recombination which is a feature of long term mtDNA evolution (Atlan and Couvet, 1993, see later). As a result each organelle genome may be regarded, for the purposes of population genetic analysis, as if it were a single haploid gene.

It is important to note that the genomes have separate mechanisms of replication, and hence the potential for accumulation of mutations at different rates during this process. The gene organisation is also radically different among the genomes. For instance the chloroplast genome is relatively conservative in structure across taxa with restricted gene duplication (Clegg *et al*, 1986). In contrast the mitochondrial genome has numerous internal repeats and is capable of substantial intrachromosomal recombination to generate subcircles of the master chromosome (Atlan and Couvet, 1993). Structural differences between mitochondrial genomes from closely related taxa may be very substantial as a consequence of complex rearrangements (Palmer, 1992). Thus the modes of evolution and the forms of intraspecific variation detectable are likely to be very different among the three plant genomes. Given the properties outlined above there are at least three reasons for anticipating that the levels of intraspecific variation for selectively neutral markers will be lower for organelle than for nuclear genomes in plants. These are outlined below.

1.2.1 Levels of variation at drift/mutation equilibrium

For selectively neutral variation, equilibrium levels of genetic diversity found in a population of size N are governed by a balance between mutation (μ per generation) increasing diversity, and genetic drift leading to its loss. Equilibrium levels of gene diversity for the haploid and diploid cases are given in Table 1.2

Table 1.2 Levels of gene diversity H_E maintained at mutation/drift equilibrium for nuclear and organelle markers in ideal populations of size N for genes with mutation rate μ per generation.

	Nuclear marker (Diploid)	Organelle marker (Haploid)
Gene diversity at equilibrium, H_E	$4N\mu/(4N\mu + 1)$	$2N\mu/(2N\mu + 1)$

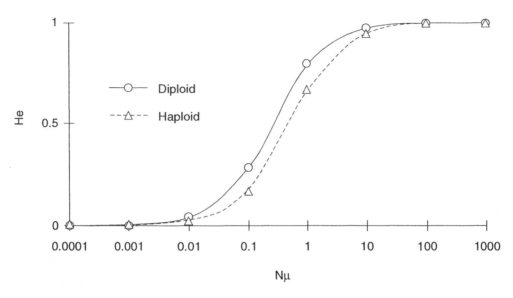

Figure 1.1 Relationship between gene diversity H_E maintained at a locus in mutation/drift equilibrium, and the product of population size N and mutation rate μ per generation. Graphs are shown for loci in both diploid and haploid genomes.

and the variation in equilibrium gene diversity with changes in $N\mu$ is illustrated in Fig. 1.1. It is clear that gene diversity rises with both population size and mutation rate, and that for an equivalent population size and mutation rate gene diversity is always lower for haploid than for diploid markers. Thus because of their haploid nature, lower levels of gene diversity are expected in organelle genomes than in diploid nuclear genomes for genes with the same rate of neutral mutation.

1.2.2 Mutation rates in plant genomes

Although we have few direct measures of mutation rates in the three plant genomes, it is possible to use data on synonymous substitution rate in coding sequences of equivalent genes in the three genomes to obtain indirect measures of these values (Wolfe *et al.*, 1987). Under the neutral theory, the rate of synonymous substitution is directly proportional only to the rate of mutation. Synonymous substitution rates differ widely among the three plant genomes. Inferred mutation rates are 3 to 5 times higher in the chloroplast than in the plant mitochondrial genome, while the inferred mutation rate in the chloroplast genome is itself only one half to one third that found in the plant nuclear genome (Wolfe *et al.*, 1987). It is important to note that synonymous mutation rates are roughly equivalent in the nuclear genomes of plants and animals, but are 5 to 10 times higher in animal mtDNA (Wolfe *et al.*, 1989). The much lower synonymous mutation rate within plant organelle genes compared with their animal counterpart has important consequences both for the levels of gene diversity anticipated, and for the interpretation of phylogenetic information derived from them.

As a consequence of these reduced rates of mutation for the organelle genes, we anticipate that lower levels of genetic variation in sequence will be maintained at equilibrium for the mitochondrial than for the chloroplast genome, and both will be lower than for gene diversity at loci in the nuclear genome (Fig. 1.1). Note that these conclusions only apply to the types of mutations used in this analysis, synonymous substitutions in coding sequences. Relative mutation rates among genomes may be very different for other forms of mutation, such as structural rearrangements or duplications, due to their very different modes of evolution and differences in selective constraints (Atlan and Couvet, 1993). To conclude, these results imply that levels of sequence variation for coding regions of chloroplast and mitochondrial genes in plants may be too low to provide useful intraspecific genetic markers for population genetic analysis, and that other regions of these genomes and/or other forms of variation may need to be targeted.

1.2.3 Reduction in organelle gene diversity due to 'selective sweeps'

The non-recombining nature of the organelle genomes compared with the more freely recombining nature of the nuclear genome has important implications for their behaviour in populations and the maintenance of genetic variation for these genomes (Maruyama and Birky, 1991). If an advantageous mutation occurs in the nuclear genome its frequency will rapidly increase over the generations. However, due to recombination it may reach fixation with very little effect on the frequency and diversity of other genes, except where they are extremely tightly linked or where the species is highly inbreeding. In contrast when an advantageous mutation in a non-recombining organelle genome sweeps to fixation it can carry all other alleles on that chromosome to fixation too. The result is a total loss of genetic variation in the organelle genome due to the 'selective sweep' (Maruyama and Birky, 1991).

The impact of selective sweeps in reducing organelle gene diversity is dependent upon the rate at which new advantageous mutations occur in these genomes. There is certainly ample evidence that mutations in organelle genomes can have large selective effects. Examples are chloroplast mutations for herbicide resistance (Mazur and Falco, 1989) and mitochondrial mutations affecting male fertility in a wide range of plants (Hanson, 1991). The effect of selective sweeps cannot therefore be ignored. However, we have too little data on the rate at which advantageous mutations arise to estimate the overall impact of selective sweeps. What we can say is that if they occur the result will be a lowering of average levels of gene diversity, and an increase in the variance of gene diversity for organelle compared with nuclear genes (Maruyama and Birky, 1991).

1.2.4 Introduction of diversity through 'organelle capture'

This discussion of organelle DNA diversity has so far considered mutation as the only means for generating intraspecific variation. However, it should be remembered that organelle diversity may potentially enter a species as the result of genetic exchange with a related taxon possessing a different organelle genome. This phenomenon has been termed organelle gene flow or genome capture and describes the apparent transfer of an organelle genome from one taxon to another when viewed on an evolutionary time scale (Rieseberg and Soltis 1991).

The mechanism responsible for organelle genome transfer between taxa depends on the mode of inheritance of the genome. For maternally inherited chloroplast and mitochondrial genomes transfer must involve repeated fertilization of the donor species by pollen from the recipient taxon. Over a number of generations of back-crossing this 'pollen swamping' (Potts and Read, 1988) will lead to replacement of the donor's nuclear genome and the creation of some plants with the organelle genomes of the donor and the nuclear genome of the recipient taxon. A population is generated which shares a common nuclear gene pool but is polymorphic for organelle genomes derived from the two parent taxa.

In situations where the organelle genome is paternally inherited (as is the case for the chloroplast genome of conifers (Neale and Sederoff, 1989)) fertilization by a donor taxon will create a plant which is hybrid for the nuclear genome, but possesses the organelle DNA of the donor taxon. If this novel organelle genome is to be integrated into the recipient species, these nuclear hybrids must contribute pollen in backcrosses with the recipient parent. Over a number of generations in which increasingly backcrossed hybrids act as pollen parents, plants will be produced which possess the recipient taxon's nuclear genomes but the donor taxon's organelle DNA. Again the creation of polymorphism for organelle genomes has occurred due to indirect transfer from a related species rather than through mutation.

The important implication is that within taxa whose origins involve interspecific hybridization or which participate in interspecific hybridization, it is likely that organelle gene diversity will be higher than in taxa where barriers to hybridization are very strict, and processes leading to interspecific organelle transfer cannot take place. Thus elevated levels of paternally inherited cpDNA diversity have been found in *Pinus* and *Picea* species where interfertile taxa meet (Wagner *et al.*, 1987; Szmidt *et al.*, 1988; Sutton *et al.*, 1994), and in *Pinus densata* which is believed to have had a hybrid origin (Wang and Szmidt, 1994). For angiosperms numerous examples of introgression of maternally inherited cpDNA in the absence of obvious nuclear introgression have been recorded (Rieseberg and Soltis, 1991; Rieseberg and Brunsfeld, 1992).

1.2.5 Conclusions and implications

This brief consideration of theory indicates that organelle genomes, because of their haploid status, their reduced rate of sequence mutation, and lack of recombination will generally show significantly lower gene sequence diversity than will nuclear genes. Moreover the variance in organelle gene diversity is likely to be large. Some species may show little or no diversity as a consequence of recent selective sweeps, while values may be much higher in species which are at mutation/drift equilibrium. Some elevation of diversity levels is expected where interspecific hybridization has occurred in the past.

These results have two implications for the use of organelle markers in studies of molecular ecology. The first is that organelle genomes are likely to provide a good source of species-specific markers simply because of the low levels of intraspecific variation in gene sequence that are predicted by theory. The second implication is that in order to detect useful levels of intraspecific organelle variation, highly mutable portions of the organelle genomes may have to be targeted.

For the chloroplast useful levels of intraspecific variation have been detected by analyzing variation in non-coding regions of the genome located outside the inverted repeats (Taberlet *et al.*, 1991; Demesure *et al.*, 1995). Compared with coding regions, these non-coding regions generally show a rather modest increase (1.5×) in the frequency of base substitution polymorphisms, but the frequency of polymorphism for small insertion/deletions is often significantly enhanced (Gielly and Taberlet, 1994).

A second useable type of intraspecific chloroplast variation takes the form of differences in the copy number of tandemly duplicated genes. Variation is generated here by elevated mutation rates giving different numbers of tandem copies of the gene (Dong and Wagner, 1993; Hipkins *et al.*, 1995). The final approach to the detection of chloroplast DNA variation involves analysis of chloroplast simple sequence repeats (SSRs or microsatellites) (Powell *et al.*, 1995; Provan *et al.*, 1999 – this volume). In these regions mutation rates are expected to be especially high, probably due to slippage during replication.

For the mitochondrial genome, attempts to detect intraspecific differences in coding sequence have often proved unfruitful (Wang *et al.*, 1996). However, analysis of variable regions within known genes that can be amplified by PCR may yield useful markers (Dawson *et al.*, 1995; Latta and Mitton, 1997). An alternative strategy is to exploit the fact that, because of the presence of numerous repeated regions within the genome, mutations involving structural rearrangements appear to be frequent (Atlan and Couvet, 1993). Analyses of intraspecific structural rearrangements in the mitochondrial genome using traditional RFLP techniques have been successful in detecting mitochondrial markers (Dong and Wagner, 1993; Samitou-Laprade *et al.*, 1993; Strauss *et al.*, 1993; Hipkins *et al.*, 1995; Sinclair *et al.*, 1998).

Where traditional population genetic methods based on allele frequencies are to be used in the analysis of intraspecific organelle variation, all of the above forms of organelle markers provide suitable data. As long as differences can be detected, the exact nature of these differences does not matter for the purposes of population genetic analysis. The only caveat to this is that care is required where mutation rates generating the variation are so high that they are of the same magnitude as migration rate (Slatkin, 1995). This may be the case for hypervariable chloroplast SSR regions (Powell *et al.*, 1995). Here the distribution of variation within and among populations may be governed more by mutation than by migration (but see Provan *et al.* (1999 – this volume) for a contrasting view). Under these circumstances analysis of genetic structure cannot be used to provide reliable insights into migration rates or genetic relatedness among populations. This result is unaffected by the method of analysis used to calculate population differentiation.

In situations where a phylogenetic approach is to be used to analyse data, only the information from characterisation of base substitutions and insertion/deletions within non-coding regions of chloroplast DNA is suitable. This is because the rate of base substitution is low enough that undetectable multiple hits are unlikely to occur, and the rate of generation of indels is not so frequent that sequence alignment is impossible. In each case the data contains information about the relatedness of the variants.

In contrast all of the other mutational processes leading to variation, generating differences in chloroplast gene repeat numbers, chloroplast SSRs and mitochondrial structure, will either generate homoplasy (same mutation generated by independent events) or variants whose evolutionary relationships cannot be readily inferred (Palmer, 1992; Demesure *et al.*, 1996). These data will therefore be unsuitable for phylogenetic analysis.

These factors impose limits on the use of plant organelle genomes for intraspecific phylogenetic analysis. The only suitable data available are those on substitutions and insertion/deletions events chiefly in non-coding regions of the chloroplast or possibly mitochondrial genome. Even in these regions of elevated mutation, the rate of substitution/insertion/deletion is too slow for recent evolutionary events (occurring over time scales of tens of thousands of years) to be investigated. This limitation needs to be more widely recognised. Phylogenetic trees based on analyses of base substitutions and indels in cpDNA are not appropriate for providing information about postglacial history spanning only the last 12,000 years (e.g. Demesure *et al.*, 1996). The events leading to the observed cpDNA differences used in the phylogeny predate the historical events being studied.

This constraint is relaxed in animal populations where the substitution rates in mitochondrial genes are an order of magnitude higher than those in the chloroplast of plants. As a consequence animal mitochondrial genes provide ideal material upon which to found an intraspecific phylogenetic approach for studying geologically recent historical events (Avise *et al.*, 1987).

We conclude that despite their generally low level of intraspecific variability, plant organelle genomes can provide markers for use in population genetic analysis. It is important to target variants for which a high mutation rate is anticipated. However this has the consequence that much of the data generated will not be suitable for intraspecific phylogenetic analysis. Where the phylogenetic relatedness of haplotypes can be reliably estimated, they may not coincide with, or be interpretable in terms of, recent population history.

1.3 Exploiting organelle markers

Given the difficulties with detecting a suitable level of organelle variation, what is the incentive for developing and scoring organelle markers in plant populations? What unique insights can they provide that cannot be delivered by analysis of nuclear markers? In the next section of this chapter we will describe three areas in which insights from analysis of organelle markers has significantly enhanced our understanding of population processes in plant populations.

1.3.1 Detecting evolutionary units and elucidating population history

Many plant populations currently possessing widespread and more or less continuous distributions have in the past been highly fragmented as a consequence of climatic change and other environmental perturbations. This is particularly true of plant species in the northern hemisphere affected by glaciation. There is currently considerable interest in reconstructing the glacial and postglacial history of these species with a view to determining their origins and their response to past climatic warming (Birks, 1989; Ferris *et al.*, 1999 – this volume). If different sections of the total population possessing different evolutionary histories (so called evolutionary units (Moritz, 1994)) can be detected, this information may be useful in the management and conservation of populations.

An analysis of genetic marker variation has the potential to contribute materially to such studies in situations where the history of population migration has left behind genetic traces in present day populations (Hewitt, 1996). As we will argue below, maternally inherited organelle markers are more likely than nuclear markers to retain information about past migration history in present day population structure. They can therefore yield insights into population history that would otherwise not be available.

In order to understand why maternally inherited organelle markers are so suitable for reconstructing past migration history, consider the cycle of fragmentation, survival in refugia, and recolonization illustrated in Fig. 1.2. In order that such events be recorded in the present day population structure, two processes must occur. Firstly genetic differentiation must take place between the two refugial populations while they are isolated at the time of maximum glaciation. Secondly the genetic differentiation established between the two parts of the population as a consequence of population history must not break down by gene flow following contact between the recolonizing populations.

The present geographic distribution of maternally inherited genetic markers is more likely to record this event in population history than nuclear markers for the following reasons. Firstly the effective population size of haploid organelle genes in a hermaphrodite, outbreeding plant species is half that of diploid nuclear genes (Birky *et al.*, 1989). In the small refugial populations differentiation by genetic drift will therefore be more marked for organelle than for nuclear genes. Secondly when the recolonizing populations meet, gene flow for the maternally inherited organelle marker will be far less substantial than for the nuclear marker. This is because maternally inherited markers are dispersed only by seed, whereas nuclear markers are dispersed by both pollen and seed (McCauley, 1995). Seed flow into an established population of plants is likely to be highly restricted, whereas pollen flow across the boundary can be very substantial (Millar, 1983; Ellstrand, 1992). As a consequence the genetic differences established at the time when populations meet will be broken down very slowly for the maternally inherited organelle marker, but will be lost rather rapidly for nuclear markers being dispersed by pollen. Thus maternally inherited organelle markers in present day populations will give a clearer picture of postglacial colonization history than will nuclear markers in the same populations.

Support for this assertion comes from a range of studies in which genetic evidence from maternally inherited markers gives a more clear cut picture of the route of postglacial recolonization than evidence from nuclear markers. Data from northwest America using chloroplast DNA variability to trace postglacial colonization is summarised by Soltis *et al.* (1997). In Europe, studies of oak (*Quercus petraea* and *Q. robur*) using a whole suite of nuclear markers indicate a significant clinal genetic change on an east–west axis based on multilocus differences in allele frequencies (Kremer and Zanetto, 1997). This is suggestive of separate eastern and western

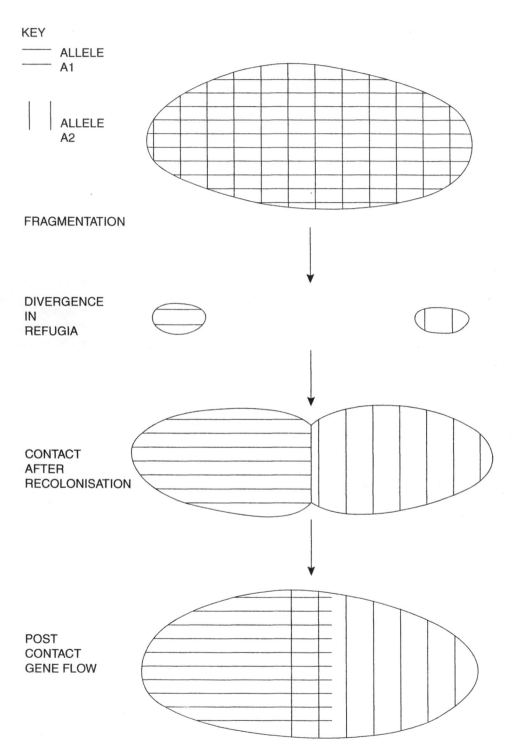

KEY

——— ALLELE
——— A1

| | ALLELE
A2

FRAGMENTATION

DIVERGENCE
IN
REFUGIA

CONTACT
AFTER
RECOLONISATION

POST
CONTACT
GENE FLOW

Figure 1.2 Genetic changes occurring during the population fragmentation and recolonization processes that are required to leave a genetic record in present day populations. Genetic differentiation by drift during periods of isolation, and limited gene flow between reunited populations from different refugia are illustrated.

origins, though no sharp boundaries are present between the regions. However for maternally inherited chloroplast markers different alleles are fixed over large areas in the east, west and centre of the European distribution, clearly indicating three origins after glaciation (Ferris *et al.*, 1993; Ferris *et al.*, 1999 – this volume). The boundaries between these regions fixed for different chloroplast alleles are sharp, indicating little gene flow by seed into areas once they have been colonized. The postglacial history is unambiguously read from the maternally inherited organelle data, and supported by the nuclear marker analysis.

In Scots pine (*Pinus sylvestris* L.) it has long been suggested on the basis of morphological analysis that populations in Fennoscandia were derived from two sources after glaciation, one colonizing from the north and east, and one from the south via Denmark (Sylven, 1916). It was further proposed that the two sources met in southern Sweden at latitude 60°N. Data from a number of types of nuclear markers give some support for this contention. Thus there are clinal changes for genetically determined monoterpene composition (Yazdani and Nilsson, 1986) at two antigen loci (Prus-Glowacki, 1983) and at a needle flavonoid locus (Lebreton *et al.*, 1990) as one moves from south to north in Sweden.

These data are compatible with invasion from two sources, but the evidence is not convincing. Analysis of mitochondrial RFLP variation on the other hand shows fixation of one mitotype in northern Fennoscandia, and fixation of a second mitotype in southern Sweden and northern Europe (Sinclair *et al.*, 1999). These data provide strong support for the dual migration hypothesis of *P. sylvestris* in Fennoscandia. The much greater differentiation for a maternally inherited organelle marker than for biparentally inherited nuclear markers, even after a possible 50 generations of contact between recolonizing populations, is as expected from population genetic theory.

Despite the obvious advantages of using maternally inherited organelle markers for tracing evolutionary history, some limitations of the approach should be made clear. The first is that an organelle genome is effectively only a single gene. The data obtained on population genetic structure from an organelle genome is therefore unreplicated. In angiosperms where both chloroplasts and mitochondria are maternally inherited, replication can be obtained by measuring patterns of population genetic differentiation for both genomes (Birky, 1995). In gymnosperms where the chloroplast genome is paternally inherited this is not possible (Hipkins *et al.*, 1994). In this case data from nuclear genes, though providing a less convincing picture, can be useful in providing independent evidence for the proposed recolonization routes, as in the cases of *Quercus* and *P. sylvestris* cited above.

The second important caution relating to the use of organelle markers for studying plant population history concerns the interpretation of intraspecific chloroplast DNA phylogenies. As outlined above the rate of base substitution and insertion/deletion even in non-coding regions of chloroplast DNA is very slow, an order of magnitude lower than for animal mitochondrial DNA (Wolfe *et al.*, 1987). The current intraspecific sequence diversity within most coding and non-coding regions of the chloroplast genome will have been established over time scales of millions of years. The current geographic distribution of this diversity of cytotypes on the other hand has been established over the last tens of thousands of years probably through an assortment of the previously evolved cytotype variation within the species. Under these circumstances the phylogenetic relationship of cytotypes in different areas is unlikely to be related in a simple way to the recent history of populations. A phylogenetic analysis will not allow a time scale to be placed on our interpretation of recent history because the phylogeny is providing information on evolutionary rather than historic events (but see Ferris *et al.*, 1999 – this volume). In these circumstances the appropriate measure of genetic relatedness that gives insights into population history is some measure of genetic distance based on analysis of gene frequencies.

1.3.2 Measuring gene flow via pollen and seed

In plants, gene flow takes place via both pollen and seed. Since the 1950s nuclear genetic markers have been used to quantify the extent of within population gene flow via pollen (Schaal, 1980). More recently paternity exclusion analysis with multiple nuclear markers has allowed us to quantify the extent of pollen flow within and between small populations with some accuracy (Ellstrand and Marshall, 1985; Meagher, 1986). However it has proved very difficult to obtain measures of gene flow via seed either by direct observation or indirect genetic methods. Our lack of understanding of this area is a major gap in our knowledge of plant population processes. The recent availability of intraspecific organelle markers and the development of appropriate theory have however provided an indirect method for estimating the extent of seed flow relative to pollen flow among populations (Ennos, 1994; McCauley, 1995).

The theory underlying this approach is relatively simple. Suppose that we have a series of plant populations joined by gene flow. Populations tend to differentiate from one another through genetic drift. Gene flow between populations tends to break down this differentiation. At equilibrium between drift and gene flow an equilibrium level of genetic differentiation measured by F_{ST} is achieved. Among the same sets of populations nuclear markers are dispersed by both pollen and seed flow. Maternally inherited organelle markers on the other hand are dispersed only via seed. As a consequence we anticipate that F_{ST} for nuclear markers ($F_{ST(B)}$) will be lower than for maternally inherited organelle markers ($F_{ST(M)}$). The degree of the discrepancy between these two measures of F_{ST} will depend on the relative rates of pollen and seed flow among populations. The greater the importance of pollen flow, the greater will be the discrepancy between the values of F_{ST}. In a simple island model for an outcrossing species the following relationship holds (Ennos, 1994):

$$\text{Pollen migration/Seed migration} = [(1/F_{ST(B)} - 1) - 2(1/F_{ST(M)} - 1)] / (1/F_{ST(M)} - 1)$$

By simultaneously measuring genetic differentiation for nuclear and maternally inherited organelle markers, the ratio of pollen to seed flow among populations can be measured for a range of plants with contrasting reproductive and dispersal modes.

Table 1.3 shows the results of applying this analysis to a range of available data. The estimated ratio of pollen to seed flow varies by two orders of magnitude among plant species in a way that is anticipated by their reproductive ecology. For oaks

Table 1.3 Modes of pollen and seed dispersal for plant species and estimated ratio of pollen to seed flow calculated from a comparison of F_{ST} values for nuclear and maternally inherited organelle markers.

Plant Species	Pollen dispersal	Seed dispersal	Pollen/Seed migration rate	Reference
Quercus spp.	Wind	Bird	196	Kremer *et al.* (1991)
Pinus contorta	Wind	Wind	28	Dong and Wagner (1993)
Argania spinosa	Insect	Ruminant	2.5	El Mousadik and Petit (1996)
Pinus sylvestris (Scotland)	Wind	Wind	18	Sinclair *et al.* (1998)
P. sylvestris (Spain)	Wind	Wind	105	Sinclair *et al.* (1999)

with wind pollination and bird dispersal of seed, the pollen/seed migration ratio is near 200, whereas for *Argania spinosa* with insect pollination and ruminant dispersal of seed the value is only 2.5.

In some cases it is possible to compare the pollen/seed ratio for the same species in two different regions. *P. sylvestris* populations in Scotland were historically widely distributed, while the same species in Spain has had a disjunct distribution at high elevation on isolated mountain ranges. Estimated ratio of pollen to seed flow is much higher in Spain (pollen/seed flow = 105) than in Scotland (pollen/seed flow = 18), perhaps reflecting the difficulties of seed dispersal between mountain ranges within Spain.

These results show that data on intraspecific organelle variation can reveal information about important biological processes that otherwise could not be obtained. Some caution is however needed when interpreting these results. There are two reasons for this. The first is that the estimates are based on a measure of $F_{ST(M)}$ derived from a single (organelle) locus. The second is that the analysis assumes that the populations are at drift/migration equilibrium. In recently founded populations this will not be true (McCauley, 1994). Further work is in progress to extend the genetic theory of differentiation for organelle markers to a variety of models of population structure, and to take into account non-equilibrium situations (McCauley *et al.*, 1996; Le Corre *et al.*, 1997; Hu and Ennos, 1997).

1.3.3 Detecting and characterising hybridization between species

The previous two applications of organelle markers make use of the fact that low levels of intraspecific variation can be detected by employing appropriate methods. The use of organelle markers in studies of hybridization exploit the fact that intraspecific diversity for organelle markers is frequently low or absent, but interspecific differences are relatively common even between closely related species. In these situations the uniparental nature of organelle inheritance not only allows hybrids to be detected, but can also reveal the maternal and paternal parentage of such hybrids.

In angiosperms maternally inherited markers have been successfully applied to identify the female parent involved in the formation of the classic allopolyploids of hybrid origin *Senecio cambrensis* (Harris and Ingram, 1992) and *Spartina anglica*

(Ferris *et al.*, 1997). In many gymnosperms the organelles show contrasting patterns of uniparental inheritance, paternal for the chloroplast and maternal for the mitochondrial genome (Hipkins *et al.*, 1994). This makes it possible to identify with certainty the male and female parents involved in first generation hybrids. For example, combined scoring of chloroplast and mitochondrial markers has allowed a detailed analysis of introgression in the Sitka spruce – interior spruce hybrid zone (Sutton *et al.*, 1994).

Our own work in this area concerns evolution within a recently formed hybrid swarm of larch. European larch, *Larix decidua*, was introduced to Scotland in the eighteenth century and was widely grown as a plantation crop on the Atholl estate in Perthshire. At the end of the nineteenth century a small stand of Japanese larch, *L. kaempferi* was planted on the same estate. In 1904 the first spontaneous F1 hybrid between these two species, *L. × eurolepis* was discovered and subsequently planted (Edwards, 1956).

The hybrid shows heterosis in growth and combines the superiority in form of *L. decidua* with the site tolerance and canker resistance of *L. kaempferi*. As a result open pollinated collections have been made from the original *L. kaempferi* introductions and thereafter from planted stands of these putative hybrids for widespread planting on the Atholl estate. Initial work on the resulting hybrid swarm is concerned with determining the composition and parentage of the larch population on the estate with the aim of understanding the early evolution of the hybrid taxon.

In order to accomplish this, samples of trees have been taken from stands planted in 1910, soon after the discovery of the hybrid, and from a stand planted in 1933. Trees have been scored simultaneously for species specific chloroplast markers (paternally inherited) and mitochondrial markers (maternally inherited) (Langdon, unpublished). Preliminary results (Table 1.4) indicate that the 1910 planting comprises mainly pure European and pure Japanese larch, with a low frequency of F1 hybrids. These have Japanese mothers and European fathers. The composition of the 1933 stand is quite different. If this represents putative first generation hybrids then three quarters of the stand are of hybrid origin with Japanese mothers and European fathers, and the remainder are pure Japanese larch.

One limitation of using organelle markers to study hybridization in this manner is that while they can provide unequivocal parentage for trees in the first generation, analysis of codominant or dominant nuclear markers is required for identifying the nature of genotypes in later generations. Indeed additional analysis of nuclear markers in the 1933 stand suggests that at least some of this material

Table 1.4 Number of plants with species specific chloroplast (paternally inherited) and mitochondrial (maternally inherited) markers found in two stands of putative hybrid larch on the Atholl estate, Perthshire (Langdon, unpublished).

Chloroplast DNA marker	Mitochondrial DNA marker	1910 stand	1933 stand
European	European	11	0
Japanese	Japanese	33	10
European	Japanese	2	30
Japanese	European	0	0

may represent F2s or backcrosses to Japanese fathers (Langdon, unpublished). Current work is focusing on the nature of the regenerating population derived from these trees.

1.4 Conclusions

We have shown above that organelle markers, because of their unusual inheritance patterns and modes of evolution, have the potential to provide novel insights into the migration history, seed dispersal and hybrid evolution of plant taxa. Such insights could not be obtained using nuclear markers alone. This in itself is a justification for expending the extra effort required to identify adequate or appropriate levels of organelle variation. However the very nature of the organelle genomes means that there are limitations on the data that they can yield. Perhaps the most restrictive characteristics of organelle genomes are their non-recombining nature and their slow rates of evolution. The former means that the information they provide is unreplicated, and the latter limits their usefulness for the phylogenetic analysis of recent population history.

These considerations imply that the most efficient way of utilising organelle markers is in combination with a parallel analysis of nuclear genetic variation. It is the comparison of data from the two sets of contrasting genomes which yields the clearest insights. Thus nuclear markers can provide replicated data to support or refute population history scenarios suggested by the geographic distribution of organelle markers. Similarly it is the comparison between organelle and nuclear population genetic structure that yields estimates of seed and pollen flow. Likewise the analysis of interspecific hybridization past the first generation demands a detailed analysis of nuclear markers, and cannot be conducted with organelle markers alone. Synthesis of information from genomes with contrasting properties has already opened up new avenues of research. Further progress will depend on the development and implementation of comprehensive population genetic theory to describe the joint behaviour of plant nuclear and organelle genomes in the presence, as well as the absence, of natural selection (Asmussen and Schnabel, 1991; Dong and Wagner, 1994; Latta and Mitton, 1997).

REFERENCES

Asmussen, M. A. and Schnabel, A. (1991). Comparative effects of pollen and seed migration on the cytonuclear structure of plant populations. I. Maternal cytoplasmic inheritance. *Genetics*, **128**, 639–654.

Atlan, A. and Couvet, D. (1993) A model simulating the dynamics of plant mitochondrial genomes. *Genetics*, **135**, 213–222.

Avise, J. C. (1994) *Molecular markers, natural history and evolution*, Chapman and Hall, New York.

Avise, J. C., Arnold, J., Ball, R. M., Bermingham, E., Lamb, T., Neigel, J. *et al.* (1987) Intraspecific phylogeography: the mitochondrial DNA bridge between population genetics and systematics. *Annual Review of Ecology and Systematics*, **18**, 489–522.

Birks, H. J. B. (1989) Holocene isochrone maps and patterns of tree-spreading in the British Isles. *Journal of Biogeography*, **18**, 103–115.

Birky, C. W. (1995) Uniparental inheritance of mitochondrial and chloroplast genes: mechanisms and evolution. *Proceedings of the National Academy of Sciences USA*, **92**, 11331–11338.

Birky, C. W., Fuerst, P. and Maruyama T. (1989) Organelle gene diversity under migration, mutation and drift: equilibrium expectations, approach to equilibrium, effects of heteroplasmic cells, and comparison to nuclear genes. *Genetics*, **121**, 613–627.

Brown, A. H. D. (1989) Genetic characterization of plant mating systems, in *Plant population genetics, breeding, and genetic resources* (eds A. H. D. Brown, M. T. Clegg, A. L. Kahler and B. S. Weir), Sinauer Associates Inc., Sunderland, Massachusetts, pp. 145–162.

Clegg, M. T., Ritland, K. and Zurawski, G. (1986) Processes of chloroplast DNA evolution, in *Evolutionary processes and theory* (eds S. Karlin and E. Nevo), Academic Press, New York, pp. 275–294.

Dawson, I. K., Simons, A. J., Waugh, R. and Powell, W. (1995) Diversity and genetic differentiation among subpopulations of *Gliricidia sepium* revealed by PCR-based assays. *Heredity*, **74**, 10–18.

Demesure, B., Sodzi, N. and Petit, R. J. (1995) A set of universal primers for amplification of polymorphic non-coding regions of mitochondrial and chloroplast DNA in plants. *Molecular Ecology*, **4**, 129–131.

Demesure, B., Comps, B. and Petit, R. J. (1996) Chloroplast DNA phylogeography of the common beech (*Fagus sylvatica* L.) in Europe. *Evolution*, **50**, 2515–2520.

Dong, J. and Wagner, D. B. (1993) Taxonomic and population differentiation of mitochondrial diversity in *Pinus banksiana* and *Pinus contorta*. *Theoretical and Applied Genetics*, **86**, 573–578.

Dong, J. and Wagner, D. B. (1994) Paternally inherited chloroplast polymorphism in *Pinus*: estimation of diversity and population subdivision, and tests of disequilibrium with a maternally inherited mitochondrial polymorphism. *Genetics*, **136**, 1187–1194.

Edwards, M. V. (1956) The hybrid larch *Larix* × *eurolepis* Henry. *Forestry*, **29**, 29–43.

El Mousadik, A. and Petit, R. J. (1996) Chloroplast DNA phylogeography of the argan tree of Morocco. *Molecular Ecology*, **5**, 547–555.

Ellstrand, N. C. (1992) Gene flow among seed populations. *New Forests*, **6**, 241–256.

Ellstrand, N. C. and Marshall, D. L. (1985) Interpopulation gene flow by pollen in wild radish, *Raphanus sativus*. *American Naturalist*, **126**, 606–616.

Ellstrand, N. C. and Roose, M. L. (1987) Patterns of genotypic diversity in clonal plant species. *American Journal of Botany*, **74**, 132–135.

Ennos, R. A. (1994) Estimating the relative rates of pollen and seed migration among plant populations. *Heredity*, **72**, 250–259.

Ferris C., Oliver, R. P., Davy, A. J. and Hewitt, G. M. (1993) Native oak chloroplasts reveal an ancient divide across Europe. *Molecular Ecology*, **2**, 337–344.

Ferris, C., King, R. A. and Gray, A. J. (1997) Molecular evidence for the maternal parentage in the hybrid origin of *Spartina anglica*. *Molecular Ecology*, **6**, 185–187.

Ferris, C., King, R. A. and Hewitt, G. M. (1999) Isolation within species and the history of glacial refugia, in *Molecular systematics and plant evolution*, (eds P. M. Hollingsworth, R. M. Bateman and R. J. Gornall), Taylor & Francis, London, pp. 20–34.

Gielly, L. and Taberlet, P. (1994) The use of chloroplast DNA to resolve plant phylogenies: noncoding versus *rbcL* sequences. *Molecular Biology and Evolution*, **11**, 769–777.

Hanson, M. R. (1991) Plant mitochondrial mutations and male sterility. *Annual Review of Genetics*, **25**, 461–486.

Harris, S. A. and Ingram, R. (1991) Chloroplast DNA and biosystematics: the effects of intraspecific diversity and plastid transmission. *Taxon*, **40**, 393–412.

Harris, S. A. and Ingram, R. (1992) Molecular systematics of the genus *Senecio* L. 1. Hybridisation in a British polyploid complex. *Heredity*, **69**, 1–10.

Hewitt, G. M. (1996) Some genetic consequences of ice ages, and their role in divergence and speciation. *Biological Journal of the Linnean Society*, **58**, 247–276.

Hipkins, V. D., Krutovskii, K. V. and Strauss, S. H. (1994) Organelle genomes in conifers: structure, evolution and diversity. *Forest Genetics*, **1**, 179–189.

Hipkins, V. D., Marshall, K. A., Neale, D. B., Rottmann, W. H. and Strauss, S. H. (1995) A mutation hotspot in the chloroplast genome of a conifer (Douglas-fir: *Pseudotsuga*) is caused by variability in the number of direct repeats derived from a partially duplicated tRNA gene. *Current Genetics*, **27**, 572–579.

Hu, X.-S. and Ennos, R. A. (1997) On estimation of the ratio of pollen to seed flow among plant populations. *Heredity*, **79**, 541–552.

Kinloch, B. B., Westfall, R. D. and Forrest, G. I. (1986). Caledonian Scots pine: origins, and genetic structure. *New Phytologist*, **104**, 703–729.

Konnert, M. and Bergmann, F. (1995) The geographical distribution of genetic variation of silver fir (*Abies alba*, Pinaceae) in relation to its migration history. *Plant Systematics and Evolution*, **196**, 19–30.

Kremer, A. and Zanetto, A. (1997) Geographical structure of gene diversity in *Quercus petraea* (Matt.) Liebl. II: Multilocus patterns of variation. *Heredity*, **78**, 476–489.

Kremer, A., Petit, R., Zanetto, A., Fougere, V., Ducousso, A., Wagren, D. *et al.* (1991) Nuclear and organelle gene diversity in *Quercus robur* and *Q. petraea*, in *Genetic variation in European populations of forest trees*, (eds G. Muller-Starck and M. Ziehe), Sauerlander's Verlag, Frankfurt, pp. 141–166.

Latta, R. G. and Mitton, J. B. (1997) A comparison of population differentiation across four classes of gene marker in limber pine (*Pinus flexilis* James). *Genetics*, **146**, 1153–1163.

Le Corre, V., Machon, N., Petit, R. J. and Kremer, A. (1997) Colonisation with long distance seed dispersal and genetic structure of maternally inherited genes in forest trees: a simulation study. *Genetical Research Cambridge*, **69**, 117–125.

Lebreton, P., Laracine-Pittet, C., Bayet, C. and Lauranson, J. (1990) Variabilité polyphenolique et systematique du pin sylvestre *Pinus sylvestris* L. *Annales Sciences Forestiere*, **47**, 117–130.

Maruyama, T. and Birky, C. W. (1991) Effects of periodic selection on gene diversity in organelle genomes and other systems without recombination. *Genetics*, **127**, 449–451.

Mazur, B. J. and Falco, S. C. (1989) The development of herbicide resistant crops. *Annual Review of Plant Physiology and Plant Molecular Biology*, **40**, 441–470.

McCauley, D. E. (1994) Contrasting the distribution of chloroplast DNA and allozyme polymorphism among local populations of *Silene alba*: implications for the study of gene flow in plants. *Proceedings of the National Academy of Sciences USA*, **91**, 8127–8131.

McCauley, D. E. (1995) The use of chloroplast DNA polymorphism in studies of gene flow in plants. *Trends in Ecology and Evolution*, **10**, 198–202.

McCauley, D. E., Stevens, J. E., Peroni, P. A. and Raveill, J. A. (1996) The spatial distribution of chloroplast DNA and allozyme polymorphisms within a population of *Silene alba* (Caryophyllaceae). *American Journal of Botany*, **83**, 727–731.

Meagher, T. R. (1986) Analysis of paternity within a natural population of *Chamaelirium luteum*. I. Identification of most-likely male parents. *American Naturalist*, **128**, 199–215.

Millar, C. I. (1983) A steep cline in *Pinus muricata*. *Evolution*, **37**, 311–319.

Moritz, C. (1994) Defining 'evolutionarily significant units' for conservation. *Trends in Evolution and Ecology*, **9**, 373–375.

Neale, D. B. and Sederoff, R. R. (1989) Paternal inheritance of chloroplast DNA and maternal inheritance of mitochondrial DNA in loblolly pine. *Theoretical and Applied Genetics*, **77**, 212–216.

Palmer, J.D. (1992) Mitochondrial DNA in plant systematics: applications and limitations, in *Molecular systematics of plants* (eds P. S. Soltis, D. E. Soltis and J. J. Doyle), Chapman and Hall, New York, pp. 36–49

Parks, C. R., Wendel, J. F., Sewell, M. M. and Qiu, Y-L. (1994) The significance of allozyme variation and introgression in the *Liriodendron tulipifera* complex (Magnoliacae). *American Journal of Botany*, **81**, 878–889.

Petit, R. J., Kremer, A., Wagner, D. B. (1993) Finite island model for organelle and nuclear genes in plants. *Heredity*, **71**, 630–641.

Potts, B. M. and Reid, J. B. (1988) Hybridization as a dispersal mechanism. *Evolution*, **42**, 1245–1255.

Powell, W. M., Morgante, M., McDevitt, R., Vendramin, G. G. and Rafalski, J. A. (1995) Polymorphic simple sequence repeat regions in chloroplast genomes: applications to the population genetics of pines. *Proceedings of the National Academy of Sciences USA*, **92**, 7759–7763.

Provan, J., Soranzo, N., Wilson, N. J., McNicol, J. W., Morgante, M. and Powell, W. (1999) The use of uniparentally inherited simple sequence repeat markers in plant population studies and systematics, in *Molecular systematics and plant evolution*, (eds P. M. Hollingsworth, R. M. Bateman and R. J. Gornall), Taylor & Francis, London, pp. 35–50.

Prus-Glowacki, W. (1983) Serological investigation of a hybrid swarm population of *Pinus sylvestris* L. × *Pinus mugo* Turra, and the antigenic differentiation of *Pinus sylvestris* L. in Sweden, in *Proteins and nucleic acids in plant systematics* (eds U. Jensen and D. E. Fairbrothers), Springer-Verlag, Berlin, pp. 352–361.

Rieseberg, L. H. and Brunsfeld, S. J. (1992) Molecular evidence and plant introgression, in *Molecular systematics of plants* (eds P. S. Soltis, D. E. Soltis and J. J. Doyle), Chapman and Hall, New York, pp. 151–176.

Rieseberg, L. H. and Soltis, D. E. (1991) Phylogenetic consequences of cytoplasmic gene flow in plants. *Evolutionary Trends in Plants*, **5**, 65–84.

Samitou-Laprade, P., Rouwendale, G. J. A., Cuguen, J., Krens, F. A. and Michaelis, G. (1993) Different CMS sources found in *Beta vulgaris* ssp. *maritima*: mitochondrial variability in wild populations revealed by a rapid screening procedure. *Theoretical and Applied Genetics*, **85**, 529–535.

Schaal, B. (1980) Measurement of gene flow in *Lupinus texensis. Nature*, **284**, 450–451.

Sinclair, W.T., Morman, J. D. and Ennos, R.A. (1998) Multiple origins for Scots pine (*Pinus sylvestris* L.) in Scotland: evidence from mitochondrial DNA variation. *Heredity*, **80**, 233–240.

Sinclair, W. T., Morman, J. D. and Ennos, R. A. (1999) The postglacial history of Scots pine (*Pinus sylvestris* L.) in Western Europe: evidence from mitochondrial DNA variation. *Molecular Ecology*, **8**, 83–88.

Slatkin, M. (1995) A measure of population subdivision based on microsatellite allele frequencies. *Genetics*, **139**, 457–462.

Soltis, D. E., Soltis, P. S. and Milligan, B. G. (1992) Intraspecific chloroplast DNA variation: systematic and phylogenetic implications, in *Molecular systematics of plants*, (eds P. S. Soltis, D. E. Soltis and J. J. Doyle), Chapman and Hall, New York, pp. 117–150.

Soltis, D. E., Gitzendanner, M. A., Strenge, D. D. and Soltis, P. S. (1997) Chloroplast DNA intraspecific phylogeography of plants from the Pacific northwest of North America. *Plant Systematics and Evolution*, **206**, 353–373.

Strauss, S. H., Hong, Y-P. and Hipkins, V. D. (1993) High levels of population differentiation for mitochondrial DNA haplotypes in *Pinus radiata, muricata* and *attenuata. Theoretical and Applied Genetics*, **86**, 605–611.

Sutton, B. C. S., Pritchard, S. C., Gawley, J. R., Newton, C. H. and Kiss, G. K. (1994) Analysis of Sitka spruce – interior spruce introgression in British Columbia using cytoplasmic and nuclear DNA probes. *Canadian Journal of Forest Research*, **24**, 278–285.

Sylven, N. (1916) Den nordsvenska tallen. *Medd. Statens Skogsforsoksanst*, **13-14** , 9–110.

Szmidt, A. E., El-Kassaby, Y. A., Sigurgeirsson, A., Aldén, T., Lindgren, D. and Hällgren, J.-E. (1988) Classifying seedlots of *Picea sitchensis* and *P. glauca* of zones in introgression using restriction analysis of chloroplast DNA. *Theoretical and Applied Genetics*, **76**, 841–845.

Taberlet, P., Gielly, L., Pautou, G. and Bouvet, J. (1991) Universal primers for amplification of three non-coding regions of chloroplast DNA. *Plant Molecular Biology*, **17**, 1105–1109.

Wagner, D. B., Furnier, G. R., Saghai-Maroof, M. A., Williams, S. M., Dancik, B. P. and Allard, R. W. (1987) Chloroplast DNA polymorphism in lodgepole and jack pines and their hybrids. *Proceedings of the National Academy of Sciences USA*, **84**, 2097–2100.

Wang, X.-R. and Szmidt, A. E. (1994) Hybridization and chloroplast DNA variation in a *Pinus* species complex from Asia. *Evolution*, **48**, 1020–1031.

Wang, X.-R., Szmidt, A. E. and Lu, M.-Z. (1996) Genetic evidence for the presence of cytoplasmic DNA in pollen and megagametophytes and maternal inheritance of mitochondrial DNA in *Pinus*. *Forest Genetics*, **3**, 37–44.

Wolfe, K. H., Li, W-H. and Sharpe, P. M. (1987) Rates of nucleotide substitution vary greatly among plant mitochondria, chloroplast and nuclear DNAs. *Proceedings of the National Academy of Sciences USA*, **84**, 9054–9058.

Wolfe, K. H., Sharp, P. M and Li, W.-S. (1989) Rates of synonymous substitution in plant nuclear genes. *Journal of Molecular Evolution*, **29**, 208–211.

Yazdani, R. and Nilsson, J.-E. (1986) Cortical monoterpene variation in natural populations of *Pinus sylvestris* in Sweden. *Scandinavian Journal of Forest Research*, **1**, 85–93.

Chapter 2

Isolation within species and the history of glacial refugia

C. Ferris, R. A. King and G. M. Hewitt

ABSTRACT

Ice-ages, the result of instabilities in the earth's climate caused by Milankovitch cycles, can have dramatic effects on species phylogeography. During the last ice-age temperate forest tree species were excluded from northern and central Europe and survived in glacial refugia in the south. Classically, fossil pollen has been used to identify refugia for species with distinctive pollen.

During periods of isolation, divergence will occur and of prime importance is the duration of isolation; this may be some 100,000 years. If refugia can be isolated for more than one glaciation then the divergence times could be much greater. Many north European tree species also exist in mountains in the south close to their putative refugial areas. For many of these species fossil pollen levels fall dramatically at the onset of glaciation and there is no direct evidence for southward migration. This opens up the possibility that refugia may remain isolated not only during the periods of glaciation but also during the warm interglacial periods.

Isolation times between refugia can be tested using molecular genetic techniques by firstly identifying the refugial genotypes within the species and secondly investigating the levels of DNA sequence divergence between them. This can be given as an estimate of time since divergence using the molecular clock hypothesis.

Here we review information on the consequences of glaciation on plant phylogeography and the processes that are likely to give rise to divergence between populations in different refugia. Recent molecular based studies on the postglacial history of native European forest tree species are reviewed.

2.1 Introduction

Effects of Pleistocene ice-ages on the phylogeography of species may be addressed in two ways. The fossil record and palaeoenvironmental sciences can be used to gain information regarding climate change. We may correlate climate change with distributions of populations, extent of range changes and rates of migration. Alternatively, we may use knowledge of the present geographic genetic structure of taxa to determine routes of migration, and rates of hybridisation, introgression and divergence. These two separate approaches are each valuable for determining the effects of glaciations upon divergence and speciation but if we combine both approaches we can gain detailed insights into the glacial/postglacial history of species.

In *Molecular systematics and plant evolution* (1999) (eds P. M. Hollingsworth, R. M. Bateman and R. J. Gornall), Taylor & Francis, London, pp. 20–34.

For many plants there is insufficient fossil data with which to work, but many genera of forest trees have distinctive pollen, abundant in fossil sediments in lake-bed cores. The fossil pollen record is well documented in the literature for European tree species such as oak and clearly indicates where refugia occurred during the last ice-age (Huntley and Birks, 1983; Bennett *et al.*, 1991). Here we present evidence from genetic structure within species to complement the fossil data and facilitate a greater understanding of the history of glacial refugia and particularly their isolation.

2.2 Pleistocene ice-ages

The earth is subject to regular changes in climatic conditions on a time scale of 20–100 kyr. These changes result from fluctuations in the earth's orbit around the sun, known as the Milankovitch cycles (Hays *et al.*, 1976). Three major cycles are evident with frequencies of 23 kyr, 41 kyr and 100 kyr (Bradley, 1985; Imbrie and Imbrie, 1979), which produce variations in the amount of solar radiation received by the earth (Berger, 1978). During the last 2.3 myr there have been several continental glaciations which are in phase with the Milankovitch cycles, thus leading to the conclusion that these cycles are the pacemaker of the glaciations (Hays *et al.*, 1976). Clearly the Milankovitch cycles have been a factor throughout the history of the earth. Oxygen isotope data from ocean sediments of foraminifera indicate climatic fluctuations in phase with these cycles predating the onset of the Pleistocene ice-ages (Shackleton *et al.*, 1988). The appearance of glaciations is most likely related to plate tectonic movements and the late Cenozoic configuration of the continents (Imbrie, 1985; Imbrie and Imbrie, 1979) or to the rise of mountains and high plateaux (Ruddiman and Raymo, 1988).

The glaciations last for some 100 kyr, separated by warm interglacials of between 10–20 kyr. Prior to the last glaciation, the Eemian interglacial lasted for 20 kyr from 135–115 kyr ago, whereupon a drop in temperature led to an advance of the ice sheets. As water became locked in the ice, the sea level dropped and new areas of coastline and landbridges emerged. As temperatures fell lower still, areas of ice cover increased leading to full glaciation. This process may be surprisingly rapid. Ice and snow reflect up to 80% of the sun's radiant energy compared to the normal 40% reflectance, thus positive feedback occurs as the ice sheets build up with the earth absorbing progressively less energy from the sun and getting colder. During glaciation the shorter orbital cycles of 23 and 41 kyr continue, producing fluctuations of temperature during the ice-age and expansion and contraction of the ice sheets. Between 21 and 18 kyr ago a very severe fall in temperature, caused presumably by a coincidence of orbital cycles, produced the most extensive advance of the glaciation. Following this, temperatures rose, the ice subsequently began to retreat, and we entered the present Holocene interglacial.

2.3 Consequences of glaciation

The effects of glaciation upon animals and vegetation have been much studied and debated. Clearly the Milankovitch cycles have had an effect on species both in ecological and evolutionary time (Bennett, 1990). Climatic changes will alter the environment and a species will either have to adapt to the change or migrate, otherwise it will die out. The Pleistocene glaciations as described above have included relatively minor climatic fluctuations in addition to the extreme changes between glacial and interglacial stages. The expansions and contractions of species' ranges in response to climate change are therefore a dynamic process. Clearly, species which can spread and colonise rapidly and/or easily over long distances will have been at an advantage during the relatively unstable conditions of the Pleistocene. It can be argued that a sudden rapid fall in temperature for as little as a century would be enough to exclude many species from the northern part of their range, while a similar increase in temperature would likewise affect the southern part of the range. Thus the range will oscillate north and south with only populations in central regions surviving (Hewitt, 1996).

During glaciation species ranges are restricted southward. Those adapted to arctic and alpine conditions will colonise the large expanses of tundra and steppe south of the ice sheets. However, those species adapted to temperate conditions will be restricted to southern refugia. Using fossil pollen evidence the refugia for many temperate species are now known (Huntley and Birks, 1983) and in particular for north European trees (Bennett *et al.*, 1991). As the temperature rose at the start of the Holocene interglacial and large expanses of Europe became suitable for colonisation, so the temperate flora rapidly migrated northward to fill previously uncolonised regions. Rates of migration are exceptionally high under these conditions (Table 2.1) and involve long-distance dispersal. As species began to reach their northern limits in Britain and northern Europe the rates of migration slowed down to give the 'normal' rates of dispersal we see today. Whilst this is true for the large European plains, when we consider any mountains encountered the situation will be different. As the temperature rises only small areas of the mountain will be suitable for colonisation and thus dispersal will be short rather than long-range.

Table 2.1 Rates of migration following the retreat of the last glaciation. Values are based on fossil pollen data (after Birks, 1989; Webb, 1986).

| Taxon | Migration rate m/yr | | Dispersal km/gen. |
	Europe	Britain	
Birch (*Betula*)	> 2000	250	2.5
Hazel (*Corylus*)	1500	500	7.5
Elm (*Ulmus*)	500–1000	550	8
Oak (*Quercus*)	150–500	350–500	7–10
Pine (*Pinus*)	1500	100–700	1–7
Alder (*Alnus*)	500–2000	500–600	5–6
Lime (*Tilia*)	300–500	450–500	10
Ash (*Fraxinus*)	200–500	50–200	1–3
Beech (*Fagus*)	200–300	100–200	4–8

2.4 Refugia, isolation and diversity

2.4.1 Glaciation

As stated above, temperate plants will survive periods of glaciation in refugial areas
to the south. A refugium can be defined as 'an area, of any size, in which a taxon
persisted, at any population density, during a cold stage' (Bennett *et al.*, 1991).
Clearly such a definition could represent a single isolated population, several populations in an isolated area (e.g. on a mountain), or several such areas covering a
large isolated geographical region. The latter is particularly relevant to southern
Europe where the topography causes large areas of similar temperate climate such
as southern Iberia, Italy and the Balkans to be isolated. For European tree taxa, a
total of 17 refugia have been identified by Huntley and Birks (1983). The most
important of these, in terms of number of species housed during the last period of
glaciation, were in the mountains of central and southern Europe. The Balkan and
Carpathian Mountains were home to 12 and 10 species repectively (Huntley
and Birks, 1983). Some taxa are thought to have survived in only a single refugium,
e.g. the evergreen oaks and *Olea europaea*, whereas others were found in up to
eight refugia, e.g. deciduous oaks, *Pinus* and *Acer* (Bennett *et al.*, 1991; Huntley
and Birks, 1983). Given that plants are the primary producers in the ecosystem,
all animals will be either directly or indirectly dependent upon them. The plants
will therefore play a major role in determining the characteristics of the communities in these refugial areas.

Refugia will be isolated if there is no gene flow between them. As stated above,
during glaciation, changes in the short orbital cycles will lead to fluctuations
in climate and subsequent expansions and contractions of species ranges. These
changes are however, not great enough to facilitate long-range gene flow between
the major refugial areas of southern Europe, i.e. between southern Spain, Italy
and the Balkans. Gene flow between these major refugia would have to involve
dispersal across several hundred kilometres of open sea, which is unlikely for many
species. However, within each of these areas there may have been some possibility
of medium-range dispersal events, although for mountain populations it is most
likely that only short-range dispersal would have occurred. Clearly the propensity
for gene flow via pollen is much greater than via seed in many species, particularly wind pollinated species. Thus geographical patterns of genetic structure may
differ between nuclear markers and maternally inherited markers such as chloroplast DNA (Ennos *et al.*, 1999 – this volume).

During glaciation, large refugial areas can thus be isolated for periods of up to
100,000 years or more and consequently genepools will diverge. Divergence will
of course be greatly facilitated where population sizes are small and there is greater
opportunity for the fixation of new mutations.

2.4.2 When the ice retreats

In contrast to the long-term relatively stable conditions during the glacial period,
the onset of an interglacial brings extreme changes in climate which result in large
range changes for many species. As the temperatures increase suitable new areas
become available for colonisation. There are generally two distinct pathways that

NORTH

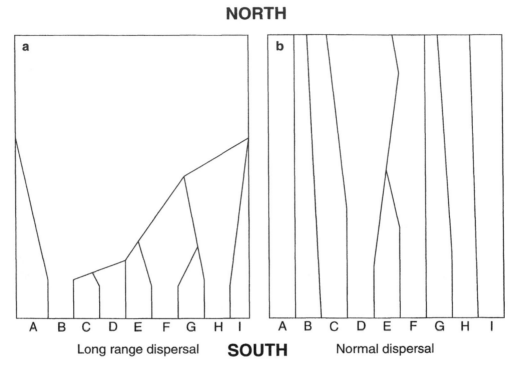

Long range dispersal   **SOUTH**   Normal dispersal

Figure 2.1 The possible effects of: a) long range and, b) normal dispersal patterns, on the genetic diversity present within northern populations after range expansion from southern refugia. In both cases nine genotypes are represented in refugial areas. Long range expansion causes a dramatic loss of diversity with in this case only type B predominating in northern areas. Normal dispersal patterns help maintain diversity with most genotypes being found in the north (after Hewitt, 1996).

species can take, with very different outcomes. In mountain areas migration will take place up the mountains and at a fairly slow rate. By contrast, on the plains vast areas become suitable for colonisation and the propensity for rapid migration is a definite advantage. The effects of speed of migration upon the levels of diversity can be seen in Fig. 2.1. Populations on the northern edge of refugia migrate quickly into the large plains of central and northern Europe with rates of anything up to 10 km per generation (Table 2.1) (Birks, 1989; Webb, 1986). Such rapid migration will involve some long-range dispersal events. Long distance migrants may produce populations ahead of the main distribution that represent only a subset of the variation of the species. Such founder effects will result in a thinning of diversity with only a limited number of genotypes migrating north (Fig. 2.1a)(Hewitt, 1996). Conversely, slow migration would not be subject to such important founder effects and most of the species diversity would be maintained (Fig. 2.1b)(Hewitt, 1996).

2.4.3 Onset of the next glaciation

It seems clear that during glaciation temperate species are restricted to refugial populations in the south where they may be isolated from each other and thus diverge. For many tree species in particular this isolation is exemplified by fossil pollen data. At the end of the ice-age the temperatures increase and species migrate up mountains in the south and onto the plains to the north. This is the natural process of postglacial succession. Again fossil pollen data has been used to follow the migration to the north.

What happens when temperatures begin to fall at the onset of the next glaciation has been little studied. For many the assumption has been that the species will migrate south as the temperatures fall. This is true for cold-adapted species such as pine and birch which migrated south during the early glacial (Bennett, 1990). However, this is not a likely scenario in the case of thermophilous species whose communities to the south consist of climax forest. Southward migration would therefore effectively entail reverse succession. In addition, the temperatures may fall extremely quickly and a simple failure of regeneration may hasten the demise of the northern populations of these temperate species. The processes that prevent southward migration are not clear but there is no indication in the fossil pollen record to support such migration for thermophilous species (Bennett et al., 1991). The populations to the north, and any diversity they do contain, are thus effectively lost and make no contribution to the refugial areas in the south. This scenario is shown in Fig. 2.2 for a single southern refugial area. In the south the mountains act as refugia to support the diversity of the species. During the cold spell the populations move down the mountains and over time mutations will occur. As the temperatures increase again the populations again migrate up the mountains and the diversity is maintained. The northern plains and hills are colonised during the warm periods by a subset of the diversity present in the south which is completely lost from the north at the early cold stage. To take the scenario further we can consider the relationship between different refugial areas. Clearly, during the warm interglacial it is possible for populations migrating from different refugia to meet and exchange genes. The interglacial period is of the order of 20,000 years and the effective generation time of a typical forest tree is a hundred years or more. Despite the ability of pollen to travel long distances in some species, it is highly unlikely that any genes could pass from one refugial area to another via the main distribution within mainland Europe. As a result the refugial areas are effectively isolated throughout the entire glacial/interglacial cycle. It is therefore possible for refugial populations of a species to have been isolated for several glaciations or even for the duration of the Pleistocene, some 2 Myr or more. In addition, if the refugial areas are effectively isolated they are vitally important for the continued survival of species. If a particular species becomes extinct in a refugial area, it can never be got back (Bennett et al. 1991), unless by the very unlikely possibility of re-colonisation from another refuge.

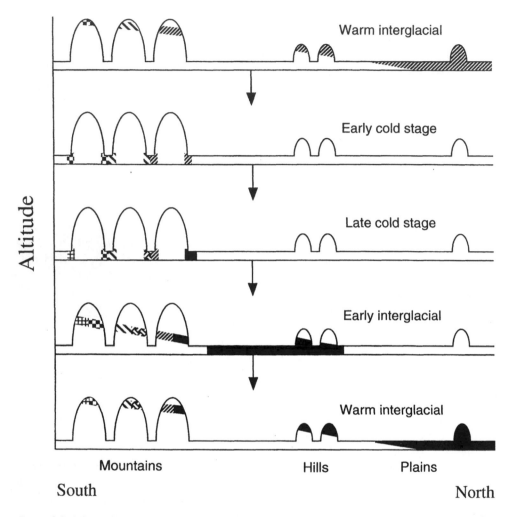

Figure 2.2 A hypothetical scenario of possible migration and divergence patterns that occur during a glacial/interglacial cycle in Europe (after Hewitt, 1996). Southern mountains represent refugial areas, while the north represents the plains of northern Europe. Shading represents different genotypes.

2.5 Some forest tree examples

2.5.1 *Quercus*

Many tree species are particularly suitable for the study of their glacial and post-glacial history due to the availability of a good fossil pollen record with which to complement studies of population genetic structure. Perhaps the best example to date is of the genus *Quercus*, which is well represented in the fossil record and for which detailed isopol maps for Europe are readily available (Huntley and Birks, 1983). About 115 kyr ago at the end of the last interglacial, the Eemian, the fossil data give no indication of a southward migration of *Quercus* as the temperatures dropped (Bennett *et al.*, 1991). Clearly, during the warm interglacial the oaks were present in mountain regions in the south of Europe from which they seeded the populations that survived the glaciation in refugia in southern Iberia, Italy and the Balkans (Bennett *et al.*, 1991; Huntley and Birks, 1983). In addition to these three main refugia, some oaks may have resided further to the east (Van Zeist and Bottema, 1977). At the start of the Holocene interglacial some 10 kyr ago, migration of oaks northward into mainland Europe was remarkably fast with maximum speeds of up to 500 m/yr representing up to 10 km per generation (Birks, 1989; Webb, 1986). Long distance dispersal events must have been common during these early stages of colonisation with a thinning of genetic diversity inevitable. Maternally inherited genomes, such as chloroplast DNA in the majority of angiosperms, are ideal for the study of glacial recolonisation for several reasons (see Ennos *et al.*, 1999 – this volume). In *Quercus*, chloroplast DNA (cpDNA) does indeed indicate that, as predicted, the diversity in the north is merely a subset of that in the south (Dumolin-Lapègue *et al.*, 1997). Population genetic studies have revealed striking patterns of genetic subdivision within Europe for the north European deciduous oaks, *Q. robur* and *Q. petraea* (Dumolin-Lapègue *et al.*, 1997; Ferris *et al.*, 1993, 1995, 1998; Petit *et al.*, 1993). The natural patterns of cpDNA variation indicate a subdivision of Europe into three main regions: western, central, and eastern (Fig. 2.3), with each region being founded from its own respective refugium (Ferris *et al.*, 1998). The cpDNA markers, obtained by direct sequencing and RFLP analysis of PCR amplified tRNA[Leu(UAA)] intron and the intergenic spacer between tRNA[Leu(UAA)] and tRNA[Phe(GAA)], define the extent to which each migration proceeded and indicate zones of contact between them. The marked boundary between the central and eastern cytotypes in southern Finland shows that there is little opportunity for gene flow via seed between the populations upon contact (Ferris *et al.*, 1998). The average sequence divergence between the eastern and the western and central cytotypes is 0.3%.

The patterns and levels of genetic variation and divergence agree well with the scenario proposed in Fig. 2.2. Diversity increases with time in the south both within and between refugia. Rapid migration at the onset of an interglacial leads to a reduced diversity in the north. Refugia are effectively isolated during both glaciation and the intervening interglacials and thus may have been isolated for many ice-ages. A divergence rate of 0.1% per Myr is given by Zurawski *et al.* (1984) as an average for cpDNA and is comparable to the rate of 0.1–0.3% per Myr found by Wolfe *et al.* (1987). Four mutations representing 0.3% divergence between refugial oak cytotypes thus represents up to 3 Myr divergence. Obviously this value is subject to many variables and potential errors and could be an overestimate, although Frascaria *et al.* (1993) have found a reduced evolutionary rate of cpDNA in the Fagaceae, the family to which *Quercus* belongs. All the above difficulties aside, the data are consistent with the refugia having been isolated for many glaciations if not for the duration of the Pleistocene, some 2.4 Myr.

2.5.2 *Fagus*

Fossil pollen data for the common beech (*Fagus sylvatica*) indicate two main refugia in the south of Europe: the Carpathian mountains and Italy (Demesure *et al.*, 1996).

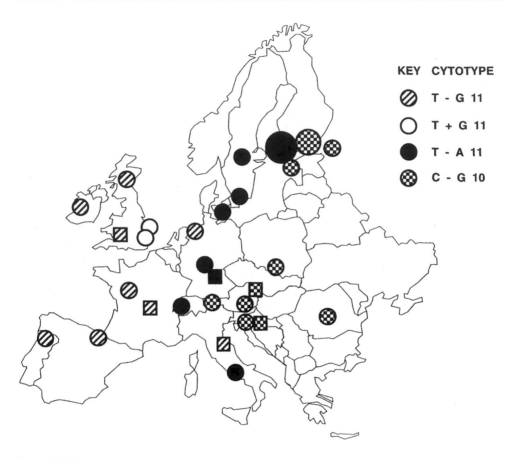

Figure 2.3 European distribution of four chloroplast DNA haplotypes in the oaks *Quercus robur* (circles) and *Q. petraea* (squares). Haplotypes were determined from sequencing 875 bases representing the tRNA$^{Leu(UAA)}$ intron and the intergenic spacer between tRNA$^{Leu(UAA)}$ and tRNA$^{Phe(GAA)}$. Symbol size is proportional to sample size (after Ferris *et al.*, 1998).

These refugia are supported by the distribution of cpDNA types throughout Europe. The phylogeographic structure of 11 haplotypes indicates the Carpathians to be a main refugium from which the north and the west of Europe were colonised. The Italian refugium gave rise to a migration which only proceeded as far as northern Italy, just south of the Alps (Demesure *et al.*, 1996). The pollen data suggest maximum rates of migration of 300 m/yr representing up to 8 km per generation (Birks, 1989; Webb, 1986). Again long distance dispersal is evident and diversity has been lost during the colonisation process. Patterns of cpDNA variation support this with most of the cpDNA diversity being in Italy and the Crimea as compared to the north and west (Demesure *et al.*, 1996). Indeed most of northern Europe is represented by a single haplotype.

2.5.3 *Alnus*

Refugia for common alder (*Alnus glutinosa*) have been proposed in the Carpathian mountains and southwest Russia, the Bay of Biscay, southern Italy, Greece (Bennett *et al.*, 1991; Huntley and Birks, 1983) and possibly Turkey (Van Zeist *et al.*, 1975). Clearly with so many refugia during the last glaciation there is the potential for high levels of diversity in the south of Europe. This has been confirmed by cpDNA studies which reveal hotspots of diversity in south-eastern Europe and Turkey (Fig. 2.4) (King and Ferris, 1998). Four cpDNA regions were PCR amplified and cut with restriction enzymes to reveal both site and size variation giving 12 haplotypes. Populations north of 45°N consist of one of two common haplotypes. One haplotype predominates in Britain and northern Europe, whilst the second is common in central regions. Of the 12 haplotyes found, 9 occur only in southern

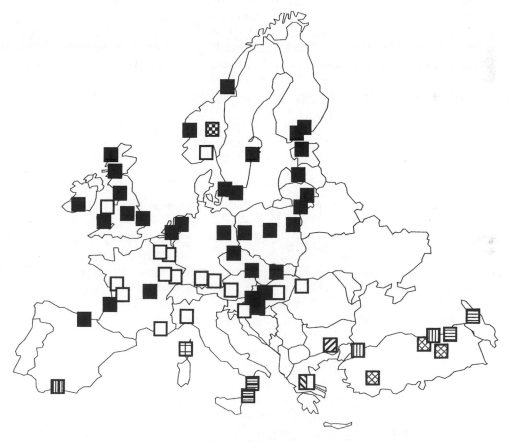

Figure 2.4 Geographical distribution of 12 cpDNA haplotypes identified in European *Alnus glutinosa* by King and Ferris (1998). Haplotypes were determined from RFLP analysis of four PCR products: tRNA$^{His(GUG)}$ – tRNA$^{Lys(UUU)}$, tRNA$^{Lys(UUU)}$ intron, tRNA$^{Ser(UGA)}$ – tRNA$^{fMet(CAU)}$ and tRNA$^{Cys(GCA)}$ – tRNA$^{Asp(GUC)}$. The haplotypes represent both size and site mutations. For polymorphic populations, shading is proportional to haplotype proportion. It can clearly be seen that two haplotypes predominate in northern Europe whilst most of the diversity within this species is held in southern Europe.

Europe; most of these are in the south-east. The decrease in cpDNA diversity from south to north can be attributed to the rapid migration at the end of the ice-age with speeds of up to 2000 m/yr (Birks, 1989). The patterns of migration in alder are similar to those of the beech with the two common haplotypes migrating north and west from the south-eastern refuge (King and Ferris, 1998).

2.5.4 Other species

If we compare the above species with genera adapted to colder climates, such as *Betula*, the patterns obtained can be very different. Birches were not restricted far enough to the south for their refugia to be isolated and thus had a large fairly continuous distribution across much of Europe during the period of glaciation (Huntley and Birks, 1983). Chloroplast DNA was found to be invariant over many *Betula* species throughout Europe, as were the ribosomal DNA ITS regions (C. Ferris, unpublished). This absence of variation in comparison with oak, beech and alder is unexpected and interesting. It may reflect a slow rate of sequence evolution in *Betula*. Clearly, other factors may at least in part be responsible for the lack of variation, including: 1) very effective seed dispersal and the propensity for very long distance gene flow; 2) the European birches may have only recently evolved; 3) birch may have been severely restricted in its distribution some time during the last few glaciations. This could have been during a previous interglacial, perhaps warmer than the present one, where birch was lost from much of southern and central Europe and thus reducing the diversity.

2.6 Divergence and speciation

Care must be taken when estimating the timing of phylogenetic events based upon DNA sequence divergence as the extent of the error on the estimate is unknown. Clearly, the information gained is useful and may give a valuable insight into the timing of evolutionary events. In general for mitochondrial DNA (mtDNA) in animals an evolutionary rate of 2% divergence per million years is accepted (Brown *et al.*, 1979), whilst for cpDNA in higher plants, a synonymous substitution rate of 0.1% per million years is taken (Zurawski *et al.*, 1984). Given that the Milankovitch cycles have a regular 20–100 kyr periodicity, it is possible to view divergence between taxa in relation to these glacial/interglacial cycles (Table 2.2). The Pleistocene ice-ages date back to about 2.5 Myr ago and thus would be represented by a mtDNA sequence divergence in higher animals of about 5%. A comprehensive review of molecular data on animal species differentiation is given by Hewitt (1996) and it is striking that most of the subspecific divergences fall within 6% while species tend to have higher values. A similar review of plant species differentiation would be most useful. For cpDNA the Pleistocene ice-ages would correspond to about 0.25% divergence. This is extremely close to the 0.29% found between populations of the north European oaks, *Q. robur* and *Q. petraea*, and may indicate that their refugial cytotypes have been isolated since about the time the ice-ages began. It is interesting to compare this value with the divergence between oaks and chestnuts, *Castanea*, another genus in the same family. Here divergence is 0.5% representing 5 Myr (C. Ferris, unpublished). If we compare the

Table 2.2 Divergence times between related pairs of taxa. Divergence time in Myr is also given in terms of number of Milankovitch range cycles. Note for comparison that divergence between *Quercus* and *Castanea* is approximately 5 Myr and that within any site the divergence between *Q. robur* and *Q. petraea* is 0.

Genus	Taxa	Divergence Myr	No. of range cycles	Reference
Quercus	populations	2.6	~ 20	Ferris *et al.*, 1998
Bombina	species	2–7	< 50	Szymura, 1993
Chorthippus	subspecies	0.5	~ 5	Lunt, 1994
Liriodendron	subspecies	1.5	10–15	Sewell *et al.*, 1996

two oak species, *Q. robur* and *Q. petraea*, for a single refugial area however, we find their cytotypes are virtually identical (Ferris *et al.*, 1998; Kremer *et al.*, 1991; Petit *et al.*, 1993). Such divergence rates of zero are likely to be due to the effects of extensive hybridisation and subsequent chloroplast capture between these two species where they co-occur (Gardiner, 1970; Rushton, 1993). We can compare these divergence values for oaks with the North American tulip tree, *Liriodendron tulipifera*, which has only 0.15% divergence between subspecies representing 1.5 Myr (Sewell *et al.*, 1996). Clearly, there may be problems in interpretation of divergence data due to factors such as hybridisation and introgression, and to differences in taxon concepts applied by plant and animal taxonomists.

The effects of glacial cycles on speciation have generally been understated or ignored by many workers, but more recently have been given more attention (Bennett, 1990; Hewitt, 1996). Major changes in climates over a 20–100 kyr cycle lead to geographical isolation and will therefore involve peripatric speciation. Gould (1985) proposed that there are processes at three distinct tiers of time which explain punctuational equilibria, the brief periods of rapid change in lineages which interrupt long periods of little change. The first tier represents ecological time up to thousands of years, the second tier represents geological time (millions of years), while the third tier represents mass extinctions with a periodicity of about 26 Myr (Gould, 1985). Bennett (1990) argued that 'changes in species that appear as a result of Darwinian natural selection are likely to be lost as a result of the reorganization of communities every 20–100 kyr. Environments (both biotic and abiotic) are not constant enough for evolution by natural selection to result in macroevolutionary change' and thus a fourth tier may be needed in Gould's (1985) 'paradox of the first tier'. Within the framework and timescale of the glaciations there is scope for any or all of the processes of speciation to occur. We must be aware however, that for plants, in particular the temperate species, the products of evolutionary processes taking place in northern Europe during an interglacial warm period will be lost if southward migration does not occur at the onset of glaciation. Refugial areas are probably isolated throughout the glacial/interglacial cycle. If a species becomes extinct from a refuge it will be lost from that refuge permanently (Bennett *et al.*, 1991). This may explain why hemlock, *Tsuga*, regularly colonised Britain in the interglacials but failed to arrive for the last three and the present Holocene (West, 1970). To enter Britain before its separation from the mainland by the English Channel, hemlock probably colonised from a refuge in southern Spain. Not only has it been lost from this western refuge it can also no longer reach Britain by colonisation from the east, and nor can it repopulate the Spanish refuge. The same may account for the loss of spruce, *Picea*, from Britain over the last and the present Holocene interglacials although it formerly colonised Britain in all previous interglacials for which data are available (West, 1970).

It is also informative to compare tree species found in the three main refugial areas during the last glaciation. The Italian and Balkan refugia were home to many tree species including oak, pine, hornbeam, fir, beech, elm and alder amongst others, whereas of these only pine and oak were evident as abundant fossil pollen in Spain (Bennett *et al.*, 1991). It may be that many of the common European tree genera once survived in the Spanish refuge but have been lost over time. It is vital, therefore, to fully understand the importance of these refugial areas both during the warm periods as well as during glaciation. For conservation purposes it would seem important to ensure the continued survival of the tree species and their associated flora and fauna presently in refuge populations in the mountains of southern Europe, particularly Spain. The survival of these communities is vital if we are to maintain the levels of biodiversity created throughout Europe over the last three million years.

REFERENCES

Bennett, K. D. (1990) Milankovitch cycles and their effects on species in ecological and evolutionary time. *Paleobiology*, **16**, 11–21.

Bennett, K. D., Tzedakis, P. C. and Willis, K. J. (1991) Quaternary refugia of north European trees. *Journal of Biogeography*, **18**, 103–115.

Berger, A. (1978) Long-term variations of calorific insolation resulting from the earth's orbital elements. *Quaternary Research*, **9**, 139–167.

Birks, H. J. B. (1989) Holocene isochrone maps and patterns of tree-spreading in the British Isles. *Journal of Biogeography*, **16**, 503–540.

Bradley, R. S. (1985) *Quaternary paleoclimatology: methods of paleoclimatic reconstruction.* Boston, Allen and Unwin.

Brown, W. M., George, M. J. and Wilson, A. C. (1979) Rapid evolution of animal mitochondrial DNA. *Proceedings of the National Academy of Sciences USA*, **76**, 1967–1971.

Demesure, B., Comps, B. and Petit, R. J. (1996) Chloroplast DNA phylogeography of the Common Beech (*Fagus sylvatica* L.) in Europe. *Evolution*, **50**, 2515–2520.

Dumolin-Lapègue, S., Demesure, B., Fineschi, S., Le Corre, V. and Petit, R. J. (1997) Phylogeographic structure of white oaks throughout the European continent. *Genetics*, **146**, 1475–1487.

Ennos, R. A., Sinclair, W. T., Hu, X.-S. and Langdon, A. (1999) Using organelle markers to elucidate the history, ecology and evolution of plant populations, in *Molecular systematics and plant evolution*, (eds P. M. Hollingsworth, R. M. Bateman and R. J. Gornall), London, Taylor & Francis, pp. 1–19.

Ferris, C., Oliver, R. P., Davy, A. J. and Hewitt, G. M. (1993) Native oak chloroplasts reveal an ancient divide across Europe. *Molecular Ecology*, **2**, 337–344.

Ferris, C., Oliver, R. P., Davy, A. J. and Hewitt, G. M. (1995) Using chloroplast DNA to trace postglacial migration routes of oaks into Britain. *Molecular Ecology*, **4**, 731–738.

Ferris, C., King, R. A., Väinölä, R. and Hewitt, G. M. (1998) Chloroplast DNA recognises three refugial sources of European oaks and shows independent eastern and western immigrations to Finland. *Heredity*, **80**, 584-593.

Frascaria, N., Maggia, L., Michaud, M. and Bousquet, J. (1993) The *rbcL* gene sequence from chestnut indicates a slow rate of evolution in the Fagaceae. *Genome*, **36**, 668-671.

Gardiner, A. S. (1970) Pedunculate and sessile oak (*Quercus robur* L. and *Quercus petraea* (Mattuschka) Liebl.). A review of the hybrid controversy. *Forestry*, **43**, 151-160.

Gould, S. J. (1985) The paradox of the first tier: an agenda for paleobiology. *Paleobiology*, **11**, 2-12.

Hays, J. D., Imbrie, J. and Shackleton, N. H. (1976) Variations in the earth's orbit: pacemaker of the ice ages. *Science*, **194**, 1121-1132.

Hewitt, G. M. (1996) Some genetic consequences of ice ages, and their role in divergence and speciation. *Biological Journal of the Linnean Society*, **58**, 247-276.

Huntley, B. and Birks, H. J. B. (1983) *An atlas of past and present pollen maps for Europe: 0-13 000 years ago.* Cambridge, Cambridge University Press.

Imbrie, J. (1985) A theoretical framework for the Pleistocene ice ages. *Journal of the Geological Society of London*, **142**, 417-432.

Imbrie, J. and Imbrie, K. P. (1979) *Ice ages: solving the mystery.* London, Macmillan.

King, R. A. and Ferris, C. (1998) Chloroplast DNA phylogeography of *Alnus glutinosa* (L.) Gaertn. *Molecular Ecology*, **7**, 1151-1161.

Kremer, A., Petit, R., Zanetto, A., Wagner, D. and Chauvin, C. (1991) Nuclear and organelle gene diversity in *Quercus robur* and *Q. petraea*, in *Genetic variation of forest tree populations in Europe*, (eds M. Ziehe and Müller-Starck), pp. 141-166. Frankfurt am Main: Sauerländer's Verlag.

Lunt, D.H. (1994) mtDNA differentiation across Europe in the meadow grasshopper *Chorthippus parallelus* (Orthoptera: Acrididae). PhD thesis, University of East Anglia, Norwich.

Petit, R. J., Kremer, A. and Wagner, D. B. (1993) Geographical structure of chloroplast DNA polymorphisms in European oaks. *Theoretical and Applied Genetics*, **87**, 122-128.

Ruddiman, W. F. and Raymo, M. E. (1988) Northern hemisphere climate regimes during the past 3Ma: possible tectonic connections. *Philosophical Transactions of the Royal Society of London*, **318B**, 411-430.

Rushton, B. S. (1993) Natural hybridization within the genus *Quercus* L. *Annales des Sciences Forestieres*, **50**, 73s-90s.

Sewell, M. M., Parks, C. R. and Chase, M. W. (1996) Intraspecific chloroplast DNA variation and biogeography of North American *Liriodendron* L. (Magnoliaceae). *Evolution*, **50**, 1147-1154.

Shackleton, N. J., Imbrie, J. and Pisias, N. G. (1988) The evolution of oceanic oxygen-isotope variability in the North Atlantic over the past three million years. *Philosophical Transactions of the Royal Society of London*, **318B**, 679-688.

Szymura, J.M. (1993) Analysis of hybrid zones with *Bombina*, in *Hybrid zones and the evolutionary process*, (ed R. G. Harrison), New York: Oxford University Press, pp. 261-389.

Van Zeist, W. and Bottema, S. (1977) Palynological investigations in western Iran. *Palaeohistoria*, **19**, 19-83.

Van Zeist, W., Woldring, H. and Stapert, D. (1975) Late quaternary vegetation and climate of south western Turkey. *Palaeohistoria*, **17**, 53-143.

Webb, T., III. (1986) Is vegetation in equilibrium with climate? How to interpret late-quaternary pollen data. *Vegetatio*, **69**, 177-187.

West, R. G. (1970) Pollen zones in the Pleistocene of Great Britain and their correlation. *New Phytologist*, **69**, 1179–1183.

Wolfe, K. H., Li, W. and Sharp, P. M. (1987) Rates of nucleotide substitution vary greatly among plant mitochondrial, chloroplast and nuclear DNAs. *Proceedings of the National Academy of Sciences USA*, **84**, 9054–9058.

Zurawski, G., Clegg, M. T. and Brown, A. H. D. (1984) The nature of nucleotide sequence divergence between barley and maize chloroplast DNA. *Genetics*, **106**, 735–749.

Chapter 3

The use of uniparentally inherited simple sequence repeat markers in plant population studies and systematics

J. Provan, N. Soranzo, N. J. Wilson,
J. W. McNicol, M. Morgante and W. Powell

ABSTRACT

The analysis of mitochondrial (mt) DNA polymorphism has revolutionised zoological population genetics and systematics. The mainly uniparental inheritance of the animal mitochondrial genome, coupled with the absence of recombination, provides a series of characters that allow a clear definition of maternal lineages. Plant mtDNA, however, has a very slow substitution rate, which coupled with frequent intramolecular recombination, reduces its utility as a general source of markers. Chloroplast DNA is a reliable source of characters for phylogenetic analyses, but low levels of variability have hampered attempts to utilise this molecule in population genetic studies, and studies involving very closely related species. However, the recent discovery of polymorphic simple sequence repeats in the chloroplast genomes of higher plants potentially provides a high-resolution assay to circumvent this problem. We have used chloroplast simple sequence repeat (cpSSR) polymorphism to study population differentiation in Scots pine (*Pinus sylvestris*). In Scots pine, the chloroplast genome is paternally transmitted and thus provides a pollen-specific marker. We have also used cpSSRs in a biosystematic context, to investigate the relationships between wild and cultivated rice (*Oryza* spp.) species. Our results show that cpSSR polymorphism provides an important tool for studies in plant population genetics and systematics.

3.1 Introduction

Mitochondrial DNA polymorphism has been used extensively to infer relationships in animal population genetic and phylogenetic studies (Moritz *et al.*, 1987; Avise *et al.*, 1987, 1989; Harrison, 1989). The uniparental inheritance of the mitochondrial genome, coupled with its rapid rate of sequence substitution and absence of recombination, has enabled high resolution analysis of maternal lineages. In plants, however, the mitochondrial genome has proved to be a more recalcitrant molecule; extremely low levels of nucleotide substitution and a high frequency of intramolecular recombination have restricted its application as a source of molecular markers for plant systematic studies (Wolfe *et al.*, 1987; Palmer *et al.*, 1992). Chloroplast DNA, on the other hand, has become a popular molecular marker for botanical phylogeneticists, and both sequencing and restriction fragment approaches have been pursued (Olmstead and Palmer, 1994). The entire chloroplast genome has

In *Molecular systematics and plant evolution* (1999) (eds P. M. Hollingsworth, R. M. Bateman and R. J. Gornall), Taylor & Francis, London, pp. 35–50.

been sequenced in several taxonomically disparate species (see Stoebe et al., 1999 – this volume) including Epifagus virginiana (Wolfe et al., 1992), Marchantia polymorpha (Ohyama et al., 1986), Nicotiana tabacum (Shinozaki et al., 1986), Oryza sativa (Hiratsuka et al., 1989), Pinus thunbergii (Wakasugi et al., 1994) and Zea mays (Maier et al., 1995). The relatively conserved gene order of the molecule and the wide availability of sequence information has facilitated the design of universal primers, and PCR based approaches are now a rapid and efficient method of generating informative data (Demesure et al., 1995; Dumolin-Lapegue et al., 1997; Taberlet et al., 1991). However, the relatively low levels of nucleotide substitution in the chloroplast genome (Wolfe et al., 1987) have limited the scope of chloroplast-based studies at the intra-specific level; for reviews see Soltis et al. (1992), McCauley (1995) and Ennos et al. (1999 – this volume). Palmer et al. (1988) stressed the need to develop technical strategies which render the analysis of cpDNA variation more useful at lower taxonomic levels.

Simple sequence repeats (SSRs: for review see Powell et al., 1996), also known as microsatellites, are a powerful new class of genetic marker that are being used extensively in animal and plant genetic analysis. The advantages of SSRs arise from their relative abundance in eukaryotic genomes, high levels of allelic diversity, co-dominance and ease of analysis using PCR. Until recently, the exploitation of PCR-based SSR length assays has been restricted to the nuclear genome. However, Powell et al. (1995a,b) have recently reported the occurrence of polymorphic mononucleotide repeats in the chloroplast genomes of soybean and pines which are analogous to nuclear microsatellites or simple sequence repeats. The publication of the complete nucleotide sequences of twelve plastid genomes (see Stoebe et al., 1999 – this volume) has provided the opportunity to design high-resolution assays based around PCR amplification of regions containing simple-sequence repeats. Such assays should provide new insights into cytoplasmic variation and allow for greater resolution than that achievable by conventional RFLPs with lower costs and greater efficiency than sequencing. The specific questions addressed by our work are summarised as follows:

1. Do SSRs occur in the chloroplast?
2. Can we detect length variation?
3. What is the molecular basis for the polymorphism?

Having addressed these points we shall then describe two case studies which deal with the extent of the polymorphism shown by cpSSRs and their value in population genetics and biosystematics.

3.2 Occurrence of cpSSRs and levels of polymorphism detected

We have searched the EMBL and GenBank nucleotide sequence databases for mononucleotide repeats of ten or more base-pairs. From six of the completely sequenced plant chloroplast genomes (Marchantia polymorpha, Pinus thunbergii, Epifagus virginiana, Nicotiana tabacum, Oryza sativa and Zea mays) and partial chloroplast sequences from other species we detected a total of 505 cpSSRs

(Table 3.1). Poly (A·T) repeats were found to be the most common, accounting for almost 99% of all SSRs found in chloroplast genomes.

Primers flanking the poly (A·T) repeat motif in the soybean *trn*M gene were used to amplify total genomic DNA from *Glycine* subgenera *Soja* and *Glycine* by PCR and a total of 10 variants was identified, ranging in size from 94bp to 103bp (Powell *et al.*, 1995b). In order to confirm that the polymorphism was due to expansion or contraction of the mononucleotide repeat region, two approaches were pursued. Firstly, an alternative pair of primers was designed which amplified a region of the *trn*M gene outwith the repeat region and subsequent PCR of the same genomic DNA samples generated a monomorphic 109bp product. Secondly, a sample of the PCR products generated using the cpSSR primers was sequenced and it was found that the length of the PCR products was directly correlated with the number of repeats in the cpSSR motif and, furthermore, the flanking sequences were found to be identical in each case (Powell *et al.*, 1995b). To confirm the maternal transmission of the cpSSRs in soybean, an artificial interspecific cross between female *Glycine soja* and male *G. max* was also examined. Variation was detected between the two parents and the progeny were monomorphic for the *G. soja* allele, demonstrating that *G. soja* was the female parent and confirming the maternal transmission of the chloroplast organelle.

The experiments outlined here have shown that SSR variability occurs in the chloroplast genome and that these repeats may be a source of polymorphic markers. The benefits of markers from cpDNA have been stressed by Ennos *et al.* (1999 – this volume) and, in particular, the largely uniparental mode of transmission of cpDNA means that cpSSRs provide seed-specific markers in angiosperms (where the chloroplast is predominantly maternally inherited) and pollen-specific markers in gymnosperms (where paternal inheritance is the rule). As with similar studies on SSRs in the non-recombining region of the human Y-chromosome (Deka *et al.*, 1996; Goldstein *et al.*, 1996; Kayser *et al.*, 1997; de Knijff *et al.*, 1997; Perez-Lezaun *et al.*, 1997a) and studies of animal mitochondrial DNA (Avise *et al.*, 1987, 1989; Moritz *et al.*, 1987; Harrison, 1989), data from multiple, physically linked cpSSR loci can be combined to give highly informative haplotypes for population genetic and biosystematic studies.

Table 3.1 Number of SSRs observed in the fully sequenced chloroplast genomes and other chloroplast sequences present in GenBank.

	Marchantia polymorpha	Pinus thunbergii	Epifagus virginiana	Nicotiana tabacum	Oryza sativa	Zea mays	Other	Total
A·T	82	19	25	39	12	30	225	485
G·C	0	1	0	0	0	2	3	6
AT·TA	5	0	1	0	0	5	1	12
TC·AG	0	0	0	0	0	1	0	1
ATT·TAA	0	0	0	0	0	1	0	1
Total	87	20	26	39	12	39	229	505
Genome (kb)	121	120	70	156	135	135		

3.3 Case studies

3.3.1 Analysis of natural populations of Scots pine (Pinus sylvestris)

The native populations of Scots pine (*P. sylvestris*) in Scotland represent the only remaining natural pine woodland in the UK. Excessive exploitation and mismanagement over the last few hundred years has led to severe depletion of the Caledonian pinewoods and recent interest in the historical, scientific and recreational aspects of these woodlands has made the conservation of Scots pine an important issue (Aldhous, 1995). Knowledge of the genetic structure of existing populations should form the basis of a rational and sustainable conservation programme. When we initiated this study, the only available molecular data on these populations was from monoterpenes and isozymes, representing five and sixteen nuclear loci respectively. No data were available from organellar genomes, nor had any data been obtained directly from DNA, as opposed to its products. DNA-based techniques, however, are now available for the molecular genetic analysis of plant populations (for review see Powell *et al.*, 1995c). We have studied levels of cpSSR variation in seven native Scottish and eight European populations of Scots pine to quantify levels of differentiation both within and between populations (Provan *et al.*, 1998). Our aims were to compare levels of polymorphism detected using cpSSRs with those obtained previously from monoterpenes and isozyme analyses and to study the relationships between native Scottish and European populations.

Analysis of 13 polymorphic cpSSRs in the seven Scottish populations revealed 174 haplotypes in the 330 individuals studied. Average diversity values (Nei, 1987) across individual loci (0.268–0.365) were comparable with earlier monoterpene (0.272–0.378) and isozyme (0.291–0.311) analyses carried out on the same populations (Kinloch *et al.*, 1986). Since there is no recombination in the chloroplast molecule, the data from the seventeen physically-linked loci were combined to give haplotypes and diversity values based on haplotype frequency ranged from 0.950 to 0.987. Around 38% of the trees studied could be individually genotyped, highlighting the power of cpSSRs for uniquely differentiating single trees.

Using five of the polymorphic cpSSR loci on the seven Scottish and eight mainland European populations, 133 different haplotypes were found in the 515 individuals studied (Fig. 3.1). From this, it can be seen that only one haplotype (haplotype 1) was found in all fifteen populations and that a large number of the woodlands contained private haplotypes (i.e. found only in a single woodland). The total within-species gene diversity value based on the five loci was 0.977, which was much higher than levels previously reported in cpSSR studies in European populations of *P. leucodermis* (0.411: Powell *et al.*, 1995a) and *P. halepensis* (0.596: Morgante *et al.*, 1997) and American *P. resinosa* populations (0.568: Echt *et al.*, 1998). *Pinus sylvestris* had a continuous distribution throughout Europe before the last glaciation whereas the other species have been highly fragmented since the Tertiary period (Mirov, 1967). This is reflected in the amounts of genetic variability partitioned between populations. An analysis of molecular variance (AMOVA) of cpSSR variability in *P. sylvestris* showed that a small, non-significant (1.48%, $P = 0.1331$) portion of the total variation existed between the Scottish and the European woodlands, which would suggest that there is no genetic difference

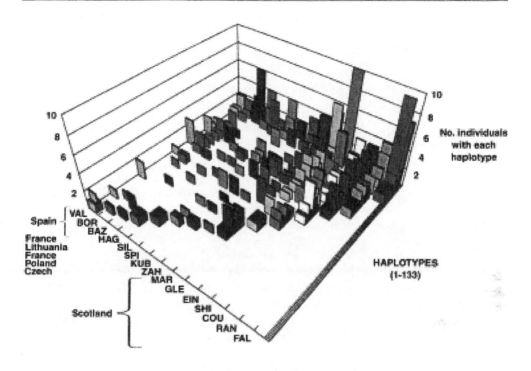

Figure 3.1 Distribution of cpSSR haplotypes in fifteen Scots pine populations. Abbreviations are as follows: VAL – Valsain; BOR – Borau; BAZ – Baza; HAG – Haguenau; SIL – Silene; SPI – Spitzberg; KUB – Kubryk; ZAH – Zahorie; MAR – Loch Maree; GLE – Glenmore; EIN – Glen Einig; SHI – Shieldaig; COU – Coulin; RAN – Rannoch; FAL – Glen Falloch.

between Scottish and European populations of Scots pine based on cpSSRs. Analysis of cpSSR data from populations of *P. leucodermis, P. halepensis* and *P. resinosa* showed a greater proportion of genetic diversity between populations (22%, 48% and 12% respectively). The lack of significant cpSSR differentiation between Scottish and mainland European populations, despite the presence of morphological differences, could be explained if the mutation rate of these markers is so high that recurrent mutations have obscured population structure. We do not believe, however, that this is the case. Previous studies in our lab on the narrow endemic *P. torreyana* suggests that although the mutation rates at cpSSR loci are higher than substitution rates in the rest of the chloroplast genome, it is unlikely that they are of the order of magnitude of migration rate (Provan *et al.*, unpublished). Thus, mutations at cpSSR loci would not appear to occur at a level which would confound analyses of population genetic structure. This is reflected in the high levels of population differentiation reported using cpSSRs in the other *Pinus* species mentioned previously.

We have demonstrated here that cpSSR polymorphism can be used to analyse diversity in the chloroplast genome of *P. sylvestris*. Ultimately, when cpSSR polymorphisms are considered along with nuclear and mitochondrial markers (Sinclair *et al.*, 1998) to give an indication of the overall history and genetic architecture of our native pinewoods, this understanding will be a valuable asset when considering legislative decisions concerning the management of existing populations of Scots pine.

3.3.2 Biosystematics of wild and cultivated rice (Oryza spp.)

Rice (*Oryza sativa*) is the world's most important food crop and is considered to be a model plant for cereal genome research (Kurata *et al.*, 1994). The conservation, enhancement and utilisation of rice biodiversity is of paramount importance. The complete chloroplast DNA sequence of rice (Hiratsuka *et al.*, 1989) provides an opportunity to explore the use of chloroplast mononucleotide repeats to generate informative assays for the detection of both intra- and interspecific chloroplast variability in cultivated rice and its wild relatives. We have used cpSSRs to study subspecies differentiation in cultivated rice as well as to examine the relationships between diploid and tetraploid *Oryza* species.

Between two and five alleles were detected at six polymorphic cpSSR loci in a range of 20 wild and cultivated *Oryza* species (Table 3.2) and Nei's diversity values for individual loci ranged from 0.07 to 0.72 in the total sample. Analysis of the joint distribution of alleles at five of these loci (the sixth only being polymorphic in a single wild accession) gave rise to 15 haplotypes in the 20 accessions studied and an overall diversity value of 0.90 (Provan *et al.*, 1996). Rice varietal classification is recognised as a pre-requisite for rational rice breeding programmes and the methods presented here will be complementary to nuclear-based assays. The sizes of amplification products generated at these five cpSSR loci will allow multiplexing of all the loci on a single polyacrylamide gel and the resulting haplotype data may be used to genotype rice accessions. The inclusion of wild *Oryza* species also demonstrates that primers designed to amplify chloroplast mononucleotide repeats in *O. sativa* can also amplify corresponding size variants in related species.

In a larger, parallel study of 53 cultivated rice (*O. sativa*) accessions (Provan *et al.*, 1996), 35 haplotypes were detected and a comparison of the variation observed in the two subspecies of *O. sativa*, ssp. *indica* and ssp. *japonica*, shows that the variation detected within the *japonica* accessions (0.898) was lower than that detected in *indica* (0.936). This reflects similar studies in subspecies diversity previously carried out using isozymes (Chen *et al.*, 1994) and nuclear RFLPs (Zhang *et al.*, 1992), although the levels of variation revealed by the cpSSR assay are much higher. The distribution of alleles at the five cpSSR loci was compared in subspecies *indica* and *japonica* to establish if the variation detected at cpSSR loci was related to differences between the subspecies. A contingency χ^2 test for independence showed significant association ($p < 0.05$) of allele size with different subspecies for four of the five cpSSR loci (Fig. 3.2), indicating that the morphological basis of subspecies differentiation is supported by the allele distributions at these cpSSR loci.

In a further study (Provan *et al.*, 1997), we have used cpSSRs to examine the relationships between wild and cultivated rice species. Nayar (1973) has described the genome classification of *Oryza* based on meiotic pairing of F_1 hybrids where different species are classified by the letters A to E. Diploids are designated AA, BB, CC etc., depending on which genome they have and tetraploids containing multiple genome types are designated by four letter codes. In this study, seven loci

Table 3.2 Allele sizes, numbers of alleles and diversity indices for six rice cpSSR loci.

Species	Name	SSR					
		OSCP3536	OSCP75969	OSCP76221	OSCP78412	OSCP80599	OSCP134510
O. sativa ssp. japonica	S201	92	82	123	178	95	140
	Primorsh	92	82	123	178	95	140
	Kosh 275	92	82	123	178	95	140
	Changpai	92	82	123	178	95	140
	Shinano Moshi	92	82	123	178	95	140
	Chacarero FA	92	82	123	178	94	139
O. sativa ssp. indica	Guizhao	92	81	123	178	97	143
	IR36	92	81	122	177	97	142
	Wu Tao Yeh Tao	92	82	123	177	96	141
	Ai Yeh Hu	92	81	123	174	96	141
	Teqing	92	82	123	178	97	142
	DEH7	92	82	123	178	95	142
	Yu Tao	92	82	123	174	97	141
	BS125	92	82	123	178	97	142
O. longistaminata		92	82	123	178	97	142
O. nivara		92	82	122	179	97	141
O. rufipogon		92	82	123	178	97	141
O. rufipogon		92	81	122	178	97	141
O. nivara/rufipogon		92	82	122	179	98	142
O. spontanea		93	80	122	177	95	142
Number of alleles		2	3	2	4	5	5
Diversity		0.070	0.397	0.375	0.540	0.570	0.720

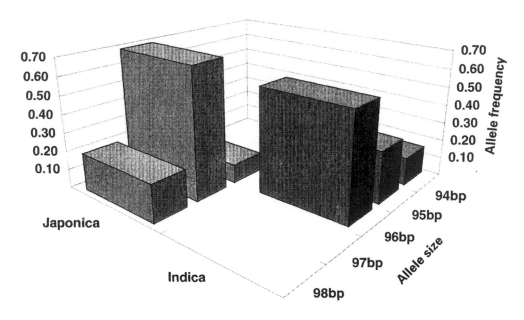

Figure 3.2 Allele distribution at locus OSCP80599 in *indica* and *japonica* subspecies of *O. sativa*.

were found to be polymorphic (three to six alleles present per locus) in 42 accessions from a total of 17 diploid and tetraploid species. The two loci that detected most differentiation between Nayar's genome types, OSCP80599 and OSCP134510 (Provan *et al.*, 1997), were used to construct a two-dimensional plot showing the distribution of alleles at these loci – effectively a two-locus haplotype (Fig. 3.3). From this, it was found that the direction of mutational events at these two loci appears to be correlated and a number of additional interesting features are apparent. The diploid species with the BB and CC genomes shared the same cpSSR haplotypes, which were not found in species with the AA genome. Furthermore, the cpSSR haplotypes suggest two genetically distinct subgroups within the BBCC genome types. The African BBCC genome species (*O. punctata*) shares a cpSSR haplotype with the BB and CC genome species including diploid *O. punctata* (BB). The tetraploid Asian BBCC genome species, *O. minuta* and *O. malampuzhaensis*, however, share cpSSR haplotypes with the diploid AA species, *O. nivara*. This is indicative of at least two separate origins of tetraploids within the BBCC complex, as highlighted by Wang *et al.* (1992), who also suggested that CCDD genome species could have arisen in a single polyploidization event. While relatively few CCDD genome accessions were included in this study, they do not show the same bi-modal distribution of alleles as was noted for the BBCC genome species. Three of the four haplotypes found in CCDD species are shared with individuals possessing the AA genome. Wang *et al.* (1992) found no evidence of this in their studies on nuclear RFLPs but a relationship between the cytoplasms of AA and CCDD genome species has been proposed by Ichikawa *et al.* (1986) based on previous studies of chloroplast markers (RFLPs and *rbc*L sequences).

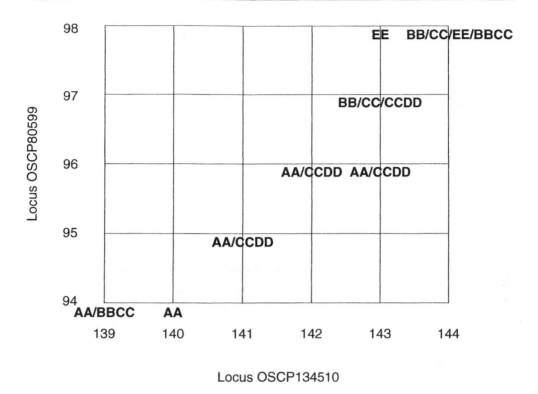

Figure 3.3 Two-dimensional representation of alleles present at rice cpSSR loci OSCP80599 and OSCP134510 in diploid and tetraploid rice genomes (after Provan *et al.*, 1997).

In conclusion, we have shown that cpSSR haplotypes provide complementary information to nuclear assays in genotyping studies in cultivated rice and its relatives. Primers designed to amplify cpSSRs in cultivated rice can also be used to examine diversity in related wild species. These factors, together with the substantial information content of cpSSRs, suggests that they will have an important role to play in genotyping and biosystematic studies, including determining the possible origins of polyploid complexes.

3.4 Problems and concerns associated with cpSSR analysis

Although cpSSRs offer many potential advantages as tools for plant systematists, they share many of the limitations associated with nuclear microsatellites. So far, cpSSRs have mainly only been studied in those species whose chloroplast genomes have been fully sequenced, or their close relatives. Attempts to design 'universal' primers which amplify polymorphic cpSSRs across higher plants have so far been unsuccessful, mainly due to non-conservation or degeneration of mononucleotide repeats across species (Provan, unpublished results). As an alternative, the complete chloroplast sequences of *Oryza sativa* (angiosperm: monocot),

Nicotiana tabacum (angiosperm: dicot), *Pinus thunbergii* (gymnosperm) and *Marchantia polymorpha* (bryophyte) were compared with the goal of identifying possible regions of SSR conservation. Genes or regions of the genome where mononucleotide repeats of ten or more repeats were found in more than one of the above species are given in Table 3.3. Depending on the taxonomic affinity of any given study species, this table could be used to highlight a few regions to sequence which would give a high probability of finding a cpSSR for which specific primers could be designed. A more extensive survey including partially sequenced chloroplast genomes is currently underway.

Data interpretation problems are primarily related to the mutational processes which occur at SSR loci (Callen *et al.*, 1993; Jarne and Lagoda, 1996; Goldstein and Pollock, 1997; Orti *et al.*, 1997). Indeed, it has been shown through theoretical and empirical studies that a detailed understanding of these processes is essential for the correct utilisation of these markers in population genetic and phylogenetic studies.

SSR mutation processes have been elucidated through both *in vitro* experiments (Henderson and Petes, 1992; Schlötterer and Tautz, 1992) and the analysis of human pedigrees (Weber and Wong, 1993). The majority of mutations at SSR loci involve the gain or loss of one or two repeats during DNA replication (Levinson and Gutman, 1987) and the distribution of allelic variants at individual loci conform largely to the stepwise mutation model (SMM) of Ohta and Kimura (1973). Larger mutations involving multiple repeat units have, however, been observed, albeit at much lower frequency (DiRienzo *et al.*, 1994).

One consequence of the bi-directional mutation process of SSRs is the potential occurrence of substantial levels of homoplasy. This occurs through the discrepancy between alleles which are identical by descent (IBD: i.e. have descended from the same ancestral state) and those which are identical in state (IIS) but are not IBD. In the simple case where an individual has gained a repeat unit, followed by a subsequent loss of a repeat unit, it will display the same allele as a similar individual which has not mutated. The extent of homoplasy in any given data set is largely dependent upon the mutation rate at individual loci, as well as the level of

Table 3.3 Genes/intergenic regions which contain cpSSRs across rice (*Oryza sativa*), tobacco (*Nicotiana tabacum*), black pine (*Pinus thunbergii*) and a liverwort (*Marchantia polymorpha*).

Region	Species			
	O. sativa	*N. tabacum*	*P. thunbergii*	*M. polymorpha*
*trn*K	Yes	Yes	Yes	
*rpo*C2	Yes	Yes		
Upstream *trn*S	Yes	Yes		
Downstream *trn*S			Yes	Yes
*trn*G		Yes	Yes	
*psb*C		Yes	Yes	
*rbc*L		Yes	Yes	Yes
*trn*R		Yes		Yes
*trn*I/*trn*A			Yes	Yes

divergence between the taxa. Given the variability of mutation rates and the inequality of ranks across taxonomic groups, predictions on the most appropriate taxonomic level for SSRs remain difficult. Studies by Estoup *et al.* (1995) using interrupted SSR motifs in the honeybee *Apis mellifera* suggest that homoplasy is most prevalent above the species/subspecies level, although notable levels of homoplasy have been observed within populations of the horseshoe crab, *Limulus polyphemus* (Orti *et al.*, 1997). The lower number of alleles generally detected in cpSSRs relative to nuclear SSRs suggests a lower mutation rate and thus the problems of homoplasy in chloroplast repeats may not be as pronounced at lower taxonomic levels as in studies utilising their nuclear counterparts. However, as levels of divergence increase, homoplasy is likely to become a problem and a recent comparison of cpRFLPs with two cpSSR loci indicated considerable size homoplasy for alleles at SSR loci within the genus *Glycine* (Doyle *et al.*, 1998).

In addition to the recurrent origins of allelic variants, other aspects of the mutational process at SSR loci require some consideration. Foremost among these are variation in mutation rates between loci and different repeat motifs, directional asymmetry of mutation and the observation of apparent constraints on allele size at SSR loci. Mutation rates at SSR loci can differ widely between different types of repeat motif (Weber and Wong, 1993; Chakraborty *et al.*, 1997), between loci (Schlötterer, 1998) and even between alleles at the same locus (Wierdl *et al.*, 1997), leading to problems in calculating divergence times. These problems are further confounded by the observations of directionality bias at SSR loci (Garza *et al.*, 1995; Zhivotovsky *et al.*, 1997), where alleles of a small size are more likely to gain repeats and alleles of a large size more likely to lose them. The main consequence of this is the limitation of allele sizes at SSR loci (Bowcock *et al.*, 1994; Goldstein *et al.*, 1995a), leading to discrepancies in the linear relationship between allele size and genetic distance/divergence. The rapid breakdown of extremely long repeat lengths at SSR loci through deletions may also be a factor in the limited range of observed allele sizes (Levinson and Gutman, 1987). For a more detailed review of the implications of such mutational processes on phylogenetic inference see Goldstein and Pollock (1997).

The above patterns of mutation at SSR loci have resulted in a requirement for the development of tailor-made statistics for microsatellite data. For example, Wright's (1951) F_{ST} is a popular approach to analyse the distribution of genetic variation between natural populations. However, F_{ST} assumes both a low mutation rate (not true for many SSR loci) and an adherence to the infinite alleles model (IAM), where each mutation can give rise to an infinite range of alleles which are considered to be equidistant. Since this may not be appropriate for the SMM, Slatkin (1995) formulated a new statistic, R_{ST}, which considers the proportions of variance in allele sizes between and within populations and is theoretically more appropriate for describing population structure based on SSR data that conform to a strict stepwise mutation model with no size constraints. However, given that most SSR loci do not follow the model exactly (DiRienzo *et al.*, 1994), there is still a great deal of debate concerning the application of such statistics to SSR data sets, particularly those in which the relative roles of migration and genetic drift are unknown (Viard *et al.*, 1996; Paetkau *et al.*, 1997; Perez-Lezaun *et al.*, 1997b; Santos *et al.*, 1997; Valsecchi *et al.*, 1997). Likewise, distances which have been formulated under the IAM, i.e. Nei's D_A distance (1987), tend to be inaccurate for SSR loci. Various distance metrics have been proposed to accommodate the SMM, including Goldstein's $\delta\mu^2$ (Goldstein *et al.*, 1995b), which is the squared difference in repeat sizes between populations averaged across loci, and Shriver *et al.*'s (1995) stepwise distance (D_{SW}), which is a modification of Nei's minimum genetic distance. Several studies have been carried out to assess the relative merits of these and other metrics (Goldstein *et al.*, 1995a; Takezaki and Nei, 1996; Perez-Lezaun *et al.*, 1997b) and in general, the more divergent the taxa under study, the more appropriate the SMM metrics ($\delta\mu^2$ and D_{SW}) become (Goldstein and Pollock, 1997).

Mutational processes occurring at SSR loci (including cpSSRs) have led to a need for the development of new statistics for the analysis of SSR data and there remains much debate about the utilisation of SSRs for phylogenetic and population analyses. Further studies comparing chloroplast and nuclear SSRs with other classes of DNA markers will be informative in shedding light on mutational processes, as well as evaluating the scope and limitations of these markers as tools in plant evolutionary biology.

3.5 Conclusions

The uniparentally inherited nature of organellar molecular markers means that they can provide seed- and pollen-specific markers for systematic and population studies. We have demonstrated a generic approach to assessing genetic variation in plants based on PCR-mediated analysis of simple sequence repeats specific to the chloroplast genome. Such analyses should allow investigation into the micro- and macro-evolutionary processes operating in plants, and in particular provide a method of detecting variation in the otherwise highly conserved chloroplast genome.

ACKNOWLEDGEMENTS

JP, NS, NJW, JWMcN and WP were supported by the Scottish Office Agriculture, Environment and Fisheries Department. Aspects of this research also received support from NATO (Grant No. CRG 940005). The authors would like to thank Pete Hollingsworth for his comments on the original draft of the manuscript.

REFERENCES

Aldhous, J. R. (ed.) (1995) *Our pinewood heritage.* The Forestry Authority, Surrey, UK.

Avise, J. C., Arnold, J., Ball, R. M., Bermingham, E., Lamb, T., Neigel, J. E. *et al.* (1987) Intraspecific phylogeography – the mitochondrial DNA bridge between population genetics and systematics. *Annual Review of Ecology and Systematics*, **18**, 489–522.

Avise, J. C., Bowen, B. W. and Lamb, T. (1989) DNA fingerprints from hypervariable mitochondrial genotypes. *Molecular Biology and Evolution*, **6**, 258–269.

Bowcock, A. M., Ruiz-Linares, A., Tomfohrde, J., Minch, E., Kidd, J. R. and Cavalli-Sforza, L. L. (1994) High resolution of human evolutionary trees with polymorphic microsatellites. *Nature*, **368**, 455–457.

Callen, D. F., Thompson, A. D., Shen, Y., Phillips, H. A., Richards, R. I., Mulley, J. C. and Sutherland, G. R. (1993) Incidence and origin of null alleles in the $(AC)_n$ microsatellite markers. *American Journal of Human Genetics*, **52**, 922–927.

Chakraborty, R., Kimmel, M., Stivers, D. N., Davison, L. J. and Deka, R. (1997) Relative mutation rates at di-, tri- and tetranucleotide microsatellite loci. *Proceedings of the National Academy of Sciences USA*, **94**, 1041–1046.

Chen, W-B., Sato, Y-I., Nakamura, I. and Nakai, H. (1994) *Indica-Japonica* differentiation in Chinese rice landraces. *Euphytica*, **74**, 195–201.

Deka, R., Jin, L., Shriver, M. D., Yu, L. M., Saha, N., Barrantes, R. *et al.* (1996) Dispersion of human Y chromosome haplotypes based on five microsatellites in global populations. *Genome Research*, **6**, 1177–1184.

deKnijff, P., Kayser, M., Caglia, A., Corach, D., Fretwell, N., Gehrig, C. *et al.* (1997) Chromosome Y microsatellites: population genetic and evolutionary aspects. *International Journal of Legal Medicine*, **110**, 134–140.

Demesure, B., Sodzi, N. and Petit, R. J. (1995) A set of universal primers for amplification of polymorphic noncoding regions of mitochondrial and chloroplast DNA in plants. *Molecular Ecology*, **4**, 129–131.

DiRienzo, A., Peterson, A. C., Garza, J. C., Valdes, A. M., Slatkin, M. and Freimer, N. B. (1994) Mutational processes of simple-sequence repeat loci in human populations. *Proceedings of the National Academy of Sciences USA*, **91**, 3166–3170.

Doyle, J. J., Morgante, M., Tingey, S. V. and Powell, W. (1998) Size homoplasy in chloroplast microsatellites of wild and perennial relatives of soybean (*Glycine* Subgenus *Glycine*). *Molecular Biology and Evolution*, **15**, 215–218.

Dumolin-Lapegue, S., Pemonge, M. H. and Petit, R. J. (1997) An enlarged set of consensus primers for the study of organelle DNA in plants. *Molecular Ecology*, **6**, 393–397.

Echt, C. S., DeVerno, L. L., Anzidei, M. and Vendramin, G. G. (1998) Chloroplast microsatellites reveal population genetic diversity in red pine, *Pinus resinosa* Ait. *Molecular Ecology*, **7**, 307–316.

Ennos, R. A., Sinclair, W. T., Hu, X.-S. and Langdon, A. (1999) Using organelle markers to elucidate the history, ecology and evolution of plant populations, in *Molecular systematics and plant evolution*, (eds P. M. Hollingsworth, R. M. Bateman and R. J. Gornall), Taylor & Francis, London, pp. 1–19.

Estoup, A., Taillez, C., Cornuet, J-M. and Solignac, M. (1995) Size homoplasy and mutational processes of interrupted microsatellites in Apidae species, *Apis mellifera* and *Bombus terrestris*. *Molecular Biology and Evolution*, **12**, 1074–1084.

Garza, J. C., Slatkin, M. and Freimer, N. B. (1995) Microsatellite allele frequencies in humans and chimpanzees, with implications for constraints in allele size. *Molecular Biology and Evolution*, **12**, 594–603.

Gepts, P. and Clegg, M.T. (1989) Genetic diversity in pearl millet (*Pennisetum glaucum* (L.)R. Br.) at the DNA sequence level. *Journal of Heredity*, **80**, 203–208.

Goldstein, D. B. and Pollock, D. D. (1997) Launching microsatellites: a review of mutation processes and methods of phylogenetic inference. *Journal of Heredity*, **88**, 335–342.

Goldstein, D. B., Ruiz-Linares, A., Cavalli-Sforza, L. L. and Feldman, M. W. (1995a) An evaluation of genetic distances for use with microsatellite loci. *Genetics*, **139**, 463–471.

Goldstein, D. B., Ruiz-Linares, A., Cavalli-Svorza, L. L. and Feldman, M. W. (1995b) Genetic absolute dating based on microsatellites and the origin of modern humans. *Proceedings of the National Academy of Sciences USA*, **92**, 6723–6727.

Goldstein, D. B., Zhivotovsky, L. A., Nayar, K., Ruiz-Linares, A., Cavalli-Sforza, L. L. and Feldman, M. W. (1996) Statistical properties of the variation at linked microsatellite loci: implications for the history of human Y chromosomes. *Molecular Biology and Evolution*, **13**, 1213–1218.

Harrison, R. G. (1989) Animal mitochondrial DNA as a genetic marker in population and evolutionary biology. *Trends in Ecology and Evolution*, **4**, 6–11.

Henderson, S. T. and Petes, T. D. (1992) Instability of simple sequence DNA in *Saccharomyces cerevisiae*. *Molecular Cell Biology*, **12**, 2749–2757.

Hiratsuka, J., Shimada, H., Whittier, R., Ishibashi, T., Sakamoto, M., Mori, M. *et al.* (1989) The complete sequence of the rice (*Oryza sativa*) chloroplast genome – intermolecular recombination between distinct transfer RNA genes accounts for a major plastid DNA inversion during the evolution of the cereals. *Molecular and General Genetics*, **217**, 185–194.

Ichikawa, H., Hirai, A. and Katayama, T. (1986) Genetic analyses of *Oryza* species by molecular markers for chloroplast genomes. *Theoretical and Applied Genetics*, **72**, 353–358.

Jarne, P. and Lagoda, P. J. L. (1996) Microsatellites, from molecules to populations and back. *Trends in Ecology and Evolution*, **11**, 424–430.

Kayser, M., de Knijff, P., Dieltjes, P., Krawczak, M., Nagy, M., Zerjal, T. *et al.* (1997) Applications of human microsatellite-based Y chromosome haplotyping. *Electrophoresis*, **18**, 1602–1607.

Kinloch, B. B., Westfall, R. D. and Forrest, G. I. (1986) Caledonian Scots pine: origins and genetic structure. *New Phytologist*, **104**, 703–729.

Kurata, N., Moore, G., Nagamura, Y., Foote, T., Yano, M., Minobe, Y. *et al.* (1994) Conservation of genome structure between rice and wheat. *BioTechnology*, **12**, 276–278.

Levinson, G. and Gutman, G. A. (1987) High frequency of short frameshifts in poly-CA/GT tandem repeats borne by bacteriophage M13 in *Escherichia coli* K12. *Nucleic Acids Research*, **15**, 5323–5338.

McCauley, D. E. (1995) The use of chloroplast DNA polymorphism in studies of gene flow in plants. *Trends in Ecology and Evolution*, **10**, 198–202.

Maier, R. M., Neckermann, K., Igloi, G. L. and Kossell, H. (1995) Complete sequence of the maize chloroplast genome – gene content, hotspots of divergence and fine-tuning of genetic information by transcript editing. *Journal of Molecular Biology*, **251**, 614–628.

Mirov, N.T. (1967) *The genus* Pinus, Ronald Press, New York.

Morgante, M., Felice, N. and Vendramin, G. G. (1997) Analysis of hypervariable chloroplast microsatellites in *Pinus halepensis* reveals a dramatic genetic bottleneck, in *Molecular tools for screening biodiversity: plants and animals*, (eds A. Karp, P. G. Issac and D. S. Ingram), Chapman and Hall, London.

Moritz, C., Dowling, T. E. and Brown, W. M. (1987) Evolution of animal mitochondrial DNA – relevance for population biology and systematics. *Annual Review of Ecology and Systematics*, **18**, 269–292.

Nayar, N. M. (1973) Origin and cytogenetics of rice. *Advances in Genetics*, **17**, 153–292.

Nei, M. (1987) *Molecular evolutionary genetics*, Columbia University Press, New York.

Ohta, T. and Kimura, M. (1973) The model of mutation appropriate to estimate the number of electrophoretically detectable alleles in a genetic population. *Genetics Research*, **22**, 201–204.

Ohyama, K., Fukuzawa, H., Kohchi, T., Shirai, H., Sano, T., Sano, S. *et al.* (1986) Chloroplast gene organization deduced from complete sequence of liverwort *Marchantia polymorpha* chloroplast DNA. *Nature* **322**, 572–574.

Olmstead, R. G. and Palmer J. D. (1994) Chloroplast DNA systematics – a review of methods and data-analysis. *American Journal of Botany*, **81**, 1205–1224.

Orti, G., Pearse, D. E. and Avise, J. C. (1997) Phylogenetic analysis of length variation at a microsatellite locus. *Proceedings of the National Academy of Sciences USA*, **94**, 10745–10749.

Paetkau, D., Waits, L. P., Clarkson, P. L., Craighead, L. and Strobeck, C. (1997) An empirical evaluation of genetic distance statistics using microsatellite data from bear (Ursidae) populations. *Genetics*, **147**, 1943–1957.

Palmer, J. D., Jansen, R. K., Michaels, H., Chase, M. W. and Manhart, J. R. (1988) Chloroplast DNA variation and plant phylogeny. *Annals of the Missouri Botanical Garden*, **75**, 1180–1206.

Palmer, J.D., Soltis, D. E. and Soltis, P. S. (1992) Large size and complex structure of mitochondrial DNA in 2 nonflowering land plants. *Current Genetics*, **21**, 125–129.

Perez-Lezaun, A., Calafell, F., Seielstad, M., Mateu, E., Comas, D., Bosch, E. *et al.* (1997a) Population genetics of Y-chromosome short tandem repeats in humans. *Journal of Molecular Evolution*, **45**, 265–270.

Perez-Lezaun, A., Calafell, F., Mateu, E., Comas, D., Ruiz-Pacheco, R. and Bertranpetit, J. (1997b) Microsatellite variation and the differentiation of modern humans. *Human Genetics*, **99**, 1–7.

Powell, W., Morgante, M., McDevitt, R., Vendramin, G. G. and Rafalski, J. A. (1995a) Polymorphic simple sequence repeat regions in chloroplast genomes: applications to the population genetics of pines. *Proceedings of the National Academy of Sciences USA*, **92**, 7759–7763.

Powell, W., Morgante, M., Andre, C., McNicol, J. W., Machray, G. C., Doyle, J. J. *et al.* (1995b) Hypervariable microsatellites provide a general source of polymorphic DNA markers for the chloroplast genome. *Current Biology*, **5**, 1023–1029.

Powell, W., Orozco-Castillo, C., Chalmers, K. J., Provan, J. and Waugh, R. (1995c) Polymerase chain reaction-based assays for the characterisation of plant genetic resources. *Electrophoresis*, **16**, 1726–1730.

Powell, W., Machray, G. and Provan, J. (1996) Polymorphism revealed by simple sequence repeats. *Trends in Plant Science*, **1**, 215–222.

Provan, J., Corbett, G., Waugh, R., McNicol, J. W., Morgante, M. and Powell, W. (1996) DNA fingerprints of rice (*Oryza sativa*) obtained from hypervariable chloroplast simple sequence repeats. *Proceedings of the Royal Society of London* **B**, **263**, 1275–1281.

Provan, J., Corbett, G., McNicol, J. W. and Powell, W. (1997) Chloroplast variability in wild and cultivated rice (*Oryza* spp.) revealed by polymorphic chloroplast simple sequence repeats. *Genome*, **40**, 104–110.

Provan, J., Soranzo, N., Wilson, N. J., McNicol, J. W., Forrest, G. I., Cottrell, J. *et al.* (1998) Gene-pool variation in Caledonian and European Scots pine (*Pinus sylvestris* L.) revealed by chloroplast simple-sequence repeats. *Proceedings of the Royal Society of London* **B**, **265**, 1697–1705.

Santos, E. J. M., Epplen, J. T. and Epplen, C. (1997) Extensive gene flow in human populations as revealed by protein and microsatellite DNA markers. *Human Heredity*, **47**, 165–172.

Schlötterer, C. (1998) Are microsatellites really simple sequences? *Current Biology*, **8**, R132–R134.

Schlötterer, C. and Tautz, D. (1992) Slippage synthesis of simple sequence DNA. *Nucleic Acids Research*, **20**, 211–215.

Shinozaki, K., Ohme, M., Tanaka, M., Wakasugi, T., Hayashida, N., Matsubayashi, T. *et al.* (1986) The complete sequence of the tobacco chloroplast genome – its gene organization and expression. *EMBO Journal*, **5**, 2043–2049.

Shriver, M.D., Jin, L., Boerwinkle, E., Deka, R., Ferrell, R. E. and Chakraborty, R. (1995) A novel measure of genetic distance for highly polymorphic tandem repeat loci. *Molecular Biology and Evolution*, **12**, 914–920.

Sinclair, W. T., Morman J. D. and Ennos, R. A. (1998) Multiple origins for Scots pine

(*Pinus sylvestris* L.) in Scotland: evidence from mitochondrial DNA variation. *Heredity*, **80**, 233–240.

Slatkin, M. (1995) A measure of population subdivision based on microsatellite allele frequencies. *Genetics*, **139**, 457–462.

Soltis, D. E., Soltis, P. S. and Milligan, B. G. (1992) Intraspecific chloroplast DNA variation: systematic and phylogenetic implications, in *Molecular systematics of plants*, (eds P. S. Soltis, D. E. Soltis and J. J. Doyle.) Chapman and Hall, New York pp. 117–150.

Stoebe, B., Hansmann, S., Goremykin, V., Kowallik, K. V. and Martin, W. (1999) Proteins encoded in sequenced chloroplast genomes: An overview of gene content, phylogenetic information and endosymbiotic transfer to the nucleus, in *Molecular systematics and plant evolution*, (eds P. M. Hollingsworth, R. M. Bateman and R. J. Gornall), Taylor & Francis, London, pp. 327–352.

Taberlet, P., Gielly, L. and Pautou, G. (1991) Universal primers for amplification of three noncoding regions of chloroplast DNA. *Plant Molecular Biology*, **17**, 1105–1109.

Takezaki, N. and Nei, M. (1996) Genetic distances and reconstruction of phylogenetic trees from microsatellite DNA. *Genetics*, **144**, 389–399.

Valsecchi, E., Palsbøll, P., Hale, P., Glockner-Ferrari, D., Ferrari, M., Clapham, P. *et al.* (1997) Microsatellite genetic distances between oceanic populations of the humpback whale (*Megaptera novaeangliae*). *Molecular Biology and Evolution*, **14**, 355–362.

Viard, F., Bremond, P., Labbo, R., Justy, F., Delay, B. and Jarne, P. (1996) Microsatellites and the genetics of highly selfing populations in the freshwater snail *Bulinus truncatus*. *Genetics*, **142**, 1237–1247.

Wakasugi, T., Tsudzuki, J., Ito, S., Nakashima, K., Tsudzuki, T. and Sugiura, M. (1994) Loss of all *NDH* genes as determined by sequencing the entire chloroplast genome of the black pine *Pinus thunbergii*. *Proceedings of the National Academy of Sciences USA*, **91**, 9794–9798.

Wang, Z. Y, Second, G. and Tanksley, S. D. (1992) Polymorphism and phylogenetic relationships among species in the genus *Oryza* as determined by analysis of nuclear RFLPs. *Theoretical and Applied Genetics*, **83**, 565–581.

Weber, J. and Wong, C. (1993) Mutation of human short tandem repeats. *Human Molecular Genetics*, **2**, 1123–1128.

Wierdl, M., Dominska, M. and Petes, T. D. (1997) Microsatellite instability in yeast: dependence on the length of the microsatellite. *Genetics*, **146**, 769–779.

Wolfe, K. H., Li, W. H. and Sharp, P. M. (1987) Rates of nucleotide substitution vary greatly among plant mitochondrial, chloroplast and nuclear DNA. *Proceedings of the National Academy of Sciences USA*, **84**, 9054–9058.

Wolfe, K. H., Morden, C. W. and Palmer, J. D. (1992) Function and evolution of a minimal plastid genome from a non-photosynthetic parasitic plant. *Proceedings of the National Academy of Sciences USA*, **89**, 10648–10652.

Wright, S. (1951) The genetic structure of populations. *Annals of Eugenics*, **13**, 323–354.

Zhang, Q., Saghai-Maroof, M. A., Lu, T. Y. and Shen, B. Z. (1992) Genetic diversity and differentiation of *indica* and *japonica* rice detected by RFLP analysis. *Theoretical and Applied Genetics*, **83**, 495–499.

Zhivotovsky, L. A., Feldman, M. W. and Grishechkin, S. A. (1997) Biased mutations and microsatellite variation. *Molecular Biology and Evolution*, **14**, 926–933.

Chapter 4

The use of RAPD data in the analysis of population genetic structure: case studies of *Alkanna* (Boraginaceae) and *Plantago* (Plantaginaceae)

K. Wolff and M. Morgan-Richards

ABSTRACT

The use of randomly amplified polymorphic DNA (RAPD) data in studies of population genetic structure and evolution is considered. Two case studies are described to exemplify the use to which RAPD data may be put. In the first, population substructuring and gene flow in an Egyptian population of *Alkanna orientalis* is investigated, the results of which indicate that the substructuring is related to topographical barriers and that gene flow over distances of more than a few metres is mediated more by flash floods than by pollinators. In the second study, analysis of RAPD variation in *Plantago major* s.l. has revealed the existence of two species: *P. major* and *P. intermedia*, despite their morphological resemblance and absence of breeding barriers. RAPD analysis is a very good starting point for studies of relationships within and among closely related species. When the limitations are kept in mind and the techniques are applied properly RAPDs will often give a quick, easily obtainable and repeatable result, even in the hands of a beginner.

4.1 Introduction

To understand why and how the population genetic structure of a species has evolved we need to know what processes and evolutionary forces have been important in its development and maintenance. Mutation, genetic drift due to finite population size, and natural selection favouring adaptations to local environmental conditions, are forces that lead to genetic differentiation of local populations. Migration or gene flow, the movement of individuals or gametes between populations, will oppose that differentiation (Slatkin, 1987). The more gene dispersal, the less differentiation there will be, assuming other factors are constant. The importance of each force and the balance between them is difficult to ascertain. Only in the exceptional case where there is clear genotype–environment covariation can we differentiate between selection and genetic drift as causes of gene frequency mosaics. If the population structure of neutral and selected characters are studied simultaneously this may shed some light on the interaction of processes.

In inbreeding or asexual plant species, additional processes such as hitchhiking, selective sweeps and background selection may play an important role in the development of genetic structure (Hedrick, 1980; Charlesworth *et al.*, 1995). These

In *Molecular systematics and plant evolution* (1999) (eds P. M. Hollingsworth, R. M. Bateman and R. J. Gornall), Taylor & Francis, London, pp. 51–73.

processes will cause loss of genetic variation in populations at selectively neutral loci due to selection for or against alleles at other loci in the genome.

Population structure is studied to answer a broad range of questions that can roughly be divided into four categories. Firstly, we can study ecotypic differentiation by investigating the influence of natural selection on population structure. Fine scale mosaics due to selection can easily occur, especially in highly selfing species. Nevo et al. (1994) and Dawson et al. (1993) have investigated the population genetic structure of the predominantly selfing species *Aegilops peregrina* and *Hordeum spontaneum*, respectively, using allozyme electrophoresis and randomly amplified polymorphic DNA (RAPD) analysis. Significant genetic differentiation was found between populations, and specific loci or fragments were associated with soil types or aridity stress. Since both species are highly inbreeding and the allozyme loci and RAPD markers are assumed to be selectively neutral, these associations between molecular and ecological factors may be caused by high levels of linkage disequilibrium between alleles at marker loci and alleles at loci affected by selection (Hastings, 1989).

Secondly, hypotheses regarding migration rates (pollen and seed), reproductive isolation of populations and the effect of historical events can be tested. There are several plant population studies that describe the effect of refugia and recolonisation after the last glaciation (e.g. Gabrielsen et al., 1997; Le Corre et al., 1997; Ferris et al., 1999 – this volume). In addition, population structure may provide insights into the mating system of a species.

Thirdly, the analysis of population structure may answer questions related to the speciation process. By looking at population structure it may become clear how different a population is from other populations of the same or a related species. This may help us understand the evolution of varieties, subspecies or ecotypes. Recent theories on local speciation show that population dynamics and metapopulation structure are important and stress the importance of an understanding of population structure (Levin, 1995; Barrett and Pannell, 1999 – this volume).

Fourthly, genetic structure has a bearing on domestication and gene conservation in small populations. Knowledge of genetic variation and its structure is valuable when studying endangered or otherwise valuable species, for instance in order to determine how to protect or to sample populations for storage of genetic information or reintroduction in the wild (Rossetto et al., 1995; Travis et al., 1996).

4.2 The tools: allozymes and DNA markers

In the late 1960s allozymes became available to examine levels of genetic variation and they have since been studied in numerous wild species. From these studies theories were developed to explain the high levels of genetic variation generally found. Variation at the population and species level and its distribution is described in various ways (Wright, 1951; Nei, 1973; Hartl and Clark, 1989). To obtain measures that are comparable among species, all loci, monomorphic as well as polymorphic, should be included.

Hamrick and Godt (1989, 1996) collated published data from allozyme studies and concluded that, in general, widespread species have high levels of variation and that the mating system is the second most important determinant of the amount and distribution of variation. Selfing species maintain as much diversity at their polymorphic loci as outcrossed species, but a much larger proportion is found among populations. As mentioned above, theoretical and simulation studies have shown that selective sweeps and background selection could contribute to the stronger population differentiation in selfing species compared to outcrossed species (Hedrick, 1980; Charlesworth *et al.*, 1995).

Although allozymes have been successfully used for hundreds of studies into population structure, in some cases they have been shown to be not the perfect tool. Their inactivity when the tissue is not fresh, and the small number of loci that are polymorphic and that can readily be visualised, may present problems. A few polymorphic loci with little variation may not be sufficient for accurate estimation of population structure.

With the development of DNA-centred techniques, new tools have become available for studying the genetic structure of populations and it seems that for most organisms and most questions a suitable technique and marker system can be developed. As DNA is much more stable than are enzymes, samples for laboratory analyses can be collected in the field more easily. For example, plant material can be dried in the field using silica gel for subsequent extraction of the DNA in the laboratory.

Amongst the early applications at the population level was a rDNA RFLP analysis by King and Schaal (1989). They detected restriction site variation in the rRNA gene in *Rudbeckia missouriensis* and calculated the distribution of genetic variation within and among populations as well as migration rates. Another technique is DNA fingerprinting, where the 33bp repeat of M13 has been used as a minisatellite hybridisation probe (e.g. Wolff *et al.*, 1994). Generally DNA fingerprinting techniques give patterns specific to individuals although, in the strongly inbreeding species *Plantago major*, only limited numbers of phenotypes were found in some populations (Wolff *et al.*, 1994).

With the introduction of PCR-based techniques a multitude of tools became available. The main PCR technique that has been applied to the study of populations is RAPD analysis (Williams *et al.*, 1990); techniques related to it include arbitrarily primed (AP) PCR (Welsh and McClelland, 1990) and anchored inter-simple sequence repeat (inter-SSR) PCR (Zietkiewicz *et al.*, 1994). The arbitrary nature of the primers means that they can be used on any species, without previous knowledge of their DNA sequences. RAPDs have been used to study genetic variation within species, to determine relationships between closely related species and genotypes within a species, to identify particular genotypes (cultivar identification) and to study clonal structure (Kresovich *et al.*, 1992; Vierling and Nguyen, 1992; Hsiao and Rieseberg, 1994).

RAPD analysis appears to be a simple and efficient tool if allozyme electrophoresis cannot answer the questions posed. But how do RAPDs and allozymes compare when both are used on the same plants in the same study? Vicario *et al.* (1995) found that both allozymes and RAPDs showed a difference between two *Abies* species and differentiation between populations. Huff *et al.* (1993) and Peakall *et al.* (1995) compared allozyme and RAPD variability in four populations of buffalo grass (*Buchloë dactyloides*) from Mexico and Texas. There was large-scale concordance between allozyme and RAPD variation patterns, with both sets of data indicating that considerable diversity is contained between the two areas, i.e. as much allozyme variation or just over half as much RAPD variation is contained within as between individual populations. Although complete congruence between allozyme and RAPD genotype data was found in a study of *Picea mariana*, use of RAPD phenotypes can lead to biased estimates of population genetic parameters (Isabel *et al.*, 1995). In two aspen (*Populus*) species and in *Hordeum spontaneum* there were higher levels of variation with RAPD analysis than with allozyme or RFLP analysis (Liu and Furnier, 1993; Dawson *et al.*, 1993). Le Corre *et al.* (1997) examined variation in RAPDs, allozymes, and chloroplast DNA of sessile oak (*Quercus petraea*), finding that RAPDs were less polymorphic than allozymes. However, although RAPDs were less variable on a per locus basis, there were more loci available, and the pattern resolved gave a better picture of the role of geography and history in explaining population structure than did allozymes. Also, with RAPDs the genetic distance between populations was correlated with geographic distance, which was not the case for allozyme or chloroplast variation. It may be that RAPDs are a better tool than allozymes since variation in the latter may not be selectively neutral and excess of heterozygotes might occur due to over-dominance. In general RAPD analysis may sometimes show higher levels of variation than allozyme electrophoresis and its ease of use and general applicability have made it very popular. Using a combination of marker types seems to be the best approach as each marker type may have a special level of discrimination.

4.2.1 Disadvantages of RAPDs

Although a large number of studies have successfully used RAPDs and related techniques to study genetic variation in plants, there are a few problems with this method. Most of these are discussed by Harris (1999 – this volume), but those that are of particular importance for studies of population structure will be considered here.

The main shortcoming is that most RAPD fragments are dominant markers: thus, only presence and absence of a band can be scored and one does not know whether a band is present in a heterozygous or homozygous condition. Therefore, allele frequencies cannot be estimated accurately and equations and computer programs traditionally used for allozyme data cannot be applied blindly to this type of data.

Another problem is that of scaling the variability. With allozyme electrophoresis there is a relative scale (admittedly crude) by which all species can be compared for their levels of variation, if all loci, polymorphic as well as monomorphic, are included in the study. With RAPDs the choice of primers and enzyme determine, for a great part, the level of variation observed. Often primers are not randomly chosen: the 'best' primers are chosen from a preliminary survey. Monomorphic as well as faint bands are generally not scored. Therefore, with RAPD analysis, comparisons between studies are difficult if not impossible. However, if the distribution of the variation is the objective of the study the absence of a general scale of variability need not be a problem.

A more general shortcoming is that although the majority of the amplified DNA originates from the nuclear DNA, the chloroplast or the mitochondrial genome can also be the source (Lorenz *et al.*, 1994). As the chloroplast and the mitochondrial DNA evolve at different rates and have a different mode of inheritance this could be a problem (Ennos, 1994; McCauley, 1995; Ennos *et al.*, 1999 – this volume; Harris, 1999 – this volume; Provan *et al.*, 1999 – this volume).

Another issue is the homology of bands. Although it has been shown, for instance by Rieseberg (1996), that products with the same mobility in a gel are in general homologous, this is not necessarily always the case and depends in part on the taxonomic scope of the study: at the population level the problem is likely to be minimal. The alleged lack of repeatability is not confirmed in most studies, at least within a laboratory, and although patterns obtained in separate laboratories may not be identical, the conclusions drawn are the same. It should be remembered that there are many large and costly mapping projects in which RAPDs are used successfully. Indeed, despite the issues raised by Harris (1999 – this volume) we believe that RAPDs represent a time- and cost-efficient method of surveying genetic variation, at least at low taxonomic levels. The sheer number of studies that have generated apparently biologically meaningful data attests to the applicability of the technique.

4.2.2 *Quantitative analysis of RAPD variation*

In order to quantify relatedness between RAPD phenotypes, a basic analysis is the calculation of similarity for which several types of equations are available (Weising *et al.*, 1995). Generally each phenotype is compared with every other in pairwise comparisons. There are numerous ways to calculate similarity, but Nei and Li's (1979) similarity is most often used. Here twice the number of fragments in common is divided by the sum of the fragments in both phenotypes. Schierenbeek *et al.* (1997) used the Euclidean distance as well as Jaccard's index, from the RAPDistance package (Armstrong *et al.*, 1994) with very similar results. The Euclidean distance is calculated as $E = n [1 - 2n_{xy}/2n]$, where n_{xy} is the number of bands shared, and n is the total number of bands scored for that primer. Other similarity estimates may take into account not only the shared presence, but also the shared absence of a band. Rossetto *et al.* (1995) used such a similarity measure in the endangered species, *Grevillea scapigera*. They calculated the simple matching coefficient, which is the number of shared absences plus the number of shared presences all divided by the total number of bands. They argued that in endangered plants it might be safer to slightly overestimate the similarity.

To describe the level and distribution of variation, several solutions to the problem of incomplete genotype information due to dominance of the fragments have been found. One of them is the use of Shannon's diversity index (King and Schaal, 1989; Chalmers *et al.*, 1992; Russell *et al.*, 1993; Yeh *et al.*, 1995). It describes the phenotypic diversity as $H = -\Sigma p_i \log_2 p_i$, where p_i is the frequency of the *i*th band. H can be calculated at several levels: H_{popx} can be calculated for each population and then averaged over populations giving H_{pop}, and H_{sp} is the corresponding value for the species. The proportion of the variation found within populations is expressed as H_{pop}/H_{sp} and the proportion distributed among populations is $(H_{sp} - H_{pop})/H_{sp}$.

Another often-used tool is the program AMOVA (analysis of molecular variance), designed by Excoffier *et al.* (1992). Although this program was originally meant for analysis of haploid data within a species it was soon used for analysis of RAPD data (Huff *et al.*, 1993; Peakall *et al.*, 1995; Yeh *et al.*, 1995; Stewart and Excoffier, 1996). Using the distances between genotypes (distance = 1 – similarity) the program produces estimates of variance components and F-statistics analogs, reflecting the correlation of diversity at different levels of hierarchical subdivision.

However, under specific circumstances one can also attempt to estimate allele frequencies. Stewart and Excoffier (1996) have modified the AMOVA analysis for use with dominant.characters by using the amount of selfing to calculate genotypic frequencies (Stewart et al., 1996). An additional possibility is to use HOMOVA (homogeneity of molecular variance) to test whether population genetic heterogeneity is significantly different among populations (Stewart and Excoffier, 1996; Stewart et al., 1996) which may be interesting from the point of view of conservation.

Lynch and Milligan (1994) gave estimators for several population genetic parameters and their sampling variance. However, in their method the inbreeding coefficient needs to be known and very large sample sizes are required to accurately estimate allele frequencies. This poses a problem in most plant species as many have mixed mating systems that are not necessarily constant over time and over populations. If Hardy–Weinberg (HW) equilibrium and random mating in populations can be assumed, allele frequencies of the recessive alleles can be calculated as the square root of the frequency of the absence of the bands if their frequency is not too low. In this last case traditional allozyme statistics and the computer program RAPDPLOT can be used (Black, 1995). This approach was used by Travis et al. (1996) who studied population variability and structure in Astragalus cremnophylax using dominant AFLP (amplified fragment length polymorphism) markers. As part of their analysis they calculated allele frequencies by assuming HW equilibrium and from the allele frequencies F_{ST} values were determined. Although the species can successfully self, the actual mating system of the species is unknown. Also, their sample sizes were too small to have any confidence in the estimates obtained. Therefore, the approach of assuming HW equilibrium may not be appropriate in this case and the alternative analyses conducted by the authors are to be preferred.

Le Corre et al. (1997) used a different and elegant strategy in Quercus petraea. From allozyme data the outcrossing rates and inbreeding coefficient were calculated and these in turn were used to calculate RAPD allele frequencies in the populations. With these frequencies H_S and G_{ST} were calculated as traditionally with allozymes analysis. Bonnin et al. (1996) used RAPDs to describe the population structure of Medicago truncatula, which is an inbreeding annual species. Additionally, the plants collected were selfed for two to three generations before being analysed. Therefore the authors assumed all genotypes were homozygous and allele frequencies could be calculated. The RAPD data were treated as haploid data using, amongst others, the GENEPOP program (Raymond and Rousset, 1995).

4.2.3 Graphical display of RAPD variation

To graph RAPD variation of several individuals or of several populations, two approaches have been used, namely, the construction of a dendrogram and the use of principal co-ordinate analysis (PCO). Huff et al. (1993) studied 48 individuals and 98 markers in buffalo grass and represented the data as a dendrogram depicting the similarity of the individuals. Yeh *et al.* (1995) used a different strategy where the differentiation of populations was visualised. They used the average Euclidean distances comparing each individual from one population with each individual from another population. Average distances among individuals of two populations were then taken as the distance between those two populations. Both distance matrices, containing distances between pairs of individuals or containing distances between pairs of populations can be used to make dendrograms, for instance with SYSTAT, Phylip or RAPDistance.

PCO is a multi-dimensional presentation of the variation observed and in some cases gives a better picture of the results than a two dimensional dendrogram, for instance in the case of hybridising species (Russell *et al.*, 1993; Nesbitt *et al.*, 1995). However, when there are many individuals and many populations, the resulting graph can be far from clear.

4.3 Case studies

Over the past few years many studies have used RAPDs to survey population structure in a large variety of organisms and many interesting and important questions have been answered. To highlight the potential of RAPDs in studies of population structure we present below two case studies: one where a test was made of the relationship between pollinator behaviour and gene flow, and another which examined the relationship between, and the level of population differentiation within, two closely related taxa.

4.3.1 Population substructure in Alkanna orientalis, an outcrossing plant

It is well known that the genetic structure of a plant species is strongly influenced by factors with an impact on gene flow, like pollination biology and mating system (Handel, 1985; Govindaraju, 1988). Heywood (1991) concluded that a high level of gene flow due to pollen dispersal has a strong influence on population structure by lowering differentiation among populations. In some insect pollinated species gene flow is often more correlated with the flight behaviour of the pollinator (pollen dispersal) than with seed movement. In order to get an impression of the relative importance of gene flow by pollen dispersal on population substructure a survey of RAPD variation was conducted in a population of the plant *Alkanna orientalis* (Boraginaceae), located near the town of St Catherine (1641 m above sea level) in the Sinai Desert, Egypt. *A. orientalis* flowers in April and its main insect visitor and pollinator is the solitary bee *Anthophora pauperata* (Willmer *et al.*, 1994; Gilbert *et al.*, 1996). Males of *A. pauperata* patrol a territory of a size ranging from 5 × 5 m to 15 × 15 m; in each territory there may be three to nine plants. It seemed, therefore, that genetic divergence between subpopulations may easily occur due to restricted gene flow.

Plants were collected from three wadis (Wadi Arbaein, Wadi Dir and Wadi Tofaha) and from the Plain of El Raha that connects them (Fig. 4.1). Wadis are riverbeds that are dry most of the year, and are isolated from each other for most of their length by mountains rising up to 2400 m. Leaf material was collected from

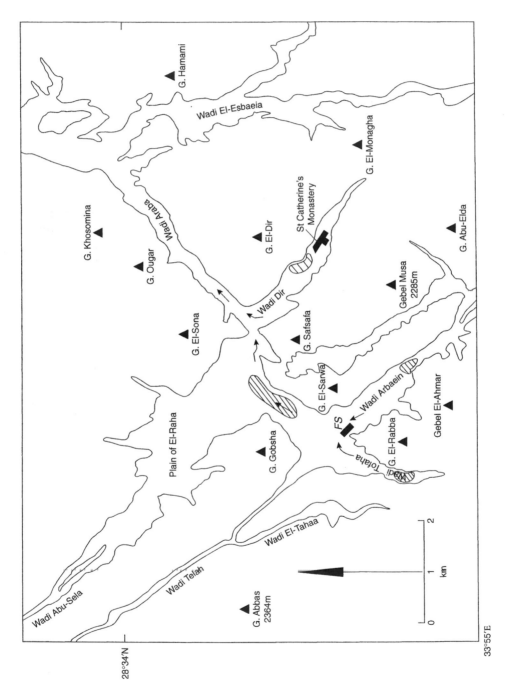

Figure 4.1 Location of the study population of *Alkanna orientalis* in the Sinai desert. Redrawn from Gilbert *et al.* (1996). The arrows represent the direction of the flash floods.

randomly chosen plants on the plain and up to 2 km into the wadis from the plain, mainly from the base of the wadis (Table 4.1). RAPD analysis was performed according to Wolff *et al.* (1995). Seven RAPD primers were used and this resulted in a presence/absence matrix of 45 fragments. The polymorphic fragments were analysed using RAPDistance, Shannon's index of diversity and AMOVA.

A considerable amount of genetic variation was detected by the seven RAPD primers used in the study. Some of the primers detected polymorphisms within, as well as between, subpopulations, whereas other primers showed clear differentiation between subpopulations. To compare levels of variation in each subpopulation and in the total population, Shannon's diversity index was calculated (Table 4.1). The subpopulations contain similar levels of diversity. The average subpopulation diversity and the total diversity were used to estimate the distribution of the variation (Table 4.2).

Genetic distances were calculated between subpopulations (Phi_{ST} values) based on Euclidean distances between all pairs of individuals, within and between populations. AMOVA performs an analysis of variance on these distances, and partitions the variation into a within and a between subpopulation component. The significance of differences between subpopulations was tested by a non-parametric permutation procedure, using 1000 permutations (Tables 4.2 and 4.3). The results were in broad agreement with those of the Shannon's diversity index analysis.

Table 4.1 Shannon's index of diversity of the subpopulations of *Alkanna orientalis*, the average subpopulation diversity and the total diversity. The number of plants in each subpopulation is given in brackets.

Subpopulation	Diversity	Level	Diversity
Plain (n = 7)	11.54	Average subpopulation	10.32
Wadi Arbaein (n = 9)	8.98	Total population	16.78
Wadi Dir (n = 16)	10.41		
Wadi Tofaha (n = 7)	10.36		

Table 4.2 Distribution of variation in *Alkanna orientalis* using Shannon's index of diversity and an analysis of variance (AMOVA).

Level	Shannon's index	AMOVA
Within subpopulations	62%	68%
Among subpopulations	38%	32%

Table 4.3 Pairwise Phi_{ST} values between subpopulations.

Subpopulation	W. Tofaha	W. Arbaein	Plain	W. Dir
W. Tofaha	—	0.29**	0.14***	0.38***
W. Arbaein		—	0.20**	0.40***
Plain			—	0.34***

** = P < 0.01, *** = P < 0.001

Depending on the statistic used, 62% (Shannon) or 68% (AMOVA) of the total variation is contained within subpopulations. The Phi_{ST} value for population differentiation was significantly different from zero (p < 0.001).

Furthermore, genetic differentiation between particular subpopulations was tested using the program AMOVA. All pairwise Phi_{ST} values were significantly different from zero (Table 4.3). Thus, all subpopulations may be considered as significantly different from each other, with the Wadi Dir subpopulation most different from the other three subpopulations, and the Plain and Wadi Tofaha subpopulations being the most similar. A Neighbour-Joining (NJ) tree was constructed from the Euclidean distances between individuals (Fig. 4.2) using RAPDistance (version 1.03), Phylip and Treeview, to represent the relationships among individuals. The tree

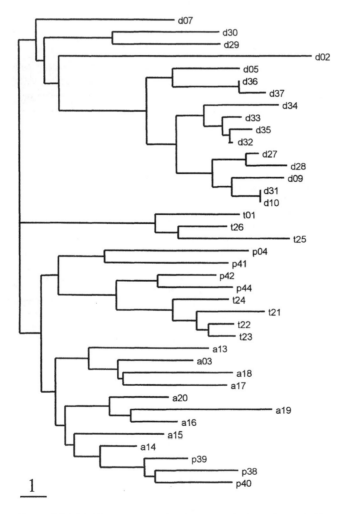

Figure 4.2 Neighbour-joining dendrogram of the individual *Alkanna orientalis* plants. Redrawn from Wolff *et al.* (1997). Scale bar represents Euclidean distance. Numbers are individual codes, prefixed by a letter indentifying which subpopulation they are from (a = Wadi Arbaein, d = Wadi Dir, t = Wadi Tofaha, p = Plain).

shows that individuals from a given subpopulation tend to cluster together and are, therefore, more genetically similar than individuals from different subpopulations. The Wadi Dir subpopulation is most clearly differentiated from the other subpopulations. Although individuals from the Wadi Tofaha and Wadi Arbaein subpopulations do not cluster together, certain individuals from the Plain subpopulation group with some Wadi Tofaha individuals and others group with Wadi Arbaein individuals. This means that the individuals from the Plain do not cluster all together, but are in two clusters. The NJ tree and the values for genetic differentiation calculated with AMOVA (Fig. 4.2 and Table 4.3) suggest that there is considerable genetic similarity between individuals in the Plain subpopulation and certain individuals within the Wadi Arbaein or Wadi Tofaha subpopulations. This is more than might be expected based on the foraging strategies of the bees, which extend over only several metres.

Instead, the population differentiation observed reflects the high mountain ridges between the wadis. Movement by bees across the ridges is unlikely and transport of pollen by *Anthophora pauperata* via the interconnecting plain also seems unlikely because of the short flight distances of the bees. The specific population substructure observed seems to be most likely caused by flash floods, which occur very regularly in Sinai at the end of winter, transporting the seeds. Wadi Tofaha is the narrowest and steepest of the three wadis and transport of seeds by water would be greatest from this subpopulation into the Plain. The Wadi Dir subpopulation is not only the most genetically isolated of the four subpopulations but also the most geographically isolated, being both distant and downstream.

This study, therefore, has shown that a survey of RAPD variation provides a good understanding of population substructure in *Alkanna orientalis*. Genetic differentiation in the population is probably due to limited pollen dispersal caused by the mountain ridges separating the wadis; and the evidence for gene flow is best explained as a result of seed transport by flash floods.

4.3.2 Population structure and differentiation of Plantago major s.l.

Populations of a species that are reproductively isolated may diverge from each other through drift and/or differential selection. If they have diverged sufficiently they may be recognised as ecotypes, subspecies or even species.

Plantago major is an almost cosmopolitan species. It has been extensively studied in the Netherlands, and much is known of its ecology, genetics, and reproductive biology (Van Dijk, 1984; Wolff, 1991a). It is a polymorphic taxon and some authors have recognised a number of subspecies within *P. major*, each divided into varieties, but others have recognised the subspecies as separate species. This has led to confusion. For example, *P. intermedia* was described as a species by Gilibert (1806). It was later reduced to a subspecies of *P. major* by Lange (1856: 107) and then to a variety of it by Pilger (1937). Pilger (1937) also described *P. major* subsp. *pleiosperma*, a taxon we regard as being synonymous with subsp. *intermedia* (the latter has nomenclatural priority at this rank). The work described in the following account starts to address the twin problems of the evolutionary and taxonomic relationships between *P. intermedia* and *P. major*.

The two taxa are capable of interbreeding in the greenhouse and, where they are sympatric, hybrids may be found. The two taxa are distinguished by a combination of morphological characters and by the habitat in which they occur (Mølgaard, 1976). *P. major* usually has more veins per leaf, a lower scape-length to spike-length ratio and fewer seeds per capsule than *P. intermedia*. However, there is some overlap of these characters. *P. major* is abundant on footpaths and rough ground. In contrast, *P. intermedia* usually grows on disturbed ground, in agricultural fields, and on sites close to the sea or on riverbanks. From extensive morphological and genetic comparisons workers in the Netherlands concluded that the two taxa were ecologically different and that within each taxon different ecotypes could be distinguished (Van Dijk, 1984, 1989). They suggested that the morphological characters that distinguish the species and ecotypes are maintained by strong selection and inbreeding in the face of gene flow.

The absence of a breeding barrier and the resemblance in morphology is matched by allozyme studies which showed that the two taxa have similar allele frequencies for all loci except three (Van Dijk and Van Delden, 1981). It is suggested that the allozyme loci that differ between the two taxa may be linked to genes controlling their respective morphologies. In contrast DNA fingerprinting as well as chloroplast RFLP analysis showed a much higher degree of differentiation between the two taxa (Wolff and Schaal, 1992; Wolff *et al.*, 1994). This paradox led us to sample widely from populations of both taxa to examine their evolutionary relationship and current genetic structure.

4.3.2.1 The study

Seeds were collected from 30 locations from ten countries in Europe as well as in Trinidad and the USA (Table 4.4). Approximately six plants from each population were grown in the greenhouse under uniform conditions. Ripe scapes were collected from each plant and the number of seeds per capsule was counted. DNA was extracted from young leaf tissue from each plant. For PCR amplification 12 RAPD primers and five anchored inter-SSR (simple sequence repeat) primers were used. Inter-SSRs are conceptually similar to RAPDs in that both use arbitrary primers; in the case of inter-SSRs the primers are targetted at inverted simple sequence repeats. In total 73 RAPD and inter-SSR markers were scored. A chloroplast DNA fragment was amplified using primer set *trn*S and *trnf*M (Demesure *et al.*, 1995). The fragment was digested with *Taq*I which revealed two haplotypes (Wolff and Morgan-Richards, 1998).

4.3.2.2 Morphology

The mean number of seeds per capsule is the character that most successfully separates the two groups. Plants with fewer than 15 seeds per capsule were named *P. major* and plants with more than 15 seeds per capsule were designated *P. intermedia*. It appeared that from most locations a single species was collected, and that plants from all British locations were *P. major*. However, two of the 112 plants collected from Britain fell within the *P. intermedia* group: one plant from Harwell and one from Aston Hill. Two *P. major* plants from Sibton had more seeds per capsule than expected given the width of their leaves (data not presented).

Table 4.4 Collection localities, dates and sample sizes of *Plantago major* (M) and *P. intermedia* (I).
Species identification based on morphology.

Country	Location	Species	Collection year	Number of plants
Britain	Aston Hill	M(one I)	1991, 1992	5, 5
	Bottesford	M	1991	4
	Chilworth	M	1990	2
	Eskdalemuir	M	1990	4
	East Malling	M	1990	3
	Harwell	M(one I)	1990	4
	ISP	M	1985, 1994	12,10
	Leatherhead	M	1990	3
	Lullington	M	1990,1991,1992	5,5,5
	Saint Osyth	M	1991,1992	3,5
	Scaftworth	M	1988,1994	5,5
	Sibton	M	1994	5
	Strath Vaich	M	1994	5
	Totley	M	1985,1994	5,5
	Wray	M	1991	5
Austria	Siebersdorf	M	1994	5
Belgium	Tervuren	M	1994	5
Germany	Braunschweig	M	1994	5
Denmark	Nybol	I	1995	5
the Netherlands	Noordpolderzijl	I	1995	5
	Tiel	I	1995	5
Switzerland	Belp	M	1994	4
Spain	Ebro	I	1990	5
	Valsain	M	1993	5
Greece	Benaki	I	1990	5
	Anavriti	I	1990	5
	Olympus	M	1993	5
USA	Los Angeles	I	1991	5
Trinidad		I	1991	5

4.3.2.3 RAPD and inter-SSR DNA fragments: discrimination of taxa

Two large groups of plants were distinguished by the NJ analysis of 73 RAPD and
inter-SSR markers using 100 plants (all *P. intermedia* plants and a selection of *P.
major* plants) (Fig. 4.3). One group, A, contains all those plants identified from
morphology as *P. intermedia*. Within this group, plants cluster with others that were
collected from the same location. The exception was one plant from Los Angeles
airport (USA) which was more similar to plants from Trinidad than to plants from
the same collection site. Within the second group, B, all the *P. major* plants cluster,
but with much less geographic structure than seen within *P. intermedia* (Fig. 4.3).
Between groups A and B lie the two plants from Sibton that resemble *P. major* except
for their greater number of seeds; see later. Fig. 4.4 is a NJ tree with additional *P.
major* plants and two *P. intermedia* plants to show the relationships of *P. major*
plants in more detail. Although for some locations the *P. major* plants from the same
site cluster together (e.g. Olympus and Belgium), for other locations the plants are
as genetically similar to plants from elsewhere as they are to adjacent plants. This
lack of geographical structure in *P. major* suggests that it represents a much more

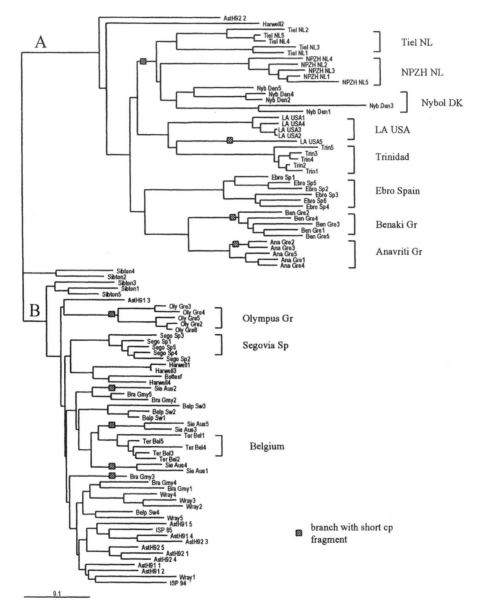

Figure 4.3 Neighbour-joining dendrogram of a hundred *Plantago major* (group B) and *Plantago inter-media* (group A) individuals using 73 RAPD and inter-SSR PCR markers. All individuals are marked according to their location of origin. The country of origin is given except for the British locations. For those locations for which multiple collection years are used the year is indicated. Scale bar represents Nei and Li's distance.

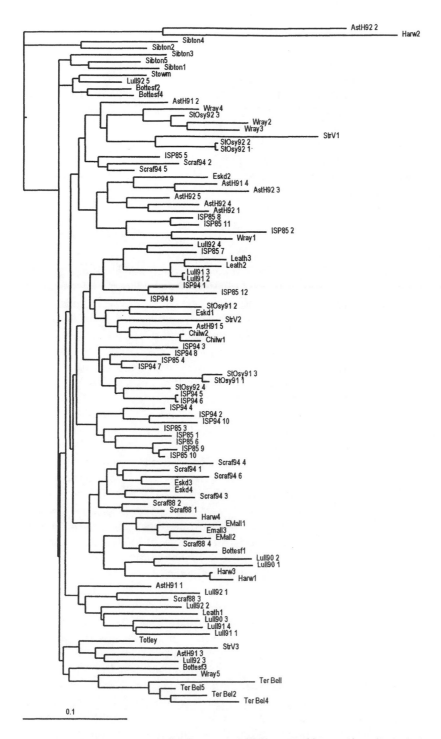

0.1

Figure 4.4 Neighbour-joining dendrogram of 98 *P. major* (Group B) and two *P. intermedia* (Group A) individuals using 73 RAPD and inter-SSR PCR markers. All individuals are marked according to their location of origin. The country of origin is given except for the British locations. For those locations for which multiple collection years are used the year is indicated. Scale bar represents Nei and Li's distance.

homogenous taxon or may indicate that we have not resolved structure that is present, due to the low number of variable markers.

All but four of the plants from Britain cluster with the *P. major* samples; two of the exceptions are from Harwell and Aston Hill. These are the same plants whose morphological characters placed them with the *P. intermedia* plants. Thus, we have clear concordance of morphology and molecular data. They did not cluster with other populations sampled but appeared to represent distinct *P. intermedia* individuals, sympatric with *P. major* in Britain.

The other two exceptions are plants from Sibton. These plants share an absence of DNA fragments at three 'loci' with *P. intermedia* and share the presence of DNA fragments with *P. intermedia* at six 'loci'. They may well be of hybrid origin and this prompted us to investigate other populations for this phenomenon.

4.3.2.4 Hybridisation

The present study indicated that there are mixed populations containing plants of both taxa (Harwell and Aston Hill) and that there are also populations that resemble *P. major*, but with some morphological traits and RAPD markers of *P. intermedia* (Sibton). To investigate how common hybridisation is we looked at potential hybrid populations, namely where habitats of the two taxa are parapatric. RAPD analysis can be used to test for the hybrid origin of a species by comparing its fragment patterns with those of the presumed parental species. Recently, Lifante and Aguinagalde (1996) found in *Asphodelus* species that several RAPD markers which distinguished two species were combined in their putative hybrid. Howard *et al.* (1997) examined in detail hybrid populations of two *Quercus* species using 14 species-specific RAPD markers which were found to reflect the morphological variation and the amount of hybridisation.

In a preliminary study, plants were collected from four sites where a *P. major* habitat bordered a *P. intermedia* habitat to see whether natural hybrids were formed and to what extent the species-specific markers were present (Wolff, unpublished). In one of the sites there were only pure *P. major* or pure *P. intermedia* plants. In the other three sites several (6–30%) of the plants turned out to be hybrids: they contained at least one marker of each species and in general the number of seeds per capsule was intermediate between both species. However, if one looks carefully at the number of species-specific markers the hybrids in general contained more *P. major*-specific than *P. intermedia*-specific RAPD and inter-SSR markers, namely 77% *P. major* markers. This could be attributed to differences in fitness, to higher outcrossing rates of *P. major*, or to maternal inheritance of some of the markers if they were of plastid or mitochondrial origin. An asymmetric contribution of markers was also found in hybrids of two *Quercus* species, and this was attributed to a possible stronger selection against markers or linked genes in one direction (Howard *et al.*, 1997).

In an earlier publication (Wolff and Morgan-Richards, 1998) the polymorphic cp PCR/RFLP fragment showed a distinction between the Dutch populations of the two species. However, in the Scottish populations the haplotypes are not restricted to one species. In the present data set it appeared that the majority of the *P. intermedia* plants possess the short fragment whereas the majority of the *P. major* plants have the longer fragment (Fig. 4.3). This may be due to an ancient polymorphism that has been retained in both species, and is not necessarily evidence for hybridisation between them.

4.3.2.5 Status of the Plantago taxa

The data show that the two taxa are genetically differentiated even in sympatry, have low gene flow and distinct population structure (Wolff and Morgan-Richards, 1998). The discrimination of the two taxa confirms the ecological and morphological differences, especially the number of seeds per capsule, between them. The overlap of morphological characters is most probably an artefact caused by the analysis of hybrids or mixed populations, but this should be confirmed in a more detailed study. Since there are not only many RAPD markers that discriminate the two taxa but also ecological and morphological differences, we consider that *P. major* and *P. intermedia* should be regarded as separate species.

There is no absolute measure of genetic difference that defines the boundary of a species (e.g. Mallet, 1995). However, comparisons with similar studies and examination of genetic evidence in tandem with other information such as the ecology, morphology and mating systems allows for an educated view of the independence of the taxa, their cohesion and their potential for separate evolutionary fates.

Van Buren *et al.* (1994) elevated the autumn buttercup, an endangered endemic in Utah, to species rank based on RAPD analysis and ecological and morphological characteristics. The *Plantago* study can also be compared with a study of Howard *et al.* (1997) in which they compared similar numbers of two *Quercus* species using RAPD analysis. The two oaks are considered to be good species that can be discriminated by morphological characters and are placed in two separate subsections of white oaks. Howard *et al.* (1997) used 700 different primers but found only 14 markers with six primers that were species specific: the two oak species are genetically extremely similar. The two *Plantago* taxa studied here have many more markers that discriminate them, in addition to diagnostic morphological and ecological characters, and therefore we believe they should be regarded as separate species.

As both *Plantago* taxa are self-fertilising species it is possible, at least in theory, that because of hitchhiking, selective sweeps and background selection (Hedrick, 1980; Charlesworth *et al.*, 1995) markers would be retained in some populations (or in a restricted species) purely due to chance. However, we found specific markers in each species, not on a local scale but on a European scale, and possibly even on a world scale. We therefore feel that the consistency over a wide geographic range indicates that each species is distinct with genetic cohesion.

4.3.2.6 Distribution of variation

In order to estimate levels of genetic differentiation, the RAPD and inter-SSR data from two groups of individuals were analysed. One group consisted of *P. intermedia* plants represented by ten populations (from Spain, Greece, USA, Trinidad, the Netherlands, Denmark and Britain), and the other group of *P. major* plants represented by nine populations (from Austria, Greece, Spain, Germany, Belgium, Switzerland and Britain). Using AMOVA greater genetic differentiation was detected among populations of *P. intermedia* than among those of *P. major* (Table 4.5). Combined AMOVA analysis of the data showed that the largest proportion of the variation was between species and the least within populations.

As mentioned earlier there were no significant differences in allozyme variation between the two species but a clear distinction between the species using a repetitive DNA probe. The RAPD and inter-SSR results take an intermediate position between the allozyme and fingerprint results with many bands shared between the two species, but also many bands that discriminate the species. At least for RAPDs the amplicons are known to be a mixture of coding and noncoding DNA (Williams *et al.*, 1993); amongst this latter group there will be stretches of repetitive DNA. Concerted evolution may cause a larger sequence similarity within than between reproductively isolated taxa: it is thus a force that, between populations, may cause a greater divergence in repetitive DNA than in single copy DNA (Elder and Turner, 1995). This might therefore be the cause of the greater divergence of the species for RAPDs than for allozymes. This is analogous to the result found by Chalmers *et al.* (1992) in a tropical tree, in which most of the RAPD variation was distributed among populations and not, as predicted by Hamrick and Godt (1996), within populations. This could be explained by the repetitiveness of at least some of the RAPD amplicons. Alternatively, some of the RAPD fragments may be of chloroplast or mitochondrial origin and these could show a greater differentiation due to their maternal inheritance.

In terms of the population subdivision in each of the two species, three possible explanations can be suggested to account for the greater differentiation observed in *P. intermedia*. These explanations are not mutually exclusive.

Firstly, a difference in rate of inbreeding could lead to distinct population genetic structure since a modest amount of gene flow reduces the level of subdivision among natural populations (Wright, 1951; Hamrick and Godt, 1996). A higher rate of outcrossing in *P. major* would result in homogenising genetic variation compared to *P. intermedia*, in which a higher rate of inbreeding would lead to increased differentiation among populations. There is empirical evidence for the role of the breeding system. Previous studies have shown that populations of *P. major* have, on average, an outcrossing rate of 14%, whereas in populations of *P. intermedia* the outcrossing rate is approximately 4% (Van Dijk *et al.*, 1988; Wolff, 1991b). This means that in *P. major* the outcrossing rate may be high enough for populations not to diverge while the gene flow in the other species may be too low to

Table 4.5 Genetic differentiation within and among *Plantago major* and *P. intermedia* using 73 random genetic markers and analysed by AMOVA.

Taxon	PHI_{ST}
P. intermedia	0.672
P. major	0.397

Variance components for both species combined (nested analysis):

Level	Percentage	PHI
Variance among species, PHI_{ST}	46.3	0.763
Variance among populations within species, PHI_{SC}	30.0	0.559
Variance within populations, PHI_{CT}	23.7	0.463

prevent population differentiation. This difference in outcrossing rate may at least explain part of the difference in population structure of the two species.

Secondly, *P. major* may have been derived from *P. intermedia* and thus represent a subset of the genetic variation within *P. intermedia* or *P. major* may have been subject to a recent bottleneck.

Thirdly, the rate of molecular evolution in *P. major* might be slower than within *P. intermedia* due to population or genome structure.

4.3.2.7 Conclusion of the Plantago study

The RAPD analysis of the population structure in *P. major* and *P. intermedia* has shown that the two taxa can easily be distinguished using RAPD markers: the separation of the two groups of populations coincides with the grouping based on the number of seeds per capsule. This means that they can be considered as two separate species, with limited gene flow between them. The markers can be used in future research into the amount of interspecific gene flow and the fate of hybrids in natural populations. The analysis showed that *P. intermedia* has a much greater degree of population differentiation than *P. major*, a finding which may be caused by a higher outcrossing rate in *P. major* and/or historical effects.

ACKNOWLEDGEMENTS

We thank Dr. R. J. Abbott and two anonymous reviewers for helpful suggestions on the manuscript and Dr. S. El-Akkad for the work on the *Alkanna* study.

REFERENCES

Armstrong, J. S., Gibbs, A. J., Peakall, R. and Weiller, G. (1994) *The RAPDistance package*. Obtainable via anonymous ftp or WWW: ftp::/life.anu.edu/au/pub/RAPDistance or http://life.anu.edu.au/molecular/software/rapd.html.

Barrett, S. C. H. and Pannell, J. R. (1999) Metapopulation dynamics and mating-system evolution in plants, in *Molecular systematics and plant evolution*, (eds P. M. Hollingsworth, R. M. Bateman and R. J. Gornall), Taylor & Francis, London, pp. 74–100.

Black, W. C. IV (1995) *FORTRAN programs for the analysis of RAPD-PCR markers in populations*. Colorado State University, Dept. of Microbiology, Ft. Collins, CO 80523.

Bonnin, I., Huguet, T., Gherardi, M., Prosperi, J-M. and Olivieri, I. (1996) High level of polymorphism and spatial structure in a selfing plant species, *Medicago truncatula* (Leguminosae), shown using RAPD markers. *American Journal of Botany*, **83**, 843–855.

Chalmers, K. J., Waugh, R., Sprent, J. I., Simons, A. J. and Powell, W. (1992) Detection of genetic variation between and within populations of *Gliricidia sepium* and *G. maculata* using RAPD markers. *Heredity*, **69**, 465–472.

Charlesworth, D., Charlesworth, B. and Morgan, M. T. (1995). The pattern of neutral molecular variation under the background model. *Genetics*, **141**, 1619–1632.

Dawson, I. K., Chalmers, K., Waugh, R. and Powell, W. (1993) Detection and analysis of genetic variation in *Hordeum spontaneum* populations from Israel using RAPD markers. *Molecular Ecology*, **2**, 151–159.

Demesure, B., Sodzi, N. and Petit, R. J. (1995) A set of universal primers for amplification of polymorphic non-coding regions of mitochondrial and chloroplast DNA in plants. *Molecular Ecology*, **4**, 129–131.

Elder, J. F. and Turner, B. J. (1995) Concerted evolution of repetitive DNA sequences in eukaryotes. *Quarterly Review of Biology*, **70**, 297–320.

Ennos, R. A. (1994) Estimating the relative rates of pollen and seed migration among plant populations. *Heredity*, **72**, 250–259.

Ennos, R. A., Sinclair, W. T., Hu, X-S. and Langdon, A. (1999) Using organelle markers to elucidate the history, ecology and evolution of plant populations, in *Molecular systematics and plant evolution*, (eds P. M. Hollingwood, R. M. Bateman and R. J. Gornall), Taylor & Francis, London, pp. 1–19.

Excoffier, L., Smouse, P.E. and Quattro, J.M. (1992) Analysis of molecular variance inferred from metric distances among DNA haplotypes: application to human mitochondrial DNA data. *Genetics*, **131**, 479–491.

Ferris, C., King, R. A. and Hewitt, G. M. (1999) Isolation within species and the history of glacial refugia, in *Molecular systematics and plant evolution*, (eds P. M. Hollingsworth, R. M. Bateman and R. J. Gornall), Taylor & Francis, London, pp. 20–34.

Gabrielsen, T. M., Bachmann, K., Jakobsen, K. S. and Brochmann, C. (1997) Glacial survival does not matter: RAPD phylogeography of Nordic *Saxifraga oppositifolia*. *Molecular Ecology*, **6**, 831–842.

Gilbert, F., Willmer, P., Semida, F., Ghazoul, J. and Zalat, S. (1996) Spatial variation in selection in a plant pollinator system in the wadis of Sinai, Egypt. *Oecologia*, **108**, 479–487.

Gilibert, J. E. (1806) *Histoire des plantes d'Europe*, 2nd ed., **1**, 125. Paris.

Govindaraju, D. R. (1988) Relationship between dispersal ability and levels of gene flow in plants. *Oikos*, **52**, 31–35.

Hamrick, J. L. and Godt, M. J. W. (1989) Allozyme diversity in plant species, in *Plant population genetics, breeding and genetic resources* (eds A. H. D. Brown, M. T. Clegg, A. L. Kahler and B. S. Weir), Sinauer, Sunderland, MA, pp. 43–63.

Hamrick, J. L. and Godt, M. J. W. (1996) Effects of life-history traits on genetic diversity in plant species. *Philosophical Transactions of the Royal Society of London, series B*, **351**, 1291–1298.

Handel, S. N. (1985) Pollen flow patterns and the creation of local genotypic variation. In: *Structure and Functioning of Plant Populations 2* (eds J. Haeck and J. W. Woldendorp), North-Holland, Amsterdam, pp. 251–265.

Harris, S. A. (1999) RAPDs in systematics – a useful methodology? in *Molecular systematics and plant evolution*, (eds P. M. Hollingsworth, R. M. Bateman and R. J. Gornall), Taylor & Francis, London, pp. 211–228.

Hartl, D. L. and Clark, A. G. (1989) *Principles of population genetics*, 2nd ed. Sinauer associates, Sunderland, MA.

Hastings, A. (1989) The interaction between selection and linkage disequilibrium in plant populations, in *Plant population genetics, breeding and genetic resources* (eds A. H. D. Brown, M. T. Clegg, A. L. Kahler and B. S. Weir), Sinauer, Sunderland, MA, pp. 163–180.

Hedrick, P. W. (1980) Hitchhiking: a comparison of linkage and partial selfing. *Genetics*, **94**, 791–808.

Heywood, J. S. (1991) Spatial analysis of genetic variation in plant populations. *Annual Review of Ecology and Systematics*, **22**, 335–355.

Howard, D. J., Preszler, R. W., Williams, J., Fenchell, S. and Boecklen, W. J. (1997) How discrete are oak species? Insights from a hybrid zone between *Quercus grisea* and *Quercus gambelii*. *Evolution*, **51**, 747–755.

Hsiao, J.-Y. and Rieseberg, L. H. (1994) Population genetic structure of *Yushania niitakayamensis* (Bambusoideae, Poaceae) in Taiwan. *Molecular Ecology*, **3**, 201–208.

Huff, D. R., Peakall, R. and Smouse, P. E. (1993) RAPD variation within and among natural populations of outcrossing buffalograss (*Buchloë dactyloides* (Nutt. Engelm.). *Theoretical and Applied Genetics*, **86**, 927–934.

Isabel, N., Beaulieu, J. and Bousquet, J. (1995) Complete congruence between gene diversity estimates derived from genotypic data at enzyme and random amplified polymorphic DNA loci in black spruce. *Proceedings of the National Academy of Sciences USA*, **92**, 6369–6373.

King, L. M. and Schaal, B. A. (1989) Ribosomal DNA variation and distribution in *Rudbeckia missouriensis. Evolution*, **43**, 1117–1119.

Kresovich, S., Williams, J. G. K., McFerson, J. R., Routman, E. J. and Schaal, B. A. (1992) Characterisation of genetic identities and relationships of *Brassica oleracea* L. via a random amplified polymorphic DNA assay. *Theoretical and Applied Genetics*, **85**, 190–196.

Lange, J. M. C. (1856) *Haandbog i den Danske flora*, 2nd ed. C.A.Reitzels, Copenhagen.

Le Corre, V., Dumolin-Lapègue, S. and Kremer, A. (1997) Genetic variation at allozyme and RAPD loci in sessile oak *Quercus petraea* (Matt.) Liebl.: the role of history and geography. *Molecular Ecology*, **6**, 519–529.

Levin, D. A. (1995) Metapopulations: an arena for local speciation. *Journal of Evolutionary Biology*, **8**, 635–644.

Lifante, Z. D. and Aguinagalde, I. (1996) The use of random amplified polymorphic DNA (RAPD) markers for the study of taxonomical relationships among species of *Asphodelus* sect. *Verinea* (Asphodelaceae). *American Journal of Botany*, **83**, 949–953.

Liu, Z. and Furnier, G.R. (1993) Comparison of allozyme, RFLP, and RAPD markers for revealing genetic variation within and between trembling aspen and bigtooth aspen. *Theoretical and Applied Genetics*, **87**, 97–105.

Lorenz, M., Weihe, A. and Börner, T. (1994) DNA fragments of organellar origin in random amplified polymorphic DNA (RAPD) patterns of sugar beet (*Beta vulgaris* L.). *Theoretical and Applied Genetics*, **88**, 775–779.

Lynch, M. and Milligan, B. G. (1994) Analysis of population genetic structure with RAPD markers. *Molecular Ecology*, **3**, 91–99.

Mallet, J. (1995) A species definition for the Modern Synthesis. *Trends in Ecology and Evolution*, **10**, 294–299.

McCauley, D. E. (1995) The use of chloroplast DNA polymorphism in studies of gene flow in plants. *Trends in Ecology and Evolution*, **10**, 198–202.

Mølgaard, P. (1976) *Plantago major* ssp. *major* and ssp. *pleiosperma*. Morphology, biology and ecology in Denmark. *Botanisk Tidskrift*, **71**, 31–56.

Nei, M. (1973) Analysis of gene diversity in subdivided populations. *Proceedings of the National Academy of Sciences USA*, **70**, 3321–3323.

Nei, M. and Li, W.-H. (1979) Mathematical model for studying genetic variation in terms of restriction endonucleases. *Proceedings of the National Academy of Sciences USA*, **76**, 5269–5273

Nesbitt, K. A., Potts, B. M., Vaillancourt, R. E., West, A. K. and Reid, J. B. (1995) Partitioning and distribution of RAPD variation in a forest tree species, *Eucalyptus globulus* (Mytaceae). *Heredity*, **74**, 628–637.

Nevo, E., Krugman, T. and Beiles, A. (1994) Edaphic natural-selection of allozyme polymorphisms in *Aegilops peregrina* at a Galilee microsite in Israel. *Heredity*, **72**, 109–112.

Peakall, R., Smouse, P. E. and Huff, D. R. (1995) Evolutionary implications of allozyme and RAPD variation in diploid populations of dioecious buffalograss *Buchloë dactyloides. Molecular Ecology*, **4**, 135–147.

Pilger, R. (1937). *Plantaginaceae. Das Pflanzenreich Regni vegetabilis conspectus. part 4269.* Neudruck, pp. 466.

Provan, J., Soranzo, N., Wilson, N. J., McNicol, J. W., Morgante, M. and Powell, W. (1999) The use of uniparentally inherited simple sequence repeat markers in plant population studies and systematics, in *Molecular systematics and plant evolution*, (eds P. M. Hollingsworth, R. M. Bateman and R. J. Gornall), Taylor & Francis, London, pp. 35–50.

Raymond, M. and Rousset, F. (1995) GENEPOP, a population genetic software for exact tests and ecumenicism. *Journal of Heredity*, **86**, 248–249.

Rieseberg, L. H. (1996) Homology among RAPD fragments in interspecific comparisons. *Molecular Ecology*, **5**, 99–105.

Rossetto, M., Weaver, P. K. and Dixon, K. W. (1995) Use of RAPD analysis in devising conservation strategies for the rare and endangered *Grevillea scapigera*. *Molecular Ecology*, **4**, 321–329.

Russell, J. R., Hosein, F., Johnson, E., Waugh, R. and Powell, W. (1993) Genetic differentiation of cocoa (*Theobroma cacao* L.) populations revealed by RAPD analysis. *Molecular Ecology*, **2**, 87–97.

Schierenbeek, K. A., Skupski, M., Lieberman, D. and Lieberman, M. (1997) Population structure and genetic diversity in four tropical tree species in Costa Rica. *Molecular Ecology*, **6**, 137–144.

Slatkin, M. (1987) Gene flow and the geographic structure of natural populations. *Science*, **236**, 787–792.

Stewart Jr, C. N. and Excoffier, L. (1996) Assessing population genetic structure and variability with RAPD data: application to *Vaccinium macrocarpon* (American cranberry). *Journal of Evolutionary Biology*, **9**, 153–171.

Stewart Jr, C. N., Rosson, G., Shirley, B. W. and Porter, D. M. (1996) Population genetic variation in rare and endangered *Iliamna* (Malvaceae) in Virginia. *Biological Journal of the Linnean Society*, **58**, 357–369.

Travis, S. E., Maschinski, J. and Keim, P. (1996) An analysis of genetic variation in *Astragalus cremnophylax* var. *cremnophylax*, a critically endangered plant, using AFLP markers. *Molecular Ecology*, **5**, 735–745.

Van Buren, R., Harper, K. T., Andersen, W. R., Stanton, D. J., Seyoum, S. and England, J. L. (1994) Evaluating the relationship of autumn buttercup (*Ranunculus acriformis* var. *aestivalis*) to some close congeners using Random Amplified Polymorphic DNA. *American Journal of Botany*, **81**, 514–519.

Van Dijk, H. (1984) Genetic variability in *Plantago* species in relation to their ecology 2. Quantitative characters and allozyme loci in *P. major*. *Theoretical and Applied Genetics*, **68**, 43–52.

Van Dijk, H. (1989) Genetic variability in *Plantago* species in relation to their ecology 4. Ecotypic differentiation in *P. major*. *Theoretical and Applied Genetics*, **77**, 749–759.

Van Dijk, H. and Van Delden, W. (1981) Genetic variability in *Plantago* species in relation to their ecology Part1: genetic analysis of the allozyme variation in *P. major* subspecies. *Theoretical and Applied Genetics*, **60**, 285–290.

Van Dijk, H., Wolff, K. and De Vries, A. (1988) Genetic variability in *Plantago* species in relation to their ecology. 3. Genetic structure of populations of *P. major*, *P. lanceolata* and *P. coronopus*. *Theoretical and Applied Genetics*, **75**, 518–528.

Vicario, F., Vendramin, G. G., Rossi, P., Liò, P. and Giannini, R. (1995) Allozyme, chloroplast DNA and RAPD markers for determining genetic relationships between *Abies alba* and the relic population of *Abies nebrodensis*. *Theoretical and Applied Genetics*, **90**, 1012–1018.

Vierling, R. A. and Nguyen, H. T. (1992) Use of RAPD markers to determine the genetic diveristy of diploid wheat genotypes. *Theoretical and Applied Genetics*, **84**, 835–838.

Weising, K., Nybom, H., Wolff, K. and Meyer, W. (1995) *DNA fingerprinting in plants and fungi.* CRC Press, Boca Raton.

Welsh, J. and McClelland, M. (1990) Fingerprinting genomes using PCR with arbitrary primers. *Nucleic Acids Research*, **18**, 7213–7218.

Williams, J. G. K., Kubelik, A. R., Livak, K. J., Rafalski, J. A. and Tingey, S. V. (1990) DNA polymorphisms amplified by aribitrary primers are useful as genetic markers. *Nucleic Acids Research*, **18**, 6531–6535.

Williams, J. G. K., Hanafey, M. K., Rafalski, J. A. and Tingey, S. V. (1993) Genetic analysis using random amplified polymorphic DNA markers. *Methods in Enzymology*, **218**, 704–740.

Willmer, P., Gilbert, F., Ghazoul, J., Zalat, S. and Semida, F. (1994) A novel form of territoriality: daily paternal investment in an anthophorid bee. *Animal Behaviour*, **48**, 535–549.

Wolff, K. (1991a) Genetic analysis of ecological relevant morphological variability in three *Plantago* species with different mating systems. *Theoretical and Applied Genetics*, **73**, 903–914.

Wolff, K. (1991b) Analysis of allozyme variability in three *Plantago* species and a comparison to morphological variability. *Theoretical and Applied Genetics*, **81**, 119–126.

Wolff, K. and Morgan-Richards, M. (1998) RAPD markers distinguish *Plantago major* subspecies. *Theoretical and Applied Genetics*, **96**, 282–286.

Wolff, K. and Schaal, B. A. (1992) Chloroplast DNA variation within and among five *Plantago* species. *Journal of Evolutionary Biology*, **5**, 325–344.

Wolff, K., Schaal, B. A. and Rogstad, S. H. (1994) Population and species variation of minisatellite DNA in *Plantago*. *Theoretical and Applied Genetics*, **87**, 733–740.

Wolff, K., Zietkiewicz, E. and Peters-Van Rijn, J. (1995) Identification of chrysanthemum cultivars and stability of fingerprint patterns. *Theoretical and Applied Genetics*, **91**, 439–447.

Wolff, K., El-Akkad, S.and Abbott, R. J. (1997) Population structure in *Alkanna orientalis* L. Boiss. (Boraginaceae) in the Sinai desert, in relation to its pollinator behaviour. *Molecular Ecology*, **6**, 365–372.

Wright, S. (1951) The genetical structure of populations. *Annals of Eugenics*, **15**, 313–354.

Yeh, F. C., Chong, D. K. X. and Yang, R. C. (1995) RAPD variation within and among natural populations of trembling aspen (*Populus tremuloides* Michx.) from Alberta. *Journal of Heredity*, **86**, 454–460.

Zietkiewicz, E., Rafalski, A. and Labuda D. (1994) Genome fingerprinting by simple sequence repeats (SSR)-anchored PCR amplification. *Genomics*, **20**, 176–183.

Chapter 5

Metapopulation dynamics and mating-system evolution in plants

S. C. H. Barrett and J. R. Pannell

ABSTRACT

Patterns of plant mating are strongly influenced by genetically determined factors such as floral design, display and phenology, self-incompatibility mechanisms, and levels of inbreeding depression. They also depend on the ecological context in which mating takes place. While the importance of pollination biology for mating-system evolution is increasingly being recognised, the demographic context of mating has received relatively little attention to date. Here, we outline metapopulation models which assess the effect of recurrent local extinction and re-colonization on the evolution and maintenance of mating-system alleles. Such metapopulation processes can alter selective pressures on patterns of sex allocation, modes of self-fertilization and the maintenance of outcrossing. Our results suggest that for species with ephemeral populations, broadscale studies of plants at the landscape level are needed. We review studies of several taxa (*Eichhornia paniculata*, *Lythrum salicaria* and *Mercurialis annua*) which display striking patterns of geographical variation in their sexual systems. Such intraspecific studies lead to insights that are not evident from the analysis of local populations alone. Moreover, we suggest that research at the metapopulation level may form a bridge between microevolutionary investigations within local populations and historical reconstruction of the evolution of reproductive traits using phylogenetic methods.

5.1 Introduction

Flowering plants display unrivalled structural diversity in their reproductive traits, which function to promote reproductive success through their influence on pollination and mating. A particularly striking feature of flowering plants is that patterns of mating vary considerably, often among closely related species. The mode of transmission of genes from one generation to the next can range from predominant self-fertilisation, through mixed mating involving different amounts of self- and cross-fertilisation, to complete outcrossing, sometimes among populations of a single species (Barrett and Eckert, 1990). Variation in mating systems has a direct influence on population genetic structure, quantitative and molecular variation and rates of evolution (Brown, 1979; Hamrick and Godt, 1990; Charlesworth, 1992; Charlesworth and Charlesworth, 1995; Liu *et al.*, 1998). A major task of evolutionary biology is to determine which selective mechanisms are responsible for

In *Molecular systematics and plant evolution* (1999) (eds P. M. Hollingsworth, R. M. Bateman and R. J. Gornall), Taylor & Francis, London, pp. 74–100.

specific changes in plant mating systems. The way plants mate is also relevant to plant systematics. This is because mating systems have a strong influence on patterns of group variability, and evolutionary changes in reproductive mode are often closely associated with the development of reproductive isolation and speciation (Stebbins, 1957; Baker, 1959). Indeed, systematists were among the first biologists to recognise that variation in 'breeding systems' have both important micro- and macroevolutionary consequences (Davis and Heywood, 1963; Ornduff, 1969; Briggs and Walters, 1984).

Current research on the evolution of plant mating systems involves three complementary approaches. Ecological and genetic studies of local populations focus on the selective mechanisms responsible for the evolution and maintenance of particular patterns of mating. Such microevolutionary studies employ experimental manipulations and the use of genetic markers as tools for investigating the proximate ecological factors that influence mating success and the fitness consequences of different mating patterns (reviewed by Barrett and Harder, 1996). Much of this empirical work has been motivated by the development of a rich theoretical literature on mating-system evolution in plants (reviewed by Morgan and Schoen, 1997a). A variety of population genetic and sex allocation models have been developed to explore the conditions that favour different levels of selfing and outcrossing, as well as the evolution of sexual systems (e.g. Charlesworth and Charlesworth, 1981; Charnov, 1982; Lande and Schemske, 1985; Holsinger, 1991; Uyenoyama et al., 1993). These two approaches have dominated the field of mating-system biology for the past two decades, and both focus on the local population as the most relevant evolutionary unit. More recently, a broader perspective has begun to emerge as the comparative biology of reproductive traits is investigated through advances in phylogeny reconstruction and molecular systematics (Donoghue, 1989; Barrett et al., 1996; Hodges, 1997; Sakai et al., 1997; Schoen et al., 1997). Using these methods it is possible to address historical questions concerned with the origin of particular mating systems. Evolutionary models that make predictions about the functional associations of traits can be tested by examining the sequence in which they appear in a phylogeny. A major issue when using these different approaches is whether the macroevolutionary patterns of interest to systematists can be satisfactorily explained by microevolutionary processes evident within local populations. A potential link between these two levels in the genealogical hierarchy is the study of geographical variation in mating systems within species (Barrett, 1995).

Most theoretical models of mating-system evolution in plants involve single populations of infinite size. Within such populations selection usually proceeds to a particular deterministic equilibrium that is governed by fitness differences among genotypes or phenotypes. However, real populations are often short-lived because demographic or environmental stochasticity results in local extinction before any deterministic equilibrium has been reached (Schaffer, 1987; Barrett and Husband, 1997). Among plants with annual or short-lived perennial life histories the longevity of single populations may often be insufficient for significant evolutionary changes in mating patterns to occur. Yet much of the strongest empirical evidence for intraspecific variation in mating systems involves short-lived species, particularly annuals that typically exhibit frequent colonisation and extinction cycles (e.g. Amsinckia – Ganders et al., 1985; Clarkia – Holtsford and Ellstrand, 1989;

Eichhornia – Barrett and Husband, 1990; *Gilia* – Schoen, 1982; *Ipomoea* – Clegg and Epperson, 1988; *Leavenworthia* – Lloyd, 1965; *Limnanthes* – McNeill and Jain, 1983; *Lupinus* – Harding *et al.*, 1974; *Mercurialis* – Pannell, 1997a; *Mimulus* – Ritland and Ritland, 1989; *Senecio* – Abbott, 1985). This association raises the question as to whether the local population is the most appropriate evolutionary unit for investigating the evolution of mating systems in these species. Here we propose that, for species with ephemeral populations, it may often be more appropriate to consider processes at the landscape level using a metapopulation approach, rather than to focus exclusively on those occurring within single populations. A similar argument based on the ephemeral nature of the local population has been made by Levin (1995) for studies of speciation.

Metapopulations are assemblages of interconnected populations existing in a balance between extinction (or more correctly local extirpation) and colonisation (Levins, 1969). The metapopulation approach provides a means of investigating ecological and evolutionary processes at a spatial scale above the local population at both landscape and regional levels. Metapopulation models depict species occupying an array of habitat patches linked by migration. The longevity of a metapopulation will depend on the relative rates of population extinction and colonisation. Since metapopulations will greatly outlive local populations, they provide greater opportunities for significant evolutionary changes, to establish within their boundaries. Despite a rapidly growing literature on metapopulation dynamics (reviewed by Hanski and Gilpin, 1997), relatively few studies have involved plant populations and even fewer have addressed issues of mating-system evolution (Husband and Barrett, 1996). Here, we ask whether application of a metapopulation perspective can provide novel insights into ecological processes governing changes in reproductive traits that are not evident from studies of local populations alone.

In this chapter we apply a metapopulation approach to two central topics in plant reproductive biology: the selection of self- versus cross-fertilisation and the evolution of combined versus separate sexes. Both problems were studied extensively by Darwin (1876, 1877) and a large literature has accumulated on these topics this century (reviewed by Richards, 1997). Here we summarise several recent empirical studies that indicate the importance of metapopulation perspectives for understanding the ecological factors responsible for regional variation in mating systems. We also outline several theoretical models of metapopulation dynamics that provide insights into these empirical observations. However, before we discuss these topics, we first consider general principles involved in plant reproduction within a metapopulation framework and ask whether the levels and direction of selection might differ between local populations and metapopulations.

5.2 General principles and levels of selection

The implications of metapopulation dynamics for mating-system evolution can be distilled into three general principles (Table 5.1). Because individual plants incapable of producing seeds by selfing cannot establish colonies on their own, metapopulation processes (i.e. local colony extinction and recolonisation) will tend to select against unisexuality and self-incompatibility. This first principle is essentially Baker's Law (Baker, 1955, 1967), considered in the context of a metapopulation (Pannell

Table 5.1 Some general principles concerning the influence of metapopulation dynamics on plant reproductive traits and their evolutionary implications.

Principle	Implication
(1) Single seeds incapable of selfing cannot found new colonies	Unisexual and self-incompatible individuals selected against
(2) Seeds but not pollen can found new colonies	Female allocation selected
(3) Ecological processes within versus among populations are distinct	Direction of selection at population and metapopulation levels may differ

and Barrett, 1998a). Traditionally, Baker's Law has been invoked to explain the higher frequency of self-compatible relative to self-incompatible species on oceanic islands (e.g. McMullen, 1987; Webb and Kelly, 1993; Barrett, 1996), but the concept applies to any situation where dispersal to a new site results in potential reproductive failure through reduced opportunities for cross-fertilisation. Baker (1955) himself drew upon observations of what we would now term metapopulation dynamics in notostracan clam shrimps to support his hypothesis, noting that cosexuality had evolved from unisexuality where populations were ephemeral and frequent recolonisation was necessary.

The second principle linking metapopulation dynamics to plant reproduction implicates selective effects on sex allocation. Although plants transmit genes to their progeny through both pollen and seeds, the recolonisation of sites in a metapopulation after local population extinction will favour increased allocation to female function, because seeds, not pollen, can found new colonies (Table 5.1). Such increased female allocation can be manifest either in the provisioning of seeds with structures for better dispersal (Olivieri *et al.*, 1990), by producing better quality seeds (Holsinger, 1986), or by simply producing more seeds (Pannell and Barrett, 1998b). As discussed below, selection on the sex allocation of cosexuals has implications for the differential success in metapopulations of selfing versus outcrossing phenotypes, cosexual versus unisexual phenotypes, and of quantitative variation in sex allocation of cosexual plants generally.

The third principle concerning the effects of metapopulation dynamics on plant reproductive traits involves the consequences of contrasting selective pressures at the population and metapopulation levels. For example, in a metapopulation there will be a trade-off between resource allocation to growth and competition versus that to reproduction and dispersal (Hamilton and May, 1977). This antagonism between selective forces acting during colonisation and population growth was termed the 'metapopulation effect' by Olivieri and Gouyon (1997). They reviewed several cases where such an effect has influenced the evolution of plant life history traits, including those affecting reproduction. For example, in *Thymus vulgaris* (Lamiaceae), cytoplasmic male-sterility is thought to be maintained at the metapopulation level even though nuclear fertility restorers might have checked its spread within local populations (Gouyon and Couvet, 1987). A recent theoretical model has shown that a joint nuclear–cytoplasmic polymorphism for male sterility can be maintained in a metapopulation with recurrent local extinction as a result of the tendency of females to disperse more seeds to surrounding populations than hermaphrodites (Couvet *et al.*, 1998). In this chapter we present several additional examples that demonstrate how selection on mating-system traits may differ within versus between populations.

Our second general principle indicates that metapopulation dynamics may select for increased seed production. This suggests that we might reconsider arguments for mating-system evolution that invoke 'reproductive efficiency' and, by implication, group selection. Such ideas were common currency in the early literature on plant evolution (Darlington and Mather, 1949; Stebbins, 1950) and have continued to hold sway (e.g. Cruden, 1977; Anderson and Stebbins, 1984; Richards, 1997; Karoly, 1994), though they seldom stand the test of rigorous analysis (Lloyd, 1965; Charlesworth, 1980). Arguments for the importance of reproductive efficiency have tended to consider fitness as a population quantity measured in terms of seed productivity (e.g. Mulcahy, 1967; Putwain and Harper, 1972). This view has largely been discredited by sex allocation theory, which recognises the equal importance of pollen and ovules as vehicles for gene transmission and which predicts the selection of equal allocation of resources to both male and female functions in outcrossing populations (Charnov, 1982; Lloyd, 1987). According to sex allocation theory, deviations from equal sex allocation are selected when marginal fitness gains through investment to either sexual function are reduced by such factors as local mate competition (Hamilton, 1967) or local resource competition (Lloyd, 1984).

In stable populations, biased sex allocation is unlikely to arise through an inefficiency of resource use rendering some populations fitter (i.e. more productive) than others. However, for species characterised by metapopulation dynamics, the concept of differential population fitness and selection between groups may appear to be more applicable. Populations comprising individuals with a greater capacity to disperse seeds will make a disproportionate contribution to the composition of a metapopulation before they eventually go extinct. Nevertheless, unless populations suffer non-random extinction probabilities as a result of traits shared by their component individuals, arguments invoking group selection are unnecessary. Couvet *et al.* (1998) explained the maintenance of a nuclear–cytoplasmic polymorphism for male sterility by appealing to group selection, because populations with mainly females grow more rapidly and disperse more seeds than those with lower female frequencies. However, we suggest that metapopulation dynamics are better regarded simply as another component of a species' ecology by which the characteristics of its constituent individuals, including the mating system, are selected. Metapopulation dynamics may thus confer fitness dividends on individuals with a female-biased sex allocation because the availability of vacant habitat patches across the landscape effectively accelerates fitness gains through female function.

The greater premium on seed (as opposed to pollen) production in a metapopulation has implications for the selection of alleles for increased selfing if selfed seeds suffer from inbreeding depression and are reduced in either quality or number. Because individuals with greater outcrossing rates may produce more seeds compared with selfers, the populations in which they occur will have a faster growth rate and hence a greater potential to export seeds to other sites in the metapopulation. This is illustrated in Fig. 5.1 which plots the expected growth rate of a population comprising predominant selfers and predominant outcrossers against the frequency of an allele for increased selfing (see legend for details). Differences in

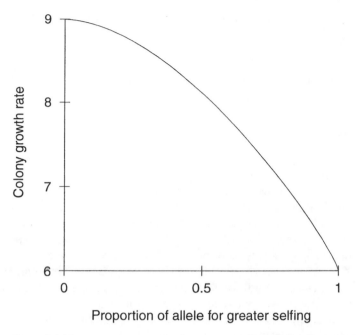

Figure 5.1 The intrinsic rate of population growth as a function of the frequency of the allele for greater selfing. Colonies with a higher frequency of the allele for greater outcrossing grow more rapidly than those for greater selfing, because the progeny of outcrossers do not suffer from inbreeding depression. In the model, inbreeding depression was 0.5. The allele for greater selfing and outcrossing caused selfing rates of 0.8 and 0.2, respectively, and plants produced 10 progeny before inbreeding depression was expressed. Populations were sufficiently large for competing self-fertilisation to be ignored. (After Pannell and Barrett, 1998b).

growth rate between colonies comprising mainly selfers and those comprising mainly outcrossers can therefore cause selection to favour outcrossing at the level of the metapopulation, even where selfing might tend to spread within individual populations (see below). Thus, the direction of selection at a mating-system locus may differ between these two levels of the ecological hierarchy.

5.3 Self- and cross-fertilisation

The comparison of self and cross-fertilisation has been considered the central topic of floral biology (Lloyd and Schoen, 1992). This is because different rates of selfing and outcrossing have profound genetic and evolutionary consequences. The effects of selfing and outcrossing on fitness through inbreeding depression and heterosis are well established (Charlesworth and Charlesworth, 1987; Ritland, 1996), and extensive data on their influence in determining patterns of genetic diversity within and among plant populations exist (Hamrick and Godt, 1990). Selection for self- versus cross-pollination often involves evolutionary modifications to aspects of floral design and display, so that shifts in mating pattern are also accompanied by changes in sex allocation (Charnov, 1982). Since resource allocation, pollination and mating determine reproductive success in an integrated manner, changes in mating system can also have important ecological and life-history consequences. Functional associations between different plant life histories and contrasting rates of self-fertilisation indicate that changes in mating patterns are usually driven by altered ecological and demographic circumstances (Barrett *et al.*, 1996; Morgan and Schoen, 1997b).

5.3.1 Theoretical background

Early models of the evolution of self- and cross-fertilisation focused mainly on genetic factors, particularly inbreeding depression resulting from the expression of deleterious recessive alleles and the transmission advantage of alleles governing the selfing rate (Fisher, 1941; Lande and Schemske, 1985). Most genetic models predicted that predominant selfing and predominant outcrossing should be alternative stable outcomes of mating-system evolution. Whereas survey data on the distribution of outcrossing rates generally supports this prediction (reviewed by Barrett and Harder, 1996), many species exhibit intermediate selfing rates, and stable mixed mating can occur when additional genetic factors are considered in theoretical models. These include biparental inbreeding, inbreeding depression caused by overdominance, and the association between mating-system alleles and viability genes (reviewed by Uyenoyama *et al.*, 1993). The evolutionary dynamics of selfing and outcrossing are also influenced by a range of ecological factors associated with pollination biology (Harder and Barrett, 1996; Holsinger, 1996). These include pollen and seed discounting, reproductive assurance and the mode of self-pollination. More recent models have explored their effects on selfing rates and have shown that these factors modify the threshold values of inbreeding depression above which selfing cannot evolve (Lloyd, 1992; Holsinger, 1996; Schoen *et al.*, 1996; Harder and Wilson, 1998).

Theories on mating-system evolution have become increasingly sophisticated through attempts to build more ecological realism into models. However, a noticeable feature of most models, whether they are primarily genetical in emphasis or involve ecological factors, is that they ignore aspects of population structure and do not consider the influence that stochastic processes might have on mating-system evolution (though see Ronfort and Couvet, 1995). Since the majority of species with wide variation in selfing rates have colonising life histories it seems likely that ecological and demographic influences on population structure are important in the evolution of their reproductive systems. Lloyd (1980) identified a range of demographic factors that might influence frequencies of self- versus cross-fertilisation, but there have been surprisingly few attempts to relate the evolution of mating systems to such ecological factors. What evidence is there that population structure is relevant to plant reproduction, and how might altered demographic circumstances influence mating-system evolution?

5.3.2 Population structure in tristylous species

Tristylous species provide convenient model systems for investigating the influence of ecological and demographic factors on the maintenance of outcrossing (Barrett,

1993). These sexual polymorphisms possess simple inheritance, and theoretical models indicate that frequency-dependent selection due to disassortative mating between the three style morphs should give rise to equal frequencies in large ideal populations (Heuch and Lie, 1985). However, real populations of tristylous plants are often small, and in species with colonising life histories frequent colonisation and extinction cycles provide opportunities for destabilisation of the polymorphism, leading to the evolution of alternate mating systems. Geographical surveys of population structure provide opportunities to investigate the maintenance of these polymorphisms under varying ecological and demographic circumstances at the landscape level. In particular, the spatial dispersion of populations across the landscape and their mutual isolation are likely to be critical in determining the local stability of the polymorphism.

Comparisons of style-morph frequency variation between native and introduced populations of tristylous *Lythrum salicaria* (Lythraceae) have provided valuable insights into the role of population structure and stochastic forces in governing the maintenance of the polymorphism (Eckert and Barrett, 1992; Eckert *et al.*, 1996). A sample of 102 populations from south-western France revealed few populations (5%) that did not contain the three style morphs. In contrast, in the introduced range in southern Ontario, Canada, where the species is rapidly invading wetlands, 23% of 102 populations sampled were missing style morphs. In both areas there was a greater likelihood that small rather than large populations were missing style morphs, implying a role for stochastic forces in morph loss. Despite the lower frequency of morph loss in France, native population sizes were significantly smaller than those in Ontario. How can this observation be reconciled with data on the relation between morph loss and population size? The answer appears to lie in differences in ecology and population structure between the two regions. French populations of *L. salicaria* occur primarily in roadside ditches associated with the agricultural landscapes of the region. The distribution of populations implies a high level of connectivity which provides opportunities for gene flow among populations. Metapopulation models indicate that levels of gene flow on the order of $m \geqslant 0.05$ can account for the maintenance of tristyly in small French populations (Eckert *et al.*, 1996). In contrast, populations in Ontario are more isolated from one another, and opportunities for missing morphs to establish in non-trimorphic populations are thus restricted. Founder effects associated with colonisation of new territory appear to be largely responsible for the higher incidence of morph loss in this region.

Similar demographic processes leading to stochastic morph loss have played a crucial role in destabilising tristyly in the annual emergent aquatic *Eichhornia paniculata* (Pontederiaceae) in northeastern Brazil (Barrett *et al.*, 1989). As in *L. salicaria*, large-scale geographical surveys of style-morph frequencies indicate that morph loss is associated with small population size. Of 167 populations sampled, 29% had at least one style morph missing, larger populations were more likely to be trimorphic than smaller ones, and there was a significant positive correlation between style-morph evenness and the harmonic mean of population size. Moreover, populations missing style morphs were significantly more vulnerable to local extinction than trimorphic populations (Husband and Barrett, 1992; Barrett and Husband, 1997). These patterns strongly implicate ecological factors associated with colonization–extinction cycles as the primary mechanism leading to the stochastic loss of style morphs from local populations. Unlike *L. salicaria*, a self-incompatible perennial, *E. paniculata* is annual and self-compatible, thus providing greater opportunities for demographic changes to exert their influence on mating patterns. Indeed, there is good evidence that the loss of style morphs from populations is accompanied by increases in selfing rate (Barrett and Husband, 1990). Self-pollinating variants occur rarely in trimorphic populations, are common in dimorphic populations, and are frequently fixed in monomorphic populations.

What evidence is there that metapopulation processes may be important in the dissolution of tristyly and the evolution of self-fertilisation in *E. paniculata*? Estimates of gene flow among 44 populations from seven spatially separated areas in northeastern Brazil indicate significant variation among areas (Husband and Barrett, 1995). Estimates of gene flow were highest among areas in which populations were exclusively trimorphic. In contrast, reduced estimates of gene flow were evident in the part of the range in which selfing populations dominate (Fig. 5.2). In this area demographic conditions differ. Populations are smaller in size, less dense and more isolated from one another than elsewhere in the range. The precise ecological factors responsible for these differences in population structure are not known, but they may be related to rainfall patterns and the availability of habitats suitable for *E. paniculata* in the zone of selfing. Nevertheless, under these altered demographic conditions populations appear to be more vulnerable to the stochastic loss of style morphs, and restricted gene flow due to population isolation restricts opportunities for the re-invasion of missing style morphs. These patterns are consistent with the view that metapopulation dynamics play a role in destabilising outcrossing and precipitating the evolution of self-fertilisation. Importantly, they could not have been revealed by focusing on a few local populations of *E. paniculata*, and it is clear that in these studies a metapopulation perspective has provided valuable new insights.

Microevolutionary forces operating at the metapopulation level may ultimately be expressed at higher taxonomic levels. This appears to be the case in *Lythrum* and *Eichhornia*, as stochastic loss of the short-styled morph has influenced pathways of floral evolution in both genera. New World members of the genus *Lythrum* (section *Euhyssopifolia*) are uniformly distylous and likely derived from a tristylous Eurasian ancestor. Comparative morphological evidence indicates that the morphs in distylous taxa correspond to the long- and mid-styled morphs of a tristylous species (Ornduff, 1979), implying loss of the short-styled morph during migration to North America. In *Eichhornia* phylogenetic reconstructions indicate multiple breakdown of tristyly, giving rise to several predominantly selfing species (Graham and Barrett, 1995; Kohn *et al.*, 1996). These small flowered homostylous species are composed of self-pollinating long- and mid-styled morphs. This pattern is evident at both the inter- and intra-specific levels, since each of the tristylous species displays predominantly selfing populations that typically lack the short-styled morph (Barrett, 1988). As discussed above, metapopulation processes resulting in stochastic loss of the short-styled morph appear to have played a key role in eliciting these patterns within the constraints imposed by the inheritance of tristyly.

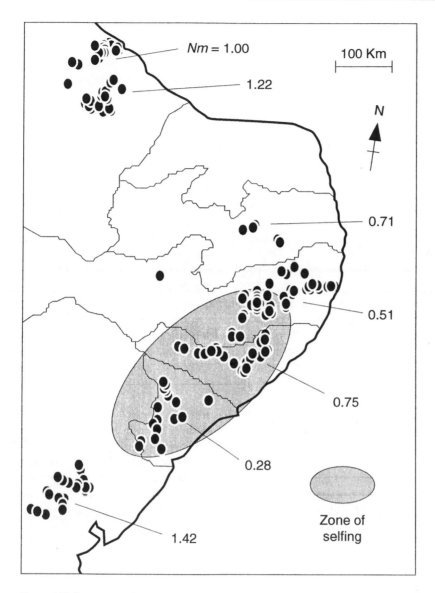

Figure 5.2 Estimates of gene flow based on polymorphism at allozyme loci among populations of *Eichhornia paniculata* from different states in northeastern Brazil. Values of *Nm* were lowest in the zone in which selfing populations occur. (After Husband and Barrett, 1995).

5.3.3 Polymorphisms for selfing and outcrossing

Populations of *E. paniculata* in the zone of selfing indicated in Fig. 5.2 are frequently dimorphic for style morph and they possess a selfing–outcrossing polymorphism. Long-styled plants exhibit high outcrossing rates whereas mid-styled plants experience considerable self-fertilisation because of genetic modifications to their stamen position. Selfing–outcrossing polymorphisms are known in other heterostylous plants (e.g. *Amsinckia* – Ganders *et al.*, 1985; *Oxalis* – Ornduff, 1972; *Primula* – Charlesworth and Charlesworth, 1979), as well as non-heterostylous taxa such as *Senecio vulgaris* (Abbott, 1985). An important question concerning these mating-system polymorphisms is whether they are evolutionarily stable, or whether selection inevitably drives one of the mating morphs to fixation. To address this problem, we have recently developed a computer simulation model that examines the effects of genetic drift and selection at various levels of the ecological hierarchy on the maintenance of a selfing–outcrossing polymorphism (Pannell and Barrett, 1998b).

In our model, we assume that there are two mating phenotypes: one that self-fertilises 80% of its ovules (the selfer); and one that selfs only 20% of its ovules (the outcrosser). Selfing occurs autonomously prior to opportunities for outcrossing; this mode of selfing is termed 'prior self-fertilisation' (Lloyd, 1992). Following prior-selfing, pollen grains dispersed in the population compete for the remaining unfertilised ovules. In large populations, almost all these ovules are outcrossed, but in small populations some are selfed through 'competing selfing' (Lloyd, 1992). This is because pollen dispersed into the population pollen pool is more likely to return to the plant that dispersed it. The mating system in our model is thus context-dependent, varying with population size and the frequency of selfing and outcrossing phenotypes (Fig. 5.3). Because of this context dependence, recently founded colonies will have higher selfing rates than larger and older colonies. Context-dependent mating has been demonstrated experimentally in bee-pollinated *E. paniculata* (Kohn and Barrett, 1994) and is likely to be particularly common in wind-pollinated species where selfing rates are sensitive to plant density (Farris and Mitton, 1984; Holsinger, 1992; and see below).

In our simulations, it was difficult to find conditions under which a selfing–outcrossing polymorphism could be maintained indefinitely, although with certain parameter values the loss of one or other of the mating-system phenotypes was very protracted. These results help to explain the overall rarity of selfing–outcrossing polymorphisms in plants, and it is possible that the known mixed-mating poly-morphisms cited above represent transitional states from predominant outcrossing to selfing. Another important outcome of our model was that with high colony turnover in a metapopulation, the allele for predominant outcrossing was fixed more often than that for selfing, even though the mating-system locus was neutral (Fig. 5.4). This lends further support for the 'metapopulation effect' discussed above (i.e. Principle 3 in Table 5.1), where selection at the metapopulation level acts in a direction not evident within single component populations. This outcome is due to the relatively greater seed output of colonies with a high frequency of outcrossers, the seeds from which did not suffer from inbreeding depression (Fig. 5.1). Thus, because only seeds can found new colonies, selection at the metapopulation level acts in favour of increased seed fertility (Principles 1 and 2, respectively, in Table 5.1).

A third finding of our simulations was that when selfed progeny suffered inbreeding depression, selfers became fixed within finite populations more often than expected. This 'fixation bias' was due to the fact that predominant outcrossers selfed more of their seeds by competing selfing than did selfers; this is analogous to the finding that geitonogamous selfing (often an important component of

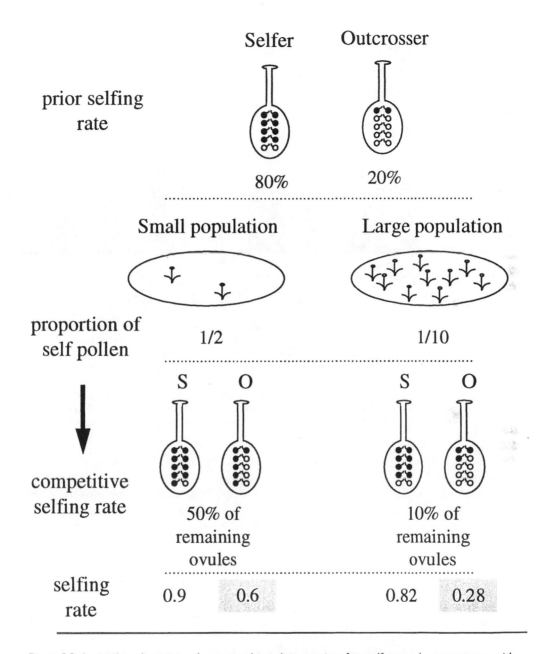

Figure 5.3 A graphic depiction of context-dependent mating for selfers and outcrossers with different prior selfing rates. In small populations, self pollen accounted for an appreciable amount of the pollen dispersed to the pollen pool, while in large populations negligible self pollen was dispersed. In small populations the rate of competing selfing is greater than in large populations. As a result, the difference in selfing rate between selfers and outcrossers is smaller than in large populations.

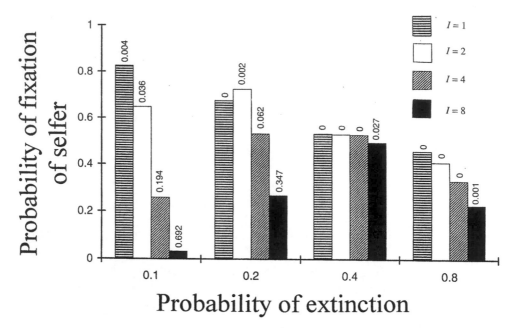

Figure 5.4 Model of a selfing–outcrossing polymorphism in a metapopulation. The probability of fixation of an allele for greater selfing in a metapopulation of 20 sites, in relation to the colony extinction rate and the mean number of immigrants into each site per generation, *I*. The number of immigrants followed a clumped distribution, such that seeds arrived at sites in groups of 5*I* with a probability of 0.2. The genotypic composition of the immigrants was determined probabilistically according to their frequency in the metapopulation. Numbers given over bars are the probability that a selfing–outcrossing polymorphism was maintained after 2000 generations. (After Pannell and Barrett, 1998b).

competing selfing) can never be selected (Lloyd, 1992). These results highlight the need to consider the context in which mating occurs (i.e. population size, density and composition) as well as the mode of self-fertilisation. Both these aspects can be influenced by population structure and metapopulation dynamics and it is clear that empirical studies of the ecological context in which mating occurs are badly needed.

5.3.4 Reproductive assurance in a metapopulation

Darwin (1876) was of the opinion that reproductive assurance was the main selective force in the evolution of selfing from outcrossing in flowering plants, and this view is supported by recent theoretical models (Schoen *et al.*, 1996). Most authors have invoked pollinator scarcity as the ecological mechanism responsible for the advantage of reproductive assurance within single populations (reviewed by Lloyd, 1980). However, in a metapopulation, phenotypes with assured reproduction are likely to be selected because they can found new colonies as single seeds. In contrast, outcrossers will fail to reproduce, irrespective of the presence or absence of pollinators, unless compatible partners reach a site concurrently (Principle 1).

As discussed earlier, Baker's Law can be viewed as a generalisation concerning the selective advantage of reproductive assurance when colonisation is frequent. Whereas it has most often been invoked to explain the high frequency of self-fertilisation in species which have undergone long-distance dispersal, the principle applies equally to the process of colonisation in a metapopulation. Although examples illustrating the principle of Baker's Law have been drawn from both plants and animals, many counter examples have been identified and its generality has been questioned (Carlquist, 1966; Carr *et al.*, 1986; Sakai *et al.*, 1995b). We might ask, for example, to what extent the success of self-incompatible colonising species vitiates the hypothesis. In an attempt to answer this question, and to quantify the selective pressure for reproductive assurance invoked by Baker's Law, we have used a metapopulation model that reframes the question in terms of colony turnover rates and plant life history traits (Pannell and Barrett, 1998a).

Our model considers the maintenance of a single phenotype in a metapopulation in which individual colonies go extinct at a rate E per year and are recolonised at a rate B. If a proportion p of the sites in the metapopulation is occupied at any point in time, then it follows that the relation $Ep = (1 - p)B$ describes the state of the metapopulation at equilibrium. Thus, for any given extinction rate we can find the colonisation rate at which a proportion p of the sites in the metapopulation remains occupied. The model assumes that B will be some function of the mean number of seeds immigrating to a site per year, and this will depend on particular aspects of demography and life history (e.g. perenniality, seed dormancy, dispersal dynamics).

As an example, consider the contrast between the maintenance of a selfer and an outcrosser in a metapopulation. Whereas only a single selfing individual is required for the founding of a new colony, at least two obligate outcrossers are needed. Thus, for outcrossers to be maintained in a metapopulation, they will have to be more proficient at dispersing seeds to vacant sites than selfers. Fig. 5.5 indicates that with greater extinction rates the seed fecundity of outcrossers required for their maintenance in a metapopulation must increase relative to that of selfers. This difference in requisite fecundities between selfers and outcrossers can be viewed as an index of the selective pressure for reproductive assurance and will depend on the site occupancy levels within the metapopulation (Fig. 5.5). In particular, as p falls towards zero, reproductive assurance becomes increasingly important and the relative fecundity tends towards infinity. This result provides an explanation for the otherwise puzzling observation that several successful sexual colonising species are in fact self-incompatible (e.g. *Centaurea solstitialis* – Sun and Ritland, 1998; *Papaver rhoeas* – Campbell and Lawrence, 1981; *Senecio squalidus* – Abbott and Forbes, 1993; *Turnera ulmifolia* – Barrett, 1978) or dioecious (e.g. *Ecballium elaterium* – Costich, 1995; *Mercurialis annua* – Pannell, 1997a; and see below), in apparent contradiction to Baker's Law. Our model indicates that selection for reproductive assurance will be weak when most of the available sites in the metapopulation are occupied, even for high colony extinction rates (Fig. 5.5). It will be strongest in situations where a species is not common across the landscape (e.g. on the periphery of its geographic distribution), and this accords well with empirical observations of the geographical distribution of selfing (reviewed by Lloyd, 1980).

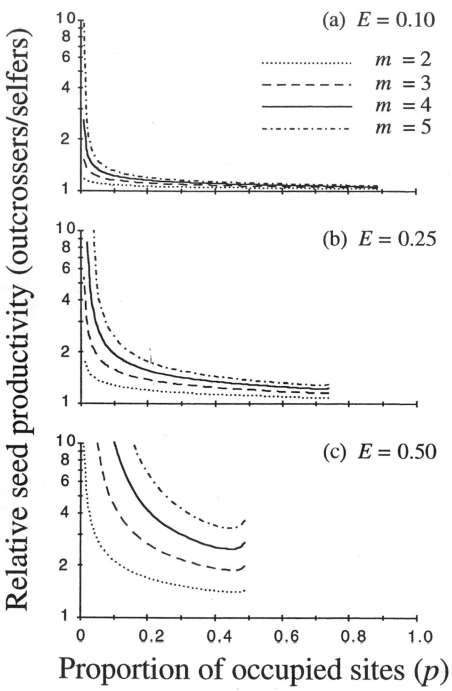

Figure 5.5 Model of reproductive assurance in a metapopulation. The number of seeds that outcrossers must produce relative to selfers to be maintained in a metapopulation in which a proportion *p* sites are occupied at equilibrium. Extinction rates for graphs a, b and c, are *E* = 0.10, 0.25 and 0.50, respectively. Different curves correspond to cases where at least *m* seeds are required for successful colonisation. Thus, selfers need only a single seed to colonise a new site (*m* = 1), whereas outcrossers require *m* > 1 seeds. Curves are truncated at the maximum possible colony site occupancy levels for the respective extinction rate. (After Pannell and Barrett, 1998a).

5.4 Combined versus separate sexes

Darwin (1877) raised the question of why some plants have evolved unisexuality. The overwhelming majority of plant species are cosexual (hermaphroditic or monoecious), and this no doubt reflects the advantages of combined versus separate sexes. Because animal-pollinated species must invest resources towards attracting and rewarding pollinators, the ability of cosexuals to share these fixed costs between male and female functions can amount to significant advantages over unisexuals, which incur the same costs for only one sexual function (Heath, 1977). Saturating male or female fitness-gain curves (or both) are also disadvantageous to unisexual plants because marginal gains in fitness are reduced when allocation to each sexual function is high. As a result, cosexuals, which divide their resources between two sexes, are more likely to be selected (Charnov, 1982)

5.4.1 Ecological and demographic associations

The evolution of combined versus separate sexes is closely associated with plant life histories. For example, dioecy is particularly common in long-lived, woody species but is rare in annuals (Renner and Ricklefs, 1995; Sakai et al., 1995a). This pattern presumably reflects, in part, the risks associated with reproductive failure for annual life histories and the costs of geitonogamy in perennials with large floral displays. Evidence for associations between life history and sexual systems are evident in the few taxa showing gender monomorphism and dimorphism at the species level. For example, in the Australian geophyte, Wurmbea dioica (Colchicaceae), plants in dioecious populations are longer lived than those in cosexual populations, even where the two sexual systems co-occur in sympatry (A. L. Case and S. C. H. Barrett, unpublished). Similar patterns occur in the emergent aquatic, Sagittaria latifolia (Alismataceae), in Ontario, Canada (S. C. H. Barrett, unpublished). Monoecious populations are commonly found in drainage ditches, stream edges and other disturbed aquatic environments and are short-lived, whereas dioecious populations are restricted to extensive marshes and more stable wetland habitats and are both long-lived and highly clonal. A particularly striking example of intraspecific variation in sexual systems occurs in Mercurialis annua (Euphorbiaceae), an annual ruderal distributed across western and central Europe and around the Mediterranean (Durand, 1963). Recent microevolutionary investigations of this species have revealed associations between ecology, demography and sexual system (Pannell, 1997a, b, c).

Throughout most of its range, M. annua is dioecious; however, its western Mediterranean populations are largely monoecious. Whereas dioecious populations tend to persist throughout the year, monoecious populations are winter annuals. In southern Spain, southern Portugal and northern Morocco, males co-occur with monoecious individuals in androdioecious populations. Monoecy has been selected in the drier, less predictable parts of the species' range, presumably as a result of selection for reproductive assurance. In contrast, males have been able to persist with cosexuals in the moister regions where populations are longer lived and more densely distributed across the landscape (Pannell, 1997a).

A theoretical model has produced results supporting a metapopulation interpretation of mating-system variation in M. annua (Pannell, 1997d). The model consists of an array of equivalent sites that suffer constant extinction rates and exchange seeds according to an island model of dispersal. As in *M. annua* (Pannell, 1997a, b), unisexuality is determined by a dominant allele, and unisexual individuals produce five times as many seeds or as much pollen as their cosexual counterparts. Simulations indicated that when rates of colony extinction were low and immigration rates high, unisexuals could be maintained at high frequencies in the metapopulation. By contrast, when colony turnover was rapid, cosexuality spreads to fixation in the metapopulation, which could not be invaded by unisexuals (Fig. 5.6). In this model, unisexuality always spreads locally on its invasion into a cosexual population, but extinction and colonisation dynamics selected against males and females at the metapopulation level. As illustrated above for the contrast between selfing and outcrossing, these results highlight the fact that selection on mating-system alleles can act in opposing directions at the metapopulation and population levels. This implies that an important component of selection on mating systems may be invisible within local populations where biologists have concentrated their efforts in estimating selection parameters. Although rates of colony extinction and recolonisation can be difficult to measure in the field, it is important to realise that processes not readily apparent in local populations may have shaped the course of mating-system evolution in significant ways.

5.4.2 Sex allocation

The theory of sex allocation (Charnov, 1982) has been a powerful tool for understanding gender variation in plants. However, most models have focused exclusively on selection acting within local populations (but see Lloyd, 1982; Ronce and Olivieri, 1997). As we have seen above, processes occurring at the metapopulation level can influence the evolutionary stability of mating systems, but the models and data we have discussed thus far have confined themselves to discrete variation: self-compatibility versus self-incompatibility; selfing versus outcrossing in general; and cosexuality versus unisexuality. Whilst each of these issues is intimately linked with questions regarding sex allocation, we now consider briefly the influence of metapopulation dynamics on quantitative variation in the allocation of resources to male and female functions.

The model we summarise here (Pannell, 1995) is similar to that described in the previous section, except that cosexuals are assumed to vary in their sex allocation rather than to have a single phenotype. For brevity, we describe the model only for an androdioecious species, though it can be applied equally to nuclear-inherited gynodioecy, which yielded similar simulation results. Sex allocation is determined by the additive interaction of co-dominant alleles at a locus unlinked to the one that determines unisexuality. The model assumes that males are more than twice as successful at fertilising ovules than cosexuals, which divide their resources equally between sexual functions. This might occur in a wind-pollinated species in which males are taller and possess inflorescences better suited to pollen dispersal than cosexuals, as found in *M. annua* (Pannell, 1997a). Simulations were conducted for an array of 500 sites interconnected by seed dispersal, with varying colony extinction and recolonisation rates.

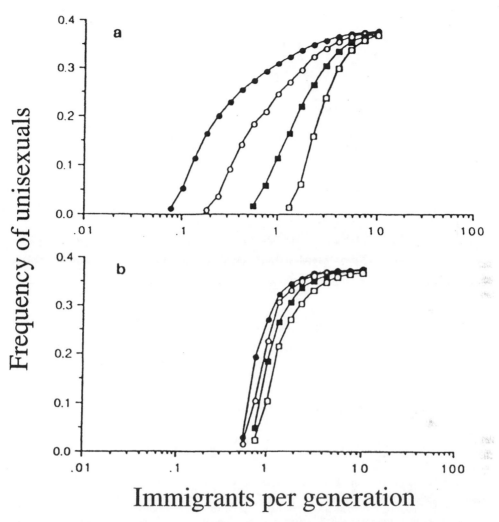

Figure 5.6 Model of combined versus separate sexes in a metapopulation. The frequency of unisexuals at equilibrium in a metapopulation of 2500 sites, plotted against the mean number of immigrants into each site per generation. Curves for four extinction rates are given: 0.05 (closed circles); 0.10 (open circles); 0.20 (closed squares); and 0.40 (open squares). Plots are given for the case of a dominant (a) and a recessive (b) allele for unisexuality. Unisexuals were assumed to produce five times as many seeds (females) or five times as much pollen (males) as cosexuals. (After Pannell, 1997d).

Simulation results differed from those for which cosexuals were assumed to have a single allocation phenotype. In particular, for any given extinction rate, males and cosexuals could be maintained together in the metapopulation only within a relatively narrow range of immigration rates. This is illustrated in Fig. 5.7 for an androdioecious metapopulation with an extinction rate of 0.5. When the mean number of immigrants to a site per generation was high, males occurred at a frequency that approached 0.5 and cosexuals were almost completely female

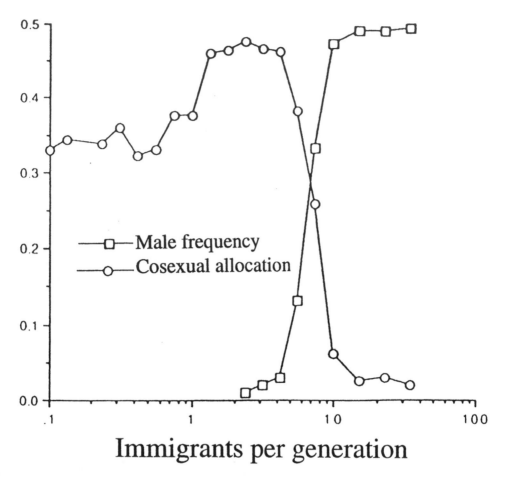

Figure 5.7 Model of sex allocation in a metapopulation. The male frequency (squares) and the relative allocation of resources to female function by cosexuals (circles) in a metapopulation of 500 sites at equilibrium, plotted against the mean number of immigrants into each site per generation, *I*. At low and high values of *I*, cosexuality and dioecy, respectively, are selected. (After Pannell, 1995).

(i.e. dioecy was stable). At intermediate immigration rates, the frequency of males fell towards zero, and cosexuals became equisexual. As long as males persisted in the metapopulation, the mean sex allocation, averaged across males and cosexuals, was approximately equal. However, when the immigration rate was low, males were lost from the metapopulation entirely, and cosexual sex allocation became strongly female biased (Fig. 5.7). As discussed above, this is because metapopulation dynamics with a high colony turnover rate favour phenotypes with increased seed productivity, because only seeds are able to found new colonies.

These results provide a further instance of the metapopulation effect, where selection acts in different directions at the local colony and metapopulation levels. Within individual populations, frequency-dependent selection favours equal sex allocation through dioecy with a sex ratio of 0.5. In contrast, at the metapopulation level, female-biased cosexuality is selected when colony turnover is rapid. This is because males and females lack reproductive assurance, and high seed productivity in cosexuals carries a premium when colonisation of vacant sites is important. The model provides a plausible explanation for the distribution of sexual systems in *Mercurialis*. All seven species of the genus are dioecious and wind-pollinated, and all but *M. annua* are perennial (Tutin *et al.*, 1968). Inflorescence architecture is highly conserved throughout *Mercurialis*, with males bearing their flowers on erect peduncles above the plant and female flowers axillary in position (Pannell, 1997a). This sexual dimorphism is probably the result of selection for optimal dispersal and receipt of pollen (Niklas, 1985), and the convex fitness–gain curves implied by it are most likely the basis for the evolutionary stability of dioecy in the genus. In western Mediterranean populations of *M. annua*, however, it appears that increased population turnover rates have led to the destabilisation of dioecy at the metapopulation level and the evolution of cosexuality. In this region, androdioecy is likely the outcome of selection for cosexuality at the landscape level, opposed by selection for dioecy in established populations due to morphological constraints in inflorescence design and the syndrome of wind-pollination.

5.5 Conclusions

The empirical data reviewed for *Lythrum*, *Eichhornia* and *Mercurialis* indicate that the intraspecific variation in mating systems displayed in these taxa is best understood by considering ecological processes operating at a landscape level. Moreover, our metapopulation models provide plausible explanations for the patterns of variation observed. These results are encouraging and argue for an increased focus on broad-scale spatial variation in ecological and demographic factors as important determinants of evolutionary changes in reproductive mode. Future work on plants will benefit if workers abandon the view of the local population as the only relevant evolutionary unit for studies of mating-system evolution and replace this by a more inclusive perspective that considers the demography and connectivity of populations and their distribution across the landscape.

Metapopulations are likely to have properties more conducive to evolutionary diversification than local populations, because of their greater longevity and the typically broader range of ecological conditions that they encompass. A considerable literature, tracing back to Sewall Wright's 'shifting balance' hypothesis, has considered the importance of population subdivision and metapopulation structure on speciation and the evolution of adaptations (e.g. Wright, 1931; Lande, 1985; Barton and Whitlock, 1997). However, little of this literature has been applied to problems in plant evolution, despite the fact that many plants have small effective population sizes and exhibit strong spatial population structure. Because gene flow, and genetic transmission generally, are governed directly by the mating system, metapopulation processes that influence reproductive mode will also affect processes of adaptive divergence and speciation. In this sense, studies at the metapopulation level may act as a conceptual bridge in the genealogical hierarchy linking micro- and macro-evolutionary enquiry.

What are the prospects that metapopulation studies will help inform systematic and evolutionary studies of plants? At the present time few workers have used these approaches presumably because of the difficulty in measuring extinction, colonisation and migration (Husband and Barrett, 1996). However, the increased availability of a broader range of genetic markers and the development of phylogeographic methods provide exciting new opportunities to examine the history of genetic exchange among populations (Avise, 1998; Schaal, 1998; Templeton, 1998). Mating system studies will undoubtedly benefit from advances in genetic-marker technology and new statistical models that enable pollen- versus seed-mediated gene flow to be measured at both local and landscape levels (Ennos 1994; Ennos et al., 1999 – this volume; McCauley, 1997; Sork et al., 1999). For species in which populations experience high colonisation and extinction rates, studies of metapopulation dynamics are likely to provide novel insights into the evolution of plant reproductive traits.

ACKNOWLEDGEMENTS

We thank Kent Holsinger, Lawrence Harder, Brian Husband and Martin Morgan for valuable discussions concerning metapopulations and plant reproduction. Research discussed in this review was supported by grants from the Natural Sciences and Engineering Research Council of Canada (NSERC) to SCHB and a studentship from the Association of Commonwealth Universities to JRP. The paper was written while JRP was a postdoctoral fellow at the University of Toronto, supported by a research grant from NSERC to SCHB.

REFERENCES

Abbott, R. J. (1985) Maintenance of a polymorphism for outcrossing frequency in a predominantly selfing plant, in *Structure and functioning of plant populations II. Phenotypic and genotypic variation in plant populations*, (eds J. Haeck and J. Woldendorp), North Holland Publishing Company, Amstrdam, pp. 277–86.

Abbott, R. J. and Forbes, D. G. (1993) Outcrossing rate and self-incompatibility in the colonising species *Senecio squalidus*. *Heredity*, **71**, 155–59.

Anderson, G. J. and Stebbins, G. L. (1984) Dioecy versus gametophytic self-incompatibility: a test. *American Naturalist*, **124**, 423–28.

Avise, J. (1998) The history and purview of phylogeography: a personal reflection. *Molecular Ecology*, **7**, 371–79.

Baker, H. G. (1955) Self-compatibility and establishment after 'long-distance' dispersal. *Evolution*, **9**, 347–49.

Baker, H. G. (1959) Reproductive methods as factors of speciation in flowering plants. *Cold Springs Harbor Symposia on Quantitative Biology*, **24**, 177–92.

Baker, H. G. (1967) Support for Baker's Law – as a rule. *Evolution*, **21**, 853–56.

Barrett, S. C. H. (1978) Heterostyly in a tropical weed: the reproductive biology of the *Turnera ulmifolia* complex (Turneraceae). *Canadian Journal of Botany*, **56**, 1713–25.

Barrett, S. C. H. (1988) Evolution of breeding systems in *Eichhornia* (Pontederiaceae): a review. *Annals of the Missouri Botanical Garden*, **75**, 741–760.

Barrett, S. C. H. (1993) The evolutionary biology of tristyly. *Oxford Surveys in Evolutionary Biology*, **9**, 283–326.

Barrett, S. C. H. (1995) Mating-system evolution in flowering plants: micro- and macroevolutionary approaches. *Acta Botanica Neerlandica*, **44**, 385–402.

Barrett, S. C. H. (1996) The reproductive biology and genetics of island plants. *Philosophical Transactions of the Royal Society of London*, B **351**, 725–33.

Barrett, S. C. H. and Eckert, C. G. (1990) Variation and evolution of mating systems in seed plants, in *Biological approaches and evolutionary trends in plants*, (ed S. Kawano), Academic Press, Tokyo, pp. 229–54.

Barrett, S. C. H. and Harder, L. D. (1996) Ecology and evolution of plant mating. *Trends in Ecology and Evolution*, **11**, 73–9.

Barrett, S. C. H. and Husband, B. C. (1990) Variation in outcrossing rates in *Eichhornia paniculata*: the role of demographic and reproductive factors. *Plant Species Biology*, **5**, 41–55.

Barrett, S. C. H. and Husband, B. C. (1997) Ecology and genetics of ephemeral plant populations: *Eichhornia paniculata* (Pontederiaceae) in northeast Brazil. *Journal of Heredity*, **88**, 277–84.

Barrett, S. C. H., Morgan, M. T. and Husband, B. C. (1989) The dissolution of a complex genetic polymorphism: the evolution of self-fertilization in tristylous *Eichhornia paniculata* (Pontederiaceae). *Evolution*, **43**, 1398–1416.

Barrett, S. C. H., Harder, L. D. and Worley, A. C. (1996) The comparative biology of pollination and mating in flowering plants. *Philosophical Transactions of the Royal Society of London*, B **351**, 1272–80.

Barton, N. H. and Whitlock, M. C. (1997) The evolution of metapopulations, in *Metapopulation dynamics: ecology, genetics and evolution*, (eds I. Hanski and M. E. Gilpin), Academic Press, San Diego, pp. 183–210.

Briggs, D. and Walters, S. M. (1984) *Plant variation and evolution*. Cambridge University Press, Cambridge.

Brown, A. H. D. (1979) Enzyme polymorphism in plant populations. *Theoretical Population Biology*, **15**, 1–42.

Campbell, J. M. and Lawrence, M. J. (1981) The population genetics of the self-incompatibility polymorphism in *Papaver rhoeas*. I. The number and distribution of *S*-alleles in families from three localities. *Heredity*, **46**, 69–79.

Carlquist, S. (1966) The biota of long distance dispersal. *Evolution*, **20**, 30–48.

Carr, G. D., Powell, E. A. and Kyhos, D. W. (1986) Self-incompatibility in the Hawaiian Madiinae (Compositae): an exception to Baker's Rule. *Evolution*, **40**, 430–34.

Charlesworth, B. (1980) Genetic constraints on the evolution of plant reproductive systems, in *Lecture notes in biomathematics: population genetics in forestry, vol. 60* (ed H. R. Gregorius), Springer-Verlag, Berlin, pp. 155–79.

Charlesworth, B. (1992) Evolutionary rates in partially self-fertilizing species. *American Naturalist*, **140**, 126–48.

Charlesworth, D. and Charlesworth, B. (1979) The maintenance and breakdown of distyly. *American Naturalist*, **114**, 499–513.

Charlesworth, D. and Charlesworth, B. (1981) Allocation of resources to male and female functions in hermaphrodites. *Biological Journal of the Linnean Society*, **15**, 57–74.

Charlesworth, D. and Charlesworth, B. (1987) Inbreeding depression and its evolutionary consequences. *Annual Review of Ecology and Systematics*, **18**, 273–88.

Charlesworth, D. and Charlesworth, B. (1995) Quantitative genetics in plants: the effect of the breeding system on genetic variability. *Evolution*, **49**, 911–20.

Charnov, E. L. (1982) *The theory of sex allocation*. Princeton University Press, Princeton, NJ.

Clegg, M. T. and Epperson, B. K. (1988) Natural selection of flower color polymorphisms in morning glory populations, in *Plant evolutionary biology*, (eds L. D. Gottlieb and S. K. Jain), Chapman and Hall, London, pp. 255–73.

Costich, D. E. (1995) Gender specialisation across a climatic gradient: experimental comparisons of monoecious and dioecious *Ecballium*. *Ecology*, **76**, 1036–50.

Couvet, D., Ronce, O. and Gliddon, C. (1998) The maintenance of nucleocytoplasmic polymorphism in a metapopulation: the case of gynodioecy. *American Naturalist*, **152**, 59–70.

Cruden, R. W. (1977) Pollen-ovule ratios: a conservative index of breeding systems in flowering plants. *Evolution*, **31**, 32–46.

Darlington, C. D. and Mather, K. (1949) *The elements of genetics*. Allen and Unwin, London.

Darwin, C. (1876) *The effects of cross- and self-fertilisation in the vegetable kingdom*. John Murray, London.

Darwin, C. (1877) *The different forms of flowers on plants of the same species*. John Murray, London.

Davis, P. H. and Heywood, V. H. (1963) *Principles of angiosperm taxonomy*. Oliver and Boyd, Edinburgh.

Donoghue, M. J. (1989) Phylogenies and the analysis of evolutionary sequences, with examples from seed plants. *Evolution*, **43**, 1137–56.

Durand, B. (1963) Le complexe *Mercurialis annua* L. *s.l.*: une étude biosystematique. *Annales des Sciences Naturelles, Botanique, Paris*, **12**, 579–736.

Eckert, C. G. and Barrett, S. C. H. (1992) Stochastic loss of style morphs from populations of tristylous *Lythrum salicaria* and *Decodon verticillatus* (Lythraceae). *Evolution*, **46**, 1014–29.

Eckert, C. G., Manicacci, D. and Barrett, S. C. H. (1996) Genetic drift and founder effect in native versus introduced populations of an invading plant, *Lythrum salicaria* (Lythraceae). *Evolution*, **50**, 1512–9.

Ennos, R. A. (1994) Estimating the relative rates of pollen and seed migration among plant populations. *Heredity*, **72**, 250–259.

Ennos, R. A., Sinclair, W. T., Hu, X.-S. and Langdon, A. (1999) Using organelle markers to elucidate the history, ecology and evolution of plant populations, in *Molecular systematics and plant evolution*, (eds P. M. Hollingsworth, R. M. Bateman and R. J. Gornall), London, Taylor & Francis, pp. 1–19.

Farris, M. A. and Mitton, J. W. (1984) Population density, outcrossing rate, and heterozygous superiority in Ponderosa pine. *Evolution*, **38**, 1151–4.

Fisher, R. A. (1941) Average excess and average effect of a gene substitution. *Annals of Eugenics*, **11**, 53–63.

Ganders, F. R., Denny, S. K. and Tsai, D. (1985) Breeding system variation in *Amsinckia spectabilis* (Boraginaceae). *Canadian Journal of Botany*, **63**, 533–8.

Gouyon, P. H. and Couvet, D. (1987) A conflict between two sexes, females and hermaphrodites, in *The evolution of sex and its consequences*, (ed. S. C. Stearns), Birkhäuser Verlag, Basel, pp. 245–61.

Graham, S. W. and Barrett, S. C. H. (1995) Phylogenetic systematics of Pontederiales: implications for breeding-system evolution, in *Monocotyledons: systematics and evolution*, (eds P. J. Rudall, P. J. Cribb, D. F. Cutler and C. J. Humphries), Royal Botanic Gardens, Kew, UK, pp. 415–41.

Hamilton, W. D. (1967) Extraordinary sex ratios. *Science*, **156**, 477–88.

Hamilton, W. D. and May, R. M. (1977) Dispersal in stable habitats. *Nature*, **269**, 578–81.

Hamrick, J. L. and Godt, M. J. W. (1990). Allozyme diversity in plant species, in *Plant population genetics, breeding, and genetic resources*, (eds A. H. D. Brown, M. T. Clegg, A. L. Kahler and B. S. Weir), Sinauer, Sunderland, pp. 43–63.

Hanski, I. and Gilpin, M. E. (1997) *Metapopulation biology: ecology, genetics, and evolution*. Academic Press, San Diego.

Harder, L. D. and Barrett, S. C. H. (1996) Pollen dispersal and mating patterns in animal-pollinated plants, in *Studies on floral evolution in animal-pollinated plants*, (eds D. G. Lloyd and S. C. H. Barrett), Chapman and Hall, New York, pp. 140–90.

Harder, L. D. and Wilson, W. G. (1998) A clarification of pollen discounting and its joint effects with inbreeding depression on mating-system evolution. *American Naturalist*, **152**, 684–95.

Harding, J., Mankinen, C. B. and Elliot, M. (1974) Genetics of *Lupinus*. VII. Outcrossing, autofertility and variability in natural populations of the Nanus group. *Taxon*, **23**, 729–38.

Heath, D. J. (1977) Simultaneous hermaphroditism: cost and benefit. *Journal of Theoretical Biology*, **64**, 363–73.

Heuch, I. and Lie, R. T. (1985) Genotype frequencies associated with incompatibility systems in tristylous plants. *Theoretical Population Biology*, **27**, 318–36.

Hodges, S. A. (1997) Rapid radiation due to a key innovation in columbines, in *Molecular evolution and adaptive radiation*, (eds T. J. Givnish and K. J. Sytsma), Cambridge University Press, Cambridge, pp. 391–405.

Holsinger, K. E. (1986) Dispersal and plant mating systems: the evolution of self-fertilization in subdivided populations. *Evolution*, **40**, 405–13.

Holsinger, K. E. (1991) Mass-action models of plant mating systems: the evolutionary stability of mixed mating systems. *American Naturalist*, **138**, 606–22.

Holsinger, K. E. (1992) Ecological models of plant mating systems and the evolutionary stability of mixed mating systems, in *Ecology and evolution of plant reproduction*, (ed. R. Wyatt), Chapman and Hall, New York, pp. 169–91.

Holsinger, K. E. (1996) Pollination biology and the evolution of mating systems in flowering plants, in *Evolutionary biology*. vol. 29, (eds M. K. Hecht, W. C. Steere and B. Wallace), Plenum Press, New York, pp. 107–49.

Holtsford, T. P. and Ellstrand, N.C. (1989) Variation in outcrossing rate and population genetic structure of *Clarkia tembloriensis* (Onagraceae). *Theoretical and Applied Genetics*, **78**, 480–8.

Husband, B. C. and Barrett, S. C. H. (1992) Genetic drift and the maintenance of the style length polymorphism in tristylous populations of *Eichhornia paniculata* (Pontederiaceae). *Heredity*, **69**, 440–9.

Husband, B. C. and Barrett, S. C. H. (1995) Estimates of gene flow in *Eichhornia paniculata*: effects of range substructure. *Heredity*, **75**, 549–60.

Husband, B. C. and Barrett, S. C. H. (1996) A metapopulation perspective in plant population biology. *Journal of Ecology*, **84**, 461–9.

Karoly, K. (1994) Dioecy and gametophytic self-incompatibility: reproductive efficiency revisited. *American Naturalist*, **144**, 677–87.

Kohn, J. R. and Barrett, S. C. H. (1994) Pollen discounting and the spread of a selfing variant in tristylous *Eichhornia paniculata*: evidence from experimental populations. *Evolution*, **48**, 1576–94.

Kohn, J. R., Graham, S. W., Morton, B., Doyle, J. J. and Barrett, S. C. H. (1996) Reconstruction of the evolution of reproductive characters in Pontederiaceae using phylogenetic evidence from chloroplast DNA restriction-site variation. *Evolution*, **50**, 1454–69.

Lande, R. (1985) The fixation of chromosomal rearrangements in a subdivided population with local extinction and recolonisation. *Heredity*, **54**, 323–32.

Lande, R. and Schemske, D. W. (1985) The evolution of self-fertilization and inbreeding depression. *Evolution*, **39**, 24–40.

Levin, D. A. (1995) Metapopulations: an arena for local speciation. *Journal of Evolutionary Biology*, **8**, 635–44.

Levins, R. (1969) Some demographic and genetic consequences of environmental heterogeneity for biological control. *Bulletin of the Entomological Society of America*, **15**, 237–40.

Liu, F., Zhang, L. and Charlesworth, D. (1998) Genetic diversity in *Leavenworthia* populations with different inbreeding levels. *Proceedings of the Royal Society*, **B 265**, 293–301.

Lloyd, D. G. (1965) Evolution of self-compatibility and racial differentiation in *Leavenworthia* (Cruciferae). *Contributions to the Gray Herbarium*, **195**, 3–134.

Lloyd, D. G. (1980) Demographic factors and mating patterns in angiosperms, in *Demography and evolution in plant populations* (ed. O. T. Solbrig), Blackwell, Oxford, pp. 209–24.

Lloyd, D. G. (1982) Selection of combined versus separate sexes in seed plants. *American Naturalist*, **120**, 571–85.

Lloyd, D. G. (1984) Gender allocation in outcrossing cosexual plants, in *Principles of plant population ecology*, (eds R. Dirzo and J. Sarukhan), Sinauer, Sunderland, MA, pp. 277–300.

Lloyd, D. G. (1987) Allocations to pollen, seeds and pollination mechanisms in self-fertilizing plants. *Functional Ecology*, **1**, 83–9.

Lloyd, D. G. (1992) Self- and cross-fertilization in plants. II. The selection of self-fertilization. *International Journal of Plant Science*, **153**, 370–80.

Lloyd, D. G. and Schoen, D. J. (1992) Self- and cross-fertilization in plants. I. Functional dimensions. *International Journal of Plant Science*, **153**, 358–69.

McCauley, D. (1997) The relative contributions of seed and pollen movement to the local genetic structure of *Silene alba*. *Journal of Heredity*, **88**, 257–63.

McMullen, C. K. (1987) Breeding systems of selected Galápagos Islands angiosperms. *American Journal of Botany*, **74**, 1694–1705.

McNeill, C. I. and Jain, S. K. (1983) Genetic differentiation studies and phylogenetic inference in the plant genus *Limnanthes* (section *Inflexae*). *Theoretical and Applied Genetics*, **66**, 257–69.

Morgan, M. T. and Schoen, D. J. (1997a) The role of theory in an emerging new plant reproductive biology. *Trends in Ecology and Evolution*, **12**, 231–4.

Morgan, M. T. and Schoen, D. J. (1997b) The evolution of self-fertilization in perennials. *American Naturalist*, **150**, 618–38.

Mulcahy, D. L. (1967) Optimal sex ratio in *Silene alba*. *Heredity*, **22**, 411–23.

Niklas, K. J. (1985) The aerodynamics of wind-pollination. *Botanical Review*, **51**, 328–86.

Olivieri, I. and Gouyon, P. H. (1997) Evolution of migration rate and other traits: the metapopulation effect, in *Metapopulation biology: ecology, genetics, and evolution*, (eds I. Hanski and M. E. Gilpin), Academic Press, San Diego, pp. 293–323.

Olivieri, I., Couvet, D. and Gouyon, P. H. (1990) The genetics of transient populations: research at the metapopulation level. *Trends in Ecology and Evolution*, **5**, 207–10.

Ornduff, R. (1969) Reproductive biology in relation to systematics. *Taxon*, **18**, 121–33.

Ornduff, R. (1972) The breakdown of trimorphic incompatibility in *Oxalis* section *Corniculatae*. *Evolution*, **26**, 52–65.

Ornduff, R. (1979) The morphological nature of distyly in *Lythrum* section *Euhyssopifolia*. *Bulletin of the Torrey Botanical Club*, **106**, 4–8.

Pannell, J. R. (1995) *Models of androdioecy and studies of* Mercurialis annua L. Unpublished D.Phil. Thesis, Oxford University.

Pannell, J. R. (1997a) Widespread functional androdioecy in *Mercurialis annua* L. (Euphorbiaceae). *Biological Journal of the Linnean Society*, **61**, 95–116.

Pannell, J. R. (1997b) Mixed genetic and environmental sex determination in an androdioecious population of *Mercurialis annua*. *Heredity*, **78**, 50–6.

Pannell, J. R. (1997c) Variation in sex ratios and sex allocation in androdioecious *Mercurialis annua*. *Journal of Ecology*, **85**, 57–69.

Pannell, J. R. (1997d) The maintenance of gynodioecy and androdioecy in a metapopulation. *Evolution*, **51**, 10–20.

Pannell, J. R. and Barrett, S. C. H. (1998a) Baker's Law revisited: reproductive assurance in a metapopulation. *Evolution*, **52**, 657–668.

Pannell, J. R. and Barrett, S. C. H. (1998b) The effect of genetic drift and metapopulation dynamics on the maintenance of a mixed-mating polymorphism in plants. *Unpublished MS*.

Putwain, P. D. and Harper, J. L. (1972) Studies in the dynamics of plant populations. V. Mechanisms governing the sex ratio in *Rumex acetosa* and *R. acetosella*. *Journal of Ecology*, **60**, 113–29.

Renner, S. S. and Ricklefs, R. E. (1995) Dioecy and its correlates in the flowering plants. *American Journal of Botany*, **82**, 596–606.

Richards, A. J. (1997) *Plant breeding systems*. Second Edition, Chapman and Hall, London.

Ritland, C. R. and Ritland, K. (1989) Variation of sex allocation among eight taxa of the *Mimulus guttatus* species complex (Scrophulariaceae). *American Journal of Botany*, **76**, 1731–9.

Ritland, K. (1996) Inferring the genetic basis of inbreeding depression in plants. *Genome*, **39**, 1–8.

Ronce, O. and Olivieri, I. (1997) Evolution of reproductive effort in a metapopulation with local extinctions and ecological succession. *American Naturalist*, **150**, 220–49.

Ronfort, J. and Couvet, D. (1995) A stochastic model of selection on selfing rates in structured populations. *Genetical Research, Cambridge*, **65**, 209–22.

Sakai, A. K., Wagner, W. L., Ferguson, D. M. and Herbst, D. R. (1995a) Biogeographical and ecological correlates of dioecy in the Hawaiian flora. *Ecology*, **76**, 2530–43.

Sakai, A. K., Wagner, W. L., Ferguson, D. M. and Herbst, D. R. (1995b) Origins of dioecy in the Hawaiian flora. *Ecology*, **76**, 2517–29.

Sakai, A. K., Weller, S. G., Wagner, W. L., Soltis, P. S. and Soltis, D. E. (1997) Phylogenetic perspectives on the evolution of dioecy: adaptive radiation in the endemic Hawaiian genera *Schiedea* and *Alsinodendron* (Caryophyllaceae: Alsinoideae), in *Molecular evolution and adaptive radiation*, (eds T. J. Givnish and K. J. Sytsma), Cambridge University Press, Cambridge, pp. 455–73.

Schaal, B. (1998) Phylogeographic studies in plants: problems and prospects. *Molecular Ecology*, **7**, 465–74.

Schaffer, M. (1987) Minimum viable populations: coping with uncertainty, in *Viable Populations for Conservation* (ed. M. Soulé), Cambridge University Press, Cambridge, pp. 69–86.

Schoen, D. J. (1982) The breeding system of *Gilia achilleifolia*: variation in floral characteristics and outcrossing rate. *Evolution*, **36**, 352–60.

Schoen, D. J., Morgan, M. T. and Bataillon, T. (1996) How does self-pollination evolve? Inferences from floral ecology and molecular genetic variation. *Philosophical Transactions of the Royal Society of London*, **B 351**, 1281–90.

Schoen, D. J., Johnston, M. O., L'Heureux, A. M. and Marsolais, J. V. (1997) Evolutionary history of the mating system in *Amsinckia* (Boraginaceae). *Evolution*, **51**, 1090–9.

Sork, V. L., Nason, J., Campbell, D. R. and Fernandez, J. F. (1999) Landscape approaches to historical and contemporary gene flow in plants. *Trends in Ecology and Evolution*, **14**, 219–24.

Stebbins, G. L. (1950) *Variation and evolution in plants*. Columbia University Press, New York.

Stebbins, G. L. (1957) Self-fertilization and population variability in the higher plants. *American Naturalist*, **91**, 337–54.

Sun, M. and Ritland, K. (1998) Mating system of yellow starthistle *(Centaurea solstitialis)*, a successful coloniser in North America. *Heredity*, **80**, 225–32.

Templeton, A. R. (1998) Nested clade analyses of phylogeographic data: testing hypotheses about gene flow and population history. *Molecular Ecology*, **7**, 381–97.

Tutin, T. G., Heywood, V. H., Burges, N. A., Moore, D. M., Valentine, D. H., Walters, S. M. and Webb, D. A. (1968) *Flora Europaea*. Cambridge University Press, Cambridge.

Uyenoyama, M. K., Holsinger, K. E. and Waller, D. M. (1993) Ecological and genetical factors directing the evolution of self-fertilization. *Oxford Surveys in Evolutionary Biology*, **9**, 327–81.

Webb, C. J. and Kelly, D. (1993) The reproductive biology of the New Zealand flora. *Trends in Ecology and Evolution*, **8**, 442–7.

Wright, S. (1931) Evolution in Mendelian populations. *Genetics*, **16**, 97–159.

Chapter 6

Identifying multiple origins in polyploid homosporous pteridophytes

J. C. Vogel, J. A. Barrett, F. J. Rumsey and M. Gibby

ABSTRACT

Polyploidy is a common mechanism of speciation in plants. In recent years molecular techniques have provided new data for the investigation of patterns and processes of polyploidy. From these data it has been inferred that genetic diversity in polyploids is mainly due to multiple origins, leading to the claim that 'the multiple origin of polyploids is the rule and not the exception'. In this paper we examine critically the premises which have been used to interpret allozyme data and that have led to the inference that multiple origin is the main cause of genetic diversity in polyploids. We argue that, at best, the evidence for multiple origin in most allozyme studies is equivocal and other explanations are possible. The best evidence for multiple origins can be obtained from studies of neo-polyploids. The difficulties in obtaining conclusive evidence from palaeopolyploids are outlined. Although this review focuses on multiple origin in homosporous pteridophytes, the principles outlined here can be applied to other organisms.

6.1 Introduction

Speciation in plants is often associated with polyploidy (Grant, 1981; Wagner and Wagner, 1980). Cytological investigations originally provided the means for unravelling polyploid complexes (Manton, 1950), but in recent years molecular techniques have been used to study polyploid evolution. Soltis and Soltis (1993) reviewed the cases of 15 pteridophyte taxa where evidence obtained from molecular studies indicates multiple origin of polyploids. Of these, only three have been published (Werth *et al.*, 1985a, b; Ranker *et al.*, 1989, 1994a, b).

Homosporous ferns have an alternation of generations with the mature diploid sporophyte forming haploid spores after meiosis. These spores germinate into short-lived, haploid gametophytes that bear the sex organs, antheridia and archegonia. Fertilisation occurs on the haploid gametophyte and the diploid sporophyte emerges from the gametophyte. Three breeding systems can be found in homosporous ferns: outbreeding, inter- and intragametophytic selfing. Intragametophytic selfing is an extreme form of self-fertilisation which occurs within a single haploid gametophyte, intergametophytic selfing is fertilisation between gametophytes derived from spores from one plant, and outcrossing is fertilisation between gametophytes derived from different plants. Allozyme studies have shown that outbreeding is common in

In *Molecular systematics and plant evolution* (1999) (eds P. M. Hollingsworth, R. M. Bateman and R. J. Gornall), Taylor & Francis, London, pp. 101–117.

diploids while inbreeding is reported to be the predominant breeding system in polyploids (Soltis and Soltis, 1989, 1992; Masuyama and Watano, 1990).

The first investigation using molecular methods and demonstrating multiple origin in vascular plants was on pteridophytes in the Appalachian *Asplenium* (Spleenwort) complex (Werth *et al.*, 1985a, b). This group of species is geographically confined to the eastern United States and consists of three diploid taxa, their three allopolyploid derivatives and various backcrosses (Wagner, 1954; Wagner *et al.*, 1993). The three ancestral diploids are morphologically distinct from each other and the polyploids have intermediate morphologies that indicate their parentage. The Appalachian *Asplenium* complex had been studied extensively cytogenetically (Wagner, 1954) and chemotaxonomically (Smith and Levin, 1963; Smith and Harborne, 1971) to determine the relationships between the taxa. The tetraploids were confirmed to be allotetraploids. Werth *et al.* (1985a, b) used allozyme electrophoresis to investigate all taxa in the complex to confirm the results obtained by the other methods. They found species-unique alleles for each of the diploid taxa and additive allozyme patterns in their derived allotetraploids. Two of the allotetraploid taxa showed allozyme polymorphism at loci that were also polymorphic in their diploid progenitors. This allozyme evidence, in conjunction with the results obtained by the cytological investigations, provided strong evidence for independent origins of two of the three allopolyploids, *A. bradleyi* and *A. pinnatifidum*.

Since this discovery many studies have employed DNA markers, allozymes (contributing most of the data), or a combination of both to reconstruct patterns of polyploid evolution (e.g. Comes and Abbott (1999 – this volume) and for a review see Soltis and Soltis (1993)). Evidence for multiple origins of polyploid plants can be obtained from chloroplast DNA (cpDNA) markers. Chloroplasts are normally uniparentally inherited and breeding experiments have demonstrated that they are transmitted maternally in most angiosperms and some pteridophytes (Birky, 1995; Vogel *et al.*, 1998). Multiple origins of polyploids may be demonstrated if the cpDNA of the diploid progenitors can be discriminated and can be recognised in different populations of the derived polyploids.

Recent studies on polyploids using molecular techniques have been the basis of the claim 'that recurrent formation of polyploid species is the rule, rather than the exception' (Soltis and Soltis, 1993). All organisms investigated are vascular plants, with the exception of the bryophyte *Plagiomnium medium* (Wyatt *et al.*, 1988, 1992), and 15 of the 46 taxa included in the review by Soltis and Soltis (1993) are homosporous ferns.

Genetic variation in polyploids can be explained in several ways:

- multiple (independent) origin;
- segregation in hybrid swarms;
- mutation, including loss-of-function mutations;
- hybridisation and introgression;
- gene flow from diploids to autopolyploids and segmental allotetraploids via unreduced gametes;
- random pairing of different genomes in autopolyploids or segmental allopolyploids;
- or combinations of the above.

In this paper we shall examine the premises that have been used to interpret data and that have led to the inference that multiple origin is the main cause of genetic diversity in polyploids. We shall argue that, at best, the evidence for multiple origin in most studies that are based on allozyme data is equivocal and other explanations are possible. Although we focus on multiple origins in homosporous pteridophytes, the principles outlined here can be applied to other organisms.

6.2 Implicit assumptions leading to the inference of multiple origins of polyploid pteridophytes

The implicit argument underlying the inference that multiple origins can explain variability in polyploid pteridophytes is shown in Fig. 6.1. Here, two diploid taxa, A and B, grow together. Their gametes form a single diploid hybrid sporophyte on which diplospores can form via restitution nuclei. A diplospore disperses, grows into a gametophyte and forms an allotetraploid sporophyte via intragametophytic selfing; such a sporophyte will be 100% homozygous at all loci. If there are any allelic differences at one or more loci in the diploid parents, these will appear in the tetraploid offspring as an electrophoretic pattern indistinguishable from that found in a diploid heterozygote. However, this pattern will be due to homozygosity at two different loci, one in each parental genome, and results from intergenomic 'fixed' heterozygosity in the tetraploid. Thus, if the parental diploid taxa are monomorphic for different alleles, or have distinguishable private alleles, allotetraploid derivatives will show corresponding patterns of fixed heterozygosity. Moreover, if individual plants of either taxon A or B (or both) are heterozygous at one or more loci, then genotypically different diploid hybrids can be produced in crosses between the same plants. If this is the normal process, and individual diploid hybrids occur independently in time and space, then every different polyploid genotype, in the absence of mutation and recombination between genotypes, must have arisen from an independent hybridisation event.

The argument appears to be based on the following two principal premises: 1) the polyploid taxa studied are allopolyploids, with disomic inheritance; and 2) the diploid hybrid and derivative allopolyploid are either capable of intragametophytic selfing or may even be intragametophytic obligate inbreeders (Ranker *et al.*, 1989, 1994a, b). Violation of either of these assumptions completely invalidates any inference made from the observations.

6.3 Genetic evidence for allopolyploidy

6.3.1 Cytological investigations

Three different forms of polyploid were defined by Stebbins (1947), with the fundamental distinction being between autopolyploids and allopolyploids. An autopolyploid is a polyploid that originates by the multiplication of one basic set of chromosomes e.g. $AA \rightarrow AA\ AA$. An allopolyploid originates by hybridisation between two different species followed by chromosome doubling, e.g. $AA \times BB \rightarrow AB$, doubling of chromosomes $\rightarrow AA\ BB$. The term 'segmental allopolyploid'

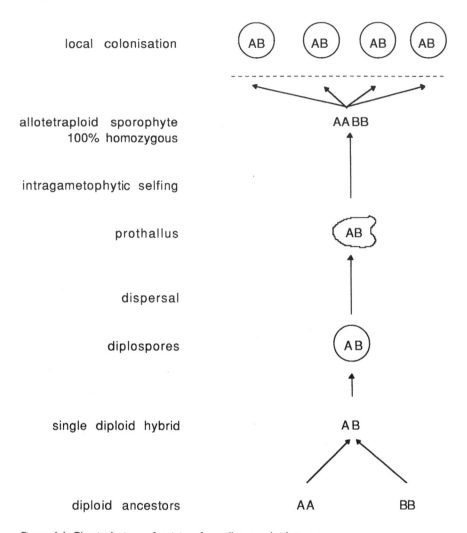

local colonisation

allotetraploid sporophyte
100% homozygous

intragametophytic selfing

prothallus

dispersal

diplospores

single diploid hybrid

diploid ancestors

Figure 6.1 Classical view of origin of an allotetraploid taxon.

was suggested for all intermediate cases where the parental diploid taxa possess some measure of chromosomal and genetic differentiation between their genomes, but are similar enough to be assigned to the same species or have a recent common ancestor.

In diploid hybrid ferns, chromosomes may fail properly to pair and segregate unevenly during anaphase 1 of meiosis, resulting in gametes with different chromosome numbers (unbalanced chromosome complements) and, ultimately, in spores that are inviable and abortive (Gibby, 1980). Sometimes, in diploid hybrids restitution occurs at the first stage of meiosis, the univalents fail to separate at anaphase 1, and only two spores result, each with the diploid chromosome number. The prothalli that they produce will be uniformly diploid, and self-fertilisation of these gametophytes will result in the production of tetraploid sporophytes. Polyploidy by restitution, following hybrid formation in which meiotic disturbances result in unreduced gametes, has been suggested as the usual origin for allopolyploid ferns (Wagner and Wagner, 1980).

Somatic tissue can also spontaneously become polyploid because of a failure in mitosis. Sporangia produced on such polyploid tissues have been observed in the wild on sterile hybrids in the genera *Woodsia* (Butters and Tryon, 1948) and *Athyrium* (Schneller and Rasbach, 1984; Rasbach *et al.*, 1991). These sporangia contained fertile spores.

Evidence for biosystematic relationships in major European and North American fern genera such as *Dryopteris* and *Asplenium* has been based on morphology and cytological investigations of pairing behaviour of chromosomes during meiosis (Manton, 1950; Lovis, 1977). To test hypotheses of ancestry and the relationships between tetraploid and diploid taxa, investigations have been carried out in Europe, using synthesised hybrids between the tetraploids and their putative ancestral diploid taxa (Shivas, 1956; Lovis, 1958; Emmott, 1963; Sleep, 1966, 1983; Vida, 1970; Brownsey, 1973; Gibby, 1977; Lovis, 1977; Reichstein, 1981).

Many studies have not followed this approach and instead have used allozyme electrophoresis to infer the allopolyploid status of the species. The presence of heterozygous patterns, as revealed by allozyme electrophoresis, combined with genetically uniform populations and an absence of segregation in breeding experiments have been used to infer that these patterns are 'fixed' heterozygotes and evidence of the allopolyploid nature of plants. Care, however, is needed in the evaluation of such data.

6.3.2 Fixed heterozygosity

The presence of 'fixed' heterozygosity cannot be used directly to infer that the taxon concerned is an allopolyploid. Fixed heterozygosity of allelic patterns is not confined solely to 'true' allopolyploid taxa, but may also be present in autopolyploids. For example, somatic polyploidy is not just restricted to diploid hybrid ferns, but can also occur in a sexual diploid, giving rise to polyploids without the involvement of aberrant meiosis. Bouharmont (1972) produced artificial tetraploids via apospory from a diploid plant of *A. ruta-muraria* subsp. *dolomiticum* from the Gorges du Verdon, southern France. These autotetraploid plants showed not only different degrees of meiotic irregularity including quadrivalents, trivalents, bivalents and univalents, but also cells with almost regular meioses showing up to 63 bivalents (full pairing would be 72 bivalents).

Thus, if a diploid taxon could produce diplospores at a low frequency then, by intragametophytic selfing, autotetraploids can be produced. Bouharmont's observations on variation in pairing behaviour suggest that natural selection among the newly formed autotetraploids would favour those showing more regular meiosis, with the formation of bivalents rather than multivalents, thus excluding random pairing of genomes (and tetrasomic inheritance), which would result in the segregation of the allozyme patterns. If the diploid plant were heterozygous for a particular enzyme system, this pattern could then become 'fixed' via a combination of intragametophytic selfing and the evolution of regular pairing.

6.3.3 Sample size and segregation analysis

Sample size is important when attempting to establish segregation ratios and the genotypic variation of a population of a tetraploid plant species. If one assumes that a single haploid spore of an autotetraploid plant with the constitution *Aa* for a particular enzyme arrives at a safe site for gametophyte development, it will germinate into a gametophyte with the constitution *Aa*. A sporophyte resulting from intragametophytic selfing would then have the constitution of *AA aa* (duplex). In the absence of any further immigrant spores, this sporophyte will be the sole founder of the local population.

Assuming tetrasomic inheritance and random-mating, according to Haldane (1930) we would expect the genotype ratio at generation one to be 1 *AAAA* : 34 *A–a–* : 1 *aaaa* and, at equilibrium, 1 *AAAA* : 14 *A–a–* : 1 *aaaa*, with chromosomal segregation; if there were chromatid segregation we would expect 9 *AAAA* : 178 *A–a–* : 9 *aaaa* at generation one, and 9 *AAAA* : 82 *A–a–* : 9 *aaaa* at equilibrium (also see Haldane (1930) for mixed random and intergametophytic mating). Assuming tetrasomic inheritance and obligate intragametophytic selfing, we would expect at generation one, 1 *AAAA* : 4 *A–a–* : 1 *aaaa* and, at equilibrium, 1 *AAAA* : 1 *aaaa*, for chromosomal segregation; and for chromatid segregation, we would expect 3 *AAAA* : 8 *A–a–* : 3 *aaaa* at generation one and 1 *AAAA* : 1 *aaaa* at equilibrium.

Therefore, in an allozyme study of an hypothetical autotetraploid population, founded by a single duplex individual, with random mating between its gametophytic progeny and between its subsequent descendants, the expected proportion of homozygous genotypes lies between about 1/20 and 3.6/20 plants, depending on how many generations have elapsed since the initial colonisation and the form of inheritance. Such rare genotypes could easily be overlooked in small samples. Even for the case which gives the most frequent occurrence of homozygotes (viz. 3.6/20), one would need a sample of more than 15 plants to stand a 95% chance of detecting at least one homozygote. The apparent uniformity of genotypes in such a population could then, misleadingly, be interpreted as fixed heterozygosity in an allotetraploid taxon.

There is a distinct possibility that populations of autotetraploids can consist predominantly of heterozygous genotypes (simplex, duplex or triplex) that cannot be reliably distinguished by gene-dosage effects on an electrophoretic gel. This together with the effects of single spore colonisation, population substructuring, small populations, genetic drift and small sample sizes, might preclude any inference that a plant is an allotetraploid on the evidence of protein electrophoresis alone.

Although heterozygotes may be fairly common during the early generations after the establishment of the population, in the case of intragametophytic selfing ultimately the population will come to equilibrium, polymorphic, but homozygous, for the two alleles present in the founding individual. When this approach is extended to more than one locus, the equilibrium population will consist of a number of different homozygous multilocus genotypes.

Furthermore, many ferns seem to be capable of single spore colonisation, via intragametophytic selfing, giving rise to genetically monomorphic populations. Therefore no segregation can be observed and no inferences about the breeding system can be made; indeed many polyploid taxa may not be (obligate) inbreeders (Vogel, unpubl.).

6.4 Studies on multiple origins in polyploid pteridophytes

Soltis and Soltis (1993) refer to three published studies of multiple origins in polyploid ferns. The study by Werth *et al.* (1985a, b) has already been described in section 6.1. In this section we examine the remaining two studies.

6.4.1 *Asplenium adiantum-nigrum* in northern America and Hawaii

Asplenium adiantum-nigrum s.l. is a tetraploid taxon, reported to have a wide distribution in Europe and western Asia, with (small) outlying populations in continental North America, Hawaii, Mt Kenya, Mauritius and southern Africa. It has been assumed that *A. adiantum-nigrum* is an allotetraploid derived from two European diploid taxa, the serpentine endemic *A. cuneifolium* and *A. onopteris* (Shivas, 1969).

Ranker *et al.* (1994a, b) carried out allozyme studies of populations of *A. adiantum-nigrum* from continental U.S.A. and Hawaii. The aim of these investigations was to test whether these small colonies of *A. adiantum-nigrum* outside the main area of distribution had been founded via long-range spore dispersal, or whether they represent relict populations of a formerly continuous distribution. It was assumed that the taxon investigated was identical to European *A. adiantum-nigrum*.

Wild populations of *A. adiantum-nigrum* were investigated at seven sites on the island of Hawaii and from three populations in continental U.S.A.: Boulder (Colorado), Zion (Utah) and Elden (Arizona). The samples included twenty plants each from the three populations in continental U.S.A. which were compared with two plants raised from spores from Chihuahua, Mexico, one plant raised from spores from the Caucasus and 236 plants from seven populations in Hawaii. Seventeen loci in 11 enzyme systems were reported to have been informative.

Samples from Elden and Zion exhibited no within- or between-population allelic variability at eight loci. Multiple-banding patterns for nine loci were interpreted as fixed heterozygosity. Furthermore, this fixed heterozygosity for the same loci was reported to have been observed in individual gametophytes, but the results were not published. The population at Boulder expressed variation at only a single locus. *A. adiantum-nigrum* from Hawaii (Ranker *et al.*, 1994a) was reported to exhibit more allozymic variability and twelve alleles were found to be unique to Hawaiian populations. Seven populations with sample sizes of between 11 and 39 specimens were investigated. The samples from Mexico and the Caucasus showed the greatest amount of genetic divergence compared with each other and relative to all other populations sampled.

Ranker *et al.* (1994b) concluded from the comparison of American, Mexican and Caucasian samples that material from different geographical regions had unique origins, resulting from at least several independent events of hybridisation and polyploidy, followed by long-distance dispersal. For the Hawaiian populations Ranker *et al.* (1994a) reported a high frequency of monomorphic loci, and that not all possible allelic combinations were observed at variable locus pairs. Ranker *et al.*

(1994a) suggested that this species is highly (if not totally) inbred and that different allelic combinations within and across locus pairs cannot be attributed to genetic recombination as would be possible in an outbreeding taxon. It was concluded that each multilocus genotype represents a distinct hybridisation (and, hence, colonisation) event, if sexual recombination is absent in these populations. No segregation analysis of the allozymes in crosses was carried out for Hawaiian material. The data were claimed to support multiple hybrid origins for the Hawaiian populations surveyed, with a minimum of three, and possibly as many as seventeen, discrete hybridisation events having produced the genetic diversity observed.

These studies (Ranker et al., 1994a, b) concentrated on non-European material and neither the allozymic variation of the putative diploid ancestral taxa, nor of European A. adiantum-nigrum was investigated. They assumed A. adiantum-nigrum to be an allotetraploid taxon, but A. adiantum-nigrum from Europe had been shown to be a segmental allotetraploid by Shivas (1969), Lovis and Vida (1969), Sleep (1983), and Cubas and Sleep (1994). Furthermore, no unequivocal evidence for the nature of the polyploidy of non-European material has yet been obtained from cytogenetic investigations. Fixed heterozygosity can only occur in allotetraploids with no autosyndetic pairing, yet fixed heterozygosity was inferred in both sets of samples from the continental U.S.A. and Hawaii, in the absence of any information on the cytological status of these populations. No investigations of the F1 generations from individual plants of A. adiantum-nigrum from the study sites have been presented and the homogeneity of allozyme patterns in the populations in U.S.A. was used to infer fixed heterozygosity of loci in the taxon; both (obligate) intragametophytic selfing and fixed heterozygosity were linked and mutually interdependent in this interpretation (Ranker et al., 1994a, b).

The A. adiantum-nigrum complex consists of diploid, tetraploid and hexaploid taxa of similar overall appearance. The study of this complex has been complicated by taxonomic confusion, and conclusions about relationships have been based either on equivocal cytogenetic evidence (Shivas, 1969; Lovis and Vida, 1969; Deschartes et al., 1978; Sleep, 1980, 1985; Jermy, 1981; Demiriz et al., 1981; Fraser-Jenkins, 1992; Rasbach in Reichstein et al., 1994; Rasbach et al., 1995) or on equivocal micromorphological evidence (Viane in Reichstein et al., 1994). The great morphological variation in the complex has variously been recognised at subspecific or varietal level (Demiriz et al., 1981; Derrick et al., 1987; Reichstein et al., 1994), adding to the confusion. Recent investigation of the A. adiantum-nigrum complex, including putative diploid ancestors, has revealed that there are at least two tetraploid taxa in Europe (Vogel et al., 1996). Our evidence from cpDNA and allozymes indicates that the serpentine endemic tetraploid taxon in western Europe and around the Mediterranean could be an allotetraploid derived from A. cuneifolium and A. onopteris, with A. cuneifolium as the maternal parent, but this is not the taxon that Ranker et al. (1994a, b) have investigated (Vogel et al., 1996). Asplenium adiantum-nigrum from non-serpentine substrates from all over Europe, the Caucasus, Macaronesia, Saudi Arabia and Hawaii has the cpDNA of A. onopteris, but so far allozymes do not link it firmly to either of the two European diploid taxa. However, the considerable amount of genetic variation within populations of both tetraploid taxa of A. adiantum-nigrum s.l. from Europe suggests that these taxa are not obligate inbreeders (Vogel, unpubl.).

The comparison of genetic identity by Ranker *et al.* (1994b) between populations of *A. adiantum-nigrum* s.l. from continental U.S.A., Mexico and the Caucasus was between different taxa with different chromosome numbers. Material from one unknown locality in continental U.S.A. (probably Boulder) was shown to be tetraploid by Shivas (1956). Ranker *et al.* (1994a, b) cited Shivas (1969) on the putative allotetraploid nature of *A. adiantum-nigrum* s.l.. However, the nature of the polyploidy of plants from the U.S.A. has never been established (Shivas, 1956, 1969). Material from Chihuahua, Mexico, was shown to be hexaploid by Rasbach *et al.* (1994) and its rank as *A. chihuahuense* was confirmed. Chromosome counts were not obtained by Ranker *et al.* (1994 a, b) for any material used in their studies and material from the Caucasus could be either diploid (*A. cuneifolium* subsp. *woronowii*) or tetraploid (*A. adiantum-nigrum* subsp. *yuanum*).

No genetic evidence is presented by Ranker *et al.* (1994a, b) linking Hawaiian or North American material to the European diploids and no cytological evidence for the allotetraploid status of any *A. adiantum-nigrum* s.l. is presented. Instead, fixed heterozygosity, the allotetraploidy, obligate inbreeding and multiple origin are inferred from each other in a circular argument. Whilst the data reported in these papers are compatible with the inference of multiple hybridisation events if the assumptions are true, all of the assumptions required to make this inference are mutually interdependent and were not tested independently.

The high level of genetic identity within the Hawaiian populations of 0.994 and of 0.902 between the Hawaiian and continental U.S.A. populations is well within the range for species of homosporous ferns (Soltis and Soltis, 1989). The allozyme data presented in the two publications would therefore have to be attributed to genetic variation in outlying populations of one, at least occasionally, outcrossing tetraploid taxon. A multiple origin for *A. adiantum-nigrum* on Hawaii or in continental U.S.A. cannot be inferred. The data presented by Ranker *et al.* (1994a) for the populations on Hawaii could be explained by the genetic diversity in a single sporangium, or a spore tetrad, derived from a single heterozygous plant – and perhaps some mutation followed by some degree of outbreeding. This explanation appears to be more parsimonious than the assumption that *A. adiantum-nigrum* evolved seventeen times in Europe and all of these taxa found their way independently via long-range spore dispersal to establish themselves in seven populations on one island in the middle of the Pacific Ocean.

6.4.2 Hemionitis pinnatifida

Hemionitis pinnatifida is a tetraploid fern distributed throughout the neotropics. Ranker *et al.* (1989) used allozyme electrophoresis to investigate the origin of the tetraploid in relation to one of its presumed ancestral diploids, *H. palmata*, and also to infer and reconstruct a missing diploid.

In naturally occurring triploid hybrids between *H. palmata* and *H. pinnatifida*, Smith and Mickel (1977) observed the formation of 30 bivalents and 30 univalents. They inferred from these observations that *H. pinnatifida* arose as a cross between diploid *H. palmata* and an unknown diploid form of *H. pinnatifida* and attributed the bivalent formation to allosyndetic pairing. Ranker *et al.* (1989) studied three populations of the tetraploid with sample sizes of n = 8, n = 17 and n = 7, as well as nine populations of the diploid *H. palmata* with sample sizes of between n = 17 and n = 81. Ten loci were scored of which five were polymorphic in the tetraploid. Comparing the genetic diversity in two unrelated *Hemionitis* species, the tetraploid *H. pinnatifida* and its presumed ancestral diploid *H. palmata*, the genetic constitution of the missing diploid was reconstructed. From the differences in enzyme profiles it was concluded that the tetraploid may have evolved *de novo* at least five times.

In triploid hybrids, at metaphase 1 the presence only of univalents has been interpreted as demonstrating that unrelated genomes are present and that the hybrid must involve an allotetraploid and an unrelated diploid, e.g. *Asplenium foreziense* (genomic formula: *ObOb FoFo*) × *Asplenium onopteris* (*OnOn*) (Sleep, 1983). However, where an equal number of bivalents and univalents are observed two interpretations are possible. In the first, pairing is between one genome from the tetraploid and the (homologous) genome from the diploid, e.g. *Asplenium adulterinum* (*TT VV*) × *Asplenium viride* (*VV*) (Lovis, 1958), with 36 bivalents and 36 univalents. The homologous genomes in the hybrid are derived from different gametes, resulting in allosyndetic pairing of chromosomes. In the second interpretation, pairing is between the two (homologous or homoeologous) genomes in the tetraploid and the chromosomes from the unrelated diploid remain as univalents, e.g. *Asplenium petrarchae* (*PePe PePe*) × *Asplenium fontanum* (*FoFo*) (Sleep, 1983). In this case, the homologous genomes are derived from the same gamete, resulting in autosyndetic pairing of chromosomes.

In the case of *Hemionitis pinnatifida*, the formation of 30 bivalents and 30 univalents in the putative triploid hybrid could either be full autosyndetic pairing between the two genomes in *H. pinnatifida*, thus suggesting it to be an autotetraploid, or it could be allosyndetic pairing between one genome of *H. pinnatifida* and the genome of *H. palmata*, suggesting the tetraploid to be an allotetraploid. However, no conclusions about the form of polyploidy in the tetraploid *H. pinnatifida* can be drawn without the necessary hybridisation experiments, as outlined by Sleep (1983).

Five different multilocus phenotypes (MLP) were observed by Ranker *et al.* (1989) in the three populations investigated, of which three were found in one population. No segregation analysis was carried out on any MLP. The allozyme data obtained from a very small sample size from the tetraploid *H. pinnatifida* (total of 32 plants from three populations) were interpreted on the assumption that this taxon is an allotetraploid. No data on genetic diversity in either the tetraploid, or in the putative ancestral taxa were presented. The circular argument was completed by the inference that tetraploid *H. pinnatifida* is an obligate inbreeder. The evidence presented in this study does not allow any conclusions to be drawn about the nature of polyploidy in the tetraploid, or about the origin of its genetic variation.

6.5 How many origins?

The number of allozyme multilocus phenotypes observed in polyploid pteridophytes has been used to assess the number of independent origins (process depicted in Fig. 6.1). While the authors of this article share the enthusiasm concerning the important role of multiple origins in pteridophyte polyploid evolution, the nature of pteridophyte reproductive biology makes it very difficult accurately to assess their number.

Our own field observations in European pteridophyte polyploid complexes have revealed that diploid hybrids are extremely rare in the wild. However, at sites where such diploid hybrids occur, these hybrids are long-lived, form repeatedly, and are normally present in some numbers. Based on these observations, we can envisage a different process (Fig. 6.2). Genetic polymorphism is present in both diploid taxa A and B. Gametes of these taxa form a swarm of diploid hybrids, each being genetically different, and each producing diplospores. Whilst it is almost necessary that a prothallus derived from a diplospore should be capable of intragametophytic selfing, it does not necessarily follow that it should be obligatorily so. Indeed, the assumption of intragametophytic selfing can be substantially relaxed if the parental diploid taxa show a fairly high, or even a promiscuous, tendency to hybridise, giving rise to a localised hybrid swarm. If the derived allotetraploid sub-population from such a hybrid swarm does not show obligate intragametophytic selfing, then any degree of outcrossing will allow reassortment and recombination to occur within the sub-population thus producing an enhanced range of genotypes compared with those produced solely by hybridisation.

Migration, as spores, from the allotetraploid population can then occur. If we assume that the probability of any migrant spore finding a suitable niche is rather small, any prothallus growing from it is unlikely to be in the company of other prothalli of the same allotetraploid and so has to undergo intragametophytic selfing, if a sporophyte is to be produced. This sporophyte will be homozygous at all loci, as will any further sporophytes produced sexually from this single plant, although they may all show intergenomic fixed heterozygosity. Furthermore, different colonies produced by migrant spores from the original hybrid swarm population will show the same effect, but for different alleles. Each of these outlying populations will most likely be genetically uniform, in contrast to the initial founder population, which was polymorphic.

Indeed, our investigations of genetic variability in European pteridophytes (*Dryopteris*, *Asplenium*, Hymenophyllaceae), in areas colonised since the last glaciation, have revealed that variation is partitioned mainly between sites and populations (Vogel *et al.*, 1999; Rumsey *et al.*, 1998; Vogel, unpubl.). Most populations of *Asplenium* in these areas of Europe show little or no genetic variation, and each site appears to have been colonised by a different multilocus phenotype (Holderegger and Schneller, 1994; Schneller and Holderegger, 1996; Vogel *et al.*, 1999; Vogel, unpubl.). Such observations can be explained by single (haploid) spore colonisation, followed by intragametophytic selfing, producing a homozygous sporophyte that subsequently builds up a genetically uniform population. Further, but rare, immigrant spores may bring in additional genotypes, leading, potentially, to recombination and segregation. However, if, by chance, only genetically uniform populations are found in a study, no inferences can be made about the origin of the genetic diversity, or the breeding system prevailing in a taxon. All plants of a population founded from a single allotetraploid sporophyte produced by intragametophytic selfing are homozygous for a single multilocus allozyme pattern whether the population is subsequently inbreeding or outbreeding. Furthermore, due to

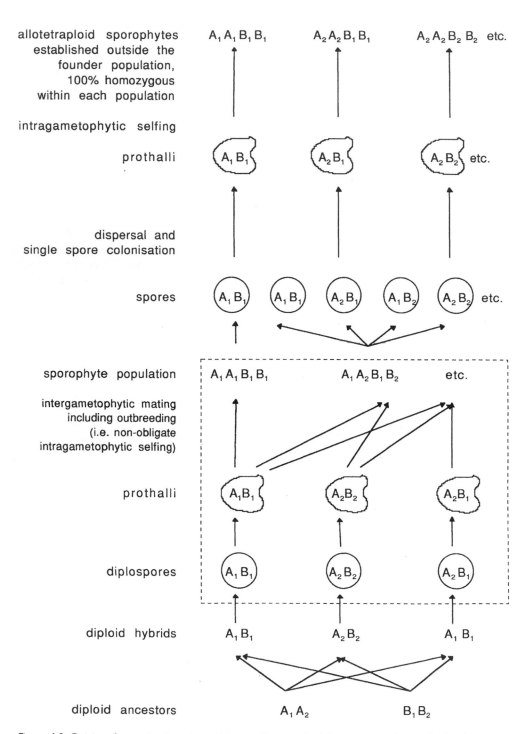

allotetraploid sporophytes established outside the founder population, 100% homozygous within each population $\quad A_1A_1B_1B_1 \qquad A_2A_2B_1B_1 \qquad A_2A_2B_2B_2$ etc.

intragametophytic selfing

prothalli $\qquad A_1B_1 \qquad A_2B_1 \qquad A_2B_2$ etc.

dispersal and single spore colonisation

spores $\qquad A_1B_1 \quad A_1B_1 \quad A_2B_1 \quad A_1B_2 \quad A_2B_2$ etc.

sporophyte population $\qquad A_1A_1B_1B_1 \qquad A_1A_2B_1B_2 \qquad$ etc.

intergametophytic mating including outbreeding (i.e. non-obligate intragametophytic selfing)

prothalli $\qquad A_1B_1 \qquad A_2B_2 \qquad A_2B_1$

diplospores $\qquad A_1B_1 \qquad A_2B_2 \qquad A_2B_1$

diploid hybrids $\qquad A_1B_1 \qquad A_2B_2 \qquad A_1B_1$

diploid ancestors $\qquad\qquad A_1A_2 \qquad\qquad B_1B_2$

Figure 6.2 Origin of genetic diversity within an allotetraploid founder population (in box).

substructuring in some fern populations, errors caused by the Wahlund Effect can be introduced into tests of Hardy-Weinberg equilibrium if such internal structuring is not recognised.

It is highly likely that initial source populations containing both diploid hybrids and F1 tetraploids are either rare, and therefore easily overlooked, or long extinct. If mixed diploid/tetraploid source populations have existed, and have been responsible for the production of tetraploid taxa, a few diploid hybrids could produce a large number of different multilocus patterns, with modest levels of outcrossing among the derived tetraploids. Thus, the apparent genetic uniformity within present-day populations of tetraploids, and the genetic differences between such populations, particularly when they are remote both in space and time from their diploid ancestors, does not allow any hypotheses about the number of original hybridisation events to be tested, in the absence of other evidence.

6.6 Conclusions

Of the three published studies on pteridophytes cited by Soltis and Soltis (1993) to support the claim that recurrent formation of polyploid species is the rule, rather than the exception, only one (Werth *et al.*, 1985a, b) provides unequivocal evidence. The inferences in the remaining two studies (Ranker *et al.*, 1989; Ranker *et al.*, 1994a, b) are based on untested assumptions, lack of independent and corroborating evidence and circular argument; most of the sample sizes of tetraploids are very small and the geographical range is not covered comprehensively. Furthermore, the genetic variation in putative ancestral diploids has either not been studied at all, or only in small sample sizes from restricted areas. In order to avoid the problems displayed by the two latter studies, we propose the following list of *conditiones sine qua non*, to explain genetic variation in polyploids:

- Evidence for multiple origin should be based on several factors and certainly on a combination of molecular techniques with other methods, e.g. morphology, cytology, ecology, breeding experiments, biogeography and natural history.
- The genetic diversity in both ancestral diploids and derived polyploids should be investigated over a large geographical range covering as many populations as possible.
- Evidence for necessary assumptions should be supported by independent experiments.

Given this set of requirements, it is clear that limits are placed on the types of polyploid that can be investigated effectively. For more recently evolved polyploids (neo-polyploids) it is more likely that the ancestral diploids will still exist, so that studies of the natural history, experimental hybridisation and tests of breeding systems can be carried out, and direct comparisons made of genetic variation and morphology. Also, with more recent polyploids, ecology and biogeography may suggest where initial hybridisation could have occurred, and independent origin might more easily be recognised in these areas.

Among palaeopolyploids, it is more likely that: a) ancestral diploids have become extinct, even where descendants of the ancestral diploids still persist; b) their genomes have diverged to such an extent from the ancestral state that cytogenetic investigation of artificial hybrids may produce ambiguous results; c) originally autopolyploid taxa have become diploidised, and thus may appear to be allopolyploids in cytogenetic studies; and d) drift and mutation, including loss of function mutation, separately or in combination, in either ancestral diploids or derived polyploids, may obscure the ancestral genetic variation.

Indeed, it may well be the case that, in the absence of any *a priori* suggestions from biogeography, ecology and geology, discrimination of neo-polyploids from palaeopolyploids may rely on the relative ease with which the requirements for testing the hypothesis of multiple origins can be satisfied.

ACKNOWLEDGEMENT

The authors would like to thank two anonymous reviewers for very helpful suggestions and Bob Press for help with the computer graphics.

REFERENCES

Birky, C. W. (1995) Uniparental inheritance of mitochondrial and chloroplast genes: mechanisms and evolution. *Proceedings of the National Academy of Sciences of the USA*, **92**, 11331–11338.

Bouharmont, J. (1972) Origine de la polyploïdie chez *Asplenium ruta-muraria* L. *Bulletin du Jardin Botanique National de Belgique*, **42**, 375–383.

Brownsey, P. J. (1973) *An evolutionary study of the* Asplenium lepidum *complex*. Ph. D. thesis, University of Leeds.

Butters, F. K. and Tryon, R. M. (1948) A fertile mutant of a *Woodsia* hybrid. *American Journal of Botany*, **35**, 132.

Comes, H. P. and Abbott, R. J. (1999) Reticulate evolution in the Mediterranean species complex of *Senecio* sect. *Senecio*: uniting phylogenetic and population-level approaches, in *Molecular systematics and plant evolution*, (eds P. M. Hollingsworth, R. M. Bateman and R. J. Gornall), Taylor & Francis, London, pp. 171–198.

Cubas, P. and Sleep. A. (1994) *Asplenium × sarniense* (Aspleniaceae, Pteridophyta) from Guernsey Islands, UK. A cytological enigma? *Fern Gazette*, **14**, 269–288.

Demiriz, H., Fraser-Jenkins, C. R., Lovis, J. D., Reichstein, T., Schneller, J. J. and Vida, G. (1981) *Asplenium woronowii* Christ (Aspleniaceae, Pteridophyta), a diploid ancestral fern new to Turkey, and the status of *Asplenium pseudofonatnum* Fomin. *Candollea*, **36**, 181–193.

Derrick, L. N., Jermy, A. C. and Paul, A. M. (1987) Checklist of European pteridophytes. *Sommerfeltia*, **6**, 1–94.

Deschartes, R., Schneller, J. J. and Reichstein, T. (1978) A tetraploid cytotype of *Asplenium cuneifolium* Viv. in Corsica. *Fern Gazette*, **11**, 343–344.

Emmott, J. I. (1963) *A cytotaxonomical investigation of the north temperate species of* Phyllitis. Ph. D. thesis, University of Leeds.

Fraser-Jenkins, C. R. (1992) The ferns and allies of the far-west Himalaya. Some additions and corrections. *Botanica Helvetica*, **102**, 143–157.

Gibby, M. (1977) *A cytogenetic and taxonomic study of the* Dryopteris carthusiana *complex*. Ph. D. thesis, University of Liverpool.

Gibby, M. (1980) Polyploidy and its evolutionary significance, in *The evolving biosphere*, (ed. P. L. Forey), British Museum (Natural History), Cambridge University Press, London, pp. 87–96.

Grant, V. (1981) *Plant speciation*, 2nd ed., Columbia University Press, New York.

Haldane, J. B. S. (1930) Theoretical genetics of autotetraploids. *Journal of Genetics*, **22**, 359–372.

Holderegger, R. and Schneller, J. J. (1994) Are small isolated populations of *Asplenium septentrionale* variable? *Biological Journal of the Linnean Society*, **51**, 377–385.

Jermy, A. C. (1981) *Asplenium cuneifolium* Viv. erroneously recorded in the British Isles. *Watsonia*, **13**, 322–323.

Lovis, J. D. (1958) *An evolutionary study of the fern* Asplenium trichomanes. Ph. D. thesis, University of Leeds.

Lovis, J. D. (1977) Evolutionary patterns and processes in ferns. *Advances in Botanical Research*, **4**, 229–415.

Lovis, J. D. and Vida, G. (1969) The resynthesis and cytogenetic investigation of × *Asplenophyllitis microdon* and × *A. jacksonii*. *British Fern Gazette*, **10**, 53–67.

Manton, I. (1950) *Problems of cytology and evolution in the Pteridophyta*, Cambridge University Press, Cambridge.

Masuyama, S. and Watano, Y. (1990) Trends for inbreeding in polyploid pteridophytes. *Plant Species Biology*, **5**, 13–17.

Ranker, T. A., Haufler, C. H., Soltis, P. S. and Soltis, D. E. (1989). Genetic evidence for allopolyploidy in the neotropical fern *Hemionitis pinnatifida* (Adiantaceae) and the reconstruction of an ancestral genome. *Systematic Botany*, **14**, 439–447.

Ranker, T. A., Floyd, S. K. and Trapp, P. G. (1994a) Multiple colonization of *Asplenium adiantum-nigrum* onto the Hawaiian Archipelago. *Evolution*, **48**, 1364–1370.

Ranker, T. A., Floyd, S. K., Windham, M. D. and Trapp, P. G. (1994b) Historical biogeography of *Asplenium adiantum-nigrum* (Aspleniaceae) in North America and implications for speciation theory in homosporous pteridophytes. *American Journal of Botany*, **81**, 776–781.

Rasbach, H., Reichstein, T. and Schneller, J. J. (1991) Hybrids and polyploidy in the genus *Athyrium* (Pteridophyta) in Europe. 2. Origin and description of two triploid hybrids and synthesis of allotetraploids. *Botanica Helvetica*, **101**, 209–225.

Rasbach, H., Reichstein, T. and Vianne, R. R. L. (1994) *Asplenium chihuahuense* (Aspleniaceae, Pteridophyta), an allohexaploid species, and the description of a simplified hybridization technique. *American Fern Journal*, **84**, 11–40.

Rasbach, H., Rasbach, K., Viane, R. R. L. and Boudrie, M. (1995) Neue Funde von zwei seltenen *Asplenium*-Hybriden in Frankreich. *Farnblätter*, **26/27**, 89–101.

Reichstein, T. (1981) Hybrids in European Aspleniaceae (Pteridophyta). *Botanica Helvetica*, **91**, 89–139.

Reichstein, T., Viane, R., Rasbach, H. and Schneller, J. J. (1994) *Asplenium adiantum-nigrum* L. subsp. *yuanum* (Ching) Viane, Rasbach, Reichstein and Schneller stat. nov. and the status of *A. woronowii* Christ. (Aspleniaceae, Pteridophyta). *Candollea*, **49**, 281–328.

Rumsey, F. J., Vogel, J. C., Russell, S. J., Barrett, J. A. and Gibby, M. (1998) Climate, colonisation, celibacy: population structure of *Trichomanes speciosum* Willd. in central Europe (Hymenophyllaceae, Pteridophyta). *Botanica Acta*, **111**, 481–489.

Schneller, J. J. and Holderegger, R. (1996) Colonization events and genetic variability within populations of *Asplenium ruta-muraria* L., in *Pteridology in perspective*, (eds J. M. Camus, M. Gibby and R. J. Johns), Royal Botanic Gardens, Kew, pp. 571–580.

Schneller, J. J. and Rasbach, H. (1984) Hybrids and polyploidy in the genus *Athyrium* (Pteridophyta) in Europe. *Botanica Helvetica*, **94**, 81–99.

Shivas, M. G. (1956) *Some problems in cytology and taxonomy in fern genera* Polypodium *and* Asplenium. Ph. D. thesis, University of Leeds.

Shivas, M. G. (1969) A cytotaxonomic study of the *Asplenium adiantum-nigrum* complex. *British Fern Gazette*, **10**, 68–80.

Sleep, A. (1966) *Some cytotaxonomic problems in the fern genera* Asplenium *and* Polystichum. Ph. D. thesis, University of Leeds.

Sleep, A. (1980) On the reported occurrence of *Asplenium cuneifolium* and *A. adiantum-nigrum* in the British Isles. *Fern Gazette*, **12**, 103–107.

Sleep, A. (1983) On the genus *Asplenium* in the Iberian Peninsula. *Acta Botanica Malacitana*, **8**, 11–46.

Sleep, A. (1985) Speciation in relation to edaphic factors in the *Asplenium adiantum-nigrum* group. *Proceedings Royal Society Edinburgh*, **86b**, 325–334.

Smith, A. R. and Mickel, J. T. (1977) Chromosome counts for Mexican ferns. *Brittonia*, **29**, 391–398.

Smith, D. M. and Harborne, J. B. (1971) Xanthones in the Appalachian *Asplenium* complex. *Phytochemistry*, **10**, 2117–2119.

Smith, D. M. and Levin, D. A. (1963) A chromatographic study of reticulate evolution in the Appalachian Aspleniums. *American Journal of Botany*, **50**, 952–958.

Soltis, D. E. and Soltis, P. S. (1989) Polyploidy, breeding systems, and genetic differentiation in homosporous pteridophytes, in *Isozymes in plant biology*, (eds D. E. Soltis and P. S. Soltis), Dioscorides Press, Portland, pp. 241–258.

Soltis, D. E. and Soltis, P. S. (1992) The distribution of selfing rates in homosporous ferns. *American Journal of Botany*, **79**, 97–100.

Soltis, D. E. and Soltis, P. S. (1993) Molecular data and the dynamic nature of polyploidy. *Critical Reviews in Plant Sciences*, **12**, 243–273.

Stebbins, G. L. (1947) Types of polyploids: their classification and significance. *Advances in Genetics*, **1**, 403–429.

Vida, G. (1970) The nature of polyploidy in *Asplenium ruta-muraria* L. and *A. lepidum* C. Presl. *Caryologia*, **23**, 525–547.

Vogel, J. C., Russell, S. J., Barrett, J. A. and Gibby, M. (1996) A non-coding region of chloroplast DNA as a tool to investigate reticulate evolution in European *Asplenium*, in *Pteridology in perspective*, (eds J. M. Camus, M. Gibby and R. J. Johns), Royal Botanic Gardens, Kew, pp. 313–329.

Vogel, J. C., Russell, S. J., Barrett, J. A. and Gibby, M. (1998) Evidence for maternal transmission of chloroplast DNA in the genus *Asplenium* (Aspleniaceae, Pteridophyta). *Botanica Acta*, **111**, 247–249.

Vogel, J. C., Rumsey, F. J., Schneller, J. J., Barrett, J. A. and Gibby, M. (1999) Where are the glacial refugia in Europe? Evidence from pteridophytes. *Biological Journal of the Linnean Society*, **66**, 23–37.

Wagner, W. H. (1954) Reticulate evolution in the Appalachian aspleniums. *Evolution*, **8**, 103–118.

Wagner, W. H. and Wagner, F. S. (1980) Polyploidy in pteridophytes, in *Polyploidy: biological relevance*, (ed. W. H. Lewis), Plenum Press, New York, pp. 103–144.

Wagner, W. H. Jr., Moran, R. C. and Werth, C. R. (1993) Aspleniaceae, in *Flora of North America*, **2**, 228–245, Oxford University Press.

Werth, C. R., Guttman, S. I. and Eshbaugh, W. H. (1985a) Electrophoretic evidence of reticulate evolution in the Appalachian *Asplenium* complex. *Systematic Botany*, **10**, 184–192.

Werth, C. R., Guttman, S. I. and Eshbaugh, W. H. (1985b) Recurring origins of allotetraploid species in *Asplenium*. *Science*, **228**, 731–734.

Wyatt, R., Odrozykoski, I. J., Stoneburner, A., Bass, H. W. and Galan, G. A. (1988) Allopolyploidy in bryophytes: multiple origins of *Plagiomnium medium*. *Proceedings National Academy of Sciences of the USA*, **85**, 5601–5604.

Wyatt, R., Odrozykoski, I. J. and Stoneburner, A. (1992) Isozyme evidence of reticulate evolution in mosses: *Plagiomnium medium* is an allopolyploid of *P. ellipticum* × *P. insigne*. *Systematic Botany*, **17**, 532–539.

Chapter 7

Population genetic structure in agamospermous plants

R. J. Gornall

ABSTRACT

Studies documenting the amount of molecular variation and its pattern of distribution within and between 'populations' of agamosperms are reviewed. Agamosperms have a similar population genetic structure on average to other clonal groups, although variation is such that it is difficult to make generalisations. Agamospermous microspecies with a narrowly-circumscribed morphological phenotype usually exhibit a low clonal diversity, comprising one or very few molecular phenotypes. Members of agamospermous groups display high genetic identities typical of those found among conspecific populations. The relative roles of sexual and asexual processes in the origin of the variability are assessed and evidence is presented to show that much of the variability can probably be attributed to the sexual ancestry of the agamospermous group, with facultatively apomictic behaviour sometimes contributing subsequently. There is insufficient evidence as yet to believe that asexual events play anything other than a minor role.

7.1 Introduction

Agamospermy is the production of seeds without the sexual fusion of male and female gametes. Agamospermous embryos can originate from sporophytic tissue by adventitious embryony (sporophytic apomixis) or from unreduced megagametophytes (gametophytic apomixis). Unreduced megagametophytes can be derived from the megasporangium wall (apospory) or from the archesporium (diplospory); in the latter case they may arise following a restitutional meiosis (*Taraxacum*-type) or a mitosis (*Antennaria*-type). Whatever pathway is involved, either a regular meiosis is avoided and the embryo is formed from an unreduced egg (parthenogenesis) or other unreduced embryo-sac cell (apogamety) or, in a few cases in *Rubus*, a synaptic, reductional meiosis occurs but the embryo is derived from a fusion between two haploid egg cells (automixis). Irrespective of its mode of origin, the agamospermous embryo derives its genome entirely from its mother. In some cases, successful embryo-formation requires pollination and the subsequent fertilisation of the polar nuclei by a male gamete to generate the endosperm – a process known as pseudogamy. In this context it is noteworthy that some agamosperms retain male fertility, undergoing regular microsporogenesis to produce viable, haploid pollen. Agamospermy can be obligate or facultative and, in the latter case, some embryos

In *Molecular systematics and plant evolution* (1999) (eds P. M. Hollingsworth, R. M. Bateman and R. J. Gornall), Taylor & Francis, London, pp. 118–138.

are produced sexually following normal meiosis and gamete fusion, whereas others are produced agamospermously. The two sorts of embryo may even occur together in the same flower, and this is particularly common in groups with apospory or sporophytic apomixis, in which the archesporium can be a precursor to a normal meiosis. The various pathways have been described in more detail by Richards (1997).

Models to account for the genetic control of agamospermy are numerous and range from those involving several genes to those requiring only one (Mogie, 1992; Koltunow, 1993; Richards, 1997). An often repeated generalisation is that most agamosperms are polyploid. Whilst this is true, relatively recently it has become clear that the statement should be refined. Thus most apomicts with sporophytic agamospermy are in fact diploid, whereas almost all those with gametophytic agamospermy (the commonest type) are polyploid (Richards, 1997). Although there is no evidence that there is a direct causal relationship, it has been suggested that the polyploid condition in gametophytic agamosperms somehow facilitates the expression of the apomixis-inducing gene or genes (Mogie, 1992). An alternative hypothesis (Nogler, 1984b; Richards, 1997), which also accounts for the occurrence of diploid sporophytic agamosperms, depends on linkage between apomixis genes and deleterious recessives which inevitably accumulate owing to the lack of recombination. Such deleterious recessives will be expressed in the haploid gametophytes of a sexual diploid, presumably resulting in their demise along with the linked apomixis genes. In a polyploid, however, deleterious recessives are masked at the gametophyte stage and both they and any linked apomixis genes are consequently preserved. (The recessives are similarly masked in the diploid embryos of sporophytic agamosperms.) The few exceptional cases of apparently bona fide gametophytic agamospermous diploids (Roy, 1995), however, suggest that the phenomenon is not clearcut.

The taxonomic consequences of agamospermous reproduction have long been recognised, resulting as it does in a myriad, morphologically more or less uniform phenotypes. The term 'agamospecies' was introduced by Turesson (1929) for those agamospermous groups whose members could be regarded on morphological, cytological or other grounds as having had a common origin. The concept was later revised and narrowed, being regarded as an apomictic parallel to that of the ecospecies (Turesson, 1943), a taxon more or less equivalent to that of a conventional species. Thus an agamospecies became 'a grouping of apomictic biotypes that constitute a morphological unit which can be more or less adequately characterised' (Davis and Heywood, 1963). The individual apomictic biotypes were originally called 'formae apomictae' (Turesson, 1929) and, later, 'agamotypes' (Turesson, 1943), analagous to the ecotypes of sexual species. Löve (1960) proposed applying the term agamospecies in its most restricted sense to the 'formae apomictae', and this narrow usage has become widespread in some groups, e.g. *Taraxacum* (Dudman and Richards, 1997). Gustafsson (1946/7: 269) preferred to use the term 'microspecies'. Although the ranks of agamospecies and microspecies are not recognised by the International Code of Botanical Nomenclature, where agamospermous microspecies can be distinguished morphologically by the skilled and practised observer, they are often accorded the official taxonomic rank of species and given a binomial, notwithstanding the knowledge that they bear little biological analogy to their sexually-reproducing counterparts. In Europe, especially, taxonomic study of the major agamospermous groups, such as *Taraxacum* and *Hieracium* in the Asteraceae and *Rubus* in the Rosaceae, has resulted in thousands of binomials being created. For example, in *Hieracium* alone over 10,000 binomials have been published (Index Kewensis CD-ROM, 1993). Although this approach has been less fashionable in the New World, North America has also had its agamosperm enthusiasts, and mention may be made of the varous treatments of *Crataegus*, in which over 1000 species were described in the early part of this century (Palmer, 1932). Other than it being a cataloguing exercise, however, this type of work did not appear to lead anywhere ('rather pointless and a wasting of print and paper', according to Winge (1938)), although Gustafsson (1946/7: 221) defended it on the grounds that, studied properly, agamosperms provide excellent material for the analysis of plant migrations and distributions. In terms of the amount of genetic variability that was contained within agamospermous groups, however, the compilations of names did not appear to provide anything other than crude estimates.

Until the advent of molecular techniques the genetic constitution of agamospermous taxa and the populations they form was virtually unknown. With little more than the morphological phenotype and some early cytological work as evidence, speculation abounded regarding their origin, genetic makeup and fate, and a certain dogma became widespread, viz:

- agamosperms are usually of hybrid origin and therefore highly heterozygous (Gustafsson, 1946/7; Stebbins, 1950)
- agamosperms are often (always?) genetically invariant lineages (Löve, 1960; Janzen, 1977)
- agamosperms are evolutionary 'dead-ends' (Darlington, 1939) and can produce new variation only on an old theme (Stebbins, 1950)

At least the last two of these assertions were intended to apply to situations in which agamospermy was obligate. Some workers, however, have questioned whether most agamosperms are ever thus, and have suggested instead that such breeding systems are usually facultative and form part of a continuum of, or alternative to, sexuality (Asker, 1979; Nogler, 1984a). If sexual reproduction does sometimes occur in agamosperms then, clearly, ideas and predictions about their genetic constitution and evolutionary potential must be revised, because even low levels of sexuality can have far-reaching effects if the genome is highly heterozygous. Such a re-evaluation was anticipated by Clausen (1954) who developed the so-called 'model-T' hypothesis in which occasional sexual reproduction allows for the formation of successful genotypes which are then mass-produced by agamospermy (Marshall and Brown, 1981).

Whatever the frequency of sex in otherwise agamospermous lineages, it is still of interest to enquire about the origin and nature of their genetic variability, and whether there is any evidence to suggest that such lineages might be active evolutionarily and, if so, by what mechanism. The purpose of the following review is to address these points as they relate to gametophytic agamosperms in particular, using the data that have accumulated from molecular studies.

7.2 Heterozygosity

Agamosperms are widely credited with possessing elevated levels of heterozygosity, for three chief reasons. Firstly they are believed typically to be of hybrid origin (Darlington, 1939). Gustafsson (1946/7: 166, 299, 302) reviewed the evidence from a variety of genera that indicates that, as a rule, agamosperms are products of hybridisation, pointing out that their sexual progenitors are frequently self-incompatible (p.166) or at least allogamous (p.302). Stebbins (1950: 406) also attributed an important role to 'hybridisation and allopolyploidy between the original sexual ancestors of the agamic complex' in the origin of agamospermous lineages. Secondly, at least gametophytic agamosperms are polyploid and it is known that levels of heterozygosity are typically higher in polyploids than in related diploids. In the case of allopolyploids this is due to fixed heterozygosity and in autopolyploids to polysomic inheritance (Soltis and Soltis, 1993). Thirdly, in the absence of recombination, mutations should accumulate, thereby further increasing levels of heterozygosity.

The prevalence of polyploidy among agamosperms, coupled with the notion of hybrid origin, has led to much attention being paid in the literature to their presumed allopolyploid status. Nevertheless, not all polyploid agamosperms are allopolyploid, at least in a taxonomic sense. This fact was pointed out by Gustafsson (1946/7), who made it clear that either intra- or inter-specific hybridisation can be involved, a point later emphasised by Stebbins (1971: 173) who noted that '. . . the genetic constitution of apomictic polyploids forms a complete spectrum from strict "autopolyploidy" to "allopolyploid" origin . . .'.

In order to examine the effect of ploidy type on levels of observed heterozygosity (Ho) in agamosperms, a comparison of allopolyploids and autopolyploids with their close sexual relatives is provided in Table 7.1. (In species with both diploid and polyploid cytodemes the latter are treated as autopolyploid; for the others I have followed the authors of the respective studies in treating them as allopolyploid.) It would appear that, making allowances for the continuous nature of the allo/autopolyploid spectrum, a higher Ho is characteristic of agamospermous allopolyploids, but not necessarily of their autopolyploid counterparts. This is somewhat surprising given that sexual autopolyploids do not usually behave this way. In the case of these particular agamospermous autopolyploids it may reflect the involvement of impoverished diploid parental genepools (Yahara et al., 1991).

The relationship between ploidy type in agamosperms and their sexual relatives in terms of the mean number of alleles per locus (A) and of the proportion of loci that are polymorphic (P) is less clear. Sometimes there is a difference in favour of the allopolyploid agamosperm, sometimes not (Table 7.1). Where there is a higher value of Ho but not of P or A, the difference between agamosperms and their sexual relatives is entirely due to the polymorphic loci being fixed as heterozygotes in the former but segregating in the latter. As we shall see, this situation has a bearing on the cause of the variability to be found in agamospermous groups.

7.3 Clonal diversity

When investigating the genetic architecture of a species, it is usual to consider the way in which any variability is partitioned within and between populations. Most of the models that provide the conceptual framework for this approach assume

Table 7.1 Observed heterozygosities in agamospermous taxa of different polyploid origins (taxonomic definition). Ho = observed heterozygosity, averaged over loci and populations; P = mean percentage of loci polymorphic per population; A = mean number of alleles per locus per population.

Taxon (no. pops)	Ho	P%	A	Reference
Allopolyploid				
Antennaria neodioica (3)	0.137 (range?)[a]	–	–	Bayer and Crawford (1986)
Antennaria parlinii (21)	0.128 (range?)[a]	–	–	Bayer and Crawford (1986)
Antennaria rosea (33)	0.229 (0.077–0.400)[a]	29 (8–43)[a]	1.4 (1.1–1.6)[a]	Bayer (1989)
Erigeron compositus (7)	0.223 (0.162–0.300)[a]	35 (20–53)	1.5 (1.3–1.8)	Noyes and Soltis (1996)
Taraxacum albidum (12)	0.684[b]	68[b]	1.8[b]	Menken and Morita (1989)
Taraxacum brachyglossum (1)	0.27[a]	27	1.3	Hughes and Richards (1988)
Taraxacum pseudohamatum (1)	0.39[a]	47	1.6	Hughes and Richards (1988)
Taraxacum unguilobum (1)	0.40[a]	40	1.4	Hughes and Richards (1988)
Autopolyploid				
Antennaria friesiana ssp. *friesiana* (4)	0.097 (0.028–0.167)	21 (14–31)	1.2 (1.1–1.3)	Bayer (1991)
Antennaria monocephala ssp. *angustata* (1)	0.063	12	1.4	Bayer (1991)
Eupatorium altissimum (5)	0.08 (0–0.15)	8 (0–15)	1.1 (1.0–1.2)	Yahara *et al.* (1991)
Panicum maximum (Ni = 313)	no difference[c]	69	2.5	Assienan and Noirot (1995)

[a] significantly more than the sexual relatives; [b] higher than sexual relatives but no difference found between agamosperms and sexuals; [c] no figures given, but no statistical test applied.

that the populations contain an infinite number of diploid, randomly-mating individuals, with a Poisson distribution of offspring per parent. Agamosperms fail to meet these criteria because they are asexual, most are polyploid, and the concept of population, in the sense of gamodeme, cannot be applied. The use of a statistic such as the heterozygosity expected under Hardy–Weinberg equilibrium or gene diversity (Hs) to quantify genetic variation is consequently inappropriate and positively misleading. More informative is the calculation of the observed average heterozygosity (Ho) because this at least allows estimates to be made of the levels of fixed heterozygosity. The two other statistics commonly used for sexual species, viz. the proportion of loci that are polymorphic (P) and the mean number of alleles per locus (A), may also be of some use, although care must be taken when calculating these statistics in a polyploid. However, in recognition of the fact that an asexually reproducing genome effectively forms one linkage group, instead of dealing in terms of alleles and individual loci, an alternative approach is to take the entire genotype as the unit of measurement. Variability can then be calculated in terms of genotypic or clonal diversity by using statistics often used in ecology (Peet, 1974; Ellstrand and Roose, 1987; Menken *et al.*, 1995).

A survey of the literature shows that two main approaches have been adopted in studies of genetic variability in agamosperms, according to whether the microspecies or the whole agamospermous aggregate has been the focus in each locality studied. In the latter case the taxonomic identities of the constituent genotypes in each local population are not addressed.

In a review of genotypic diversity in 21 clonally-reproducing plant species, Ellstrand and Roose (1987) summarised data that included four agamosperms and one apogamous fern (in none of which was sexual reproduction an obvious alternative), but did not treat the data from aggregate-based studies separately from those which distinguished the microspecies taxonomically. In the last ten years or so, a great deal more data relating specifically to agamosperms have become available, and in the following review I have divided them according to the type of study that produced them, i.e. distinguishing a microspecies from an aggregate approach. In some cases it is unclear whether an aggregate is actually involved or whether the taxon concerned is more narrowly defined, albeit not necessarily as narrowly as a microspecies; such cases are separated from the others.

Summary statistics for each taxon have mostly been calculated by me from the data provided in the original articles. Genetic diversity within microspecies was estimated by the proportion of clones detected (Ng/Ni, where Ng is the number of clones detected among a total of Ni individuals, pooling data over all populations). Clonal diversity and equitability statistics were also calculated for each microspecies as a whole from data pooled over localities. Clonal diversity was estimated by Gini's (complement of Simpson's) Index, $Hg = 1 - \Sigma g_i^2$, corrected for finite sample sizes, where g is the frequency of the i th genotype; clonal equitability or evenness, $Eg = 1/(1 - Hg)Ng$. Diversity within populations was expressed in terms of: a) the mean number of genotypes per population; b) the proportion of populations that are multi-clonal; c) clonal diversity; and d) clonal equitability. Diversity between populations was expressed as the mean number of populations per genotype and as the proportion of genotypes that are widespread (i.e. those occurring in at least 75% of the populations) or private (i.e. those restricted to one population).

Care must be exercised when interpreting the data because differences in experimental methodology, viz. in sampling strategy, sample size and perhaps also in taxonomic expertise, and in the number and variability of the loci examined, will clearly have an effect on estimates of genetic variability. In terms of the molecular markers chosen, most studies have relied on isozymes although alternatives with potentially greater resolving power, such as minisatellite DNA and RAPD markers, have recently been used. As far as the taxa investigated are concerned, much emphasis has been placed on the microspecies of the large apomictic genera, *Rubus*, *Hieracium* and *Taraxacum*, with a limited set of data available also from one study on *Sorbus*.

7.3.1 Microspecies

Data are provided for microspecies of *Rubus* and *Sorbus* (Table 7.2), *Hieracium* (Table 7.3) and *Taraxacum* (Table 7.4). As noted already, potentially there are considerable problems in making comparisons using these data and indeed there is a significant correlation in each of the three large datasets (*Rubus*, *Hieracium* and *Taraxacum*) between sample size and the number of genotypes detected ($p < 0.01$, Spearman Rank-order Correlation). Despite the potential difference in resolution associated with the different sample sizes and methods of study, the data that are available appear to show a broadly similar pattern. In all four genera many of the microspecies are uniclonal (Fig. 7.1), at least as far as can be ascertained from the molecular markers studied (additional sampling and the use of more discriminatory markers may well change this picture). In *Rubus* and *Sorbus* each microspecies contains an average of 2.1 and 1.4 genotypes, respectively (Table 7.2), in *Hieracium* the average is 1.8 (Table 7.3) and in *Taraxacum* 5.0 (Table 7.4). The

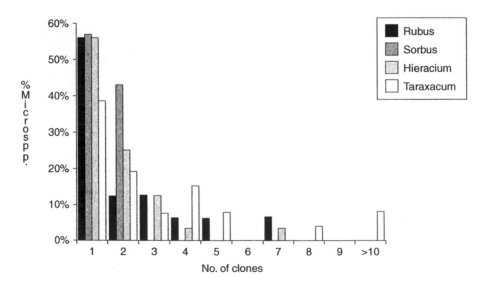

Figure 7.1 Distribution of genetic variability, as measured by the number of clones detected, among microspecies of the genera *Rubus*, *Sorbus*, *Hieracium* and *Taraxacum*.

Table 7.2 Genotypic diversity statistics for microspecies of *Rubus* and *Sorbus*. Geog = geographical area studied (Da, Denmark; En, England; Ge, Germany; Su, Sweden). Marker = molecular marker (iz, isozymes; mc, microsatellite DNA; mn, minisatellite DNA). Mk/loci = no. of markers (enzymes or DNA primers or probes) / no. of gene loci. Npop = no. of populations. Ni = total no. of individuals. Ng = total no. of genotypes detected. Hg = overall clonal diversity (multiclonal spp. only), calculated from data pooled over populations. Eg = overall clonal evenness (multiclonal spp. only), calculated from data pooled over populations. Ng/pop = average no. of genotypes per population (calculated for populations with more than one individual sampled). Mult-g pop = proportion of populations that are multiclonal. Hg/pop = mean clonal diversity per population (where Ng > 1). Eg/pop = mean clonal evenness per population (where Ng > 1). Pop/g = mean number of populations per genotype (calculated where Npop > 1). Wide g = proportion of genotypes that are widespread (calculated where Npop > 1 and Ng > 1). Priv g = proportion of genotypes that are private (calculated where Npop > 1 and Ng > 1).

Rubus	Geog	Marker	Mk/loci	Npop	Ni	Ng	Ng/Ni	Hg	Eg	Ng/pop	Mult-g pop	Hg/pop	Eg/pop	Pop/g	Wide g	Priv g	Ref.*
grabowskii	Su	mn	1/?	7	?	1	?	na	na	1.0	0.00	na	na	7.0	na	na	(a)
gracilis	Ge	mn	1/?	3	3	1	0.333	na	na	na	na	na	na	3.0	na	na	(b)
hartmanii	Su	mc,mn	2/22	7	8	1	0.125	na	na	1.0	0.00	na	na	7.0	na	na	(c)
infestus	Ge,Su	mc,mn	2/26	5	13	2	0.154	0.28	0.7	1.0	0.00	na	na	2.5 (1–4)	0.50	0.50	(c)
insularis (a)	Su	mc,mn	2/26	5	6	2	0.333	0.53	1.0	1.0	0.00	na	na	2.5 (2–3)	0.00	0.00	(c)
insularis (b)	Da,Ge,Su	mn	1/?	8	12	3	0.250	0.32	0.49	1.0	0.00	na	na	2.7 (1–6)	0.33	0.67	(b)
muenteri/scheutzii	Ge,Su	mc,mn	2/24	8	13	1	0.077	na	na	1.0	0.00	na	na	8.0	na	na	(c)
nessensis	Da,Ge,Su	mn	1/?	19	32	4	0.125	0.29	0.35	1.2 (1–2)	0.20	0.33	0.75	5.0 (1–15)	0.25	0.50	(b)
n. ssp. scissoides	Ge,Su	mn	1/?	4	6	2	0.333	0.33	0.75	1.0	0.00	na	na	2.0 (1–3)	0.50	0.25	(b)
pedemontanus	Su	mn	1/?	6	?	1	?	na	na	1.0	0.00	na	na	6.0	na	na	(a)
pensilvanicus	USA	mn	1/?	2	22	7	0.318	0.80	0.72	3.5 (2–5)	1.00	0.88 (0.76–1)	0.91 (0.83–1)	1.0	0.00	1.00	(d)
plicatus	Da,Ge,Su	mn	1/?	20	28	5	0.179	0.38	0.32	1.6 (1–3)	0.40	1.00	1.0	4.6 (1–17)	0.20	0.40	(b)
polyanthemus	Ge,Su	mc,mn	2/27	3	3	1	0.333	na	na	na	na	na	na	3.0	na	na	(c)
pseudopallidus	Su	mc,mn	2/23	4	6	1	0.167	na	na	1.0	0.00	na	na	4.0	na	na	(c)
scissus	Ge,Su	mn	1/?	8	13	3	0.231	0.50	0.67	1.0	0.00	na	na	2.7 (1–5)	0.00	0.33	(b)
septentrionalis (a)	Su	mc,mn	2/25	2	4	1	0.250	na	na	1.0	0.00	na	na	2.0	na	na	(c)
septentrionalis (b)	Da,Su	mn	1/?	2	5	1	0.200	na	na	1.0	0.00	na	na	2.0	na	na	(b)
vestervicensis	Su	mn	1/?	1	?	1	?	na	na	1.0	0.00	na	na	na	na	na	(a)
Average						**2.1**	**0.227**	**0.43**	**0.62**	**1.2**	**0.10**	**0.74**	**0.89**	**3.8**	**0.22**	**0.46**	

Sorbus	Geog	Marker	Mk/loci	Npop	Ni	Ng	Ng/Ni	Hg	Eg	Ng/pop	Mult-g pop	Hg/pop	Eg/pop	Pop/g	Wide g	Priv g	Ref.*
anglica	En	iz	1/?	8	14	2	0.143	0.26	0.68	1.0	0.00	na	na	4.0 (2–6)	0.50	0.00	(e)
devoniensis	En	iz	1/?	6	25	2	0.080	0.51	1.00	1.0	0.00	na	na	3.0 (1–5)	0.50	0.50	(e)
intermedia	En	iz	1/?	2	2	1	0.500	na	na	1.0	0.00	na	na	2.0	na	na	(e)
porrigentiformis	En	iz	1/?	5	13	2	0.154	0.15	0.60	1.0	0.00	na	na	2.5 (1–4)	0.50	0.50	(e)
rupicola	En	iz	1/?	4	22	1	0.045	na	na	1.0	0.00	na	na	4.0	na	na	(e)
subcuneata	En	iz	1/?	5	27	1	0.037	na	na	1.0	0.00	na	na	5.0	na	na	(e)
vexans	En	iz	1/?	5	19	1	0.053	na	na	1.0	0.00	na	na	5.0	na	na	(e)
Average						**1.4**	**0.145**	**0.31**	**0.76**	**1.0**	**0.00**	**–**	**–**	**3.6**	**0.50**	**0.33**	

* (a) Kraft et al. (1995); (b) Kraft et al. (1996); (c) Kraft and Nybom (1995); (d) Nybom and Schaal (1990); (e) Proctor et al. (1989).

Table 7.3 Genotypic diversity statistics for microspecies of *Hieracium*. Geog = geographical area studied (En, England; He, Switzerland; No, Norway; Sc, Scotland; Su, Sweden; Wa, Wales). Marker = marker used (iz, isozymes). Mk/loci = no. of markers (enzymes) / no. of gene loci. Npop = no. of localities (populations). Ni = total no. of individuals. Ng = total no. of genotypes detected. Hg = overall clonal diversity (multiclonal spp. only), calculated from data pooled over populations. Eg = overall clonal evenness (multiclonal spp. only), calculated from data pooled over populations. Ng/pop = average no. of genotypes per population (calculated for populations with more than one individual sampled). Mult-g pop = proportion of populations that are multiclonal. Hg/pop = mean clonal diversity per population (where Ng > 1). Eg/pop = mean clonal evenness per population (where Ng > 1). Pop/g = mean number of populations per genotype (calculated where Npop > 1). Wide g = proportion of genotypes that are widespread (calculated where Npop > 1 and Ng > 1). Priv g = proportion of genotypes that are private (calculated where Npop > 1 and Ng > 1). All data from Gornall et al. (unpublished).

Hieracium	Geog	Marker	Mk/loci	Npop	Ni	Ng	Ng/Ni	Hg	Eg	Ng/pop	Mult-g/pop	Hg/pop	Eg/pop	Pop/g	Wide g	Priv g
alpinum	He,No,Sc,Su	iz	7/12	8	121	4	0.033	0.67	0.75	1.4 (1–3)	0.25	0.34 (0.09–0.59)	0.68 (0.55–0.81)	2.8 (1–7)	0.25	0.50
atraticeps	Sc	iz	7/12	1	8	1	0.125	na	na	1.0	0.00	na	na	na	na	na
backhousei	Sc	iz	7/12	2	30	1	0.033	na	na	1.0	0.00	na	na	2.0	na	na
calenduliflorum	Sc	iz	7/12	7	119	3	0.025	0.42	0.58	1.3 (1–2)	0.33	0.03	0.51	2.7 (1–5)	0.33	0.33
calvum	Sc	iz	7/12	3	12	1	0.083	na	na	1.0	0.00	na	na	3.0	na	na
completum	Sc	iz	7/12	2	34	2	0.059	0.06	0.53	1.5 (1–2)	0.50	0.07	0.54	1.50 (1–2)	0.50	0.50
eximium	Sc	iz	7/12	4	46	1	0.022	na	na	1.0	0.00	na	na	4.0	na	na
globosiflorum	Sc	iz	7/12	3	4	2	0.500	0.50	1.0	1.0	0.00	na	na	1.5 (1–2)	0.00	0.50
graniticola	Sc	iz	7/12	1	5	1	0.200	na	na	1.0	0.00	na	na	na	na	na
hanburyi	Sc	iz	7/12	4	14	3	0.214	0.28	0.46	1.5 (1–2)	0.50	0.22	0.64	1.67 (1–3)	0.33	0.67
hanb. f. pusillum	Sc	iz	7/12	2	3	1	0.333	na	na	1.0	0.00	na	na	2.0	na	na
holosericeum	En,Sc,Wa	iz	7/12	6	336	2	0.006	0.01	0.50	1.0	0.00	na	na	3.0 (1–5)	0.50	0.50
insigne	Sc	iz	7/12	1	4	1	0.250	na	na	1.0	0.00	na	na	na	na	na
insigne f. celsum	Sc	iz	7/12	4	5	2	0.400	0.40	0.83	1.0	0.00	na	na	2.0 (1–3)	0.50	0.50
kennethii	Sc	iz	7/12	1	5	1	0.200	na	na	1.0	0.00	na	na	na	na	na
larigense	Sc	iz	7/12	1	2	2	1.000	1.00	1.00	2.0	1.00	1.00	1.00	na	na	na
leptodon	Sc	iz	7/12	1	5	1	0.200	na	na	1.0	0.00	na	na	na	na	na
macrocarpum	Sc	iz	7/12	6	26	3	0.115	0.22	0.43	1.2 (1–2)	0.20	0.50	1.00	2.33 (1–5)	0.33	0.67
marginatum	Sc	iz	7/12	2	3	1	0.333	na	na	1.0	0.00	na	na	2.0	na	na
memorabile	Sc	iz	7/12	5	13	3	0.231	0.29	0.47	1.4 (1–2)	0.40	1.00	1.00	2.33 (1–5)	0.33	0.67
milesii	Sc	iz	7/12	3	6	2	0.333	0.33	0.75	1.0	0.00	na	na	1.5 (1–2)	0.00	0.50
mundum	Sc	iz	7/12	2	2	1	0.500	na	na	na	0.00	na	na	2.0	na	na
notabile	Sc	iz	7/12	1	3	1	0.333	na	na	1.0	0.00	na	na	na	na	na
pensum	Sc	iz	7/12	1	3	1	0.333	na	na	1.0	0.00	na	na	na	na	na
perscitum	Sc	iz	7/12	1	7	2	0.286	0.29	0.70	2.0	1.00	0.29	0.70	na	na	na
probum	Sc	iz	7/12	2	20	2	0.100	0.19	0.62	1.5 (1–2)	0.50	0.22	0.64	1.5 (1–2)	0.50	0.50
pseudocurvatum	Sc	iz	7/12	1	3	1	0.333	na	na	1.0	0.00	na	na	na	na	na
pseudopetiolatum	Sc	iz	7/12	2	11	1	0.091	na	na	1.0	0.00	na	na	2.0	na	na
subglobosum	Sc	iz	7/12	1	2	1	0.500	na	na	1.0	0.00	na	na	na	na	na
subgracilentipes	En	iz	7/12	2	11	1	0.091	na	na	1.0	0.00	na	na	2.0	na	na
tenellum	Sc	iz	7/12	1	3	1	0.333	na	na	1.0	0.00	na	na	na	na	na
tenuifrons	Sc	iz	7/12	8	69	7	0.101	0.74	0.55	1.3 (1–2)	0.33	0.12 (0.08–0.15)	0.57 (0.55–0.59)	1.4 (1–3)	0.00	0.71
Average						1.8	0.241	0.39	0.66	1.2	0.16	0.38	0.73	2.2	0.30	0.55

Table 7.4 Genotypic diversity statistics for microspecies of *Taraxacum*. Geog = geographical area studied (En, England; Cz, Czechoslovakia; Da, Denmark; Ja, Japan). Marker = marker used (iz, isozymes). Mk/loci = no. of markers (enzymes) / no. of gene loci. Npop = no. of localities (populations). Ni = total no. of individuals. Ng = total no. of genotypes detected. Hg = overall clonal diversity (multiclonal spp. only), calculated from data pooled over populations. Eg = overall clonal evenness (multiclonal spp. only), calculated from data pooled over populations. Ng/pop = average no. of genotypes per population (calculated for populations with more than one individual sampled). Mult-g pop = proportion of populations that are multiclonal. Hg/pop = mean clonal diversity per population (where Ng > 1). Eg/pop = mean clonal evenness per population (where Ng > 1). Pop/g = mean number of populations per genotype (calculated where Npop > 1). Wide g = proportion of genotypes that are widespread (calculated where Npop > 1 and Ng > 1). Priv g = proportion of genotypes that are private (calculated where Npop > 1 and Ng > 1).

Taraxacum	Geog	Marker	Mk/loci	Npop	Ni	Ng	Ng/Ni	Hg	Eg	Ng/pop	Mult-g pop	Hg/pop	Eg/pop	Pop/g	Wide g	Priv g	Ref*
albidum	Ja	iz	10/19	12	109	2	0.018	0.02	0.51	1.1 (1–2)	0.11	0.20	0.63	6.5 (1–12)	0.50	0.50	(d)
ancoriferum	Cz	iz	6/8	2	3	1	0.333	na	na	1.0	0.00	na	na	2.0	na	na	(a)
atactum	?	iz	1/?	5	5	1	0.200	na	na	na	na	na	na	5.0	na	na	(e)
boerkmanii	?	iz	1/?	3	3	1	0.333	na	na	na	na	na	na	3.0	na	na	(e)
brachyglossum (a)	En	iz	3/≥6	1	12	3	0.250	0.62	0.88	3.0	1.00	0.62	0.88	na	na	na	(b)
brachyglossum (b)	En	iz	10/15	1	38	1	0.026	na	na	1.0	0.00	na	na	na	na	na	(c)
fulviforme	En	iz	3/≥6	1	12	2	0.167	0.53	1.00	2.0	1.00	0.53	1.00	na	na	na	(b)
hamatiforme	?	iz	1/?	8	8	5	0.625	0.79	0.93	na	na	na	na	1.6 (1–3)	0.00	0.80	(e)
hamatum	?	iz	1/?	8	8	4	0.500	0.64	0.70	na	na	na	na	2.0 (1–4)	0.00	0.75	(e)
hamiferum	?	iz	1/?	2	2	2	1.000	1.00	1.00	na	na	na	na	1.0	0.00	1.00	(e)
hollandicum	Cz	iz	6/8	7	231	4	0.017	0.03	0.26	1.4 (1–4)	0.14	0.13	0.29	2.5 (1–7)	0.25	0.75	(a)
kernianum	?	iz	1/?	2	2	1	0.500	na	na	na	na	na	na	2.0	na	na	(e)
lacistophyllum	En	iz	3/≥6	1	41	4	0.098	0.65	0.71	4.0	1.00	0.65	0.71	na	na	na	(b)
norstedtii	En	iz	3/≥6	1	9	2	0.222	0.56	1.00	2.0	1.00	0.56	1.00	na	na	na	(b)
obliquum	Da?	iz	7/?	2	64	1	0.016	na	na	1.0	0.00	na	na	2.0	na	na	(f)
olivaceum	Cz	iz	6/8	1	30	8	0.267	0.71	0.44	8.0	1.00	0.71	0.44	na	na	na	(a)
pauciloburn	Cz	iz	6/8	2	2	1	0.500	na	na	1.0	0.00	na	na	2.0	na	na	(a)
proximum	En	iz	3/≥6	2	2	1	0.500	na	na	1.0	0.00	na	na	2.0	na	na	(b)
pseudohamatum (a)	?	iz	1/?	18	18	4	0.222	0.73	0.93	na	na	na	na	4.5 (1–14)	0.00	0.25	(e)
pseudohamatum (b)	En	iz	10/15	45	45	3	0.067	0.28	0.46	3.0	1.00	0.28	0.46	na	na	na	(c)
quadrans	?	iz	1/?	14	14	3	0.214	0.28	0.46	na	na	na	na	4.7 (1–11)	0.33	0.67	(e)
rubicundum	Da?	iz	7/?	1	16	5	0.313	?	?	5.0	1.00	?	?	na	na	na	(f)
subalpinum	Cz	iz	6/8	1	5	1	0.200	na	na	1.0	0.00	na	na	na	na	na	(a)
subnaevosum	En	iz	3/≥6	1	10	1	0.100	na	na	1.0	0.00	na	na	na	na	na	(b)
tortilobum	Da?	iz	7/?	20	334	15	0.045	0.60	0.17	3.3 (1–7)	0.90	0.56 (0.27–0.90)	0.79 (0.51–1.00)	4.3 (1–19)	0.13	0.53	(f)
uliginosum	Cz	iz	6/8	2	28	1	0.036	na	na	1.0	0.00	na	na	2.0	na	na	(a)
unguilobum (a)	En	iz	3/≥6	1	8	2	0.250	0.43	0.88	2.0	1.00	0.43	0.88	na	na	na	(b)
unguilobum (b)	En	iz	10/15	1	30	1	0.033	na	na	1.0	0.00	na	na	na	na	na	(c)
vindobonense	Cz	iz	6/8	4	87	64	0.736	0.98	0.68	16.0 (10–28)	1.00	0.79 (0.56–0.97)	1.00	1.0	0.00	1.00	(a)
Average					87	5.0	0.269	0.55	0.69	2.9	0.51	0.50	0.73	2.9	0.14	0.69	

* (a) Battjes *et al.* (1992); (b) Ford and Richards (1985); (c) Hughes and Richards (1989); (d) Menken and Morita (1989); (e) Mogie (1985); (f) Van Oostrum *et al.* (1985).

higher figure in the last case is inflated by the extreme value for *T. vindobonense* in which the unusually high number of 64 genotypes was detected. The low clonal diversity within most microspecies reflects a close link between their morphological and molecular phenotypes. This is highlighted at least in *Hieracium*, where those microspecies that contain phenotypes with small but consistent differences display correlated molecular variation, e.g. in the cases of *Hieracium calenduliflorum* and *H. tenuifrons*, where distinct morpho-geographical races are also distinguished by their molecular phenotypes (Shi *et al.*, 1996; Stace *et al.*, 1997).

In terms of the proportion of genotypes distinguishable, microspecies of the three large genera show remarkably similar mean values (*Rubus*, 0.227; *Hieracium*, 0.241; *Taraxacum*, 0.269; with *Sorbus* only slightly lower at 0.145; Tables 7.2–7.4). These figures are also mirrored by the estimates of overall clonal diversity, Hg (Tables 7.2–7.4).

In terms of the frequency distribution of genotypes, many microspecies that are multiclonal appear to consist of a common, sometimes widespread genotype, with an additional 1–4 others (rarely more) present at relatively low frequencies. These 'satellite' genotypes usually differ from the principal genotype by one allele at only one or two loci. They are usually restricted to single populations, as is shown by the overall evenness values which generally are below about 0.7, and by the values obtained for the mean number of populations per genotype and by the lower proportion of widespread compared with private genotypes (Tables 7.2–7.4). For this last ratio, *Sorbus* has a higher mean value than the other genera but the figures may be distorted by the fact that they are derived from only one enzyme system (peroxidase) and use of additional markers may well change the picture.

Within a population, the mean number of genotypes ranges from an average of 1.0 in *Sorbus*, through 1.2 in both *Hieracium* and *Rubus*, to 2.9 in *Taraxacum*. Many microspecies in any given locality appear to be uniclonal, the proportion of multiclonal populations varying from none in *Sorbus* to 10% and 16% in *Rubus* and *Hieracium* respectively, and up to 51% in *Taraxacum* (Tables 7.2–7.4). Overall, therefore, variability in agamospermous microspecies tends to be distributed largely between rather than within populations.

Such differences as there are in the patterns of variation between the three large genera are not clearly associated with the type of agamospermy present, and the similarity between *Rubus* and *Hieracium* is unexpected, although possible methodological biases should be borne in mind. *Rubus* is aposporous and facultatively agamospermous, and therefore might be expected to produce novel variants among the offspring at a higher frequency than *Hieracium*. *Hieracium* is diplosporous and meiosis is avoided completely, hence new variation should in theory be produced only by asexual means. In the case of *Rubus*, the evidently high degree of clonality seen in many of the microspecies may owe more to its considerable powers of vegetative reproduction than to its agamospermous breeding system (Nybom, 1988). In *Taraxacum*, whose agamosperms are diplosporous but carry out a restitutional meiosis, the slightly higher levels of variability could be attributable not only to asexual processes such as point mutations and somatic recombination, but also to facultative agamospermy, involving male fertility and the wider occurrence of gene exchange with sexual diploids (Menken *et al.*, 1995). None of these explanations, however, takes into account the variability inherited as a result of the sexual ancestry of the various agamosperms, and the extent to which this can explain the patterns observed is taken up in a later section.

7.3.2 Aggregates and others

The data accruing from studies of aggregates of agamospermous species and other agamic taxa that have not been divided into microspecies are presented in Table 7.5. The general pattern of variation is similar to that found from studies of microspecies, except that the levels of variability within populations tend to be higher, especially in the case of the aggregates. The fact that many populations of agamospermous aggregates contain more genetic variation than do individual microspecies should come as no surprise (Van Oostrum et al., 1985). Indeed, it has long been known that in certain aggregates the populations can contain dozens of microspecies (Richards, 1997). The variation between taxa in many of the statistics, however, is large. This is mirrored in part by their different morphological complexities, with some of the taxa representing large, variable, agamic complexes, e.g. *Antennaria rosea* and *Taraxacum officinale*, whereas others are much more homogeneous, e.g. *Limonium cavanillesii*, which may approach a microspecies in its definition.

The available evidence indicates that, within a population of an agamospermous aggregate, the constituent clones are closely related, with genetic identities between individuals calculated as greater than 0.9 in the cases of *Antennaria* (Bayer, 1989) and *Erigeron* (Noyes and Soltis, 1996). Similarly, within an agamic aggregate and irrespective of locality, the constituent microspecies can also be genetically closely related to a degree equivalent to that typical of conspecific populations (Hughes and Richards, 1989). Leaving aside the contentious issue of appropriate taxonomic rank for the apomictic genotypes (Cowan et al., 1998; Dickinson, 1998; Hörandl, 1998; Kirschner, 1998; Stace, 1998), the chief question raised by the findings on population genetic structure is how to explain the origin of the variability found. Two arguments are current in the literature with regard to this, hinging on the extent to which sexual or asexual processes have been involved in agamosperm origins.

7.4 Origins of clonal diversity

Various explanations have been proposed to account for the variability found in agamosperms. The amount and pattern generated at the time(s) of origin may be altered later by sexual or asexual events. The patterns we see may also be determined by ecological or geographical factors, as well as various evolutionary processes affecting populations, such as drift, migration and selection. Several studies to date have highlighted how difficult it is to disentangle these factors. For example, in trying to explain the exceptionally high level of variation found in the microspecies *Taraxacum vindobonense* (64 genotypes), Battjes et al. (1992) invoked mutation, autosegregation, habitat variation, facultative agamospermy, gene exchange with sexual diploids, and polyphyletic origin as possibilities. Carefully designed experiments using matched sets of taxa are needed in order to compare and contrast the effects of the different factors outlined above.

Table 7.5 Genotypic diversity statistics for agamospermous groups where sampling is based on an agamic aggregate or other taxon rather than a particular microspecies. Geog = geographical area studied (Ge, Germany; N.Am., North America). Marker = marker used (iz, isozymes; cy, cytology; mp, morphology). Mk/loci = no. of markers (enzymes or DNA primers) / no. of gene loci. Npop = no. of populations. Ni = total no. of individuals. Ng = total no. of genotypes detected. Ng/pop = average no. of genotypes per population (calculated for populations with more than one individual sampled). Mult-g pop = proportion of populations that are multiclonal. Hg/pop = mean clonal diversity per population (where Ng > 1). Eg/pop = mean clonal evenness per population (where Ng > 1). Pop/g = mean number of populations per genotype (calculated where Npop > 1). Wide g = proportion of genotypes that are widespread (calculated where Npop > 1 and Ng > 1). Priv g = proportion of genotypes that are private (calculated where Npop > 1 and Ng > 1).

Taxon	Geog	Marker	Mk/loci	Npop	Ni	Ng	Ng/Ni	Ng/pop	Mult-g pop	Hg/pop	Eg/pop	Pop/g	Wide g	Priv g	Ref.*
Aggregates															
Antennaria rosea	N.Am	iz	10/19	63	1065	192	0.180	3.5 (1–11)	0.73	?	?	1.1 (range?)	0.00	0.89	(b)
Arabis holboellii	USA	iz	4/9	3	904	40	0.044	18.3 (6–27)	1.00	0.73 (0.40–0.92)	?	1.4 (1–3)	0.13	0.78	(k)
Erigeron annuus	USA	iz	4/?	3	300	17	0.057	14.0 (13–15)	1.00	0.87 (0.85–0.97)	0.90 (0.88–0.91)	2.4 (1–3)	0.59	0.12	(e)
E. compositus	USA	iz	15/15	7	231	24	0.104	3.4 (1–10)	0.71	0.49 (0.00–0.87)	0.61 (0.00–1.00)	1.0	0.00	1.00	(h)
Sorbus latifolia	Ge	iz	5/?	1	23	10	0.435	10.0	1.00	?	?	na	na	na	(a)
Taraxacum officinale (a)	USA	iz	6/?	3	284	4	0.014	3.7 (3–4)	1.00	0.47 (0.36–0.57)	0.61 (0.50–0.73)	2.7 (2–3)	0.75	0.00	(l)
T. officinale (b)	USA	iz,cy,mp	6/?	22	518	50	0.097	5.0 (1–13)	0.91	0.57 (0.17–0.89)	0.49 (0.10–0.83)	2.1 (1–19)	0.02	0.66	(f)
T. officinale (c)	?	mp	4/?	1	177	20	0.113	20.0	1.00	0.85	0.86	na	na	na	(n)
T. officinale (d)	England	iz,mp	?/?	1	97	18	0.186	18.0	1.00	0.99	0.56	na	na	na	(o)
T. sect. Ruderalia	Europe	iz	2/3	11	374	57	0.152	11.8 (6–17)	1.00	0.82 (0.71–0.89)	0.54 (0.26–0.85)	2.3 (range?)	0.05	0.63	(g)
Others															
Antennaria friesiana															
ssp. friesiana	N.Am	iz	9/16	4	?	?	?	3.3 (2–5)	1.00	?	?	?	?	?	(c)
A. monocephala															
ssp. angustata	N.Am	iz	9/16	1	?	3	?	3.0	1.00	?	?	na	na	na	(c)
Eupatorium altissimum	USA	iz	5/13	5	144	4	0.028	1.0	0.00	na	na	1.3 (1–2)	0.00	0.75	(m)
Limonium cavanillesii	Spain	RAPD	11/131	1	29	1	0.034	1.0	0.00	na	na	na	na	na	(i)
Limonium dufourii	Spain	RAPD	12/124	6	165	44	0.267	8.5 (2–14)	1.00	?	?	1.2 (1–3)	0.00	0.86	(j)
Pellaea andromedifolia (apogamous)	USA	iz	7/8	3	23	2	0.087	1.0	0.00	na	na	1.5 (1–2)	0.00	0.50	(d)

* (a) Aas et al. (1994); (b) Bayer (1990); (c) Bayer (1991); (d) Gastony and Gottlieb (1985); (e) Hancock and Wilson (1976); (f) Lyman and Ellstrand (1984); (g) Menken et al. (1995); (h) Noyes and Soltis (1996); (i) Palacios and González-Candelas (1997a); (j) Palacios and González-Candelas (1997b); (k) Roy (1993); (l) Solbrig and Simpson (1974); (m) Yahara et al. (1991); (n) Nilsson (1947); (o) Ford and Richards (1985).

The first question to ask, however, is to what extent the ancestry of an agamo-spermous group can explain its constituent variability. The potential amount of variability depends on whether the origin of asexuality in such groups is mono- or polyphyletic, and whether the parental sexual genepool was genetically depauperate or not. A second question is whether the variability in the group has been increased subsequently primarily by sexual or by asexual means. There are conflicting reports in the literature over the degree to which asexual processes are involved, and in the following two sections I shall examine to what extent sexuality or asexuality can explain the patterns of variability found.

7.4.1 Sexual origins

Large amounts of variability would be expected in an apomictic group if the sexual parents were divergent and heterozygous and if they formed agamospermous indi-viduals on numerous occasions by hybridisation. Smaller amounts would be expected if the sexual parents were less divergent, more homozygous and hybridised very few times or even only once. Either way, at least initially, the amount and nature of the variability in the agamospermous group should be directly related to that of its sexual parents. In this section I shall review some of the studies that have produced molecular data comparing the genetic composition of agamosperms and their sexual progenitors.

Following a study of the *Antennaria rosea* agamic complex in North America, Bayer (1990) found that the factor best correlated with diversity was proximity to related sexual species. Indeed in a series of investigations (Bayer and Crawford, 1986; Bayer, 1989, 1990, 1991), it was concluded that the variability displayed in a variety of agamospermous *Antennaria* taxa in North America was best explained on the basis of polyphyletic origins from extant sexual progenitors. Values of Nei's genetic identity (unbiased) between agamosperm and sexual relatives were typically greater than 0.9, indicating that the origins have been recent with little time for the accumulation of mutations; indeed only one novel allele was detected in *A. rosea* (Bayer, 1989). In the case of *A. media*, some populations contain a mixture of obligate and facultative apomicts and sexuals (Bayer *et al.*, 1990), making the origin of variability in this species scarcely a puzzle.

A similar picture emerged from a study of sexual and agamospermous taxa in the *Erigeron compositus* group in western North America (Noyes and Soltis, 1996). Here, the finding of high genetic identities between asexual taxa and their sexual putative progenitors led the authors to suggest that the variation in the former was best explained on the basis of inheritance from the latter, although crossing between facultative agamosperms was admitted as a possibility. Crossing between agamo-sperms and sexual diploids, however, was regarded as unlikely because of the geographical and ecological isolation of the agamosperms.

Based on an investigation of extensive clonal diversity within and between popu-lations containing diploid and triploid plants of *Taraxacum* section *Ruderalia* in central and western Europe, Menken *et al.* (1995) provided isozyme evidence from which they inferred the recurrent formation of the triploid agamosperms. The suggested mechanisms were diploid × diploid matings involving an unreduced gamete, and fertilization of diploid agamosperm eggs by haploid pollen derived from either diploids or male-fertile agamospermous triploids. Furthermore, on the basis of isozyme evidence, bidirectional gene flow was inferred to occur between asexual triploid and sexual diploid plants growing in the same or nearby populations. The evidence rests on the very large proportion of shared alleles between the two ploidy levels; also, both diploids and triploids simulate Hardy–Weinberg equilibrium and together in populations they display a marked genetic homogeneity suggesting that they form a cohesive evolutionary unit. Mutation was offered as an explanation to account for 'satellite' clones that differ from their putative parents by single alleles. (It is worth adding a note of caution here because not all diploids are necessarily ancestral to the agamospermous polyploids with which they exchange genes; there is evidence that some diploids may be derived from the latter by means of diploid/polyploid cycles, e.g. *Taraxacum* (Richards, 1997), and will as a consequence show the same high genetic identities as are found in relationships of the reverse kind (Hughes and Richards, 1989).)

The examples so far have involved groups exhibiting considerable clonal diversity. That a sexual ancestry can also account for cases in which the agamosperm displays low levels of variability is shown by *Eupatorium altissimum*. In this species, which contains both sexual and asexual races, the low clonal diversity and observed heterozygosity in the agamosperms are matched by the relatively impoverished genetic constitution of the sexual populations. Furthermore, all alleles in the agamospermous populations are also found in the sexual populations, with genetic identities in pairwise comparisons ranging from 0.994–1.000 (Yahara *et al.*, 1991). The authors concluded that the agamospermous populations had an autopolyploid origin from their sexual progenitors, but did not comment on how many times this may have happened. The finding of three clones suggests that multiple origins might be involved, although a cross between heterozygous parents might also produce the same result. Irrespective of whether agamospermy originated once or on several occasions in this species, it is clear that a sexual ancestry from a restricted genepool can account for the low levels of variability encountered.

In the Pteridophyta also, studies of isozyme diversity among triploid apomictic fern clones have produced evidence of polyphyletic origins from sexual progenitors, whether the mechanism be autoploid or alloploid (Haufler *et al.*, 1985; Suzuki and Iwatsuki, 1990). (See also Vogel *et al.* (1999 – this volume) for a discussion of multiple origins of polyploid pteridophytes.)

These studies suggest that the amount and pattern of variability in an agamosperm group can be explained as a direct consequence of its sexual ancestry. Nevertheless, the possible role of asexual processes in generating not only variability within microspecies (Gustafsson, 1946/7: 183–198) but also entirely new lineages of agamosperms has been suggested (Richards, 1997). To what extent such asexually generated variation can occur and be of evolutionary consequence is taken up in the next section.

7.4.2 Asexual origins

Odd clones within a microspecies, involving differences in one allele at one or two isozyme loci, have been attributed to mutation by other workers in addition to Menken *et al.* (1995), e.g. Battjes *et al.* (1992) and Shi *et al.* (1996). Similarly, the small differences in DNA-fingerprint profiles between some morphologically related microspecies of *Taraxacum* are also believed to be the result of mutation (Heusden *et al.*, 1991). Cytologically, aneuploid variants are sometimes found that are possibly of asexual origin, e.g. in *Hieracium macrocarpum* (Stace *et al.*, 1997). Furthermore, morphological variants of agamospermous microspecies have been described occasionally, either as varieties or as formae, giving the impression that they have been produced asexually from a parent genotype. It is not clear, however, whether whole groups of microspecies have ever been produced by such means although there are suggestions that this may be so (Richards, 1996, 1997; Mogie and Ford, 1988; Kirschner and Stepanek, 1994).

Richards (1996) and Kirschner and Stepanek (1996) have reviewed evidence that agamospecies of *Taraxacum* can generate variability asexually. Possible mechanisms include: a) point mutations; b) somatic recombination, perhaps sometimes enhanced by transposon activity; and c) disjunctional accidents resulting in polyploid, aneuploid or haploid offspring. The evidence for asexually-generated variation has accumulated from observations of morphological, cytological and molecular variation among progeny arrays produced from agamospermous mothers (Gustafsson, 1946/7; Malecka, 1973; Richards, 1996). Such variability should in theory be a result of asexual processes, such as those listed above. However, the origins of the variability described in some of the studies may be open to doubt, primarily because it is not clear whether the experimental mothers were neutered (emasculated and with stigmas removed) before being allowed to set seed. The possibility of rare sexual events therefore cannot be excluded and may explain some of the variation found within the families. A notable early exception is the work of Ostenfeld (1921) who raised families from neutered plants of *Hieracium* s.s. and within which he recovered non-maternal phenotypes at low frequency. Similarly, King and Schaal (1990) used neutered mothers to produce two families in which non-parental rDNA and Adh-1 restriction fragment genotypes occurred. In one of the families all 41 plants were deviant and, in the other, one out of 26 sibs presented a novel genotype. The authors attributed this variability to mutational events occurring in the first case at an early stage of development in the floral meristem and in the second case at a later stage during the formation of the aberrant seed.

In a study designed to distinguish explicitly between the roles of mutation and multiple sexual origins as explanations for agamosperm variability, King (1993) investigated nuclear ribosomal DNA and chloroplast DNA markers in a range of North American and European populations of *Taraxacum*. By combining the data from rDNA and cpDNA restriction fragment profiles, a pattern emerged that unambiguously demonstrated not only the multiple origin of asexual polyploids and microspecies by hybridisation but also that this process accounted for most of the variability seen. Mutation only contributed to the diversity in the agamospermous taxa to a minor extent.

A second line of evidence that has been used in support of the idea that asexual reproduction can generate new agamospermous groups rests on the inferred monophyly of such groups. At least three such cases have been proposed: *Taraxacum* section *Hamata* (Mogie and Richards, 1983), species of *Taraxacum* section *Palustria* (Heusden *et al.*, 1991; Kirschner and Stepanek, 1996) and the *Limonium binervosum* aggregate (Ingrouille and Stace, 1985), although there is now some doubt as to the strength of the evidence relating to the monophyly of the first two (Richards, pers. comm.).

Whilst all the above mentioned groups may well be monophyletic, it does not follow that the common ancestor in each case was agamospermous. In other words, the monophyletic status of each group of taxa could pre-date the establishment of agamospermy. Under this scenario, agamospermy could have become established in an ancestral sexual genepool on numerous occasions and in different localities. In many cases, however, the sexual ancestor has become extinct and therefore there are no comparative data by which to link the agamosperm group with its putative progenitors. It may be difficult, even impossible, in such cases to distinguish between sexual and asexual origins as explanations for the variability. Some light might be shed on the problem, however, by adopting an indirect approach and looking at the *pattern* of variation exhibited. In outcrossing sexual groups the relationships between genotypes will be reticulate, whereas in agamospermous groups they will be dichotomously hierarchical if variability has been generated asexually (Richards, 1996). A phylogenetic analysis can be used to distinguish these two situations, given appropriate data. If genotypes were originally produced by cross-fertilisation events then their phylogenetic tree should be largely unresolved. In contrast, if they have evolved by asexual means then the analysis should result in a dichotomously hierarchical tree. In a recent study of 36 British taxa belonging to *Hieracium* sect. *Alpina* (Stace *et al.*, 1997) the pattern of variation among them was examined by means of RAPD markers (the homologies of several of which were confirmed by Southern hybridisations). Of the 181 amplification products (marker bands) produced by 17 primers, 116 were informative in that they were present in more than one but not all taxa, 32 were present in only one taxon and 33 were common to all. A phylogenetic analysis using parsimony (unpublished) produced 126 shortest trees each of 514 steps; the strict consensus tree was almost unresolved, suggesting a great deal of reticulate variation in the data. A study of the same taxa but using ITS2 sequence data revealed 34 different sequences determined by 25 variable sites out of the total of 222 (Stace *et al.*, 1997); phylogenetic analysis produced a totally unresolved strict consensus tree (unpublished). Notwithstanding the low level of polymorphism in the ITS2 sequence data, the relationships between the microspecies appear to be largely, if not entirely, reticulate. If occasional sexuality is dismissed as an unlikely explanation because in *Hieracium* s.s. reproduction is by mitotic diplospory and, furthermore, members of sect. *Alpina* in Britain are male-sterile, then the ultimate cause of much of the variation among the microspecies would appear to be attributable to their sexual ancestry.

A largely unresolved question concerns the mechanism behind the presumed widespread extinction of sexual progenitors in such obligately agamospermous groups as *Hieracium* sect. *Alpina*. In the context of similar large north-temperate agamic complexes, lack of pollinators in the extreme post-glacial environments may well have been a crucial factor that rendered the self-incompatible (or at least outbreeding) sexual progenitors unfit (Stace *et al.*, 1997). Indeed, Bierzychudek (1987) has shown that sexual populations are at up to a four-fold disadvantage in circumstances where pollination is unpredictable and sexual mothers set submaximum amounts of seed.

7.4.3 Conclusion

Although there is evidence that asexual processes have generated variation on a minor scale, especially within microspecies, there are as yet no convincing data to support the view that they have spawned whole new groups of agamospermous taxa. At least as far as gametophytic agamosperms are concerned most of the variability found, whether it be present at a high level or a low level, can be satisfactorily explained as a direct consequence of the mode of sexual ancestry, with facultatively apomictic behaviour subsequently making an important contribution in some groups.

ACKNOWLEDGEMENTS

The part of this work relating to *Hieracium* sect. *Alpina* was funded by NERC grant GR3/8743A. I am also grateful to Prof. C. A. Stace, A. J. Richards and an anonymous referee for commenting on a draft of this chapter.

REFERENCES

Aas, G., Maier, J., Baltisberger, M. and Metzger, S. (1994) Morphology, isozyme variation, cytology, and reproduction of hybrids between *Sorbus aria* (L.) Crantz and *S. torminalis* (L.) Crantz. *Botanica Helvetica*, **104**, 195–214.

Asker, S. (1979) Progress in apomixis research. *Hereditas*, **91**, 231–240.

Assienan, B. and Noirot, M. (1995) Isozyme polymorphism and organisation of the agamic complex of the *Maximae* (*Panicum maximum* Jacq., *P. infestum* Anders., and *P. trichocladum* K. Schum.) in Tanzania. *Theoretical & Applied Genetics*, **91**, 672–680.

Battjes, J., Menken, S. B. J. and den Nijs, H. J. C. M. (1992) Clonal diversity in some microspecies of *Taraxacum* sect. *Palustria* (Lindeb. fil.) Dahlst. from Czechoslovakia. *Botanischer Jarhbucher Systematische*, **114**, 315–328.

Bayer, R. J. (1989) Patterns of isozyme variation in *Antennaria rosea* (Asteraceae: Inuleae) polyploid agamic complex. *Systematic Botany*, **14**, 389–397.

Bayer, R. J. (1990) Patterns of clonal diversity in the *Antennaria rosea* (Asteraceae) polyploid agamic complex. *American Journal of Botany*, **77**, 1313–1319.

Bayer, R. J. (1991) Allozymic and morphological variation in *Antennaria* (Asteraceae: Inuleae) from the low arctic of northwestern North America. *Systematic Botany*, **16**, 492–506.

Bayer, R. J. and Crawford, D. J. (1986) Allozyme divergence among five diploid species of *Antennaria* (Asteraceae: Inuleae) and their allopolyploid derivatives. *American Journal of Botany*, **73**, 287–296.

Bayer, R. J., Ritland, K. and Purdy, B. G. (1990) Evidence of partial apomixis in *Antennaria media* (Asteraceae: Inuleae) detected by the segregation of genetic markers. *American Journal of Botany*, **77**, 1078–1083.

Bierzychudek, P. (1987) Pollinators increase the cost of sex by avoiding female flowers. *Ecology*, **68**, 444–447.

Clausen, J. (1954) Partial apomixis as an equilibrium system in evolution. *Caryologia*, **6** (suppl.), 469–479.

Cowan, R., Ingrouille, M. J. and Lledó, M. D. (1998) The taxonomic treatment of agamosperms in the genus *Limonium* Mill. (Plumbaginaceae). *Folia Geobotanica*, **33**, 353–366.

Darlington, C. D. (1939) *The evolution of genetic systems*, Oliver & Boyd, Edinburgh.

Davis, P. H. and Heywood, V. H. (1963) *Principles of angiosperm taxonomy*, Oliver & Boyd, Edinburgh.

Dickinson, T. A. (1998) Taxonomy of agamic complexes in plants: a role for metapopulation thinking. *Folia Geobotanica*, **33**, 327–332.

Dudman, A. A. and Richards, A. J. (1997) *Dandelions of Great Britain and Ireland*, BSBI, London.

Ellstrand, N. C. and Roose, M. L. (1987) Patterns of genotypic diversity in clonal plant species. *American Journal of Botany*, **74**, 123–131.

Ford, H. and Richards, A. J. (1985) Isozyme variation within and between *Taraxacum* agamospecies in a single locality. *Heredity*, **55**, 289–291.

Gastony, G. J. and Gottlieb, L. D. (1985) Genetic variation in the homosporous fern *Pellaea andromedifolia*. *American Journal of Botany*, **72**, 257–267.

Gustafsson, Å. (1946/7) *Apomixis in higher plants*, Gleerup, Lund.

Hancock, J. F. and Wilson, R. E. (1976) Biotype selection in *Erigeron annuus* during old field succession. *Bulletin of the Torrey Botanical Club*, **103**, 122–125.

Haufler, C. H., Windham, M. D., Britton, D. M. and Robinson, S.J. (1985) Triploidy and its evolutionary consequences in *Cystopteris protrusa*. *Canadian Journal of Botany*, **63**, 1855–1863.

Heusden, A. W. V., Rouppe van der Voort, J. and Bachmann, K. (1991) Oligo(-GATA) fingerprints identify clones in asexual dandelions (*Taraxacum*, Asteraceae). *Fingerprint News*, **3**, 13–15.

Hörandl, E. (1998) Species concepts in agamic complexes: applications in the *Ranunculus auricomus* complex and general perspectives. *Folia Geobotanica*, **33**, 335–348.

Hughes, J. and Richards, A. J. (1988) The genetic structure of populations of sexual and asexual *Taraxacum* (dandelions). *Heredity*, **60**, 161–171.

Hughes, J. and Richards, A. J. (1989) Isozymes, and the status of *Taraxacum* (Asteraceae) agamospecies. *Botanical Journal of the Linnean Society*, **99**, 365–376.

Ingrouille, M. J. and Stace, C. A. (1985) Pattern of variation of agamospermous *Limonium* (Plumbaginaceae) in the British Isles. *Nordic Journal of Botany*, **5**, 113–125.

Janzen, D. J. (1977) What are dandelions and aphids? *American Naturalist*, **111**, 586–589.

King, L. M. (1993) Origins of genotypic variation in North American dandelions inferred from ribosomal DNA and chloroplast DNA restriction enzyme analysis. *Evolution*, **47**, 136–151.

King, L. M. and Schaal, B. A. (1990) Genotypic variation within asexual lineages of *Taraxacum officinale*. *Proceedings of the National Academy of Sciences of the USA*, **87**, 998–1002.

Kirschner, J. (1998) A note on the taxonomy of agamic complexes. A reply to Tim Dickinson. *Folia Geobotanica*, **33**, 333–334.

Kirschner, J. and Stepanek, J. (1994) Clonality as part of the evolution process in *Taraxacum*. *Folia Geobotanica & Phytotaxonomica*, **29**, 265–275.

Kirschner, J. and Stepanek, J. (1996) Modes of speciation and evolution of the sections in *Taraxacum*. *Folia Geobotanica & Phytotaxonomica*, **31**, 415–426.

Koltunow, A. M. (1993) Apomixis: embryo sacs and embryos formed without meiosis or fertilization in ovules. *Plant Cell*, **5**, 1425–1437.

Kraft, T. and Nybom, H. (1995) DNA fingerprinting and biometry can solve some taxonomic problems in apomictic blackberries (*Rubus* subgen. *Rubus*). *Watsonia*, **20**, 329–343.

Kraft, T., Nybom, H. and Werlemark, G. (1995) *Rubus vestervicensis* (Rosaceae) – its hybrid origin revealed by DNA fingerprinting. *Nordic Journal of Botany*, **15**, 237–242.

Kraft, T., Nybom, H. and Werlemark, G. (1996) DNA fingerprint variation in some blackberry species (*Rubus* subg. *Rubus*, Rosaceae). *Plant Systematics & Evolution*, **199**, 93–108.

Löve, A. (1960) Biosystematics and the classification of apomicts. *Feddes Repertorium*, **62**, 136–148.

Lyman, J. C. and Ellstrand, N. C. (1984) Clonal diversity in *Taraxacum officinale* (Compositae), an apomict. *Heredity*, **53**, 1–10.

Malecka, I. (1973) Problems in the mode of reproduction in microspecies of *Taraxacum* section *Palustria* Dahlstedt. *Acta Biologica Cracoviensia, series Botanica*, **16**, 37–84.

Marshall, D. R. and Brown, A. H. D. (1981) The evolution of apomixis. *Heredity*, **47**, 1–15.

Menken, S. B. J. and Morita, T. (1989) Uniclonal population structure in the pentaploid obligate agamosperm *Taraxacum albidum* Dahlst. *Plant Species Biology*, **4**, 29–36.

Menken, S. B. J., Smit, E. and den Nijs, H. J. C. M. (1995) Genetical population structure in plants: gene flow between diploid sexual and triploid asexual dandelions (*Taraxacum* section *Ruderalia*). *Evolution*, **49**, 1108–1118.

Mogie, M. (1985) Morphological, developmental and electrophoretic variation within and between obligately apomictic *Taraxacum* species. *Biological Journal of the Linnean Society*, **24**, 207–216.

Mogie, M. (1992) *The evolution of asexual reproduction in plants*, Chapman & Hall, London.

Mogie, M. and Ford, H. (1988) Sexual and asexual *Taraxacum* species. *Biological Journal of the Linnean Society*, **35**, 155–168.

Mogie, M. and Richards, A. J. (1983) Satellited chromosomes, systematics and phylogeny in *Taraxacum* (Asteraceae). *Plant Systematics & Evolution*, **141**, 219–229.

Nilsson, N. H. (1947) Totale inventierung der microtypen eines minimiareals von *Taraxacum officinale. Hereditas*, **33**, 119–142.

Nogler, G. A. (1984a) Gametophytic apomixis, in *Embryology of angiosperms*, (ed. Johri, B. M.), Springer-Verlag, Berlin, pp. 475–518.

Nogler, G. A. (1984b) Genetics of apospory in apomictic *Ranunculus auricomus*. V. Conclusion. *Botanica Helvetica*, **94**, 411–422.

Noyes, R. D. and Soltis, D. E. (1996) Genotypic variation in agamospermous *Erigeron compositus* (Asteraceae). *American Journal of Botany*, **83**, 1292–1303.

Nybom, H. (1988) Apomixis versus sexuality in blackberries (*Rubus* subg. *Rubus*, Rosaceae). *Plant Systematics & Evolution*, **160**, 207–218.

Nybom, H. and Schaal, B. A. (1990) DNA 'fingerprints' reveal genotypic distributions in natural populations of blackberries and raspberries (*Rubus*, Rosaceae). *American Journal of Botany*, **77**, 883–888.

Ostenfeld, C. H. (1921) Some experiments on the origin of new forms in the genus *Hieracium* sub-genus *Archieracium. Journal of Genetics*, **11**, 117–122.

Palacios, C. and González-Candelas, F. (1997a) Lack of genetic variability in the rare and endangered *Limonium cavanillesii* (Plumbaginaceae) using RAPD markers. *Molecular Ecology*, **6**, 671–675.

Palacios, C. and González-Candelas, F. (1997b) Analysis of population genetic structure and variability using RAPD markers in the endemic and endangered *Limonium dufourii* (Plumbaginaceae). *Molecular Ecology*, **6**, 1107–1121.

Palmer, E. J. (1932) The *Crataegus* problem. *Journal of the Arnold Arboretum*, **13**, 342–362.

Peet, K. (1974) The measurement of species diversity. *Annual Review of Ecological Systematics*, **5**, 285–307.

Proctor, M. C. F., Proctor, M. E. and Groenhof, A. C. (1989) Evidence from peroxidase polymorphism on the taxonomy and reproduction of some *Sorbus* populations in south-west England. *New Phytologist*, **112**, 569–575.

Richards, A. J. (1996) Genetic variability in obligate apomicts of the genus *Taraxacum. Folia Geobotanica & Phytotaxonomica*, **31**, 405–414.

Richards, A. J. (1997) *Plant breeding systems*, 2nd ed., Chapman & Hall, London.

Roy, B. A. (1993) Patterns of rust infection as a function of host genetic diversity and host density in natural populations of the apomictic crucifer, *Arabis holboellii*. *Evolution*, **47**, 111–124.

Roy, B. A. (1995) The breeding systems of six species of *Arabis* (Brassicaceae). *American Journal of Botany*, **82**, 869–877.

Shi, Y., Gornall, R. J., Draper, J. and Stace, C. A. (1996) Intraspecific molecular variation in *Hieracium* sect. *Alpina* (Asteraceae), an apomictic group. *Folia Geobotanica & Phytotaxonomica*, **31**, 305–313.

Solbrig, O. T. and Simpson, B. B. (1974) Components of regulation of a population of dandelions in Michigan. *Journal of Ecology*, **62**, 473–486.

Soltis, D. E. and Soltis, P. S. (1993) Molecular data and the dynamic nature of polyploidy. *Critical Reviews in Plant Sciences*, **12**, 243–273.

Stace, C. A. (1998) Species recognition in agamosperms – the need for a pragmatic approach. *Folia Geobotanica*, **33**, 319–326.

Stace, C. A., Gornall, R. J. and Shi, Y. (1997) Cytological and molecular variation in apomictic *Hieracium* sect. *Alpina*. *Opera Botanica*, **132**, 39–51.

Stebbins, G. L. (1950) *Variation and evolution in plants*, Columbia University Press, New York.

Stebbins, G. L. (1971) *Chromosomal evolution in higher plants*, Edward Arnold, London.

Suzuki, T. and Iwatsuki, K. (1990) Genetic variation in agamosporous fern *Pteris cretica* L. in Japan. *Heredity*, **65**, 221–227.

Turesson, G. (1929) Zur Natur und Begrenzung der Arteneinheiten. *Hereditas*, **12**, 323–334.

Turesson, G. (1943) Variation in the apomictic species of *Alchemilla vulgaris* L. *Botaniska Notiser*, **1943**, 413–427.

Van Oostrum, H., Sterk, A. A. and Wijsman, H. J. W. (1985) Genetic variation in agamospermous microspecies of *Taraxacum* sect. *Erythrosperma* and sect. *Obliqua*. *Heredity*, **55**, 223–228.

Vogel, J. C., Barrett, J. A., Rumsey, F. J. and Gibby, M. (1999) Identifying multiple origins in polyploid homosporous pteridophytes, in *Molecular systematics and plant evolution*, (eds P. M. Hollingsworth, R. M. Bateman and R. J. Gornall), Taylor & Francis, London, pp. 101–117.

Winge, O. (1938) The genetic aspect of the species problem. *Proceedings of the Linnean Society of London*, **1937-1938**, 231–238.

Yahara, T., Ito, M., Watanabe, K. and Crawford, D.J. (1991) Very low genetic heterozygosities in sexual and agamospermous populations of *Eupatorium altissimum* (Asteraceae). *American Journal of Botany*, **78**, 706–710.

Chapter 8

Monophyly, populations and species

J. I. Davis

ABSTRACT

Monophyly is a relationship that is applicable in hierarchic descent systems. In such systems there are basal elements (e.g. each of a set of copies of a non-recombining gene, or each of a set of species that do not exchange genes with each other), and each of these elements has a single line of ancestry that can be traced backward through time, via points of common ancestry with other elements, to the point of common ancestry of all elements in the system. Because each basal element has just one line of ancestry, descent relationships of all basal elements in the system can be depicted in a phylogenetic diagram, with each node representing the point of divergence of two or more lines of descent, and with the single point of most-recent common ancestry for any specified pair of basal elements situated at one node. Also, for any three specified elements, there is either a single point from which their lines of ancestry diverge (i.e. they are equally closely related to each other), or more often two of the elements are more closely related to each other than either of them is to the third. Willi Hennig proposed two definitions of mono-phyly: the first defines a monophyletic group as one that consists of an ancestor and all of its descendants, whereas the second defines a monophyletic group as one in which all members are more closely related to each other than any of them is to any element that is not in the group. In hierarchic descent systems these two definitions are congruent in meaning, and consequently if a group such as Liliopsida or Compositae is monophyletic under either of these definitions, it is also mono-phyletic under the other. In non-hierarchic descent systems (i.e. reticulate systems, such as exist among individual sexually reproducing organisms), Hennig's defini-tions are not congruent. Attempts to redefine monophyly in order to apply it within reticulate descent systems (e.g. to the sexually reproducing individuals of a species) have been frustrated by this dissociation of common ancestry and prox-imity of relationship in such systems. In response to these difficulties, some proponents of the applicability of monophyly in non-hierarchic descent systems have proposed modifications of Hennig's definitions, or have attempted to shift the focus of monophyly from the elements of descent to the characters borne by those elements (i.e. by redefining monophyly in terms of descent relationships among the genes borne by organisms, rather than in terms of descent relation-ships among the organisms *per se*). These attempts have led to a series of contra-dictory conceptual stances.

In *Molecular systematics and plant evolution* (1999) (eds P. M. Hollingsworth, R. M. Bateman and R. J. Gornall), Taylor & Francis, London, pp. 139–170.

My Ability of distressing you during my Life would be a species of revenge to which I should hardly stoop under any circumstances.

Jane Austen, *Lady Susan*

8.1 Introduction

It is axiomatic that systematists use evidence to discern relationships among groups of organisms. This being so, however, it remains the responsibility of individual systematists to decide what kinds of relationship are of interest, and to identify appropriate sources of evidence concerning those relationships. With the ascendance of phylogenetic systematics has come an intense focus on phylogenetic relationships – specifically on the discovery of monophyletic groups – and on shared homologues, i.e. synapomorphy, as evidence thereof (e.g. Hennig, 1966; Nelson and Platnick, 1981). More recently, with the rise of molecular systematics, there has been an increasing availability of new sets of characters, and an increased focus on certain problems that are of particular importance with respect to these characters (e.g. gene duplication, concerted evolution), but the basic equation – synapomorphy as evidence of phylogenetic structure – still stands. However, it is crucial that systematists distinguish between characters and character-bearers, and that they recognize that patterns of shared characters need not be congruent with phylogenetic history. These distinctions, in turn, are inextricably linked to the meaning of monophyly.

Although systematists have spoken of monophyly for more than 100 years (for discussions of its use by Haeckel and other pre-Hennigian authors see Hull, 1988 and Farris, 1990), the monophyly concept proposed by Hennig in 1966 was more rigorous than those of earlier authors. The increased precision embodied in Hennig's monophyly concept has since been recognized almost universally in systematics, even by those who object to monophyly as a necessary attribute of higher taxa (e.g. Stuessy, 1990; Sosef, 1997; Knox, 1998). The restrictiveness of Hennig's monophyly concept, in contrast with the less precise monophyly concepts that preceded it, is perhaps seen most clearly in the fact that he simultaneously characterized paraphyly – a relationship unrecognised prior to his time – as a form of deviation from monophyly (Hennig, 1966; Farris, 1991). Although some authors objected to the restrictiveness of Hennig's monophyly concept (e.g. Ashlock, 1971, 1972), Hennig's view eventually prevailed, and paraphyletic groups, like polyphyletic groups, are now recognised as unambiguously non-monophyletic. Once paraphyly came to be seen as a form of non-monophyly, the distinction between paraphyly and polyphyly became controversial, and a debate arose over this subject, in which the importance of characters and genealogy were discussed (Ashlock, 1971, 1972; Nelson, 1971; Platnick, 1977; Oosterbroek, 1987; Farris, 1974, 1990, 1991). Farris, using the convention of 'group membership characters' (which he later called 'identifiers'), affirmed the precision of Hennig's original paraphyly concept, and the terms monophyly, paraphyly, and polyphyly have fallen into fairly standard usage.

Even so, cladograms are mathematical constructs that need not be interpreted as reflections of descent relationships (Nelson and Platnick, 1981). Thus, monophyly (and the associated terms polyphyly and paraphyly) can be viewed from at least two perspectives that are complementary but distinct, these being cladistic relationship and phylogenetic relationship. In the first instance, a set of objects (i.e. terminals or taxa or any other elements of analysis), as depicted on a particular cladogram, can be determined to be monophyletic or non-monophyletic simply by reference to their placement on that cladogram. In this case, to assert that a group is monophyletic is to state that cladistic relationships among the members of that group satisfy a set of criteria, such as those specified by Farris (1974, 1991). Similarly, it can be stated unequivocally that a given set of terminals as represented by rows of data in a real or hypothetical data matrix, rather than in a cladogram, does or does not constitute a monophyletic group, with the criterion being that cladistic analysis (or some other specified form of data analysis), if conducted, would resolve that group as mono-phyletic. If the matrix is changed, the status of the group as monophyletic or non-monophyletic may change as well. In contrast, definitions of monophyly in terms of evolutionary or phylogenetic relationships are statements about history, and thus hypotheses regarding past events. In this context, to assert that a group of taxa con-stitutes a monophyletic group is to express a hypothesis regarding common ancestry or proximity of historical relationship among members of the group (see below). Although a difference can be recognized between cladistic monophyly and phyloge-netic monophyly, empirical studies conventionally use cladistic structure as the basis of hypotheses of evolutionary relationship, and discussions of these two aspects of monophyly thus are often commingled. The principal focus of the present paper is on monophyly as a descent relationship among groups of organisms, and the tree-like diagrams in this paper represent phylogenetic relationships.

While the distinction between monophyly and non-monophyly is well established for higher-level relationships, there has been an ongoing debate on the question of whether monophyly is applicable to relationships among individual organisms or populations. The problem can be expressed quite simply: if it is appropriate to describe groups such as the seed plants or the Compositae as monophyletic, is it also appropriate to speak of Joseph and Rose Kennedy and all of their descendants as a monophyletic group? Although it is possible to pursue this question strictly in terms of abstract descent relationships, without reference to the rules by which taxa are recognised, the debate over monophyly among organisms and populations has, in fact, been entangled with a debate over species concepts. The roots of this linkage are not obscure: Hennig (1966) presented two definitions of monophyly (see below), and in both of them he specified the relationship in terms of relationships among species; moreover, he argued that non-monophyletic groups of species should not be recognised as taxa. Those who have suggested that monophyly is applicable to relationships among organisms and populations (e.g. Donoghue, 1985; de Queiroz and Donoghue, 1988; Baum, 1992) have coupled their arguments to an assertion that all taxa, including species, should be monophyletic. Having asserted that mono-phyly is a necessary condition for the recognition of a species, and having observed that Hennig restricted the application of his monophyly concept to groups of species, these authors have placed themselves in the position of explaining how a mono-phyly concept is applicable to individual species, and how systematists can determine that a putative species, such as a local population of plants with a distinct set of characteristics, is or is not monophyletic. Ultimately, these authors have asserted that an integrated population of organisms, differentiable from all other organisms by any number of morphological or molecular characteristics, and reproductively isolated from all other groups of organisms, may not be a species (and in many cases may not be assignable to any species) if it fails a test of monophyly. Although my goal in the present paper is to discuss monophyly per se, there is, by necessity, discussion of the manner in which this concept impinges upon the related problem of species definitions.

Apart from the question of how species should be recognised in biological systematics (i.e. what criteria should be employed when determining whether a particular group of organisms is called a 'species'), it is useful to recall the more general usage of the word 'species', whereby one may speak of 'species' of minerals, 'species' of currency or Jane Austen's 'species' of revenge. In this general usage, as found in any dictionary of the English language, a species is simply a particular 'kind of thing' that is distinct in some way from other kinds of things (Mayr, 1982). Throughout most of the history of taxonomy, morphological differences usually served as criteria of species status, at least operationally. However, for at least the past hundred years, systematists have spoken of biological species, i.e. species that are groupings of organisms, including but not limited to the 'Biological Species Concepts,' or BSC (Dobzhansky, 1935, 1937a, b; Mayr, 1942, 1965; Grant, 1981; and Avise, 1994) in terms other than morphological diagnosability. Features such as evolutionary divergence, the ability to interbreed, and adaptation to a particular environment have figured in most of the species concepts proposed in the 20th century. Although criteria other than physical similarity are invoked in most modern discussions of biological species, it should be recalled that the species of biology also are 'species' in the general sense, i.e. different kinds of things. As discussed below, reference to this related meaning of 'species' helps to clarify the relationship of species concepts to monophyly concepts.

Another problem that is related to monophyly, and to species concepts, is the remarkably challenging question of how to define a 'population' (e.g. Nelson and Platnick, 1981). Because evolutionary change occurs in a population setting, discussions of evolutionary theory, including descent relationships (and thus monophyly), are innately connected to concepts of real and ideal populations, including such categories as metapopulations (e.g. Barrett and Pannell, 1999 – this volume).

A final matter associated with discussions of descent relationships, including monophyly, is the distinction between actual descent relationships and evidence of these relationships, typically in the form of character distributions among organisms and populations. It is possible to speak of descent relationships in theoretical terms, without reference to evidence, but if one's goal is to improve and extend a scientific research programme, it is important that one also describes the sorts of observations that are appropriate in the analysis of descent relationships, and the manner in which those observations should be interpreted in empirical studies. To an ever increasing extent, the evidence that is applied in empirical studies in systematic biology takes the form of DNA nucleotide sequences; the focus of the present volume on *molecular* systematics emphasises this category of evidence. If molecular systematics is to be a successful field of inquiry, it is crucial that the distinction be maintained between patterns of shared characters (e.g. patterns of co-occurrence of alleles at a locus by various individual organisms and populations) and the descent relationships of the organisms and populations in which those characters occur.

In the following pages I first discuss the paramount importance and utility of Hennig's monophyly concept. I further argue that Hennig was wise to restrict this concept to hierarchic systems, and that attempts to dilute and alter the meaning of monophyly, inspired by the desire to apply it to non-hierarchic systems, have been ill-founded, and have fostered confusion by erroneously suggesting that the monophyly concept itself is flawed. Finally, I briefly discuss the potential for incongruence between gene trees and species trees.

8.2 Hierarchic and reticulate descent systems

Contributors on all sides of the debate regarding the proper species concept for phylogenetic systematics (a debate that is only part of a much more general discussion of appropriate species concepts in biology) acknowledge that species are groups of organisms that lie approximately at the boundary between reticulate and hierarchic descent. In reticulate descent systems (e.g. sexual populations), lines of descent diverge, unite, and diverge again in succession. In hierarchic descent systems, in contrast, lines of descent, once they have diverged, do not meet again (Hennig, 1966). To state that species lie somewhere near the boundary between hierarchic and reticulate descent is to state, in general terms, that descent relationships among species are predominantly or exclusively hierarchic (reticulate species origins, as in polyploid systems (e.g. Vogel *et al.*, 1999 – this volume), are set aside for purposes of this discussion), while descent relationships within species (i.e. among elements within species, such as organisms or populations) are reticulate.

Cladistics is an appropriate means for the analysis of hierarchic descent systems because the lines of descent can be represented by a tree-like diagram (i.e. a cladogram) in which branches diverge sequentially, with no two lines, once they have diverged, ever meeting again. Such a system can be viewed, retrospectively, from the position of any of the end-products of the descent history (i.e. from the position of any of the fundamental elements of descent, the 'species' of the system). One feature that is evident from this vantage point is that the line of ancestry of each species runs from it to the root of the tree, or conversely, every species is connected to the root by just one line of descent (Fig. 8.1a). A second feature of such systems is that the lines of ancestry of any two contemporaneous species (the case of ancestral species, is here set aside) meet at just one point in the tree, which is their single point of most recent common ancestry (Fig. 8.1a). Prior to the time of most recent common ancestry (i.e. between this point and the root of the tree), the ancestry of the two species is identical, and after this time (i.e. between this point and each species) the ancestry of the two species is completely different. Thus, the line of descent from the root of the tree to the point of divergence represents the entire history of common ancestry of two species, and within this region, the point of divergence is the single unique point of most-recent common ancestry of the two species. Just as there is a point of most-recent common ancestry for any two species, there is also a point of most-recent common ancestry for any set of three or more species, this being the first point at which the line of descent of any member of the group diverges from that of any other member.

A third feature of hierarchic descent systems follows directly from the first two. For any three species there are three possible pairwise combinations (i.e. for species A, B, and C, the three pairs are A and B, A and C, and B and C), and there is, of course, a single point of most recent common ancestry for each of these pairs.

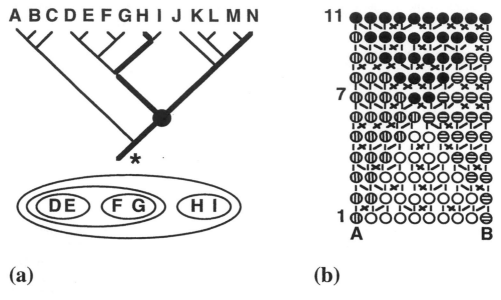

(a)   **(b)**

Figure 8.1 (a) Phylogenetic relationships among 14 terminals, A–N, depicted in a phylogenetic diagram (top), and among six of the same terminals, depicted as a Venn diagram (bottom). Each terminal has a solitary line of ancestry that leads to it from the root (*). In the phylogenetic diagram, lines of ancestry of two terminals, H and N, are represented by thicker lines. The single point of most recent common ancestry of H and N is marked with a solid circle; prior to that point their ancestry is identical. Nesting of monophyletic groups is depicted in the Venn diagram; groups DE and FG are subsets of group DEFG, and group DEFG is a subset of group DEFGHI. (b) Mating history of a population of sexually reproducing individuals through eleven generations, with lines of descent depicted for two individuals (A and B) in generation 1. Individual A is depicted as a circle with vertical bars, individual B is depicted as a circle with horizontal bars. Descendants of each of these reference individuals (groups A and B, respectively) are marked like their respective ancestors, except that individuals that are descendants of both reference individuals are depicted as solid circles. In generation 7, only two individuals are descendants of both reference individuals, but by generation 11 (and in all succeeding generations, not illustrated) all individuals are descendants of both reference individuals.

If the lines of descent of the three species diverge at a single point, the three share that single point of most recent common ancestry. Otherwise, one of the pairs has a point of common ancestry that is more recent than that of either of the other two pairs. As Hennig and many other authors have stated, the two species that share a more recent common ancestor are by definition *more closely related* to each other than either of them is to the third.

A fourth feature of hierarchic descent systems is nesting. As discussed in greater length below, Hennig defined monophyly with explicit reference to hierarchic descent systems, and one of his definitions specified that a monophyletic group consists of all descendants of a particular common ancestor. Nesting is a quality of monophyletic groups, specifically the quality that if any two different monophyletic groups are specified, and if any species are shared by the two, then one of the two groups must be wholly included in the other (Fig. 8.1a). The same idea, phrased differently, is that if any two different monophyletic groups share any species in common, only one of the groups includes species that are not members of the other. If two different monophyletic groups do have some species in common, the less inclusive group is nested within the more inclusive group. Note that this relationship also can be expressed with reference to points of most-recent common ancestry: if two different monophyletic groups are specified, group A is nested within group B if the point of most-recent common ancestry of group B lies along the path between the root of the tree and the point of most-recent common ancestry of group A.

The conditions just described are intrinsic to hierarchic descent systems. Although each of these conditions can apply to some groups in a reticulate descent system, none of the conditions is a general attribute of such systems. First, elements in a reticulate descent system have multiple lines of ancestry. For example, in a population of sexually reproducing organisms, each individual has two parents, each of which also has two parents, and so forth. Thus, each individual has multiple lines of ancestry that branch backward through time, rather than a single line of ancestry that leads back to a solitary root.

Second, the lines of ancestry of two individuals can meet at more than one point. Two individuals may have the same mother (i.e. a point of common ancestry one generation earlier), but they may also have different fathers, and yet have the same paternal grandfather (i.e. a point of common ancestry two generations earlier along a different line of ancestry). Thus, even when it is possible to identify a single most-recent common ancestor of two particular descendants (in this case, the mother of the two individuals), it does not follow that all prior lines of ancestry are identical, for the same two individuals may have different lines of ancestry for any number of generations farther back along other lines.

Third, although three individuals in a reticulate descent system may have different lines of ancestry, it does not follow that two of these individuals are more closely related to each other than either of them is to the third. For example, individuals A and B in a sexual population may have the same mother, while individuals A and C have the same father. Although all three of these individuals have different lines of ancestry, no particular pair of them is more closely related to each other than either of them is to the third.

Fourth, nesting of groups of related individuals is not a general feature of reticulate descent systems, because two different groups of closely related individuals may share some members, yet each of the groups may also include some individuals that are not members of the other. Consider group A, which consists of individual A and all of its descendants (in a population of sexually reproducing organisms), and group B, which consists of individual B and all of its descendants (Fig. 8.1b). Initially, groups A and B are mutually exclusive. However, the offspring of a mating between a member of group A and a member of group B belong to both groups, as do all descendants of those offspring, but each group also may continue to include individuals that are not members of the other. The circumstances of this example, where there is some overlap in membership between groups A and B, but neither is nested within the other, may persist indefinitely.

In summary, each of the individual elements of descent in a reticulate descent system (e.g. the individual organisms in a sexually reproducing population) have multiple lines of ancestry. Consequently, the lines of ancestry of two different elements may meet many times in different ancestors when traced backward. Also, for sets of three elements, all with different lines of ancestry, there may be no two that are more closely related to each other than either is to the third. Finally, although it is possible to identify a group of elements that constitute the complete set of descendants of a particular individual or breeding pair, and although some such groups are subsets of others, there are also groups that overlap only partially in membership, such that neither is a subset of the other.

8.3 Hennig's monophyly concept

As noted above, Hennig (1966) proposed a monophyly concept that was more rigorous than those of earlier authors. In the same pages he discussed biological descent relationships, and indicated that relationships within species are reticulate, while relationships among them – phylogenetic relationships – are hierarchic. Thus, he established a link between species concepts and the overall theory of phylogenetic systematics. He also proposed two definitions of monophyly. Confusion can arise if one considers these two definitions in isolation from his related discussion of species and hierarchic descent; if the linkage between these concepts is acknowledged, however, it becomes clear that the two definitions specify a single monophyly concept. The first of the two definitions (Hennig, 1966: 73) is:

> A monophyletic group is a group of species descended from a single ('stem') species, and which includes all species descended from this stem species.

This definition, which I have called *Hennig's monophyly Definition One* (Davis, 1997), expresses a relationship that was later identified as *common ancestry* (de Queiroz and Donoghue, 1990). What is intrinsic to this definition, however the meaning of 'species' is construed, is the recognition of monophyly in terms of descent relationships among a group of elements, where a monophyletic group is the entire set of descendants of one ancestral element. What is not completely apparent from the definition itself is the context in which it was presented. In the pages that preceded his presentation of this definition, Hennig discussed hierarchic and reticulate descent systems, and indicated that species, under the terms of the species concept that he used, are groups that are related to each other hierarchically. Hence, Hennig indicated that 'species' are elements of a hierarchic descent system, and that a monophyletic group consists of all of the descendants of one such element. Whether or not one accepts Hennig's exact notion of what a species is, his intention is unmistakable; he restricted his monophyly concept to groups of elements that are hierarchically related to each other.

Hennig's second definition (1966: 73), presented on the same page as his first, is:

> A monophyletic group is a group of species in which every species is more closely related to every other species than to any species that is classified outside this group.

In this second definition, which I have called *Hennig's monophyly Definition Two* (Davis, 1997), Hennig described a second attribute of monophyletic groups. He called this quality *kinship*, but it has since been referred to by others as *exclusivity of relationships*, or simply *exclusivity* (de Queiroz and Donoghue, 1990). Borrowing from both sources, I suggest *exclusivity of kinship* as an appropriate descriptive term for this relationship.

In Definition Two, as in Definition One, Hennig again described monophyly as a kind of relationship that exists among species, so again, regardless of Hennig's precise meaning of 'species', his determination that this concept should be restricted in application to hierarchic descent systems is clear. As discussed below, Hennig's two definitions specify a unified monophyly concept, for in hierarchic descent systems the two definitions are congruent in meaning.

8.4 'Species' and monophyletic groups in hierarchic descent systems

As noted above, Hennig's references to *species* in his definitions of monophyly have tended to link discussions of monophyly *per se* to the perennial debate over species concepts in biological systematics. One way to examine this matter is to adopt, temporarily, the general meaning of 'species' as referring to distinct 'kinds of things', with the restriction, however, that it refers to basal elements in hierarchic descent systems. Under this interpretation, each of the various alleles of a non-recombining gene, such as those of the chloroplast-encoded gene *rbc*L, is a 'species' of that gene. Variant forms of the entire chloroplast genome, whether they differ from others by single nucleotide substitutions or by substantial structural mutations, also could be regarded as 'species' of that genome. If each allele of *rbc*L (i.e. each distinct kind of gene copy, differing from other kinds by some number of nucleotide or length differences) is recognised as a 'species,' and if descent relationships among alleles are strictly hierarchic, a new 'species' arises whenever a mutation causes a new allele to appear (assume that there cannot be multiple origins of identical alleles), and the origin of a new allele can be regarded as a 'speciation' event.

Suppose that a mutation occurs, and that the resultant new allele – a new 'species' of *rbc*L – is designated allele A. This allele may replicate indefinitely without change, until any number of copies are in existence, but eventually, as additional mutations occur, a set of new alleles that are descendants of allele A may originate. Each mutation event can be regarded as a 'speciation' event. Because the descent system is hierarchic, the four qualities described above as pertaining to hierarchic systems will apply. Eventually, a hierarchic phylogenetic structure will develop, in which a group of 'species' consisting of allele A plus all alleles descended from it (each of them also a 'species') constitute what may be called a monophyletic group. This group is monophyletic under Hennig's Definition One, for it consists precisely of one 'species' and all of its descendant 'species'. This group also is monophyletic under Hennig's Definition Two, which refers to kinship, because 'species' A is an ancestor of all members of the group and is not an ancestor of any non-member (the question of whether 'species' A should be considered an ancestor of itself will not be considered here, but for the present purposes it need only be accepted that any 'species' is more closely related to itself than it is to any of its ancestors, or to any descendants of its ancestors that are not also descendants of itself). Because 'species' A is an ancestor of all members of the specified group of alleles, the most recent common ancestor of any two members of the group is either 'species' A or some descendant of it. In contrast, 'species' A is not an ancestor of any 'species' that is not in the group, so the most recent common ancestor of any 'species' pair that consists of one member of the group and one non-member must be an ancestor of A, not A or any of its descendants. Hence, the relationship (in terms of relative recency of common ancestry) between any member and any non-member is more distant (the last common ancestor is an ancestor of A) than the relationship between any two members (the last common ancestor is A or one of its descendants).

Two key points that are relevant to a discussion of monophyly are evident in the foregoing example. First, in hierarchic descent systems, where the four conditions described earlier apply, Hennig's two definitions are congruent, for any group that conforms to either of the definitions also conforms to the other. Thus, despite the fact that Hennig framed two definitions of monophyly, the monophyly concept to which they refer is unambiguous and internally consistent. Second, when describing relationships in hierarchic descent systems it is possible to refer to the basal elements as 'species', and to specified groups of these 'species' as either monophyletic or non-monophyletic. Some readers may regard the application of these terms outside the usual realm in which biologists use them as unacceptable, simply on the basis of non-traditional usage, but barring that objection, the existing relationships are sufficiently similar in their attributes to render such usage unambiguous.

If the system described in the previous example is re-examined, but mutations are eliminated from consideration, the possibility of extending the use of 'species' to individual gene copies is evident. Just as in a system of distinct alleles, there exists a hierarchic descent system, and each new element (i.e. each gene copy) arises from the subdivision of a previously existing element. Any number of indistinguishable gene copies can exist, and if each of these copies is called a 'species', each replication event can be called a 'speciation' event. Similarly, the individual cells in a plant, if they have originated strictly by a series of mitotic divisions beginning with the first division of a zygote or spore, can be described in terms of 'species' (the individual cells), 'speciation' events (the mitotic events), and monophyletic groups of cells. The basic model also applies to any system of unicellular organisms in which cell division events occur and cell fusion events do not. It should be noted, however, that in systems of indistinguishable basal elements (e.g. systems of bacterial cells or identical gene copies), it can be argued that the generalised meaning of 'species' does not apply, since there are not different 'kinds of things', just different basal elements with different histories. If there are no character differences to distinguish among subsets of gene copies or cells, and no other record of history (e.g. positions of individual cells in a multicellular organism, coupled with knowledge of the developmental programme), only an omniscient observer can identify the 'species' or monophyletic groups thereof.

If it is accepted that the terms 'species' and 'monophyly' can be applied to systems of hierarchic descent, involving replicating genes and cells, are these terms also applicable above the level of the cell, where the elements of descent are organisms and populations? If there is no sexual reproduction, use of the terms 'species', 'speciation', and 'monophyly' to refer to clonal populations of organisms is not problematic. However, as described above, descent relationships among individual sexually-reproducing organisms in a population are reticulate rather than hierarchic. A system of this sort lacks the qualities that are intrinsic to hierarchic systems, and Hennig's monophyly definitions are inapplicable (see further discussion below). Similarly, a system of local populations loosely connected by occasional episodes of gene flow between them is also a reticulate descent system. Some local populations may have been founded by migration of individual organisms, but others may have been founded by migrants from two or more local populations, and possibly augmented afterward by additional migration events. In a system of this sort, limited and local instances of hierarchic structure may be identifiable, but they are disrupted as later dispersal events occur. Systems of this sort – metapopulations – have come under increased scrutiny by ecologists and evolutionary biologists (e.g. Levins, 1969, 1970; McCauley *et al.*, 1995; Levin, 1995; Husband and Barrett, 1996; Krohne, 1997; Scheiner and Rey-Benayas, 1997; Vrijenhoek, 1997; Barrett and Pannell, 1999 – this volume), for the metapopulation model appears to be realistic, and it may indeed be an accurate model for the understanding of effective population sizes, adaptation, and the origin and subsequent survival of species.

In terms of the distinction between hierarchic and reticulate descent systems, metapopulations represent an infinitely graded series of intermediate states, but it seems appropriate to recognise them, in general, as reticulate systems that harbour temporary and local pockets of hierarchy. In this manner, the relationships among local populations within a metapopulation are similar to those that exist among individual organisms within a local population. Even in a panmictic population there are temporary and local regions of hierarchic descent, for although mating may occur at random, any particular mating event that results in two or more offspring yields a set of full siblings that are more closely related to each other than any of them is to any other individual in the population. If two of these siblings mate with each other, some degree of hierarchy is preserved. Thus, there are vestiges of hierarchy even in an ideal, random-mating population, and in any natural system there is a greater tendency for the development of local hierarchic structure. Natural metapopulations can be expected to exhibit a range of structures, and some systems surely will be found to have attributes that closely approach those of a strictly hierarchic system.

When gene exchange among local populations in a metapopulation system approaches zero, each local population reproduces and evolves in isolation from others. Similarly, when the magnitude of gene exchange among metapopulations approaches zero, the local populations within each metapopulation may continue to be united as a single genetic community, but each metapopulation acts as an independently reproducing and evolving unit. In such cases, descent relationships are hierarchic at some levels of organisation below that of the individual organism (e.g. among cell lines, alleles and gene copies), reticulate at some higher levels of organisation (e.g. among organisms within local populations, and among local populations within metapopulations), and hierarchic at some even higher levels of organisation (e.g. among metapopulations). For this reason, any discussion of hierarchy or reticulation must specify a particular level of focus. If the focus is on descent relationships among metapopulations that have ceased to exchange genes with each other, and that originate only by the fission of previously existing metapopulations, then it is appropriate to recognise the descent system as hierarchic. The disintegration of one metapopulation system into two or more such systems may be a slow and gradual process, with periods of near isolation among systems followed by episodes of gene flow that reunite them. Also, an equilibrium situation might exist for extended periods of time, during which periods of gene flow are followed repeatedly by periods of isolation. However, in some cases at least, a point eventually is reached at which one metapopulation has become fragmented into two or more metapopulations, among which all gene exchange ceases, so that each survives and evolves as a separate genetic community. At this point, the overall system can be recognised as hierarchic, and the 'species' of this system are the separate metapopulations, within each of which there is gene flow among subpopulations, and among which there is no gene exchange. Speciation has occurred.

Before the term 'metapopulation' came into use, and before the rise of modern phylogenetics, with its inherent focus on hierarchic descent, investigators from a wide range of traditions in biology had long considered and analysed systems of this sort. The literature of evolutionary biology, specifically in areas such as biosystematics, pollination biology and population biology, is replete with discussions of local populations as the arenas in which most mating and gene exchange occurs, and of the role of dispersal among such populations of pollen and propagules (and in the case of many animals, the dispersal of adults) as the means by which local populations are linked into larger genetic communities. Also, there is a long tradition of attentiveness to the limits of gene flow (e.g. Ehrlich and Raven, 1969), whether caused by geographic isolation or by intrinsic isolating barriers (Mayr, 1942, 1965; Grant, 1981). This entire field of inquiry is, of course, species biology, and if the metapopulation concept is extended to include the most widely distributed and tenuously connected systems of populations, it differs little if at all from many traditional species concepts. Different investigators will hold contrasting opinions on the precise point at which two different population systems should be recognised as different species, but virtually every modern species concept incorporates the basic idea that gene flow occurs within and among populations of the same species, and is either rare or absent among species.

In light of these considerations, the relationship developed by Hennig between species concepts and phylogenetics is readily understandable. Hennig worked during the period in which biosystematics was at its zenith (Richter and Meier, 1994). Systematic biology was largely focused on problems of gene flow, species boundaries and speciation, and systematists were engaged in the development of explicit, repeatable methods for the analysis of these problems (e.g. studies of interfertility). In the midst of this climate, Hennig developed a theory of relationships above the level of individual species, and he proposed explicit, repeatable methods for the analysis of this sort of system. Thus, he went to great lengths to establish that the upper limit of gene flow – also the upper limit of population and species biology – was the lower limit of phylogenetics, and that the distinction between hierarchic and reticulate descent was the key factor in defining the critical boundary between these complementary areas of evolutionary biology. However, species biology was dominated at the time by the BSC, in which the focus is on the *potential limits* of gene flow, rather than on historic *patterns* of gene flow.

8.5 Phylogenetic structure within 'species', phylogeography and the Phylogenetic Species Concept

Consistent with the distinction between the potential limits of gene flow and historic pattern, and between reticulate and hierarchic descent, Rosen (1978, 1979) noted that fully differentiated population systems, and phylogenetic relationships among them, could be identified within species as recognised under the BSC. If there is recoverable phylogenetic structure within a species, that species is an aggregation of less inclusive minimal phylogenetic elements (Bremer and Wanntorp, 1979). Furthermore, it is possible for such an aggregation to be paraphyletic, in which case it is neither a basal evolutionary element nor is it a set of most-closely-related evolutionary units. Even if a 'species' of this sort is a monophyletic set of basal elements, it still includes two or more discrete evolutionary elements with different histories, so phylogenetic history is lost when several of these elements are treated as only a single element or species. There is no requirement, of course, for the species of biology to be basal evolutionary elements, but biologists who favour the BSC should be prepared to explain why distinct elements of the phylogenetic system, including paraphyletic groups of such elements, should be grouped into solitary 'species', simply because, in the absence of an intrinsic reproductive isolating barrier, it is possible that they *might* at some future time, under some set of circumstances, become merged into a single reproductive community. Many persons would be surprised if astronomers were to declare that two distinct stars should be recognised as one because they had the potential to meet and fuse at some time in the future, whatever the probability might be of this actually occurring, yet proponents of the BSC, on precisely the same basis, declare that two distinct populations, completely out of genetic contact, perhaps occurring on opposite sides of the Earth, belong to the same 'species'. Rosen argued that the minimal elements of the phylogenetic system should be recognised as separate species, and he proposed an initial version of the Phylogenetic Species Concept (PSC) that was based on that reasoning. Other investigators have subsequently discussed and further developed the idea (Cracraft, 1983, 1989; Nelson and Platnick, 1981; Nixon and Wheeler, 1990; Davis and Nixon, 1992), and from these discussions there has emerged a family of related phylogenetic species concepts, united in their recognition of species as basal elements of hierarchic descent systems.

The present paper is not about the PSC *per se*, but several salient points should be mentioned with regard to the present focus on monophyly. First, because a phylogenetic species is a basal element of phylogenetics, there is no phylogenetic (i.e. hierarchic) structure within it. Indeed, the potential for phylogenetic structure within biological species was the principal criticism initially levelled at the BSC by Rosen (1978, 1979) and others (e.g. Bremer and Wanntorp, 1979). It should be noted, however, that although a phylogenetic species does not have internal phylogenetic structure, it need not be a monomorphic unit, for it can include any number of local populations, and any amount of genetic polymorphism can exist within it (Davis and Manos, 1991; Davis and Nixon, 1992). Descent relationships among sexually reproducing organisms are reticulate, and as noted above, even a reticulate system is expected to exhibit temporary and local regions of hierarchy. Thus, investigators who share an interest in phylogenetic species may hold different opinions regarding the point at which relationships properly should be recognised as reticulate rather than hierarchic. Despite this potential for differences in implementation, however, the general aim of the PSC is to recognise the boundary between reticulate and hierarchic descent, and in empirical studies, to delimit species such that there is reticulate descent within them, and hierarchic descent among them.

Because phylogenetic species are the minimal *elements* of phylogenetic systems, the minimal phylogenetic *event* is the separation of one previously existing (i.e. ancestral) phylogenetic species into two or more daughter species. This event, which represents the division of one phylogenetic element into two or more, is both cladogenesis and speciation, and as such it highlights the critical position occupied by phylogenetic species at the interface between phylogenetics and population biology (including population genetics).

As for monophyly, it is consistent with the recognition of phylogenetic species as the minimal elements of phylogenetic systems that an individual phylogenetic species is *not* monophyletic. This is a critical point, for if a species is monophyletic there must be phylogenetic structure within it, reflecting a history of hierarchic descent among some kind of lesser elements. Indeed, if a putative phylogenetic species is discovered to have an internal phylogenetic structure, it should, by definition, be recognised as a set of two or more species. With this in mind, the development of the field known as 'infraspecific phylogeography', i.e. the analysis of phylogenetic structure within 'species', typically with the aid of molecular characters (Avise *et al.*, 1987), can be seen either as an over-reaching of evidence (i.e. misinterpretation of hierarchic structure in a gene tree as evidence of phylogenetic structure among the populations from which the genes were sampled), or as evidence of actual phylogenetic structure within conventionally recognised species. In the latter case, adoption of the PSC would necessitate re-evaluation of the status of any 'species' with internal phylogenetic structure, and a change in its status from a species to a group of species. It is not coincidental that some aspects of phylogeography so closely resemble Rosen's observations, for he was similarly impressed with the potential for phylogenetic structure within conventionally recognised species. Rosen's response, and that of others who have since adopted the PSC, was to argue that any species of this sort had been delimited too broadly, to reject the BSC because it allowed for the recognition of intraspecific phylogenetic structure, and to suggest that systematics and other branches of biology are better served by a species concept in which the minimal phylogenetic elements are recognised as 'species'. The implicit response of proponents of infraspecific phylogeography, many of whom continue to work within the framework of the BSC, has been to continue to accept the BSC, and to acknowledge the potential for recoverable phylogenetic structure within 'species', some of them demonstrably paraphyletic.

8.6 Monophyletic species concepts with populations as phylogenetic elements

Although increasing numbers of systematists have rejected the BSC, not all of them have regarded the PSC as a preferred alternative. Two additional responses to the potential for phylogenetic structure within conventionally recognised 'species' are particularly significant in the present context because they represent attempts to apply the monophyly concept, or something like it, at the level of individual species. The first response was to propose that all species should be monophyletic. Following the lead of Rosen (1978, 1979), Cracraft (1983) and others who had developed the PSC, Donoghue (1985), Mishler (1985) and de Queiroz and Donoghue (1988) agreed that paraphyletic species (as allowed under the BSC) are unacceptable, but contrary to the approach taken by proponents of the PSC, the latter authors argued that individual species should be monophyletic. This assertion immediately raised the question of whether the monophyly concept is applicable at this level, and if so, how it should be implemented. Under the terms of the Monophyletic Species Concept (MSC) developed by these authors (actually a family of related species concepts, just as there is a family of phylogenetic species concepts), there is either an implicit or an explicit assertion that some kind of unit is the minimal appropriate element of phylogenetics. The MSC differs from the PSC, however, in declaring that this unit is something other than a species, for if a species is to be a monophyletic group, it must be composed of two or more elements, each of which is less inclusive than a species. It follows from the insistence that every species be monophyletic, that phylogenetic analysis (of some sort) is deemed appropriate for some kind of element that is less inclusive than a species, for it is only through the analysis of relationships among such elements that a monophyletic group could be resolved and recognised as a species. Thus, when the implications of each different version of the MSC are considered, it is important to examine the kinds of infraspecific units that are considered the appropriate basal elements of phylogenetics.

One difficulty that immediately arises with any MSC is that, whatever sorts of units are recognised as basal phylogenetic elements, some of these units may not fall within a minimal monophyletic group (Fig. 8.2a). Because the MSC is based on the assertion that the basal elements of phylogenetics should not themselves be recognised as species, and because some of those elements (perhaps most of them) cannot be assigned to minimal monophyletic groups, they cannot be assigned to species. Thus, adoption of the MSC creates the need for ad hoc rules for the accommodation of basal elements that cannot be assigned to minimal monophyletic groups. Proponents of the MSC have repeatedly emphasised the importance of a 'metaspecies' convention for this purpose (e.g. Donoghue, 1985; de Queiroz and Donoghue, 1988; Baum and Shaw, 1995). Under this convention, organisms that do not fall within minimal monophyletic groups are placed into groups known as metaspecies, and although the organisms of a metaspecies belong to more inclusive monophyletic groups (e.g. genera and families), they do not belong to any species. Thus, a genus might include 30 species and 50 metaspecies. In terms of the general meaning of 'species', a metaspecies can be a viable population that is distinct from all other species and metaspecies, but it is not a 'kind of thing'.

As for the operational mechanics of the MSC, application of any version of the MSC requires that a practitioner conduct a phylogenetic analysis, using infraspecific elements of some sort as terminals, in order to resolve monophyletic groups of these elements, and circumscribe these groups as species. In contrast, application of the PSC involves species delimitation prior to phylogenetic analysis. Despite this fundamental difference – phylogenetic analysis prior to species delimitation versus species

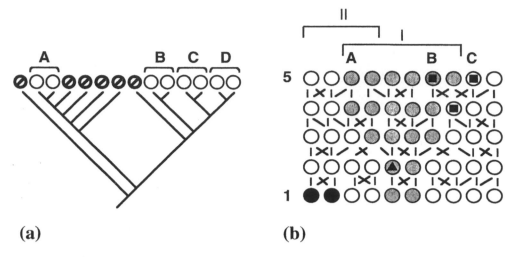

Figure 8.2 (a) Hypothetical phylogenetic structure for 14 terminals (circles), with four minimal monophyletic groups demarcated (groups A–D, each consisting of two terminals, depicted as open circles). Six other terminals, marked with slashes, do not fall within minimal monophyletic groups; if species are defined as minimal monophyletic groups, these six terminals cannot be assigned to species. Also, if species are defined as minimal monophyletic groups, the set of terminals within each such group (e.g. the two in group B) constitutes a single species, even if they are genetically distinct and reproductively isolated (see text). (b) Mating history for a population of sexually reproducing individuals through five generations. One mating pair in generation 1, and all descendants of this pair (i.e. group I, a 'monophyletic' group of organisms under the Donoghue/Baum monophyly concept) is depicted as shaded circles, and its membership in generation five is signified with bracket I. The ancestral mating pair of another 'monophyletic' group (group II) is depicted as filled circles, and the extent of this group in generation 5 is signified with bracket II. Group I includes individuals A and B, but not individual C. The last common ancestor of individuals A and B is marked with a filled triangle. Individuals B and C are marked with filled squares, as is their last common ancestor, an individual in generation 4. Individual B, which is a member of group I, is less closely related to individual A, which is also a member of group I, than to individual C, which is not a member of this group.

delimitation prior to phylogenetic analysis – it is conceivable that, under some interpretations, the distinction between these approaches could be at least partly semantic. This potential is evident when it is recognised that a version of the MSC could be proposed in which the elements regarded as phylogenetic species by proponents of the PSC were recognised (presumably under some other name) as minimal phylogenetic elements, and in which minimal monophyletic groups of such elements were recognised as 'species'. In that case, every 'species' under the terms of the MSC would be a monophyletic group of two or more 'species' under the terms of the PSC. Indeed, this potential similarity between the PSC and the MSC is evident in the statement by de Queiroz and Donoghue (1988) that populations are the appropriate minimal elements of phylogenetics, and that species, each consisting of two or more such populations, are the minimal monophyletic groups. As noted by those authors (de Queiroz and Donoghue, 1988, pp. 325–326),

Populations themselves, in contrast with their component organisms, may show a branching pattern of relationship to one another. Indeed, using populations as terminal taxa will potentially yield the finest possible resolution of phylogenetic relationships among sexually reproducing organisms. Populations, therefore, have a special role as 'basal units' in the phylogenetic systematics of organisms.

This focus on populations, as recommended by de Queiroz and Donoghue, requires clarification, specifically regarding the precise meaning of 'population', and the manner in which those who conduct empirical studies should identify such populations. For instance, it is not clear whether de Queiroz and Donoghue would argue that it is appropriate to use different local populations with no discerned character differences among them as separate terminals in phylogenetic analyses. These questions aside, though, there are similarities between the approach they favour, and those favoured by proponents of the PSC, and this similarity is evident in the focus on populations as arenas of gene exchange, on the potential for phylogenetic structure among populations, and on the recognition that descent relationships among organisms within a population are reticulate rather than hierarchic. Systematists working under the terms of this version of the MSC could potentially conduct the same sorts of studies that have been recommended by proponents of the PSC, these being the initial delimitation of 'populations' (perhaps corresponding to the species recognised under the PSC, as noted above) by means other than phylogenetic analysis, followed by phylogenetic analysis of relationships among these 'populations'.

Despite the similarities that exist between the PSC and the population-based MSC proposed by de Queiroz and Donoghue (1988), several particular problems in the MSC are confronted by anyone seeking to adopt that approach. Although the population-based MSC treats 'populations' (of whatever sort) as minimal phylogenetic elements, it also requires that every taxon be a monophyletic group of two or more such elements; the obvious implication of these two conditions is that a species can be recognised only if it comprises two or more populations (Fig. 8.2a). This situation could be avoided if a convention were adopted under which solitary populations, or at least some of them, were allowed to be recognised as species, but to adopt such a convention would be to contradict the basic precepts that underlie the population-based MSC, for if populations are regarded as the minimal elements of phylogenetics, but a unique population can be recognized as a species, then monophyly is not the criterion of species status.

The denial of species status for solitary populations is a facet of the 'metaspecies' problem, for both of these difficulties derive from the axiom that all species must be monophyletic. If a minimal phylogenetic element is not regarded as being monophyletic, it cannot be recognised as a species under the MSC. Under the monophyly criterion, a single local population that consists of organisms that are morphologically distinct from all other known organisms can never be recognised as a species. Conversely, if two such populations were found to be sister groups, they would have to be recognised as members of the same species, regardless of the number of characters by which they differed from each other, for each by itself would not be monophyletic, though together they would constitute a minimal monophyletic group (Fig. 8.2a). This example illustrates a key element of all monophyletic species concepts, which is that differentiation of groups is by itself immaterial to species status; all versions of the MSC, by virtue of their insistence that species be monophyletic, force any two closest relatives to be conspecific, regardless of how distinct those relatives are from each other.

Although fusion of completely distinct sister populations into a single species may seem undesirable to many practising systematists, it is precisely what de Queiroz and Donoghue (1988) proposed. In considering the possibility that multiple criteria of species status might be employed, these authors explicitly rejected 'disjunctive' species concepts, wherein two or more species-defining criteria would be applied, with a species being recognised if any one of the specified conditions were satisfied. Thus, if one selects the monophyly criterion, it is the imperative of this version of the MSC that observations of fixed genetic differences among sister populations, including those that confer reproductive isolation, be disregarded in the delimitation of species (Fig. 8.2a).

On the question of species monophyly *per se*, another point also should be noted. Donoghue (1985), following Rosen (1978, 1979), Bremer and Wanntorp (1979), and others, criticised the BSC on the grounds that it allows for the possibility of phylogenetic structure within species, and that in so doing it hinders the analysis of phylogenetic relationships by failing to subdivide the world's biota as finely as possible. However, the population-based MSC shares precisely this property with the BSC, for it specifically *demands* that there be phylogenetic structure within every species. The PSC, in contrast, rejects monophyly as a criterion of species status, and allows for the recognition of a species that consists of a single population.

As indicated by the foregoing discussion, it is possible to formulate a monophyletic species concept that resembles the PSC in its recognition of populations as the critical basal elements of phylogenetics (witness the previously quoted passage from de Queiroz and Donoghue, 1988). Whether or not descent relationships among 'populations' are hierarchic or reticulate depends on the precise meaning that one attaches to 'population', but in its acknowledgement that populations occupy a boundary region between these two kinds of descent, the population-based MSC, as described by de Queiroz and Donoghue, has an appropriate focus on populations. Thus, in spite of the difficulties just discussed, which arise from the demand for monophyly at the species level, it is possible to frame a monophyletic species concept that is compatible with Hennig's monophyly concept. Indeed, this must be so, for all that is required for that to be accomplished is that Hennig's monophyly concept be adopted (with an appropriate change in terminology, so that his 'species' become minimal phylogenetic elements), and that minimal monophyletic groups be called species. However, there are other versions of the MSC that are incompatible with Hennig's monophyly concept.

8.7 Monophyletic species concepts with organisms as phylogenetic elements

In Donoghue's (1985) initial formulation of the MSC he recognised neither species nor populations as the minimal elements of phylogenetics. Rather, Donoghue proposed the use of individual organisms as the basal elements in phylogenetic

analysis. In his words (Donoghue, 1985: 177), 'If one wishes to resolve phylogenetic relationships as far as possible – to find the smallest monophyletic groups of
organisms – then it seems reasonable that individual organisms should be used as
terminal taxa in analyzing relationships.' On the same page he continued:
'Organisms should be placed into more and less inclusive monophyletic groups
using shared derived characters as evidence, just as species, genera, or families are
united on this basis.' And finally, 'Although Hennig (1966) and others . . . restrict
the use of monophyly to groups of species, I can see no reason why the concept
cannot be and should not be applied more generally to any group at any level that
contains all and only the descendants of a common ancestor (de Queiroz *et al.*,
MS in prep.).'

It should be noted, particularly with reference to the last of the quoted passages,
that Donoghue's implied definition of monophyly is simply common ancestry
without exclusivity of kinship, as represented by Hennig's monophyly Definition
One but stripped of any reference to hierarchic descent. This monophyly concept
was endorsed more recently by Baum (1992: 1), who further specified what he
regarded as two acceptable meanings for common ancestor: 'here I use the term
"monophyly" to refer to all the descendants of a single organism or breeding pair,
rather than the narrower definition preferred by some.' The 'narrower definition'
to which Baum alludes is, of course, that of Hennig; it is narrower because it specifies both common ancestry and exclusivity of kinship.

The Donoghue/Baum redefinition of monophyly abandons two important conditions. First, these authors drop the restriction of monophyly to hierarchic descent
systems. Second, they set aside Hennig's Definition Two, and accept Hennig's
Definition One (as applied in any system, hierarchic or not) as a sufficient definition of monophyly. The importance of Hennig's Definition Two is complex, for in
hierarchic descent systems Definition One *is*, in fact, a sufficient definition of monophyly. In such systems, as described above, a group that consists of an ancestral
element and all of its descendants is also a group of most-closely-related elements.
However, Hennig's twin monophyly definitions become disengaged in non-hierarchic descent systems, and the decision by Donoghue and Baum to abandon the
restriction to hierarchic systems *and* Definition Two, thereby focusing exclusively
on common ancestry, represents a severe dilution of the concept, as is evident
in the following examples. First, consider a population of sexually reproducing
organisms. Using the Donoghue/Baum monophyly definition, many different 'monophyletic' groups, with varying degrees of overlapping membership, can be identified
within the population (Fig. 8.2b; see also Davis, 1997). Thus, nesting is not a characteristic of 'monophyletic' groups under the Donoghue/Baum definition. Indeed,
as discussed earlier in the present paper (see Fig. 8.1b), all four of the characteristic features of hierarchic descent systems are lacking.

A 'monophyletic group' of individual organisms, as defined by Donoghue and
Baum in terms of common ancestry alone, need not consist of a group of closest
relatives (Fig. 8.2b). This disengagement of common ancestry from exclusivity of
kinship is a direct result of the fact that individuals in reticulate descent systems
have multiple lines of ancestry that diverge backward through time. Because this
is so, an individual may be a member of a specified 'monophyletic' group (e.g. individual B in Fig. 8.2b), but be more closely related to contemporaneous relatives

that are not members of the group than to other contemporaneous relatives that are members of the group (e.g. individuals C and A, respectively, in Fig. 8.2b). The Donoghue/Baum 'monophyly' definition specifies monophyly in terms of ancestry backward through one particular path, to a particular organism or mating pair, and ignores all others.

A notable consequence of the arbitrary definition of 'monophyly' in terms of a specific line of ancestry is that a group can maintain its 'monophyletic' status even as genes flow uninterruptedly into the group (Fig. 8.3a). A 'monophyletic' group, once specified, can absorb genes from immigrants, and still be 'monophyletic', because the lines of ancestry of those immigrants are ignored. Indeed, the genetic makeup of a population can be altered to any extent by the introduction of genes from other lineages, while the group retains its status as 'monophyletic'. Under this usage, the term 'monophyly', which has always specified the unity of a lineage, comes to be applied to a group that has absorbed genes from other lineages that existed as separate groups at the time that the 'monophyletic' group was initiated.

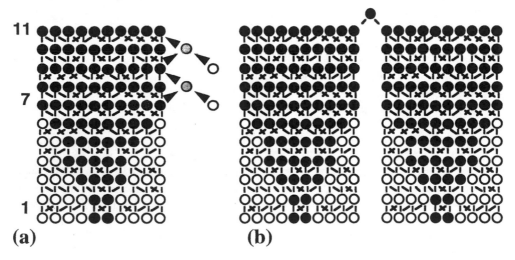

(a)   (b)

Figure 8.3 (a) Mating history for a population of sexually reproducing individuals through eleven generations, with two individuals in population 1, and all of their descendants (a 'monophyletic' group under the Donoghue/Baum monophyly concept), depicted as filled circles; all individuals in generation 7 and in all later generations are descendants of these two individuals. In generation 7 one individual of the population mates with an immigrant from another population, and the solitary offspring of this mating later mates with another individual of the population. A similar episode of immigration and mating occurs in generation 9. The offspring of these two mating events involving immigrants are members of the 'monophyletic' group, as are the offspring of any mating event in which either of the parents is a member of the 'monophyletic' group. (b) Two 'monophyletic' populations (under the Donoghue/Baum monophyly concept) and a single organism derived from a cross between one member of one population and one member of the other. The single offspring of this specified cross belongs to both 'monophyletic' groups; if it is assigned to one of the groups, that group can be recognised as a 'monophyletic' species, but the remaining members of the other group cannot.

The Donoghue/Baum 'monophyly' concept therefore is arbitrary, and Donoghue's species concept, which is derived from it, is similarly arbitrary. For example, consider two groups that are both 'monophyletic' under this definition, and a mating event that takes place between members of the two groups, and yields one or more offspring (Fig. 8.3b). These offspring are members of both 'monophyletic' groups, but the two groups together do not constitute a single 'monophyletic' group. Because the offspring of the specified cross trace their ancestry to both 'monophyletic' groups, either of the groups circumscribed so as to include the offspring of the mating between them can be recognised as a minimal 'monophyletic' group, i.e. a 'species'. However, if one of the groups is recognised as such, the other group cannot be, unless the offspring of the intergroup cross are simultaneously assigned to both groups. Because a species, under Donoghue's organism-based MSC, is a 'monophyletic' group of organisms, either of these two groups can be recognised as a species, but not both, and the decision as to which one should be recognised is arbitrary. As soon as one is designated a species, the other becomes a metaspecies.

As represented in Fig. 8.3b, this situation might appear to be a case of two fairly distinct species and one problematic hybrid individual, in which case it might appear to be the troublesome sort of situation that is encountered in the practical application of almost any species concept, whenever gene exchange between two groups occurs rarely. However, in a natural system of local populations, or even within one undifferentiated population, there are virtually infinite numbers of groups of organisms that might be recognised as 'monophyletic', on the basis of descent from arbitrarily chosen breeding pairs that lived any number of generations earlier. Innumerable overlapping 'monophyletic' sets of individuals exist at any time, and under Donoghue's definition any number of these might conditionally be recognised as species only if others are not (unless, again, an individual can belong to two or more species).

An additional complication arises with respect to character change. Donoghue (1985: 179) originally framed his species concept in a manner that required character change: 'Speciation would be the process of origination of a separate lineage characterised by a new trait.' More recently, however, with co-author Baum, he has argued that species concepts should not be based on character evolution, and specifically that any adherence to what they called 'character-based' species concepts, which link speciation to character change, 'is fundamentally a decision not to engage in the scientific enterprise' (Baum and Donoghue 1995: 571). Instead, Baum and Donoghue argued, scientists should use 'history-based' species concepts that are not contingent on character change. Systematists should consider the implications of this pronouncement, in light of Donoghue's (1985) recommendation that species should be defined as 'monophyletic' groups of individual organisms. Donoghue's organism-based MSC, stripped of any requirement for character change during the speciation process, raises serious theoretical and empirical difficulties, for it would seem to require the recognition of arbitrary 'monophyletic' groupings without reference to character information. Further questions also arise from the requirement by his organism-based MSC for a metaspecies convention, or for some other method to account for organisms that are unassignable to the 'monophyletic' groups of organisms that are to be called 'species'.

Donoghue's (1985) redefinition of monophyly might have been little more than a footnote in the literature of systematics, in light of his subsequent repudiation of this stance, with co-author de Queiroz (discussed above), in the assertion that individual organisms are inappropriate basal elements in phylogenetics (de Queiroz and Donoghue, 1988). However, Baum's revival of the idea in 1992 initiated a new line of argumentation, ultimately leading to a species concept based on gene coalescence (see below), and like the MSC this is important in the context of the present discussion of monophyly. As noted above, Baum (1992) followed Donoghue (1985) in rejecting Hennig's restriction of monophyly to hierarchic systems, and instead endorsed a monophyly concept in which the set of descendants of any organism or mating pair is recognised as a monophyletic group (i.e. a diluted version of Hennig's Definition One). He also noted that this modified monophyly concept lacks rigor, but rather than reject it, and return to Hennig's original concept, he instead invoked Hennig's Definition Two (exclusivity of kinship), which he attributed to Donoghue (1985) and de Queiroz and Donoghue (1990) as an 'important new concept' (Baum, 1992: 1). Baum then defined a species as 'the smallest *exclusive* monophyletic group' (Baum, 1992, his emphasis). In other words, he followed Donoghue in adopting a non-rigorous redefinition of monophyly, based on a diluted version of part of Hennig's original concept, and then discovered that a more rigorous concept emerges when the two original components of Hennig's monophyly concept are reunited.

Baum's reinvention of monophyly under the name 'exclusive monophyly' differs from Hennig's concept only in its intended application in non-hierarchic systems – that is, to groups of individual organisms. In this, as in his limitation of 'monophyly' to only part of Hennig's definition, he followed Donoghue (1985). However, Baum and Shaw (1995) soon discovered that Baum's (1992) organism-based MSC has the drawback of identifying small groups of closely related individuals within panmictic populations as species. A set of full siblings is, indeed, an 'exclusive monophyletic group', for the organisms in such a set are simultaneously the complete set of descendants of a mating pair (i.e. they are 'monophyletic' under the Donoghue/Baum definition) and more closely related to each other than any of them is to any other organism (i.e. they are united by exclusivity of kinship). Hence, Baum's organism-based MSC identifies trivially small sets of organisms as 'species'.

8.8 Genealogical species

Baum and Shaw (1995) observed that Baum's (1992) organism-based MSC has the undesirable quality of identifying sets of siblings within panmictic populations as 'species', and proposed the genealogical species concept (GSC). The GSC, like the various organism-based versions of its predecessor, the MSC, embodies a relationship that resembles monophyly in some ways and that is intended to be applied at the level of individual organisms. Unlike the MSC, which is based on common ancestry (and thus related to Hennig's monophly Definition One), the GSC is based on a variant of Hennig's monophyly Definition Two, exclusivity of kinship. However, the GSC also incorporates a new version of exclusivity of kinship. Baum and Shaw (1995) abandon proximity of descent relationships among *organisms* – the relationship that had allowed sets of siblings in panmictic populations to be recognised as 'species' under Baum's organism-based MSC – and instead base the GSC on proximity of relationship among *characters* (i.e. gene copies) carried by a group of organisms. Thus, under their definition, an 'exclusive' group of *organisms* is not a group of organisms that are more closely related to each other than any of them is to an organism that is not a member of the group. Rather, it is a group of organisms in which every copy of every gene carried by a member of the group is more closely related to every copy of every gene found in another organism in the group than to any copy of any gene found in any organism that is not in the group. For the purposes of the GSC, Baum and Shaw also adopted the definition of 'gene' that is used in coalescence analysis (e.g. Hudson, 1990), under which a gene is a non-recombining portion of the genome. For instance, if it is assumed that recombination does not occur among copies of the chloroplast genome, this entire genome can be regarded as a coalescent 'gene'. Within the nuclear genome, one exon or portion of an exon of an enzyme-encoding gene might be recognised as a 'gene' and elsewhere a single base-pair might be a 'gene'.

When genes are defined as non-recombining genomic segments, descent relationships among all copies of any particular 'gene' are hierarchic, by definition. This being so, Hennig's monophyly concept, without modification, is eminently suited to the purpose of identifying monophyletic sets of gene copies, and this would seem to be the reason that Baum and Shaw (1995) have adopted Hennig's monophyly concept in unmodified form. Because of the internal consistency of Hennig's monophyly concept, a monophyletic set of gene copies is an exclusive set of gene copies. Thus, following a series of redefinitions of 'monophyly', Baum and Shaw have, in part, adopted Hennig's monophyly concept. Unfortunately, their newly defined version of 'exclusivity' introduces another set of problems.

As described elsewhere in greater detail (Davis, 1997), adoption of the GSC represents a dramatic shift in focus from the descent history of organisms or populations in which characters are found to the characters themselves, for Baum and Shaw define 'exclusivity' of a group of organisms as a condition that pertains when all gene copies found in that group correspond to a specified criterion (coalescence), regardless of the descent history of the organisms themselves. It is widely understood by systematists that lineage sorting events can influence the distribution of gene copies among fully-differentiated population systems or species, so that an accurately reconstructed gene tree based on gene copies sampled from representatives of the various populations or species is incongruent with actual phylogenetic relationships among the groups of organisms in which the various alleles occur (e.g. Tateno *et al.*, 1982; Neigel and Avise, 1986; Pamilo and Nei, 1988; Harrison, 1991; Doyle, 1992, 1995; Maddison, 1995, 1997; Doyle and Davis, 1998). Under these conditions, there is 'gene tree/species tree incongruence'. Indeed, gene tree/species tree incongruence is more than an abstract possibility, for several cases have been described in which relationships among groups of species (using species concepts other than the GSC) are inconsistent with relationships among sets of alleles on a gene tree that has been constructed from gene copies sampled from those species (Figueroa *et al.*, 1988; Lawlor *et al.*, 1988; Mayer *et al.*, 1988; McConnell *et al.*, 1988; Ioerger *et al.*, 1990; Clark and Kao, 1991; Gaut and Clegg, 1993; Ayala *et al.*, 1994; Richman *et al.*, 1996; Comes and Abbott, 1999 – this volume).

Under Baum and Shaw's redefinition of 'exclusivity', however, where the exclusivity of kinship of a group of organisms is defined strictly in terms of the gene trees themselves, gene tree/species tree incongruence implies 'non-exclusivity' of the species involved (Davis, 1997, Fig. 4 and associated text). It should be recalled, however, that exclusivity of kinship, as originally described in Hennig's monophyly Definition Two, is a characteristic of all monophyletic groups. Thus, any group that is monophyletic also has the quality of exclusivity of kinship, under standard usage. However, under Baum and Shaw's redefinition of 'exclusivity', a group that is demonstrably monophyletic (again, under Hennig's monophyly concept) may not be 'exclusive' (under Baum and Shaw's definition). This occurs because a group of two or more phylogenetic species is monophyletic, and thus conforms to both of Hennig's definitions, but if there is any gene tree/species incongruence involving any species in that group and any species not in that group, the group is not 'exclusive' under Baum and Shaw's redefinition, and thus the multispecies monophyletic group is not inclusive enough to be recognized as a genealogical species. Once again, Hennig's monophyly concept has been discarded in attempts to discover a way to define monophyly, or something like it, in such a way as to be applicable to groups of organisms. In this case, the concept is discarded in an attempt to force all incongruence among gene trees to be intraspecific, with the stated goal of allowing relationships among the gene copies found in organisms to supersede relationships among the organisms themselves.

By basing their species definition on a definition of 'exclusivity' that derives from gene trees, rather than from the history of diversification of organisms or populations, Baum and Shaw (1995) operationally expand the limits of what they call a 'species' until all incongruence among genes is encompassed within 'species'. In so doing they succeed in defining gene tree/species tree incongruence out of existence, for any observed incongruence of this sort forces the amalgamation of previously recognised 'species' into a single, more inclusive 'species'. The overall effect of Baum and Shaw's redefinition of 'exclusivity', and their description of a 'species' concept based on it, is to expand the limits of 'species' to the point that a discoverable phylogenetic structure, with any number of recognisable monophyletic groups within it, is encompassed within a group that is designated a single 'species'. As described elsewhere (Davis, 1997), application of Baum and Shaw's GSC would result in the incorporation of much of the Solanaceae into a single 'species', and in the aggregation of humans, chimps and gorillas into a single 'species'. With this in mind, it should be recalled that the potential for intraspecific phylogenetic structure was one of the principal criticisms originally levelled at the BSC by phylogenetic systematists; the various arguments of Rosen (1978, 1979), Bremer and Wanntorp (1979), and Donoghue (1985), concerning the flaws of species concepts that allow for intraspecific phylogenetic structure, are also potent criticisms of the GSC.

It should also be noted that in addition to the abandonment of descent relationships among organisms as a criterion of species status in favour of relationships among the genes they carry, the GSC, like the MSC in its various manifestations, does not escape the need for a metaspecies convention, for under its terms there will be sets of organisms, populations and even species (under any conventional definition) that do not fall within any genealogical species. Baum and Shaw (1995: 301) acknowledged this, and noted that under their definition 'species have inherently fuzzy boundaries.' They argued, however (same page), that 'this fuzziness is not an attribute of the concept itself, . . . but arises in the application of the concept to real organisms.'

A final point regarding the GSC deserves mention. Baum and Donoghue (1995) described a distinction between what they called 'history-based' and 'character-based' species concepts. In order to illustrate this distinction they evaluated one species concept from each category. It was their opinion that the PSC belongs in the category of 'character-based' species concepts, and they condemned the entire category as unscientific. In contrast, they assigned the GSC to the category of 'history-based' concepts, and argued that systematists should prefer this class of species concepts. I have addressed this matter elsewhere (Davis, 1997), but it is appropriate to remember, with Baum and Donoghue's distinction between 'character-based' and 'history-based' species concepts in mind, that the GSC forces recoverable phylogenetic structure into the confines of a single 'species' to satisfy the goal of forcing all incongruence among gene trees to be intraspecific, regardless of the history of divergence of the population systems within which the genes occur. The placement of humans, chimps and gorillas into a common 'species' in order to avoid gene tree/species tree incongruence is difficult to explain except in terms of a decision to favour character patterns over history, and it therefore seems appropriate to recognize the GSC as a 'character-based' species concept. Operationally, the GSC defines 'species' in terms of a character pattern – gene tree consensus – rather than on the basis of history.

8.9 Monophyly and molecular systematics

Molecular characters are employed by systematists in a variety of ways, but the principal pattern is for gene trees to be used to infer species trees. Even in analyses of higher level relationships, where only one representative species may be sampled from each of a large number of families, relationships among the exemplar species are inferred from the gene trees that are constructed, and conclusions regarding relationships among the families represent extrapolation from the inferred species tree. How, then, does the matter of hierarchy, and with it the matter of monophyly, relate to the gene tree/species tree problem? This can be examined by looking at descent relationships among gene copies and the manner in which they are distributed among the carriers of those gene copies – organisms – at a series of levels of organisation, beginning with local populations.

Within a local population, non-recombining molecular characters behave as classical Mendelian characters. If the potential for parallel origins of alleles is set aside, the shared occurrence of a particular allele by two individuals indicates that both are descendants of an ancestor that also had that allele. The veracity of this basic assertion seems to have been the basis for Donoghue's (1985) suggestion that systematists should use individual organisms as terminals in phylogenetic analyses, with the goal of detecting 'monophyletic' groups of organisms. This line of reasoning breaks down, however, when the converse relationship is considered; although presence of an allele implies common ancestry, absence of that allele does not imply a more distant descent relationship. An organism that lacks a particular allele may be closely related to individuals that have it, and if the specified allele is at a biparentally inherited nuclear locus, an individual that lacks it may be parent, sibling, or offspring of one that has it (Mendel, 1866). If the specified allele is at a uniparentally inherited locus there are fewer possibilities. A particularly important case of uniparental inheritance is the plastid genome, which is usually maternally inherited in angiosperms. In this case complete and half-siblings that share the same seed parent will carry the same allele, but an individual still may not share the same allele as its pollen parent. Thus, the relationship between common history and patterns of shared characters in populations, i.e. 'traits', *sensu* Nixon and Wheeler (1990) is tenuous. This by itself would tend to negate the presumption that phylogenetic analysis should be conducted in such systems. In light of the general inapplicability of the monophyly concept within populations, as discussed above, it should be clear that phylogenetic analysis of relationships among organisms within populations is unjustified.

Note that the preceding example refers only to patterns of shared alleles, not to gene tree structure. As weak as the association is between shared traits and shared organismal histories within sexual populations, it is weaker still between descent relationships among alleles on a gene tree, and descent relationships among the organisms in which those alleles occur. Once again, there is the general problem that phylogenetic structure among organisms is not to be expected, nor is the monophyly concept meaningful at this level. However, there is yet another difficulty with phylogenetic inference within a population on the basis of gene tree structure, and that is a result of the fact that plesiomorphic and apomorphic alleles are polymorphisms in the population (e.g. Doyle, 1995; Davis, 1997). Gene tree structure can be inferred from samples collected within a single population only if two or more alleles co-occur as polymorphisms within that population. However, if they do co-occur, an individual with an apomorphic allele can have descendants that have a plesiomorphic allele, so descent relationships among alleles on the gene tree are not reflective of descent relationships among the organisms.

What, then, of relationships among populations? Here the conditions begin to resemble those described by Hennig (1966), later termed 'Hennig's model' (Nixon and Wheeler, 1992; Davis and Nixon, 1992), under which synapomorphies are indicative of shared history. One condition is that there be a phylogenetic structure (i.e. hierarchic descent relationships) among the elements of the analysis, that is, among the populations. As discussed above, the degree to which this condition pertains depends on the meaning of 'population'. If this term denotes a local grouping of organisms, among whose members most but not all gene exchange occurs, then there is gene exchange occurring among 'populations', and descent relationships among them are reticulate. If, instead, 'population' refers to groups that do not exchange genes with other such groups (i.e. all gene exchange occurs within populations), then the various local populations within a metapopulation are properly recognized as members of a single 'population' within which there should be no expectation of phylogenetic structure. As indicated above, metapopulations, at the extreme, are approximately as inclusive as phylogenetic species, for this is the point at which descent relationships among 'populations' are hierarchic. As for genic relationships, it is the case among local populations within a metapopulation system, as well as among individual organisms within a local population, that apomorphic and plesiomorphic alleles can co-occur as intrapopulational polymorphisms. Therefore, relationships among alleles on a gene tree, as sampled from a set of local populations within such a system, should not be expected to be congruent with descent relationships among these populations. Indeed, for local populations that occasionally exchange genes, and thus are parts of a larger population system, the situation is much like the case described above for organisms within a population; hierarchic structure is lacking, and the monophyly concept does not apply.

Finally, what of the detection of phylogenetic structure among 'populations' that have ceased to exchange genes? At this point descent relationships among 'populations' are hierarchic, individual populations can be recognized as phylogenetic species, and sets of two or more populations can be monophyletic. If, again, the possibility of parallel evolution of identical alleles (i.e. the same allele) is set aside, all new alleles that arise within populations are unique. As different combinations of alleles become fixed in the various populations, they provide the means for the diagnosis of species. Once these unique character combinations become fixed as autapomorphies of individual species, they can serve as synapomorphies of daughter species that arise as the individual species becomes subdivided. Thus, another critical aspect of Hennig's model is fulfilled, this being that all descendants of an ancestor have all of its characters, either in original or modified form (Nixon and Wheeler, 1992; Davis and Nixon, 1992). As indicated above, this condition, like that of hierarchic descent among the bearers of characters, is lacking when the elements of analysis are individual organisms or local populations that are in occasional genetic contact with other local populations. The failure of character combinations to reflect phylogenetic structure in those systems therefore is a result of deviation from both of the key conditions specified in Hennig's model.

Unfortunately, there are further barriers to phylogenetic inference using gene trees, even when the elements of the analysis are phylogenetic species. One difficulty arises from the fact that phylogenetic species may be diagnosable either by solitary alleles that are mutually exclusive (e.g. allele A fixed in one species, allele B fixed in another) or by mutually exclusive allele pools, e.g. alleles A and C in one species, alleles B and D in another, with the two alleles that occur in each species not constituting a monophyletic group on the gene tree (Davis, 1997). In the latter case, one of two daughter species derived from species A/C may be fixed for allele A and the other for allele C; these are sister species, but they do not have sister alleles. The range of potential difficulties for a systematist working with such a group of species can be imagined by the reader.

The potential for incongruence between gene trees and species trees is a substantial challenge for systematics. One approach to this challenge is to eliminate it wherever it is observed, by expanding the limits of 'species' to the point at which all incongruence among gene trees is intraspecific (i.e. to adopt the GSC). Another approach is to accept the fact that there will be incongruence between gene trees sampled from the minimal phylogenetic elements known as phylogenetic species, any number of which may lie within the limits of a genealogical species. This is the boundary zone between systematics and population biology. Analyses of patterns of allelic polymorphism within and among populations and closely related species will be instrumental in the exploration of diversification patterns and processes at this level.

8.10 Conclusions

The elegance of Hennig's monophyly concept is perhaps most evident in its resilience to redefinition by those who have declared that individual species must be monophyletic. That goal has inspired the various attempts that have been made to redefine monophyly in such a way as to facilitate its application to groups of organisms, yet each of these redefinitions has its own fatal flaws. Attempts to develop a monophyly concept that is applicable to non-hierarchic systems have led to approaches that trivialize the concept. Under these redefinitions 'monophyly' variously refers to intersecting sets of organisms within populations, or to sets of siblings. An alternative approach, the redefinition of monophyly so as to refer to the distribution of characters, rather than to the history of the bearers of these characters, leads to the recognition of minimal 'exclusive' groups – genealogical 'species' that are unnatural amalgamations of independently evolving population systems, and hence species with internal phylogenetic structure. Thus, the various redefinitions of monophyly all hit wide of the mark, either by being too inclusive or not inclusive enough.

Systematists should not be surprised by these outcomes. Definitions based only on shared common ancestry among sexual organisms will inevitably yield groups with overlapping contents, precisely because lines of descent intersect in sexual systems. Conversely, definitions based only on proximity of relationship among sexual organisms yield minimal groups that are no more inclusive than sets of siblings, or sets of siblings plus their parents, because these are the minimal genealogical groups in sexual systems. The tendency for the GSC to define unnaturally inclusive groupings as basal elements is similarly unsurprising. This is a direct result of its focus on character histories, for if lineage sorting can occur, minimal groups defined in terms of unique sets of gene copies will often have internal phylogenetic structure. Thus, the GSC identifies groups that are united by a quality that is of dubious importance to evolutionary biologists. Avise and Wollenberg (1997) seemed to acknowledge as much when they suggested that majority-rule consensus among gene trees, rather than strict consensus, might be a more useful criterion of species status. This modification might, indeed, result in the delimitation of somewhat less inclusive 'species' than would the original method of Baum and Shaw. However, the ad hoc nature of this proposal also confirms the arbitrary quality of the GSC in general; there are several other kinds of consensus trees, and any of them could be used as an arbiter of 'species' status. All of them, however, would in some cases identify groups with internal phylogenetic structure.

The principal advantage of the GSC seems to be convenience. What is so dangerous about this approach is that it abandons the basic elements of evolutionary biology – gene evolution, patterns of mating in populations – in favour of a purely mechanical, character-based, and operationalist approach to species definition. Baum and Donoghue (1995: 561) criticised the focus of the PSC on population variation patterns, arguing that any such approach 'implies knowledge of patterns of gene flow in nature.' However, to focus on populations, as difficult as it is to recognise natural populations, and as cumbersome as it is to sample them adequately, implies only that patterns of gene flow in nature are recognised as important and deserving of study. It is easier, of course, to sample genes from organisms, construct gene trees, and compute consensus trees of those gene trees, without regard to population membership, than it is to analyse patterns of genetic variation in natural populations. And surely, coalescence theory can be employed in the analysis of patterns of gene flow among natural populations (e.g. Slatkin and Maddison, 1990; Hudson *et al.*, 1992). However, the GSC was not proposed as a means for estimating patterns of gene flow, with relationships among populations the stated target of analysis. Rather, it was presented as a *definition* of species. As long as character histories are given priority over population histories, it is safe to predict that unnatural groupings of convenience will be delimited.

Hennig developed phylogenetic systematics during the golden age of biosystematics. Systematists of his time recognised the fundamental importance of populations as the arenas in which gene exchange occurs, and they understood that reticulate descent patterns are to be expected within local populations. Moreover, they realised that occasional episodes of gene flow among local populations serve to connect those populations into larger genealogical systems, and that the cessation of gene flow between population systems is a crucial element of speciation. The BSC, developed during that era, reflects much of this understanding, but because it defines species in terms of the loss of *potential* for gene flow, rather than on the actuality of cessation of gene flow, and genetic divergence, it will recognise two population systems as conspecific after gene exchange between them has ceased and they have diverged from each other in any number of characters. In contrast, Hennig emphasised history; he inspired the current focus of systematics on phylogenetics, and he recognised that the cessation of gene flow between population systems is the minimal phylogenetic event. If a successful synthesis of phylogenetics and biosystematics is to emerge, practitioners will need to take full account of the distinction between reticulate and hierarchic descent, and acknowledge the key role of population systems as basal phylogenetic elements. If molecular systematics is to succeed, it will be by focusing on molecular characters as *evidence* of descent relationship among character bearers, where the character bearers are individual organisms, populations and multi-population systems. Hennig's definitive monophyly concept is uniquely suited to these objectives.

REFERENCES

Ashlock, P. D. (1971) Monophyly and associated terms. *Systematic Zoology*, **20**, 63–9.

Ashlock, P. D. (1972) Monophyly again. *Systematic Zoology*, **21**, 430–8.

Avise, J. C. (1994) *Molecular markers, natural history and evolution*. Chapman & Hall, London.

Avise, J. C. and Wollenberg, K. (1997) Phylogenetics and the origin of species. *Proceedings of the National Academy of Sciences USA*, **94**, 7748–55.

Avise, J. C., Arnold, J., Ball, R. M., Bermingham, E., Lamb, T., Neigel, J. E., Reeb, C. A. and Saunders, N. C. (1987) Intraspecific phylogeography: the mitochondrial DNA bridge between population genetics and systematics. *Annual Review of Ecology and Systematics*, **18**, 489–522.

Ayala, F. J., Escalante, A., O'huigin, C. and Klein, J. (1994) Molecular genetics of speciation and human origins. *Proceedings of the National Academy of Science USA*, **91**, 6787–94.

Barrett, S. C. H. and Pannell, J. R. (1999) Metapopulation dynamics and mating-system evolution in plants, in *Molecular systematics and plant evolution*, (eds P. M. Hollingsworth, R. M. Bateman and R. J. Gornall), Taylor & Francis, London, pp. 74–100.

Baum, D. A. (1992) Phylogenetic species concepts. *Trends in Ecology and Evolution*, **7**, 1–2.

Baum, D. A. and Donoghue, M. J. (1995) Choosing among alternative 'phylogenetic' species concepts. *Systematic Botany*, **20**, 560–73.

Baum, D. A. and Shaw, K. L. (1995) Genealogical perspectives on the species problem, in *Experimental and molecular approaches to plant biosystematics*, (eds P. C. Hoch and A. G. Stephenson), Missouri Botanical Garden, St. Louis, pp. 289–303.

Bremer, K. and Wanntorp, H. -E. (1979) Geographic populations or biological species in phylogeny reconstruction? *Systematic Zoology*, **28**, 220–4.

Clark, A. G. and Kao, T. -H. (1991) Excess nonsynonymous substitution at shared poly-morphic sites among self-incompatibility alleles of Solanaceae. *Proceedings of the National Academy of Science USA*, **88**, 9823–7.

Comes, H. P. and Abbott, R. J. (1999) Reticulate evolution in the Mediterranean species complex of *Senecio* sect. *Senecio*: uniting phylogenetic and population-level approaches, in *Molecular systematics and plant evolution*, (eds P. M. Hollingsworth, R. M. Bateman and R. J. Gornall), Taylor & Francis, London, pp. 171–98.

Cracraft, J. (1983) Species concepts and speciation analysis. *Current Ornithology*, **1**, 159–87.

Cracraft, J. (1989) Speciation and its ontology: the empirical consequences of alternative species concepts for understanding patterns and processes of differentiation, in *Speciation and its consequences*, (eds D. Otte and J. A. Endler), Sinauer Associates, Inc., Sunderland, Massachusetts, pp. 28–59.

Davis, J. I. (1997) Evolution, evidence, and the role of species concepts in phylogenetics. *Systematic Botany*, **22**, 373–403.

Davis, J. I. and Manos, P. S. (1991) Isozyme variation and species delimitation in the *Puccinellia nuttalliana* complex (Poaceae): an application of the phylogenetic species concept. *Systematic Botany*, **16**, 431–445.

Davis, J. I. and Nixon, K. C. (1992) Populations, genetic variation, and the delimitation of phylogenetic species. *Systematic Biology*, **41**, 421–35.

de Queiroz, K. and Donoghue, M. J. (1988) Phylogenetic systematics and the species problem. *Cladistics*, **4**, 317–38.

de Queiroz, K. and Donoghue, M. J. (1990) Phylogenetic systematics or Nelson's version of cladistics? *Cladistics*, **6**, 61–75.

Dobzhansky, T. (1935) A critique of the species concept in biology. *Philosophy of Science*, **2**, 344–55.

Dobzhansky, T. (1937a) Genetic nature of species differences. *American Naturalist*, **71**, 404–20.

Dobzhansky, T. (1937b) *Genetics and the origin of species*. Columbia University Press, New York.

Donoghue, M. (1985) A critique of the biological species concept and recommendations for a phylogenetic alternative. *Bryologist*, **88**, 172–81.

Doyle, J. J. (1992) Gene trees and species trees: molecular systematics as one-character taxonomy. *Systematic Botany*, **17**, 144–63.

Doyle, J. J. (1995) The irrelevance of allele tree topologies for species delimitation, and a non-topological alternative. *Systematic Botany*, **20**, 574–88.

Doyle, J. J. and Davis, J. I. (1998) Homology in molecular phylogenetics: a parsimony perspective, in *Molecular systematics of plants II: DNA sequencing*, (eds D. E. Soltis, P. S. Soltis and J. J Doyle), Kluwer, London, pp. 101–31.

Ehrlich, P. R. and Raven, P. H. (1969) Differentiation of populations. *Science*, **165**, 1228–32.

Farris, J. S. (1974) Formal definitions of paraphyly and polyphyly. *Systematic Zoology*, **23**, 548–54.

Farris, J. S. (1990) Haeckel, history, and Hull. *Systematic Zoology*, **39**, 81–8.

Farris, J. S. (1991) Hennig defined paraphyly. *Cladistics*, **7**, 297–304.

Figueroa, F., Günther, E. and Klein, J. (1988) MHC polymorphism predating speciation. *Nature*, **335**, 265–7.

Gaut, B. S. and Clegg, M. T. (1993) Molecular evolution of the *Adh1* locus in the genus *Zea*. *Proceedings of the National Academy of Science USA*, **90**, 5095–9.

Grant, V. (1981) *Plant speciation*, ed. 2. Columbia University Press, New York.

Harrison, R. G. (1991) Molecular changes at speciation. *Annual Review of Ecology and Systematics*, **22**, 281–308.

Hennig, W. (1966) *Phylogenetic systematics*. University of Illinois Press, Urbana.

Hudson, R. R. (1990) Gene genealogies and the coalescent process. *Oxford Surveys in Evolutionary Biology*, **7**, 1–44.

Hudson, R. R., Slatkin, M. and Maddison, W. P. (1992) Estimation of levels of gene flow from DNA sequence data. *Genetics*, **132**, 583–9.

Hull, D. L. (1988) *Science as a process*. University of Chicago Press, Chicago.

Husband, B. C. and Barrett, S. C. H. (1996) A metapopulation perspective in plant population biology. *Journal of Ecology*, **84**, 461–9.

Ioerger, T. R., Clark, A. G. and Kao, T. -H. (1990) Polymorphism at the self-incompatibility locus in Solanaceae predates speciation. *Proceedings of the National Academy of Science USA*, **87**, 9732–5.

Knox, E. B. (1998) The use of hierarchies as organizational models in systematics. *Biological Journal of the Linnean Society*, **63**, 1–49.

Krohne, D. T. (1997) Dynamics of metapopulations of small mammals. *Journal of Mammalogy*, **78**, 1014–26.

Lawlor, D. A., Ward, F. E., Ennis, P. D., Jackson, A. P. and Parham, P. (1988) HLA-A and B polymorphisms predate the divergence of humans and chimpanzees. *Nature*, **335**, 268–71.

Levin, D. A. (1995) Metapopulations: an arena for local speciation. *Journal of Evolutionary Biology*, **8**, 635–644.

Levins, R. A. (1969) Some demographic and genetic consequences of environmental heterogeneity for biological control. *Bulletin of the Entomological Society of America*, **15**, 237–40.

Levins, R. A. (1970) Extinction, in *Some mathematical problems in biology*, (ed M. Gerstenhaber), American Mathematical Society, Providence, Rhode Island, pp. 77–107.

Maddison, W. P. (1995) Phylogenetic histories within and among species, in *Experimental and molecular approaches to plant biosystematics*, (eds P. C. Hoch and A. G. Stephenson), Missouri Botanical Garden, St. Louis, pp. 273–87.

Maddison, W. P. (1997) Gene trees in species trees. *Systematic Biology*, **46**, 523–36.

Mayer, W. E., Jonker, M., Klein, D., Ivanyi, P., van Seventer, G. and Klein, J. (1988) Nucleotide sequences of chimpanzee MHC class I alleles: evidence for trans-specific mode of evolution. *EMBO Journal*, **7**, 2765–74.

Mayr, E. (1942) *Systematics and the origin of species*. Columbia University Press, New York.

Mayr, E. (1965) *Animal species and evolution*. Harvard University Press, Cambridge, Massachusetts.

Mayr, E. (1982) *The growth of biological thought*. Harvard University Press, Cambridge, Massachusetts.

McCauley, D., Raveill, J. and Antonovics, J. (1995) Local founding events as determinants of genetic structure in a plant metapopulation. *Heredity*, **75**, 630–6.

McConnell, T. J., Talbot, W. S., McIndoe, R. A. and Wakeland, E. K. (1988) The origin of MHC class II gene polymorphism within the genus *Mus*. *Nature*, **332**, 651–4.

Mendel, G. (1866) Versuche über Pflanzen-Hybriden. *Verhandlungen des Naturforschenden Vereines in Brünn*, **4**, 3–47.

Mishler, B. D. (1985) The morphological, developmental, and phylogenetic basis of species concepts in bryophytes. *Bryologist*, **88**, 207–14.

Neigel, J. E. and Avise, J. C. (1986) Phylogenetic relationships of mitochondrial DNA under various demographic models of speciation, in *Evolutionary processes and theory*, (eds. S. Karlin and E. Nevo), Academic Press, New York, pp. 515–34.

Nelson, G. J. (1971) Paraphyly and polyphyly: redefinitions. *Systematic Zoology*, **20**, 471–2.

Nelson, G. J. and Platnick, N. I. (1981) *Systematics and biogeography: cladistics and vicariance*. Columbia University Press, New York.

Nixon, K. C. and Wheeler, Q. D. (1990) An amplification of the phylogenetic species concept. *Cladistics*, **6**, 211–23.

Nixon, K. C. and Wheeler, Q. D. (1992) Extinction and the origin of species, in *Extinction and Phylogeny*, (eds. M. J. Novacek and Q. D. Wheeler), Columbia University Press, New York, pp. 119–43.

Oosterbroek, P. (1987) More appropriate definitions of paraphyly and polyphyly, with a comment on the Farris 1974 model. *Systematic Zoology*, **36**, 103–8.

Pamilo, P. and Nei, M. (1988) Relationships between gene trees and species trees. *Molecular Biology and Evolution*, **5**, 568–83.

Platnick, N. I. (1977) Paraphyletic and polyphyletic groups. *Systematic Zoology*, **26**, 195–200.

Richman, A. D., Uyenoyama, M. K. and Kohn, J. R. (1996) Allelic diversity and gene genealogy at the self-incompatibility locus in the Solanaceae. *Science*, **273**, 1212–16.

Richter, S. and Meier, R. (1994) The development of phylogenetic concepts in Hennig's early theoretical publications (1947–1966). *Systematic Biology*, **43**, 212–21.

Rosen, D. E. (1978) Vicariant patterns and historical explanation in biogeography. *Systematic Zoology*, **27**, 159–88.

Rosen, D. E. (1979) Fishes from the uplands and intermontane basins of Guatemala: revisionary studies and comparative geography. *Bulletin of the American Museum of Natural History*, **162**, 267–376.

Scheiner, S. M. and Rey-Benayas, J. M. (1997) Placing empirical limits on metapopulation models for terrestrial plants. *Evolutionary Ecology*, **11**, 275–88.

Slatkin, M. and Maddison, W. P. (1990) Detecting isolation by distance using phylogenies of genes. *Genetics*, **126**, 249–60.

Sosef, M. S. M. (1997) Hierarchical models, reticulate evolution and the inevitability of paraphyletic supraspecific taxa. *Taxon*, **46**, 75–85.

Stuessy, T. (1990) *Plant taxonomy: the systematic evaluation of comparative data*. Columbia University Press, New York.

Tateno, Y., Nei, M. and Tajima, F. (1982) Accuracy of estimated phylogenetic trees from molecular data. I. Distantly related species. *Journal of Molecular Evolution*, **18**, 387–404.

Vogel, J. C., Barrett, J. A., Rumsey, F. J. and Gibby, M. (1999) Identifying multiple origins in polyploid homosporous pteridophytes, in *Molecular systematics and plant evolution*, (eds P. M. Hollingsworth, R. M. Bateman and R. J. Gornall), Taylor & Francis, London, pp. 101–17.

Vrijenhoek, R. C. (1997) Gene flow and genetic diversity in naturally fragmented metapopulations of deep-sea hydrothermal vent animals. *Journal of Heredity*, **88**, 285–293.

Chapter 9

Reticulate evolution in the Mediterranean species complex of *Senecio* sect. *Senecio*: uniting phylogenetic and population-level approaches

H. P. Comes and R. J. Abbott

ABSTRACT

Studies of reticulate evolution among plant taxa continue to be phylogenetically oriented, with an overall emphasis on pattern rather than process. Recently, however, the number of analyses has increased that are using phylogenies as a framework to derive microevolutionary hypotheses about natural hybridization and introgression in various species complexes. These hypotheses can then be tested by using data from extant populations. Here, this dual approach is illustrated by phylogenetic and population-level analyses involving the Mediterranean species complex of *Senecio* sect. *Senecio* (Asteraceae), using chloroplast (cp) DNA and rDNA (ITS) sequence information, allozymes, and RAPDs as nuclear data sets. The phylogenetic approach has allowed us to evaluate the extent and overall pattern of reticulation within the species complex. In at least two instances cpDNA *and* ITS capture, accompanied by differing amounts of introgression of the rest of the nuclear genome, has resulted in the phylogenetic alignment of taxonomically disparate species. In addition, a striking example of differential introgression of cytoplasmic and nuclear (allozymic) elements has been identified by detailed population-level examinations of two parapatric species of *Senecio* from the Near East allowing inferences to be made about the microevolutionary processes that may have led to this phenomenon. We conclude that an integration of phylogenetic and population-level approaches is critical to understanding the role of hybridization and introgression in the plant kingdom, and especially among species at an early stage of divergence from each other.

9.1 Introduction

Molecular phylogenetic reconstruction involving the use of chloroplast (cp) DNA has been widely applied to studies of natural hybridization and introgression in plants; for reviews see Rieseberg and Soltis (1991), Rieseberg and Wendel (1993) and Arnold (1997). Numerous studies on problems related to reticulate evolution have used this phylogenetic approach, which is particularly powerful when cpDNA variation is assayed in conjunction with other nuclear genetic markers, thus allowing examination of non-concordance among independently inherited uniparental and biparental data sets (Givnish and Sytsma, 1997). In particular, the tradition of examining different phylogenetic trees based on cytoplasmic and nuclear characters

In *Molecular systematics and plant evolution* (1999) (eds P. M. Hollingsworth, R. M. Bateman and R. J. Gornall), Taylor & Francis, London, pp. 171–198.

has proven a sensitive means to test for ancient or more recent hybridization, and for falsifying hypotheses about homoploid and polyploid hybrid speciation (Soltis and Soltis, 1989; Rieseberg *et al.*, 1990a; Arnold *et al.*, 1991; Rieseberg, 1991; Wendel *et al.*, 1991; Wolfe and Elisens, 1994; Sang *et al.*, 1995; Smith *et al.*, 1996; Allan *et al.*, 1997). In addition, the phylogenetic approach has provided insights into the frequency and evolutionary importance of chloroplast capture through introgressive hybridization (Larson and Doebley, 1993; Roelofs and Bachmann, 1995; Soltis and Kuzoff, 1995; Wolfe and Elisens, 1995), often in the apparent absence of nuclear introgression (Whittemore and Schaal, 1991; Cruzan *et al.*, 1993); see Rieseberg and Wendel (1993) and Soltis and Soltis (1995) for reviews. Possible reasons for the difference between the degree of interspecific cytoplasmic and nuclear gene flow have been summarized by Rieseberg and Soltis (1991).

Surprisingly, however, there have been few detailed molecular population-level analyses of natural hybridization among plant taxa: *Pinus* (Wheeler and Guries, 1987; Edwards-Burke *et al.*, 1997); *Helianthus* (Rieseberg *et al.*, 1988, 1995); *Ipomopsis* (Wolf and Soltis, 1992); *Iris* (Cruzan and Arnold, 1993); *Carpobrotus* (Gallagher *et al.*, 1997), especially when compared with the case of animals; see Barton and Hewitt (1985), Harrison (1990) and Arnold (1992, 1997) for reviews. This might reflect a major interest of botanists in the systematic implications of natural hybridization and the testing for polyploidy, and only a minor interest in the processes of natural hybridization and the formation of barriers to reproduction, viz. speciation (Arnold, 1997). However, an additional factor of importance is that whereas mitochondrial DNA variation within and among populations has been routinely used as a powerful tool in investigating natural hybridization among animal taxa (Dowling *et al.*, 1989; Forbes and Allendorf, 1991; Zink, 1994; Schneider, 1996), the analogous application of cpDNA in plants has been hampered by the view that the chloroplast genome is highly conserved and rarely displays the level of variation necessary for population studies (e.g. Banks and Birky, 1985; Neale *et al.*, 1988). Undoubtedly, the extent and magnitude of intraspecific cpDNA variation varies within and among species, and sometimes the utility of cpDNA may lie at higher taxonomic levels (Jordan *et al.*, 1996; Allan *et al.*, 1997). Nonetheless, intraspecific cpDNA variation has long been observed in many plant species (see Harris and Ingram (1991) and Soltis *et al.* (1992) for reviews), and has provided useful insights into evolutionary processes within and among populations: *Trifolium* (Milligan, 1991); *Eucalyptus* (Byrne and Moran, 1994); *Quercus* (Dumolin-Lapègue *et al.*, 1997; Ferris *et al.*, 1999 – this volume); *Argania* (El Mousadik and Petit, 1996); *Fagus* (Demesure *et al.*, 1996); *Phacelia* (Levy *et al.*, 1996); *Aquilegia* (Strand *et al.*, 1996); and *Senecio* (Comes and Abbott, 1998). Moreover, recent analyses of cpDNA microsatellite variation suggest that levels of intraspecific cpDNA variation may often be high in plants (Provan *et al.*, 1999 – this volume), e.g. *Glycine* (Powell *et al.*, 1996), *Oryza* (Provan *et al.*, 1997); *Pinus* (Powell *et al.*, 1995; Cato and Richardson, 1996).

Taken together, these newer advances and technologies should offer considerable prospect for the use of intraspecific cpDNA variation for population-level studies of interspecific hybridization and introgression among plant taxa. In this regard, the joint application of cytoplasmic and nuclear markers appears to be crucial because molecular systematic data clearly suggest that the chloroplast genome may sometimes be exchanged more freely than nuclear genes (see above). This potentially could lead to erroneous microevolutionary hypotheses when the chloroplast genome is used as a sole source of inference for overall levels of interspecific gene flow. In any case, a major advantage of the joint application of cytoplasmic and nuclear markers is the prospect of disentangling the relative contributions of seed and pollen dispersal to overall levels of interpopulational/interspecific gene flow (Hu and Ennos, 1997; Ennos *et al.*, 1999 – this volume).

As shown by Arnold's (1997) account, molecular phylogenetic reconstruction has been and will undoubtedly continue to be a powerful tool for inferring natural hybridization and introgression among plant taxa, for testing various hypotheses about hybrid speciation, and for estimating frequencies of differential capture of cytoplasmic and nuclear elements. Nonetheless, population-level analyses appear essential to our understanding of the microevolutionary processes of hybridization, and only the complementation of both historical- *and* contemporary-oriented analyses will help to determine the relative importance of non-recurrent and recurrent hybridization events on the molecular patterns observed.

In this chapter, we illustrate the utility of such a dual approach as part of an integrative research programme aimed at unravelling the molecular phylogeography of Mediterranean species of *Senecio* sect. *Senecio*, Asteraceae (*sensu* Alexander, 1979). This group serves as a model system for the study of reticulate evolution because it comprises a tractable number of closely related species exhibiting considerable interfertility, widespread sympatry or parapatry, and documented natural hybridization (Alexander, 1979; Kadereit, 1984). Moreover, several of its members have been popular models for the examination of key processes in plant evolution, not least the origin, establishment and maintenance of diploid and polyploid species (Abbott, 1992). In this report we begin with a brief summary of our study system. We then describe preliminary results of a phylogenetic approach to discern the extent and overall pattern of reticulation in the species complex by examining at least 14 species for restriction site and length variation in the chloroplast genome, and nuclear sequence variation of the internal transcribed spacer region (ITS) of the 18S-26S nrDNA cistron (see Hershkovitz *et al.* (1999 – this volume) for a review). In addition, variation in RAPD markers, the majority of which can be considered as being representative of the nuclear genome (Lorenz *et al.*, 1994; Thormann *et al.*, 1994; Harris, 1999 – this volume), was analysed among a subset of species to allow further comparison with the cpDNA and ITS data sets. We then summarize a population-level analysis conducted to determine the actual extent and level of cytoplasmic and nuclear gene flow across a particular species contact zone of *Senecio* in the Near East by examining differences in cpDNA and allozyme variation within and among populations (Comes and Abbott, 1999). First results of our phylogenetic reconstruction indicate that at least in two instances cpDNA *and* ITS capture has resulted in the phylogenetic alignment of taxonomically disparate species. In addition, a striking example of differential introgression of cytoplasmic and nuclear (allozymic) elements is identified for the parapatric species pair from the Near East, allowing inferences to be made about the microevolutionary processes that may have led to this phenomenon.

9.2 Phylogenetic systematics of the Mediterranean species complex of *Senecio* sect. *Senecio*

Certain annual and perennial species of *Senecio* sect. *Senecio* (Jeffrey *et al.*, 1977; Jeffrey, 1979) mainly from countries bordering the Mediterranean have long been recognized to form a group of variously closely related taxa based on morphological and experimental evidence (Alexander, 1979; Kadereit, 1984). However, several species of the complex are also more widely distributed throughout temperate regions of central and northern Europe, and the type species, the tetraploid *S. vulgaris* (2n = 4x = 40), is cosmopolitan in distribution. Following the latest revision by Alexander (1979) the group consists of approximately 23 diploid and related tetraploid species, and three hexaploid species. Notably, with a few exceptions, the tetraploids and hexaploids have a strong inbreeding tendency, associated with shortness or absence of outer ray florets, whereas the diploids tend to have long ray florets and are outbreeding (Alexander, 1979). The group is well known for its amenability to experimental studies of genecology and plant evolution, but is notorious for its taxonomic complexity (Walters, 1964). At least part of the taxonomic confusion within the group might be attributed to relatively high levels of species inter-fertility, the origin of polyploid, and probably also diploid species following interspecific hybridization, the likelihood of continued hybridization and introgression between some of its members, and/or the sharing of ancestral similarities across species. Much of the confusion within the species complex has been clarified or fully resolved by more recent investigations into the origin and evolutionary history of the diploid *S. squalidus* (Abbott *et al.*, 1995; Abbott and Milne, 1995) and several polyploid species: *S. vulgaris* (Weir and Ingram, 1980; Kadereit, 1984; Ashton and Abbott, 1992; Harris and Ingram, 1992a; Comes *et al.*, 1997); *S. nebrodensis* and *S. viscosus* (Kadereit *et al.*, 1995; Purps and Kadereit, 1998); *S. teneriffae* (Lowe and Abbott, 1996). However, comparatively little is known about the phylogenetic relationships among the majority of the widespread diploid taxa within the species group (*S. flavus, S. gallicus, S. glaucus, S. leucanthemifolius, S. squalidus, S. vernalis*), various geographically restricted diploid (*S. aethnensis, S. chrysanthemifolius, S. hesperidium, S. petraeus*) and polyploid endemics (*S. aegyptius, S. hoggariensis, S. massaicus*), and the relationship of all these species to a supposedly monophyletic group of glandular-hairy tetraploids, comprising *S. nebrodensis, S. lividus, S. sylvaticus*, and *S. viscosus* (Kadereit *et al.*, 1995). Similarly, detailed investigations into aspects of interspecific hybridization and introgression have focused largely on events in the British Isles, involving *S. squalidus* and *S. vulgaris*, their stabilized tetraploid introgressant *S. vulgaris* var. *hibernicus*, and the allohexaploid derivative *S. cambrensis* (see Abbott (1992) and Abbott and Lowe (1996) for reviews). Recently, however, Liston and Kadereit (1995) have produced a molecular phylogeny for the Namibian/Saharo-Arabian diploid *S. flavus* and its tetraploid relative *S. mohavensis* from North America, based on parsimony analysis of cpDNA restriction sites. Significantly, they postulated that the dramatic divergence of cpDNA genotypes found between *S. flavus* subsp. *flavus* and subsp. *breviflorus* most likely resulted from past introgressive hybridization, probably involving an unknown Mediterranean taxon closely related to *S. squalidus*. For all that, a robust phylogenetic hypothesis for the entire group is still elusive, and the overall degree to which reticulate events have affected the species complex is largely unexplored, thus highlighting the need for further research.

9.3 Phylogenetic approach to screen for natural hybridization and introgression

9.3.1 Material and Methods

Most material for molecular analysis was obtained from greenhouse plants raised from seed collected in the field. For a very few accessions silica gel-dried material was used. Voucher specimens and seed samples of collections are stored at the St Andrews (**STA**) herbarium.

9.3.1.1 DNA extraction

Total genomic DNAs for cpDNA, ITS and RAPD analysis were isolated from leaf tissue according to Abbott *et al.* (1995). Resulting extracts were cleaned via an RNase A step (Boehringer Mannheim) followed by ammonium acetate precipitation (Comes *et al.*, 1997). Some DNA samples, including all those from silica dried materials, were purified on caesium chloride/ethidium bromide gradients (Abbott *et al.*, 1995).

9.3.1.2 CpDNA analysis

The major part of our cpDNA analysis involved conventional Southern hybridization techniques. Methods for cpDNA restriction enzyme digestion, fragment separation, DNA transfer, and hybridization were as described elsewhere (Harris and Ingram, 1992b; Comes *et al.*, 1997). Filters were hybridized with cloned fragments of ^{32}P- or digoxigenin-labelled cpDNA from *Lactuca sativa* (Jansen and Palmer, 1987), and hybridized probes were visualized via autoradiography or chemiluminescence. A limited number of taxa, embracing much of the morphological diversity within the species complex (marked with an asterisk in Table 9.1), was initially screened for cpDNA restriction site and length variation over a wide range of probe–enzyme combinations (15 probes × 18 enzymes; Abbott *et al.*, 1995; Abbott, unpubl. data). This survey resolved a total of 7 polymorphisms (Table 9.2). The subsequent widescale analysis (Table 9.1) was limited largely to those probe–enzyme combinations that had identified the 7 polymorphisms in the pilot study, but also included two additional enzymes for probes pLsC 8–11 (i.e. *Hin*DIII and *Xba*I). Each polymorphism or character was divided into two character states (Table 9.2). In several instances the exact cause of polymorphisms could not be determined in that they were due to the presence/absence of a single band.

In addition, a survey of RFLP variation was conducted within the PCR-amplified region of the chloroplast *trnK* gene for 39 indvidual plants from 15 species (Table 9.1), using a slightly modified amplification protocol from Schwarzbach and Kadereit (1995). The sequences of the primers used were: *trnK*-2621/*trnK*-11: 5'-AAC TAG TCG GAT GGA GTA G-3', and 5'-CTC AAC GGT AGA GTA CTC G-3' (Arnold *et al.*, 1991; Allan *et al.*, 1997). The resulting PCR products (2560bp) were restricted with three enzymes (Table 9.2) shown to yield polymorphisms in an initial survey of 19 accessions from 13 representative species, and using a series of 16 enzymes (*Apa*I, *Bam*HI, *Bcl*I, *Bgl*II, *Bst*OI, *Cfo*I, *Dra*I, *Hpa*II, *Hinf*I, *Eco*RI, *Eco*RV, *Pst*I, *Pvu*II, *Rsa*I, *Sst*I, *Taq*I).

Table 9.1 Species groups of *Senecio* and the outgroup used in the cpDNA restriction analysis together with their geographic origin. The number of populations and the total number of individuals surveyed per species are also indicated separately for Southern hybridization analyses (Southern) and for PCR-RFLP analyses using the *trnK* gene (PCR-*trnK*). More detailed information on the localities and sample sizes of each population studied will be presented elsewhere.

Senecio species[a]	*Geographic origin*	*Populations/ total sample size*	
		Southern	PCR-trnK
DIPLOIDS (2n = 2x = 20)			
* *aethnensis*	Sicily	4/6	–
* *chrysanthemifolius*	Sicily	7/13	–
flavus	Morocco, Israel, Egypt	7/9	1/1
gallicus	France, Iberian Peninsula, Tunisia	14/72	1/1
glaucus	Morocco, Israel, Egypt	23/152	2/2
hesperidium	Morocco (Massa)	3/6	–
leucanthemifolius	Morocco, Italy, Corsica, Greece	16/29	1/1
petraeus	Spain (Sierra de las Nieves)	1/1	1/1
rodriguezii	Spain (Mallorca)	1/1	1/1
* *squalidus*	British Isles, Germany, Italy, Balkans, Greece	23/30	5/5
* *vernalis*	Germany, Greece, Turkey, Israel	14/99	1/1
TETRAPLOIDS (2n = 4x = 40)			
aegyptius	Egypt	2/3	1/1
lividus	Iberian Peninsula, Corsica, Turkey	6/10	1/1
* *nebrodensis*	Spain	6/12	6/13
sylvaticus	British Isles, France, Spain	5/5	2/2
viscosus	Norway, British Isles, Central and Eastern Europe, Turkey	6/7	5/7
vulgaris	British Isles, Jersey (Channel Islands), Germany, France, Iberian Peninsula, Italy, Sicily, Cyprus, Greece, Israel	26/30	1/1
malacitanus (outgroup)	Morocco	1/1	1/1

[a] Asterisks refer to taxa initially screened for RFLP variation over 18 enzymes × 15 *Lactuca sativa* cpDNA probes.

The genealogical relationships among cpDNA haplotypes were estimated under unweighted parsimony using PAUP 3.1.1 (Swofford, 1993). The analysis used the heuristic search algorithm with the following options in effect: MULPARS (holding multiple equally parsimonious trees), TREE BISECTION RECONNECTION (TBR) swapping, and SIMPLE addition of haplotypes. One thousand replicates of RANDOM haplotype entries were also performed, but no additional trees were found. As in previous intraspecific studies in Mediterranean *Senecio* (Comes *et al.*, 1997; Comes and Abbott, 1998), character states were polarized as ancestral or derived by outgroup comparison with *S. malacitanus*; this is a long-lived perennial species that is not a member of the Mediterranen species complex of *Senecio* sect. *Senecio* (Alexander, 1979). To assess the amount of support for individual clades, bootstrap values from 100 replicates were calculated using the PAUP HEURISTIC option with CLOSEST addition sequence of haplotypes.

Table 9.2 Chloroplast DNA polymorphisms detected in the *Senecio* material. Character states were polarized as ancestral (0) or derived (1) by outgroup comparison with *S. malacitanus*.

Poly-morphism[b]	Enzyme	Probe/ Product[c]	Mutation (fragment size in kb)[a]		Type
			0	1	
* 1	*Pvu*II	4 + 5	14.0	9.1 + 4.9	site gain
* 2	*Hae*III	4 + 5	–	–0.045	deletion
3	*Hae*III	4 + 5	1.5	absent	site
4	*Hae*III	4 + 5	–	+0.035	insertion
5	*Hae*III	4 + 5	–	–0.04	deletion
6	*Hae*III	4 + 5	–	+0.02	insertion
7	*Cla*I	6	11.1	absent	site
* 8	*Cla*I	6	–	–0.33/0.35	deletion
* 9	*Cla*I	6	3.1+(0.2)	3.3	site loss
10	*Cla*I	6	–	–0.7	deletion
* 11	*Pst*I	1–3	2.8	absent	site
12	*Pst*I	1–3	–	–0.05	deletion
* 13	*Cfo*I	6	8.4	7.0 + 1.4	site gain
14	*Cfo*I	6	–	+0.12	insertion
* 15	*Eco*RI	8–11	2.0+(0.4)	2.4	site loss
16	*Eco*RI	8–11	–	–0.06	deletion
17	*Eco*RI	8–11	4.9+1.1	6.0	site loss
18	*Eco*RI	8–11	–	–0.1	deletion
19	*Hin*DIII	8–11	3.5	absent	site
20	*Hin*DIII	8–11	10.9	4.2 + 2.5 + (4.2)	site gain
21	*Xba*I	8–11	–	+0.7	insertion
22	*Xba*I	8–11	absent	6.5	site
23	*Hpa*II	*trn*K	2.6	2.4 + 0.2	site gain
24	*Hpa*II	*trn*K	2.4	2.0 + 0.4	site gain
25	*Bst*OI	*trn*K	1.5	1.1 + 0.4	site gain
26	*Bst*OI	*trn*K	–	+0.14	insertion
27	*Cfo*I	*trn*K	–	+0.02	insertion

[a] Dashed lines indicate no equivalent band detected. Fragment sizes in brackets refer to presumed but undetected bands.
[b] Asterisks refer to seven RFLP polymorphisms resolved by an initial wide screen of seven *Senecio* taxa with 18 enzymes × 15 *Lactuca sativa* cpDNA probes.
[c] Probe designation follows Harris and Ingram (1992a, b) and Abbott *et al.* (1995).

9.3.1.3 nrDNA (ITS) sequence analysis

The ITS analysis presented here for 19 accessions from 14 species is a subset of a more complete analysis for Mediterranean species of sect. *Senecio* taken from the ITS data base of Comes and Abbott (in preparation). The complete study encompasses five additional endemic and/or polyploid species (*S. cambrensis, S. hesperidium, S. petraeus, S. rodriguezii, S. teneriffae*), and thus includes all but a small number of geographically highly restricted species in the species complex (e.g. *S. chalureaui, S. hoggariensis, S. massaicus*).

The entire ITS1–5.8S-ITS2 region was PCR amplified from total genomic DNA using the primers 'ITS4' and 'ITS5' (White *et al.*, 1990) and the protocol detailed in Comes and Abbott (in preparation). Double-stranded PCR products were purified using the PROMEGA Wizard Kit (Promega Corporation, Madison, WI), and then used as the template in a cycle sequencing reaction using the ABI PRISM™ Dye Deoxy Terminator Cycle Sequencing Kit (Perkin-Elmer, Foster City, CA), with AmpliTaq® DNA polymerase FS. Primers for cycle-sequencing included the two primers used for amplification, and a single-stranded cycle sequencing protocol (Comes and Abbott, in preparation) was followed. Unincorporated dye terminators etc. were removed by ethanol precipitation, and DNA samples were loaded on a gel of an ABI PRISM 377 DNA automated sequencer. Raw sequencing data analysis was performed using AutoAssembler™ and Sequence Navigator™ (Perkin-Elmer, Foster City, CA). Sequences of the two ITS regions (+5.8S) were aligned by CLUSTAL (Higgins *et al.*, 1992), with minor manual adjustments.

Phylogenetic trees were generated from the above data set by Fitch parsimony and the heuristic search strategy in PAUP, with the following options in effect: MULPARS, TBR swapping, STEEPEST DESCENT, and 500 replications of RANDOM sequence entries. In these searches, *S. malacitanus* served as outgroup again, and gaps (indels) were scored as missing data. Coding of informative indels was also explored, though this procedure generally increased the length of several branches and had no effect on the overall conclusions reached in the present context (data not shown). A majority-rule consensus of all most parsimonious trees was calculated, and bootstrap values were estimated from 100 replicate HEURISTIC analyses with CLOSEST addition sequence.

9.3.1.4 RAPD analysis

RAPD variation was screened over a subset of DNA samples examined for cpDNA and/or ITS variation. Details of the PCR amplification chemistry and the 17 Operon ten-mer primers used (OPH-01–09, OPH-11–15, OPH-17–19; Operon Technologies, Inc., Alameda, CA) will be presented elsewhere. Forty-five amplification cycles were performed following a profile of 94 °C for 30 sec, 36 °C for 45 sec, and 72 °C for 1.5 min. Amplified band states (362) were scored as present (1) or absent (0). The pairwise RAPD distances between samples were measured by using the DISTANCE MATRIX option implemented in PAUP. These distances are identical to those also produced by the Euclidean metric of Excoffier *et al.* (1992), $E = n[1 - (n_{xy}/n)]$, where n_{xy} is the number of bands shared by two individuals, and n is the total number of polymorphic bands. A neighbour-joining tree (Saitou and Nei, 1987) was computed from the values of the distance matrix using the routine NEIGHBOR in PHYLIP 3.5 (Felsenstein, 1993) with *S. malacitanus* chosen as outgroup. These analyses were repeated by calculating genetic distances from similarity coefficents implemented in NTSYS-pc 1.80 (Rohlf, 1993): (i) Jaccard's coefficient $(n_{xy}/n - m_{xy})$, which explicitly excludes the number of shared absence of bands (m_{xy}), and (ii) the simple matching coefficient $(n_{xy} + m_{xy}/n)$, which includes shared absence. Because all coefficients gave very similar results, only genetic distances based on the Euclidean metric are presented.

Although RAPDs may be considered an independent source of genetic information (Smith *et al.*, 1996; Allan *et al.*, 1997), there are well-known theoretical concerns about the use of RAPD analysis in phylogeny reconstruction, not least major criticisms regarding the inheritance, character independence, and homology of RAPD markers (Harris, 1995; van de Zande and Bijlsma, 1995; Harris, 1999 – this volume). Our results, therefore, will be based on the following assumptions: (1) each RAPD marker comprises a single dominant locus with alleles being either present (amplified) or absent (non-amplified); (2) the majority of RAPD markers are randomly distributed throughout the nuclear genome; and (3) comigrating bands represent homologous amplified products. Currently, apart from traditional morphological classification, the RAPD evidence is our only substantial source of evidence concerning the nuclear relationships of a larger number of species in sect. *Senecio*.

9.3.2 Results and Discussion

The cpDNA analysis resolved a total of 27 polymorphisms over all the material surveyed (Table 9.2), and these polymorphisms were ordered into 16 haplotypes (see also *Appendix*). The single most parsimonious tree generated from these data is illustrated in Table 9.3. The distribution of cpDNA haplotypes among species is summarized in the same table, and provides a broad overview of the cpDNA composition of major groups in the Mediterranean *Senecio* complex.

Our survey indicates that intraspecific cpDNA polymorphism is rather common within the group and that there is considerable sharing of cpDNA haplotypes, particularly among the widespread diploid taxa (Table 9.3). It is noteworthy that much of this cpDNA diversity also embraces the variation present in the endemic diploids, and that found previously within the putative auto- or allotetraploid derivative species *S. vulgaris* (Comes *et al.*, 1997). In regard to the glandular tetraploid taxa, these are characterized by a rather distinct set of cpDNAs (*D* and *M*) that do not form a monophyletic group in the cpDNA tree. Only *S. nebrodensis* appears to be polymorphic for cpDNA, whereas *S. viscosus* and *S. lividus*/*S. sylvaticus* are monomorphic for haplotypes *D* and *M*, respectively. Finally, the diploid *S. flavus* is polymorphic for two radically different sets of cpDNAs (*N* vs. *F* and *K*), with haplotype *F* being also present within the group of widespread diploid taxa. Overall, the magnitude of intraspecific cpDNA variation, together with the pattern of shared cpDNA haplotypes, precludes any accurate phylogenetic reconstruction of species relationships at this time.

Fig. 9.1 presents the majority-rule consensus of two equally parsimonious trees of 19 accessions from 14 species of the complex based on a cladistic analysis of nrDNA ITS sequence variation. Two allohexaploid species analysed, *S. cambrensis* and *S. teneriffae*, showed additivity at the variable sites in their ITS sequences, and, therefore, were omitted from the analysis. The ITS tree indicates that the diploid *S. vernalis* forms a separate clade with the tetraploid *S. vulgaris* (bootstrap value of 83%), whereas the majority of widespread diploid taxa, plus the Sicilian endemics *S. aethnensis* and *S. chrysanthemifolius*, form another well supported though largely unresolved clade (bootstrap value of 95%). The position of both of these subclades to each other and to the glandular tetraploid taxa is problematic because of the conflict created by several poorly (< 50%) supported branches (Fig. 9.1).

There are two instances in which both the ITS phylogeny and the cpDNA data contrast strongly with suspected species relationships based on traditional

Table 9.3 The single most parsimonious tree found under unweighted parsimony analysis of cpDNA haplotypes, and the distribution of these haplotypes among 17 species of the Mediterranean *Senecio* complex. Outgroup taxon was *S. malacitanus* (haplotype *P*). The tree had a length of 28 steps, CI (consistency index, excluding uninformative characters) = 0.923, and RI (retention index) = 0.946. Numbers above the branches refer to cpDNA polymorphisms identified in Table 9.2. All branches had bootstrap values of greater than 60% based on 100 replicates, except for the two marked by arrows (50–60%). Numbers within the table indicate the number of individuals per haplotype for each taxon surveyed. Haplotypes *A–H* correspond to those reported in previous studies (Abbott *et al.*, 1995; Kadereit *et al.*, 1995; Comes *et al.*, 1997; Comes and Abbott, 1998). Abbreviations of species are as follows: (aeg) *aegyptius*; (aeth) *aethnensis*; (chry) *chrysanthemifolius*; (flav) *flavus*; (gall) *gallicus*; (glauc) *glaucus*; (hesp) *hesperidium*; (leuc) *leucanthemifolius*; (liv) *lividus*; (mal) *malacitanus*; (nebr) *nebrodensis*; (pet) *petraeus*; (rod) *rodriguezii*; (squa) *squalidus*; (sylv) *sylvaticus*; (vern) *vernalis*; (visc) *viscosus*; (vulg) *vulgaris*.

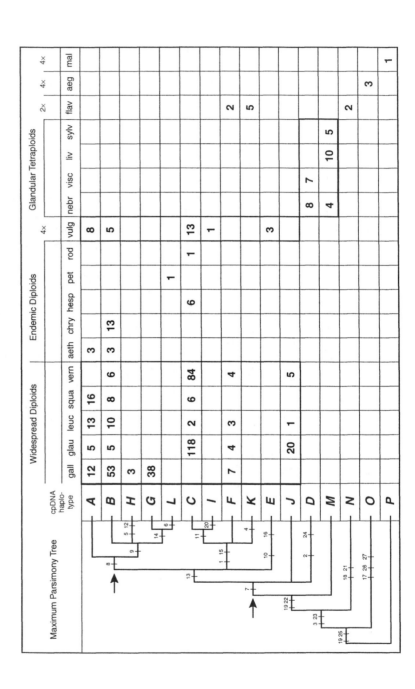

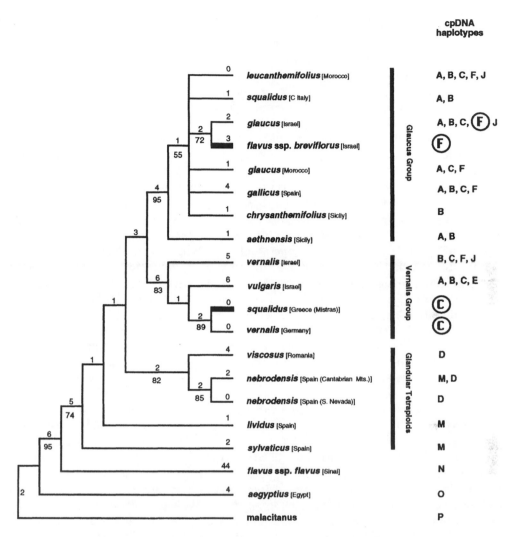

Figure 9.1 Majority-rule consensus of two equally parsimonious trees of 19 accessions from 14 species of the Mediterranean *Senecio* complex obtained under Fitch parsimony using ITS sequence data (ITS1–5.8S-ITS2) with indels scored as missing data. Outgroup taxon was *S. malacitanus*. The equal-weighted trees had a length of 119 steps, CI = 0.689 (excluding uninformative characters), and RI = 0.843. Numbers of nucleotide substitutions are shown above the branches, and bootstrap percentages from 100 replicates are shown below. Branches without bootstraps received less than 50%. Chloroplast DNA haplotypes found within species/accessions are also given for comparison. See text for further explanation.

morphological classification (indicated by solid branches in Fig. 9.1). Thus, (i) the short-rayed form of *S. flavus* (subsp. *breviflorus*), and (ii) *S. squalidus* from Greece (Mistras, Peloponnisos) possess ITS sequences and cpDNA haplotypes which otherwise are characteristic of species they co-occur with. *Senecio flavus* subsp. *breviflorus* shares a very similar ITS sequence and an identical cpDNA haplotype (*F*) with *S.*

glaucus from Israel, though both taxa are morphologically clearly distinct. Likewise, *S. squalidus* from Mistras has the same ITS sequence and the same cpDNA haplotype (*C*) as *S. vernalis* from Continental Europe (Germany). In this case, however, more detailed morphometric analyses have revealed that overall phenotype material from the Mistras population not only overlaps that of *S. squalidus* from Mt Olympus, Greece, (i.e. genuine *S. squalidus*), but also that of *S. vernalis* from Germany (King, 1994). Overall, the most likely explanation for different species sharing similar or identical ITS sequences and identical cpDNA haplotypes is transfer (or capture) of both of these molecules mediated by introgressive hybridization. *Senecio glaucus* and *S. vernalis*, respectively, must have served as the maternal parent, and in both cases homogenization among ITS repeats (Hillis and Dixon, 1991; Hillis *et al.*, 1991) obviously occurred in the direction of the maternal parent. Since we uncovered no additivity of ITS sequences within individuals of either *S. flavus* subsp. *breviflorus* or *S. squalidus* from Mistras we might conclude that both of these introgressive hybridization events were not *very* recent, and that concerted evolution (*sensu* Dover, 1982) was operating fast enough to homogenize ITS repeats within individuals and populations. In the case of *S. flavus* subsp. *breviflorus* this view of past introgression is supported by the observation that the originally captured chloroplast genome of *S. glaucus* has subsequently acquired one additional length mutation (see haplotype *K* in Table 9.3).

At least for the species pair *S. flavus/S. glaucus* there is unequivocal evidence based on RAPD markers that reticulation was largely unaccompanied by introgression of the rest of the nuclear genome. A neighbour-joining tree based on the results of the survey of RAPD variation over 9 species (18 accessions) of the Mediterranean *Senecio* complex is illustrated in Fig. 9.2. Overall, the RAPD phenogram is completely congruent with suspected species relationships based on morphological characters, and is also consistent with intraspecific groupings. However, it is in strong conflict with the cpDNA and ITS data sets with respect to *S. flavus*, such that the radiate subsp. *breviflorus* clusters now with the non-radiate subsp. *flavus*, and not with *S. glaucus* from Israel. (*Senecio squalidus* from Mistras and *S. vernalis* were not available for RAPD analysis).

Taken as a whole, our phylogenetic approach has demonstrated (i) the occurrence of introgressive hybridization, i.e. cpDNA and ITS capture, among species of phylogenetically rather different clades as identified by the ITS data; and also (ii) a striking example of an apparent lack of nuclear introgression, except for nrDNA ITS sequences. Thus, *S. flavus* subsp. *breviflorus* and *S. squalidus* from Mistras have captured the cpDNA and the ITS sequence of another co-occurring species, whereby in the case of the former, this capture evidently occurred without any or only a small change in RAPD or morphological phenotype. The presence of short marginal ray florets in *S. flavus* subsp. *breviflorus* may be regarded as such a small change in phenotype following nuclear introgression (see also Liston and Kadereit, 1995). Contrastingly, based on the morphometric evidence available, *S. squalidus* from Mistras is likely to have been more strongly impacted by the introgression of nuclear genes.

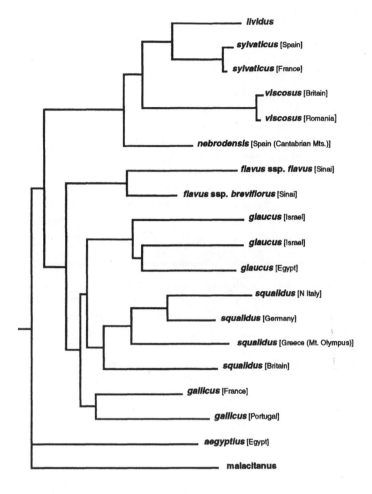

Figure 9.2 Neighbour-joining tree of 18 accessions from 9 species of the Mediterranean *Senecio* complex based on their estimated genetic distances for 362 RAPD products obtained with 17 primers. The tree was rooted using *S. malacitanus* as an outgroup. The bar indicates a genetic distance of 0.01.

9.4 Population-level study of interspecific hybridization and introgression

9.4.1 The study species

Our decision to choose *Senecio vernalis* and *S. glaucus* for this population study was governed by the fact that the two species (i) are highly interfertile (Alexander, 1979; Kadereit, 1983; Comes, unpubl. data); (ii) share four out of five distinct cpDNA haplotypes (Table 9.3); (iii) belong to two different, well-supported subclades as evidenced by the phylogenetic ITS sequence data (Fig. 9.1); and

(iv) form a distinctive zone of parapatric distribution in the Israel area as illustrated in Fig. 9.3. This makes them ideal candidates for studying the effects of differential cytoplasmic (cpDNA) and nuclear gene flow across a species contact zone, which most likely represents a zone of secondary contact rather than primary intergradation as suggested by the ITS data. In the present context, cytoplasmic gene flow implies movement of seed, since cpDNA in *Senecio* is inherited through the maternal parent (Harris, 1990; Harris and Ingram, 1992b).

Senecio vernalis and *S. glaucus* from the Near East vary little in life history and other phenotypic characteristics. In Israel, main flowering occurs from early winter through spring, with considerable overlap of flowering time between *S. vernalis* and *S. glaucus* (Feinbrun-Dothan, 1978; Comes and Abbott, pers. obs.). Both species are insect-pollinated, highly outcrossing and no obvious differences in primary floral morphology or pollinator preferences appear to exist (Feinbrun-Dothan, 1978; Alexander, 1979). Like many other members of the *Asteraceae* they possess specialized adaptations to facilitate long-distance dispersal, such that achenes bear a pappus of feathery hairs and become mucilaginous and sticky after moistening, thus being capable of moving both through the air and by hitchhiking on humans and animals. However, the species differ dramatically in their geographical distribution and ecological requirements in Israel (Fig. 9.3). Here, *S. vernalis* is one of the most important components of man-made habitats of the 'mesic' Mediterranean phytogeographic region (*sensu* Zohary, 1973), encompassing humid, sub-humid and semi-arid climatic regions (see also Fig. 9.4). By contrast, *S. glaucus* is typically associated with either coastal dune systems, or arid regions of both the steppic Irano-Turanian and the true desert Saharo-Arabian ecological zones of the Near East (Fig. 9.3). Areas of close proximity or sympatry include disturbed areas next to the Mediterranean Sea, as well as the Dead Sea (Comes and Abbott, pers. obs., 1995).

9.4.2 Material and Methods

For our cpDNA and allozyme survey, we sampled 11 Israeli populations of *S. vernalis* and 14 populations of *S. glaucus*, 11 from Israel and three from Egypt (Fig. 9.4). All populations analysed for allozyme variation were also included in the survey of restriction site and length variation of cpDNA, with the exception of two *S. vernalis* populations (abbreviated NY and Q in Fig. 9.4).

9.4.2.1 CpDNA analysis

The cpDNA survey involved DNA extraction methods and filter hybridization techniques with cloned probes of digoxigenin-labelled cpDNA from *Lactuca sativa* as indicated earlier. In a preliminary screen, one or two individuals from each of the populations (except M'AS) were analysed, using 6 of the 8 probe–enzyme combinations listed in Table 9.2, thereby excluding pLsC 8–11/*Hin*DIII and Xba I. Only those three probe–enzyme combinations that detected polymorphisms either within or among populations (pLsC 1–3/*Pst*I, pLsC 6/*Cla*I, pLsC 6/*Cfo*I) were retained for the full survey of 87 individuals of *S. vernalis* and 138 individuals of *S. glaucus*. Sample sizes of 10 individuals per population were almost equivalent across all populations.

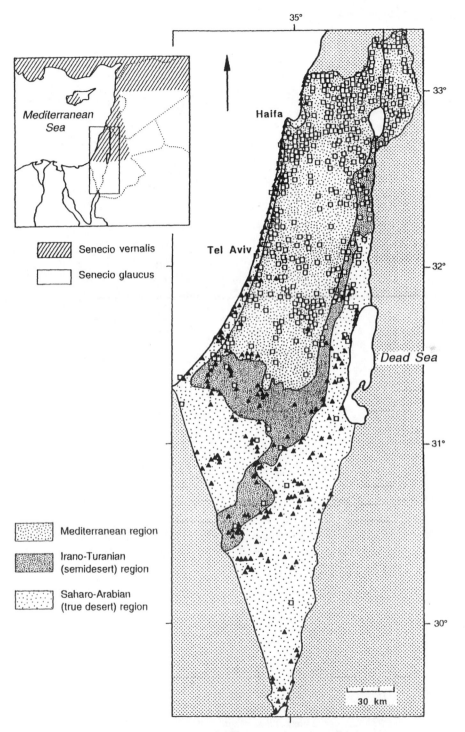

Figure 9.3 Major environmental belts (adapted from Zohary, 1973) and distribution of *S. vernalis* (squares) and *S. glaucus* (triangles) in the Israel area. Symbols indicate known populations based on locality information kindly provided by the Israel Plant Information Center (ROTEM), Hebrew University of Jerusalem. The overall, approximate ranges of *S. vernalis* and *S. glaucus* in the Eastern Mediterranean, shown in the inset, are modified from Alexander (1979) and Meusel and Jäger (1992).

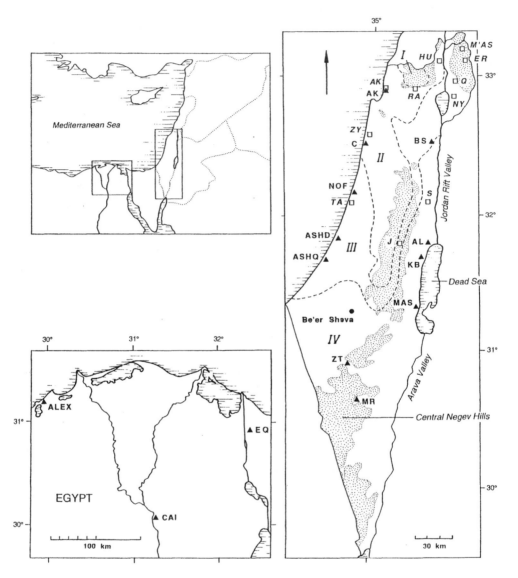

Figure 9.4 Localities of populations of *S. vernalis* (squares) and *S. glaucus* (triangles) collected in Israel and Egypt (only *S. glaucus*). The dashed lines mark the four climatic regions of Israel (I-IV) defined as (I) humid, (II) subhumid, (III) semiarid, and (IV) arid. Areas at 500m elevation or higher are indicated by stippling. Abbreviations of localities are as follows: (AK) Akko; (AL) Almog; (ALEX) Alexandria; (ASHD) Ashdod; (ASHQ) Ashquelon; (BS) Bet-She'an; (C) Caesarea; (CAI) Cairo; (EQ) El Quantara; (ER) El Rom; (HU) Hule Plain/Zomet Gonen; (J) Jerusalem; (KB) Khirbet Mezin; (M'AS) Mas'ada/Golan; (MAS) Massada; (MR) Mizpe Ramon; (NOF) Nof Yam; (NY) Nahal Yehudiyya; (Q) Quazrin; (RA) Rama; (S) Sartaba; (TA) Tel Aviv; (ZT) Zomet Telalim; (ZY) Zikhron Ya'aqov.

9.4.2.2 Allozyme analysis

Allozyme variation was assayed at eight presumptive loci (*Aat-3*; β-*Est-3*; *Idh-1*; *G3pd-1*, *Pgi-1*, *Pgi-2*; *Pgm-1* and *Pgm-2*) using horizontal starch gel electrophoresis. Details of isozyme extraction, electrophoretic techniques and the genetic interpretation of isozymes have been described elsewhere (Comes and Abbott, 1999). Allozyme variation was assayed for a total of 330 individuals for *S. vernalis* and 433 individuals of *S. glaucus*, with an average sample size per population of approximately 30 individuals. The Cavalli-Sforza chord distance (Cavalli-Sforza and Edwards, 1967; Swofford and Selander, 1989) was used to estimate the genetic distances among populations. This metric is considered to be relatively insensitive to variation in evolutionary rates not only among loci but also among evolutionary lineages (Swofford *et al.*, 1996).

9.4.3 Results and Discussion

The cpDNA survey of a total of 225 individuals of *S. vernalis* and *S. glaucus* in Israel and Egypt resolved altogether five haplotypes (*A*, *B*, *C*, *F*, and *J*). These haplotypes differed by 1–5 mutations (Table 9.3), most of which were due to site mutations rather than insertions/deletions (Table 9.2). One haplotype, *C*, was common and widespread among populations (Fig. 9.5), although more than 70% of all populations surveyed in both species were polymorphic for cpDNA. Haplotype *C*, together with haplotypes *B*, *F*, and *J*, were shared not only among neighbouring populations and non-neighbouring populations of the same species, but also among those belonging to both species (Fig. 9.5). Given that the two species occur in two quite distinct subclades in the ITS phylogeny (see above), it is most likely that the similarities in cpDNA between them result from interspecific seed flow, rather than the retention of an ancestral cpDNA polymorphism that predates species divergence. An inferred estimate of interspecific seed flow ($Nm_{(o)}$) was derived from the average of pairwise θ values for cpDNA among pairs of populations from the two species, and corrected for essentially haploid, organellar loci (Weir and Cockerham, 1984; Crow and Kimura, 1970; Birky *et al.*, 1983). This estimate was well above one, $Nm_{(o)} = 1.32$, suggesting a considerable amount of interspecific (cytoplasmic *and* nuclear) gene flow via seed (e.g. Wright, 1931; Slatkin, 1985).

The allozyme survey resolved altogether six polymorphic loci (*Aat-3*; β-*Est-3*; *Idh-1*; *G3pd-1*; *Pgi-2*; and *Pgm-2*) out of the eight loci examined. Although both species lacked fixed, diagnostic alleles, there were substantial differences in allele frequencies at polymorphic loci. A UPGMA phenogram based on chord distances between populations separated the two species into two distinct clusters, corresponding strictly to their taxonomic, i.e. morphological status (Fig. 9.6). These patterns of allozymic divergence are consistent with our view gained from the ITS tree that *S. vernalis* and *S. glaucus* are not undergoing primary divergence in the Near East, but instead are forming a zone of secondary contact. The unbiased estimate of interspecific nuclear gene flow calculated from hierarchical *F*-statistics (Wolf and Soltis, 1992) is very low, $Nm_{(n)} = 0.253$, indicating that *S. vernalis* and *S. glaucus* are almost fully isolated in terms of nuclear encoding loci.

CpDNA haplotypes

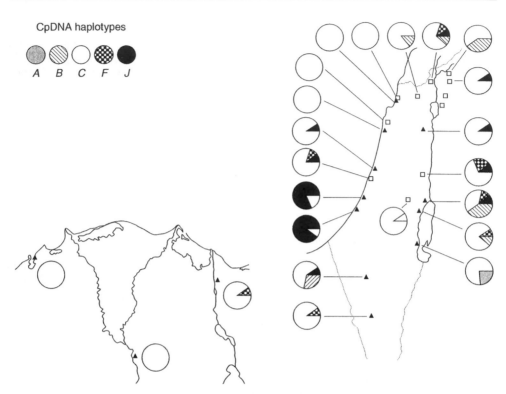

Figure 9.5 Geographic distribution and relative frequencies of five cpDNA haplotypes in sample populations of *S. vernalis* (squares) and *S. glaucus* (triangles). Pies represent samples of eight to ten individuals per site. Sample populations are identified in Fig. 9.4.

Overall, if interpreted in terms of contemporary gene flow, the observed contrast between low cytoplasmic and high nuclear (allozymic) subdivision found among *S. vernalis* and *S. glaucus* is somewhat puzzling as the relatively high level of inferred interspecific seed flow ($Nm_{(o)} = 1.32$) would be expected to prevent the occurrence of both cytoplasmic and nuclear genetic differentiation between the species. One solution to this paradox could arise from the possibility that nuclear genetic differentiation is maintained by selection favouring particular alternative allozymes (or linked nuclear genes) in the different habitats occupied by *S. vernalis* and *S. glaucus* (mesophytic/xerophytic). Alternatively, the observed pattern could result from the rare but chronic or infrequent mass immigration of seeds into 'foreign' sets of recipient populations. In such cases, the nuclear DNA of the female donor plants would be rapidly eliminated in repeated back-cross generations, while the corresponding non-recombinant cpDNA is more likely to persist. Such a process, especially if followed by rapid population growth and/or further intraspecific cpDNA migration (Rieseberg and Soltis, 1991), could lead to the accumulation of 'foreign' cpDNA haplotypes over time, which would contribute to an overestimation of ongoing levels of interspecific seed flow in terms of $Nm_{(o)}$.

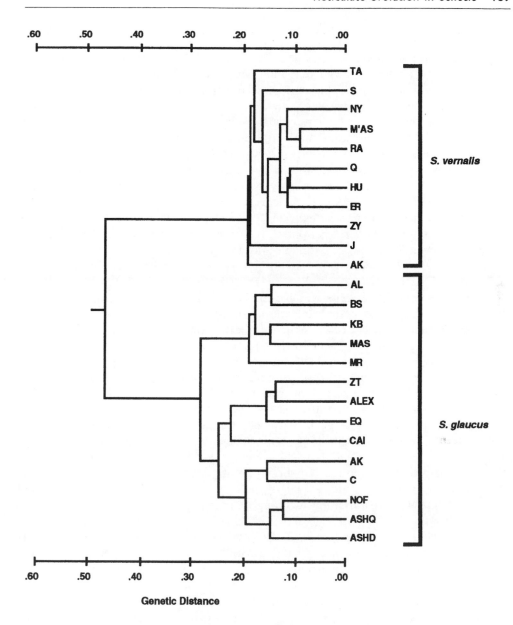

Figure 9.6 A UPGMA tree based on Cavalli-Sforza and Edwards (1967) chord distances, illustrating genetic relationships among populations of *S. vernalis* and *S. glaucus*. Refer to Fig. 9.4 for locality abbreviations.

9.5 Conclusions

The principal aim of the work reported here was to investigate a group of plant species of supposedly recent divergence to determine (i) the occurrence of inter-specific gene flow; and (ii) the long term effects of such gene flow on the evolution and maintenance of species differences. The results obtained are of broad evolutionary significance given that interspecific gene flow is not uncommon in the plant kingdom, especially among species at an early stage of divergence from each other.

Our cpDNA analysis showed that many species of the Mediterranean complex of sect. *Senecio* are polymorphic and share the same range of cpDNA haplotypes (Table 9.3). This may be due either to reticulation and/or the differential sorting of chloroplast genomes from a polymorphic ancestor (Neigel and Avise, 1986). Thus, our cpDNA results argue that traditional phylogenetic methods relying on a hierarchical pattern of lineage divergence have major limitations in inferring species relationships in recently evolved and/or hybridizing plant groups because morphologically well differentiated species may be aligned erroneously in a phylogenetic tree constructed from cpDNA data. Congruencies between cpDNA trees and organismal phylogenies are not to be expected (see Doyle, 1992), and other approaches that do not assume a hierarchical pattern of differentiation should be invoked (e.g. Slatkin and Maddison, 1989; Crandall and Templeton, 1996).

Nuclear rDNA restriction profiles and ITS sequences may provide more accurate information on the evolutionary history and species relationships (e.g. Rieseberg, 1991; Sang *et al.*, 1995; Soltis and Kuzoff, 1995), although again, due to ancient or more recent episodes of hybridization resulting in partial nuclear capture and (un)even homogenization of ITS sequences, confusion may often persist (Rieseberg *et al.*, 1990b; Sang *et al.*, 1995). In the present study, the ITS tree is clear on the taxonomic status of *S. vernalis*. This species is placed in a well-supported separate clade from the other widespread diploid taxa, including *S. leucanthemifolius* (Fig. 9.1). Traditionally treated at specific rank (e.g Feinbrun-Dothan, 1978), *S. vernalis* has also been recognized as a variety of *S. leucanthemifolius* because of the 'lack of reliable morphological characters' for separating the two taxa, and high pollen fertility of their artificial hybrids (Alexander, 1979). Based on the ITS sequence data, it seems clear that Alexander's (1979) treatment is unnatural, and that the polyphyletic group created should be abandoned. Thus, similarities in lifeform, habitat preferences, and leaf characteristics between *S. leucanthemifolius* and *S. vernalis* provide an intriguing example of convergence and/or the retention of ancestral traits.

In contrast, the ITS tree indicates that two co-occurring, but morphologically clearly distinct taxa, i.e. *S. flavus* subsp. *breviflorus* and *S. glaucus* (from Israel), are to be considered as sister groups, rather than being placed in different clades. *Senecio flavus* subsp. *flavus* is distinguished from the *S. glaucus/S. flavus* subsp. *breviflorus* clade by numerous autapomorphic mutations in the ITS sequence. It is concluded that subsp. *breviflorus* obtained its ITS sequence from *S. glaucus* through interspecific hybridization. The lesson learnt from this result is that a phylogenetic approach based on a large number of nuclear genes is more likely to provide an accurate reflection of overall organismal phylogeny when hybridization leads to cytoplasmic and partial nuclear gene capture. This conclusion is reinforced by the RAPD tree (Fig. 9.2), which placed the two subspecies of *S. flavus* together, and separate from *S. glaucus*.

Our population-level analysis has shown that a combined analysis of cpDNA and isozyme variation between two species that form a secondary zone of contact is instructive in regard to the maintenance of species differences despite introgression of cytoplasmic genomes via seed. This was demonstrated for *S. glaucus* and *S. vernalis* that form a distinctive zone of parapatric distribution in the Near East by differing in their ecogeographical regimes. The two species appear to maintain their phenotypic integrity, as reflected by a high genetic distance for allozymes (Fig. 9.6), yet seem to have exchanged chloroplast genomes in a rampant though elusive fashion. What is of central relevance here is that retention of ancestral cpDNAs appears untenable as an explanation for the sharing of haplotypes because of the need for four (out of five) different cpDNA haplotypes to have persisted through two diverging phylogenetic lineages. Nonetheless, we are unsure about the levels of ongoing cytoplasmic gene flow across allozymically defined species boundaries: either our inferred estimates of $Nm_{(o)}$ are correct, and there is selection acting against 'foreign' nuclear (allozymic) but not cytoplasmic DNA; or the two species are ecologically well isolated, and our $Nm_{(o)}$ estimates imply erroneously high levels of interspecific seed flow due to the sporadic exchange of cpDNA via seed with large homogenizing effects on cytoplasmic population structure over time. The finding of extremely low levels of hybrid genotypes among the offspring of the two species at one sympatric site, Akko (AK in Fig. 9.4), favours the second of these hypotheses. Further clarification of the allozyme/cpDNA paradox in *S. vernalis* and *S. glaucus* awaits reciprocal transplant experiments of natural selection, and experimental and computer simulation studies on the long-term consequences of sporadic cpDNA immigration.

Despite the fact that phylogenetic and population-genetical evidence both point to the conclusion that contemporary and recent transfer of cpDNA or nrDNA may have a lasting effect on the genetic constitution of a species, the establishment and maintenance of morphologically distinct forms in sect. *Senecio* is unlikely to have been strongly impacted by the recurrent gene exchange of molecular marker genes that may be selectively neutral. Nevertheless, it is important to consider that current lack of evidence for extensive nuclear introgression among diploid species of *Senecio* (i.e. *S. flavus/S. glaucus* and *S. vernalis/S. glaucus*) does not rule out the possibility of very subtle patterns of past and present introgressive hybridization of adaptively significant genes or gene complexes, not least the establishment of novel genotypes and the origin of new derivative species following interspecific hybridization. Significantly, the population of *S. squalidus* from Mistras is a likely candidate for such a scenario (King, 1994; present study), and there is also increasing evidence that British *S. squalidus*, derived from material introduced from Sicily, might be a product of diploid hybrid speciation involving the Sicilian endemics *S. aethnensis* and *S. chrysanthemifolius* (Abbott and Milne, 1995). While stressing these points, the significance of polyploidy in the founding of new lineages during the early and more recent organismal evolution of the Mediterranean species complex of sect. *Senecio* remains undisputed (Kadereit, 1984; Abbott, 1992; Abbott and Lowe, 1996).

Our final conclusion is that an integration of phylogenetic and population-level approaches is critical to understanding the role of hybridization and introgression in plant speciation and evolution. Phylogenetic evidence helps target the histories of reticulating lineages, and assists in reducing the probability of confounding common ancestry, parallel evolution, and reticulation. In turn, population studies are needed to test the likelihood of the historical scenarios posed, and serve as a fundamental basis of more process-oriented hypotheses about introgressive hybridization that may be tested experimentally.

ACKNOWLEDGEMENTS

We wish to thank Amanda Gillies, Kirsten Wolff and David Forbes for expert technical advice and assistance at the St Andrews laboratory. We are especially grateful to Mark Chase and his colleagues for allowing us to conduct the ITS sequencing at the Jodrell Laboratory, Royal Botanic Gardens, Kew, and for useful comments and advice. HPC owes a special debt of gratitude to A. Sofiman Othman for his generous assistance at Kew. We also thank Avi Shmida for access to data on the distribution of *S. vernalis* and *S. glaucus* in Israel; the many colleagues who helped in collecting seed material of *Senecio*; Joachim Kadereit for critically reading an earlier draft of the manuscript; and Anke Berg for drawing the maps in Figs 9.3 to 9.5. This research was funded by a grant from the NERC (GR3/09053) to RJA.

REFERENCES

Abbott, R. J. (1992) Plant invasions, interspecific hybridization and the evolution of new plant taxa. *Trends in Ecology and Evolution*, **7**, 401–05.

Abbott, R. J. and Lowe, A. J. (1996) A review of hybridization and evolution in British *Senecio*, in *Compositae: systematics*. Proceedings of the International Compositae Conference, Kew, 1994, (eds D. J. N. Hind and H. J. Beentje), Royal Botanic Gardens, Kew, vol. 1, pp. 679–89.

Abbott, R. J. and Milne, R. I. (1995) Origins and evolutionary effects of invasive weeds, in *Weeds in a changing world*, BCPC Symposium Proceedings No. 64, British Crop Protection Council, pp. 53–64.

Abbott, R. J., Curnow, D. J. and Irwin, J. A. (1995) Molecular systematics of *Senecio squalidus* L. and its close diploid relatives, in *Advances in Compositae systematics*, (eds D.J.N. Hind, C. Jeffrey and G.V. Pope), Royal Botanic Gardens, Kew, pp. 223–37.

Alexander, J. C. M. (1979) Mediterranean species of *Senecio* sections *Senecio* and *Delphinifolius*. *Notes of the Royal Botanical Garden Edinburgh*, **37**, 387–428.

Allan, G. J., Clark, C. and Rieseberg, L. H. (1997) Distribution of parental DNA markers in *Encelia virginensis* (Asteraceae: Heliantheae), a diploid species of putative hybrid origin. *Plant Systematics and Evolution*, **205**, 205–21.

Arnold, M. L. (1992) Natural hybridization as an evolutionary process. *Annual Review of Ecology and Systematics*, **23**, 237–61.

Arnold, M. L. (1997) *Natural hybridization and evolution*, Oxford University Press, Oxford.

Arnold, M. L., Buckner, C. M. and Robinson, J. J. (1991) Pollen mediated introgression and hybrid speciation in Louisiana irises. *Proceedings of the National Academy of Sciences of the USA*, **88**, 1398–1402.

Ashton, P. and Abbott, R. J. (1992) Isozyme evidence and the origin of *Senecio vulgaris* (Compositae). *Plant Systematics and Evolution*, **179**, 167–74.

Banks, J. A. and Birky, C. W., Jr. (1985) Chloroplast DNA diversity is low in a wild plant, *Lupinus texensis*. *Proceedings of the National Academy of Sciences of the USA*, **82**, 6950–54.

Barton, N. and Hewitt, G. M. (1985) Analysis of hybrid zones. *Annual Review of Ecology and Systematics*, **16**, 113–48.

Birky, C. W., Jr., Maruyama, T. and Fuerst, P. (1983) An approach to population and evolutionary genetic theory for genes in mitochondria and chloroplasts, and some results. *Genetics*, **103**, 513–27.

Byrne, M. and Moran, G. F. (1994) Population divergence in the chloroplast genome of *Eucalyptus nitens. Heredity*, **73**, 18–28.

Cato, S. A. and Richardson, T. E. (1996) Inter- and intraspecific polymorphism at chloroplast SSR loci and the inheritance of plastids in *Pinus radiata* D. Don. *Theoretical and Applied Genetics*, **93**, 587–592.

Cavalli-Sforza, L. L. and Edwards, A. W. F. (1967) Phylogenetic analysis: models and estimation procedures. *Evolution*, **21**, 550–70.

Comes, H. P. and Abbott, R. J. (1998) The relative importance of historical events and gene flow on the population structure of a Mediterranean ragwort, *Senecio gallicus* (Asteraceae). *Evolution*, **52**, 355–67.

Comes, H. P. and Abbott, R. J. (1999) Population genetic structure and gene flow across arid versus mesic environments: a comparative study of two parapatric *Senecio* species from the Near East *Evolution*, **53**, 36–54.

Comes, H. P., Kadereit, J. W., Pohl, A. and Abbott, R. J. (1997) Chloroplast DNA and isozyme evidence on the evolution of *Senecio vulgaris* (Asteraceae). *Plant Systematics and Evolution*, **206**, 375–92.

Crandall, K. A. and Templeton, A. R. (1996) Applications of intraspecific phylogenetics, in *New uses for new phylogenies*, (eds P.H. Harvey, A.J. Leigh Brown, J. Maynard Smith and S. Nee), Oxford University Press, Oxford, pp. 81–99.

Crow, J. F. and Kimura, M. (1970) *An introduction to population genetics theory.* Burgess Publishing Company, Minneapolis, Minnesota.

Cruzan, M. B. and Arnold, M. L. (1993) Ecological and genetic associations in an *Iris* hybrid zone. *Evolution*, **47**, 1432–45.

Cruzan, M. B., Arnold, M. L., Carney, S. E. and Wollenberg, K. R. (1993) cpDNA inheritance in interspecific crosses and evolutionary inference in Louisiana irises. *American Journal of Botany*, **80**, 344–50.

Demesure, B., Comps, B. and Petit, R. J. (1996) Chloroplast DNA phylogeography of the common beech (*Fagus sylvatica* L.) in Europe. *Evolution*, **50**, 2515–20.

Dover, G. A. (1982) Molecular drive: a cohesive mode of species evolution. *Nature*, **299**, 111–116.

Dowling T. E., Smith, G. R. and Brown, W. M. (1989) Reproductive isolation and introgression between *Notropis cornutus* and *Notropis chrysocephalus* (Family *Cyprinidae*): comparison of morphology, allozymes, and mitochondrial DNA. *Evolution*, **43**, 620–34.

Doyle, J. J. (1992) Gene trees and species trees: molecular systematics as one-character taxonomy. *Systematic Botany*, **11**, 373–91.

Dumolin-Lapègue, S., Demesure, B., Fineschi, S., Le Corre, V. and Petit, R. J. (1997) Phylogeographic structure of white oaks throughout the European continent. *Genetics*, **146**, 1475–87.

Edwards-Burke, M. A., Hamrick, J. L. and Price, R. A. (1997) Frequency and direction of hybridization in sympatric populations of *Pinus taeda* and *P. echinata* (Pinaceae). *American Journal of Botany*, **84**, 879–86.

El Mousadik, A. and Petit, R. J. (1996) Chloroplast DNA phylogeography of the argan tree of Morocco. *Molecular Ecology*, **5**, 547–55.

Ennos, R. A., Sinclair, W. T., Hu, X-S. and Langdon. A. (1999) Using organelle markers to elucidate the history, ecology and evolution of plant populations, in *Molecular systematics and plant evolution*, (eds P. M. Hollingsworth, R. M. Bateman and R. J. Gornall), Taylor & Francis, London, pp. 1–19.

Excoffier, L., Smouse, P. E. and Quattro, J. M. (1992) Analysis of molecular variance inferred from metric distances among DNA haplotypes: application to human mitochondrial DNA restriction data. *Genetics*, **86**, 991–1000.

Feinbrun-Dothan, N. (1978) *Flora Palaestina, vol. 3*, Israel Academy of Science and Humanities, Jerusalem.

Felsenstein, J. (1993) *PHYLIP 3.5*. Distributed by the author. University of Washington, Seattle.

Ferris, C., King, R. A. and Hewitt, G. M. (1999) Isolation within species and the history of glacial refugia, in *Molecular systematics and plant evolution*, (eds P. M. Hollingsworth, R. M. Bateman and R. J. Gornall), Taylor & Francis, London, pp. 20–34.

Forbes, S. H. and Allendorf, F. W. (1991) Associations between mitochondrial and nuclear genotypes in cutthroat trout hybrid swarms. *Evolution*, **45**, 1332–49.

Gallagher, K. G., Schierenbeck, K. A. and D'Antonio, C. M. (1997) Hybridization and introgression in *Carpobrotus* spp. (Aizoaceae) in California II. Allozyme evidence. *American Journal of Botany*, **84**, 905–11.

Givnish, T. J. and Sytsma, K. J. (eds) (1997) *Molecular evolution and adaptive radiation*, Cambridge University Press, Cambridge.

Harris, S. A. (1990) *Molecular systematic studies in some members of the genus* Senecio *L. (Asteraceae)*. Ph.D. Thesis, University of St Andrews.

Harris, S. A. (1995) Systematics and randomly amplified polymorphic DNA in the genus *Leucaena* (Leguminosae, Mimosoideae). *Plant Systematics and Evolution*, **197**, 195–208.

Harris, S. A. (1999) RAPDs in systematics – a useful methodology? in *Molecular systematics and plant evolution*, (eds P. M. Hollingsworth, R. M. Bateman and R. J. Gornall), Taylor & Francis, London, pp. 211–28.

Harris, S. A. and Ingram, R. (1991) Chloroplast DNA and biosystematics: the effects of intraspecific diversity and plastid transmission. *Taxon*, **40**, 393–412.

Harris, S. A. and Ingram, R. (1992a) Molecular systematics of the genus *Senecio* L., II: The origin of *S. vulgaris* L. *Heredity*, **69**, 112–21.

Harris, S. A. and Ingram, R. (1992b) Molecular systematics of the genus *Senecio* L., I: Hybridization in a British polyploid complex. *Heredity*, **69**, 1–10.

Harrison, R. G. (1990) Hybrid zones: windows on evolutionary process. *Oxford Surveys in Evolutionary Biology*, **7**, 69–128.

Hershkovitz, M. A., Zimmer, E. A. and Hahn, W. J. (1999) Ribosomal DNA sequences and angiosperm systematics, in *Molecular systematics and plant evolution*, (eds P. M. Hollingsworth, R. M. Bateman and R. J. Gornall), Taylor & Francis, London, pp. 268–326.

Higgins, D. G., Bleasby, A. J. and Fuchs, R. (1992) CLUSTAL V: improved software for multiple sequence alignment. *Computer Applications in the Biosciences* (CABIOS), **8**, 184–91.

Hillis, D. M. and Dixon, M. T. (1991) Ribosomal DNA: molecular evolution and phylogenetic inference. *Quarterly Review of Biology*, **66**, 411–53.

Hillis, D. M., Moritz, C., Porter, C. A. and Baker, R. J. (1991) Evidence for biased gene conversion in concerted evolution of ribosomal DNA. *Science*, **251**, 308–10.

Hu, X.-S. and R. A. Ennos (1997) On estimation of the ratio of pollen to seed flow among plant populations. *Heredity*, **79**, 541–52.

Jansen, R. K. and Palmer, J. D. (1987) Chloroplast DNA from lettuce and *Barnadesia* (Asteraceae): structure, gene localisation and characterisation of a large inversion. *Current Genetics*, **11**, 553–64.

Jeffrey, C. (1979) Generic and sectional limits in *Senecio* (Compositae) II. Evaluation of some recent studies. *Kew Bulletin*, **34**, 49–58.

Jeffrey, C., Halliday, P., Wilmot-Dear, M. and Jones, S. W. (1977) Generic and sectional limits in *Senecio* (Compositae) I. Progress report. *Kew Bulletin*, **32**, 47–67.

Jordan, W. C., Courtney, M. W. and Neigel, J. E. (1996) Low levels of infraspecific genetic variation at a rapidly evolving chloroplast DNA locus in North American duckweeds (Lemnaceae). *American Journal of Botany*, **83**, 430–39.

Kadereit, J. W. (1983) *Experimental study and revision of* Senecio *section* Obaejacae. Ph.D. Thesis, University of Cambridge.

Kadereit, J. W. (1984) The origin of *Senecio vulgaris* (Asteraceae). *Plant Systematics and Evolution*, **145**, 135–53.

Kadereit, J. W., Comes, H. P., Curnow, D. J., Irwin, J. A. and Abbott, R. J. (1995) Chloroplast DNA and isozyme analysis of the progenitor-derivative species relationship between *Senecio nebrodensis* and *S. viscosus* (Asteraceae). *American Journal of Botany*, **82**, 1179–85.

King, R. A. (1994) *Hybridization and ecotypic differentiation in southern European* Senecio *(Asteraceae): a preliminary study*. B.Sc. (Hons.) Dissertation, University of St Andrews.

Larson, S. R. and Doebley, J. (1993) Restriction site variation in the chloroplast genome and nuclear ribosomal DNA of *Tripsacum* (Poaceae): phylogeny and rates of sequence evolution. *Systematic Botany*, **19**, 21–34.

Levy, F., Antonovics, J., Boynton, J. E. and Gillham, N. W. (1996) A population genetic analysis of chloroplast DNA in *Phacelia*. *Heredity*, **76**, 143–55.

Liston, A. and Kadereit, J. W. (1995) Chloroplast DNA evidence for introgression and long distance dispersal in the desert annual *Senecio flavus* (Asteraceae). *Plant Systematics and Evolution*, **197**, 33–41.

Lorenz, M., Weihe, A. and Börner, T. (1994) DNA fragments of organellar origin in random amplified polymorphic DNA (RAPD) patterns of sugar beet (*Beta vulgaris* L.). *Theoretical and Applied Genetics*, **88**, 775–79.

Lowe, A. J. and Abbott, R. J. (1996) Origins of the new allopolyploid species *Senecio cambrensis* and its relationship to the Canary Island endemic *Senecio teneriffae* (Asteraceae). *American Journal of Botany*, **83**, 1365–72.

Meusel H. and Jäger, E. J. (1992) *Vergleichende Chorologie der Zentraleuropäischen Flora*, Gustav Fischer Verlag, Jena.

Milligan, B. G. (1991) Chloroplast DNA diversity within and among populations of *Trifolium pratense*. *Current Genetics*, **19**, 411–16.

Neale, D. B., Saghai-Maroof, M. A., Allard, R. W., Zhang, Q. and Jorgensen, R. A. (1988) Chloroplast DNA diversity in populations of wild and cultivated barley. *Genetics*, **120**, 1105–10.

Neigel, J. E. and Avise, J. C. (1986) Phylogenetic relationships of mitochondrial DNA under various demographic models of speciation, in *Evolutionary processes and theory*, (eds E. Nevo and S. Karlin), Academic Press, New York, pp. 515–34.

Powell, W., Morgante M., McDevitt, R., Vendramin, G. G. and Rafalski, J. A. (1995) Polymorphic simple sequence repeat regions in chloroplast genomes: applications to the population genetics of pines. *Proceedings of the National Academy of Sciences of the USA*, **92**, 7759–7763.

Powell, W., Morgante M., Doyle J. J., McNicol, J. W., Tingey, S. V. and Rafalski, J. A. (1996) Genepool variation in genus *Glycine* subgenus *Soja* revealed by polymorphic nuclear and chloroplast microsatellites. *Genetics*, **144**, 793–803.

Provan, J., Corbett, G., McNicol, J. W. and Powell, W. (1997) Chloroplast DNA variability in wild and cultivated rice (*Oryza* spp.) revealed by polymorphic chloroplast simple sequence repeats. *Genome*, **40**, 104–110.

Provan, J., Soranzo, N., Wilson, N. J., McNicol, J. W., Morgante, M. and Powell, W. (1999) The use of uniparentally inherited simple sequence repeat markers in plant population studies and systematics, in *Molecular systematics and plant evolution*, (eds P. M. Hollingsworth, R. M. Bateman and R. J. Gornall), Taylor & Francis, London, pp. 35–50.

Purps, D. M. L. and Kadereit, J. W. (1998) RAPD evidence for a sister group relationship of the presumed progenitor-derivative species pair *Senecio nebrodensis* and *S. viscosus* (Asteraceae). *Plant Systematics and Evolution*, **211**, 57–70.

Rieseberg, L. H. (1991) Homoploid reticulate evolution in *Helianthus* (Asteraceae): evidence from ribosomal genes. *American Journal of Botany*, **78**, 1218–37.

Rieseberg, L. H. and Soltis, D. E. (1991) Phylogenetic consequences of cytoplasmic gene flow in plants. *Evolutionary Trends in Plants*, **5**, 65–84.

Rieseberg, L. H. and Wendel, J. F. (1993) Introgression and its consequences in plants, in *Hybrid zones and the evolutionary process* (ed. R. G. Harrison), Oxford University Press, Oxford, pp. 70–109.

Rieseberg, L. H., Soltis, D. E. and Palmer, J. D. (1988) A molecular reexamination of introgression between *Helianthus annuus* and *H. bolanderi* (Compositae). *Evolution*, **42**, 227–38.

Rieseberg, L. H., Carter, R. and Zona, S. (1990a) Molecular tests of the hypothesized hybrid origin of two diploid *Helianthus* species (Asteraceae). *Evolution*, **44**, 1498–1511.

Rieseberg, L. H., Beckstrom-Sternberg, S. and Doan, K. (1990b) *Helianthus annuus* ssp. *texanus* has chloroplast DNA and nuclear ribosomal RNA genes of *Helianthus debilis* ssp. *cucumerifolius*. *Proceedings of the National Academy of Sciences of the USA*, **87**, 593–97.

Rieseberg, L. H., Desrochers, A. M. and Youn, S. J. (1995) Interspecific pollen competition as a reproductive barrier between sympatric species of *Helianthus* (Asteraceae). *American Journal of Botany*, **82**, 515–19.

Roelofs, D. and Bachmann, K. (1995) Chloroplast and nuclear DNA variation among homozygous plants in a population of the autogamous annual *Microseris douglasii* (Asteraceae, Lactuceae). *Plant Systematics and Evolution*, **196**, 185–94.

Rohlf, F. J. (1993) NTSYS-pc: *Numerical taxonomy and multivariate analysis system, version 1.80.* Exeter Software, Setauket, New York.

Saitou, N. and Nei, M. (1987) The neighbor-joining method: a new method for reconstructing phylogenetic trees. *Molecular Biology and Evolution*, **4**, 406–25.

Sang, T., Crawford, D. J. and Stuessy, T. F. (1995) Documentation of reticulate evolution in peonies (*Paeonia*) using internal transcribed spacer sequences of nuclear ribosomal DNA: implications for biogeography and concerted evolution. *Proceedings of the National Academy of Sciences of the USA*, **92**, 6813–17.

Schneider C. J. (1996) Distinguishing between primary and secondary intergradation among morphologically differentiated populations of *Anolis marmoratus*. *Molecular Ecology*, **5**, 239–49.

Schwarzbach, A. E. and Kadereit, J. W. (1995) Rapid radiation of North American desert genera of the Papaveraceae: evidence from restriction site mapping of PCR-amplified chloroplast fragments. *Plant Systematics and Evolution* (Suppl.), **9**, 159–70.

Slatkin, M. (1985) Gene flow in natural populations. *Annual Review of Ecology and Systematics*, **16**, 393–430.

Slatkin, M. and Maddison, W. P. (1989) A cladistic measure of gene flow inferred from the phylogenies of alleles. *Genetics*, **123**, 603–13.

Smith, J. F., Burke, C. C. and Wagner, W. L. (1996) Interspecific hybridization in natural populations of *Cyrtandra* (Gesneriaceae) on the Hawaiian Islands – evidence from RAPD markers. *Plant Systematics and Evolution*, **200**, 61–77.

Soltis, D. E. and Kuzoff, R. K. (1995) Discordance between nuclear and chloroplast phylogenies in the *Heuchera* group (Saxifragaceae). *Evolution*, **49**, 727–42.

Soltis, D. E. and Soltis, P. S. (1989) Allopolyploid speciation in *Tragopogon*: insights from chloroplast DNA. *American Journal of Botany*, **76**, 1119–24.

Soltis, P. S. and Soltis, D. E. (1995) Plant molecular systematics: inferences of phylogeny and evolutionary processes, in *Evolutionary biology*, (eds M. K. Hecht, R. J. MacIntyre and M. T. Clegg), Plenum Press, New York, vol. 28, pp. 139–94.

Soltis, D. E., Soltis, P. S. and Milligan, B. G. (1992) Intraspecific chloroplast DNA variation: systematic and phylogenetic implications, in *Molecular systematics of plants*, (eds P. S. Soltis, D. E. Soltis and J. J. Doyle), Chapman & Hall, New York, pp. 117–50.

Strand, A. E., Milligan, B. G. and Pruitt, C. M. (1996) Are populations islands? Analysis of chloroplast DNA in *Aquilegia*. *Evolution*, **50**, 1822–29.

Swofford, D. L. (1993) *PAUP: Phylogenetic Analysis Using Parsimony, version 3.1.1.* Illinois Natural History Survey, Champaign, Illinois.

Swofford, D. L. and Selander, R. B. (1989) *BIOSYS-1: A computer program for the analysis of allelic variation in population genetics and biochemical systematics, version 1.7.* University of Illinois, Urbana, Illinois.

Swofford, D. L., Olsen, G. J., Waddell, P. J. and Hillis, D. M. (1996) Phylogenetic inference, in *Molecular systematics*, (eds D. M. Hillis, C. Moritz and B. K. Mable), Sinauer Associates, Sunderland, Massachusetts, pp. 407–514.

Thormann, C. E., Ferreira, M. E., Camargo, L. E. A., Tivang, J. G. and Osborn, T. C. (1994) Comparison of RFLP and RAPD markers to estimating genetic relationships within and among cruciferous species. *Theoretical and Applied Genetics*, **88**, 973–80.

van de Zande, L. and Bijlsma, R. (1995) Limitations of the RAPD technique in phylogeny reconstruction in *Drosophila*. *Journal of Evolutionary Biology*, **8**, 645–56.

Walters, S. M. (1964) *Senecio rupestris* Waldst. & Kit. and *S. squalidus* L. *Proceedings of the Botanical Society of the British Isles*, **5**, 382.

Weir, B. S. and Cockerham, C. C. (1984) Estimating *F*-statistics for the analysis of population structure. *Evolution*, **38**, 1358–70.

Weir, J. and Ingram, R. (1980) Ray morphology and cytological aspects of *S. cambrensis* Rosser. *New Phytologist*, **86**, 237–41.

Wendel, F. W., Stewart, J. McD. and Rettig, J. H. (1991) Molecular evidence for homoploid reticulate evolution among Australian species of *Gossypium*. *Evolution*, **45**, 694–711.

Wheeler, N. C. and Guries, R. P. (1987) A quantitative measure of introgression between lodgepole and jack pines. *Canadian Journal of Botany*, **65**, 1876–85.

White, T. J., Bruns, T., Lee, S. and Taylor, J. (1990) Amplification and direct sequencing of fungal ribosomal RNA genes for phylogenetics, in *PCR protocols*, (eds M. A. Innis, D. H. Gelfand, J. J. Sninsky and T. J. White), Academic Press, New York, pp. 315–22.

Whittemore, A. T. and Schaal, B. A. (1991) Interspecific gene flow in oaks. *Proceedings of the National Academy of Sciences of the USA*, **88**, 2540–44.

Wolf, P. G. and Soltis, P. S. (1992) Estimates of gene flow among populations, geographic races, and species in the *Ipomopsis aggregata* complex. *Genetics*, **130**, 639–47.

Wolfe, A. D. and Elisens, W. J. (1994) Nuclear ribosomal DNA restriction-site variation in *Penstemon* section *Peltanthera* (Scrophulariaceae): an evaluation of diploid hybrid speciation and evidence for introgression. *American Journal of Botany*, **81**, 1627–35.

Wolfe, A. D. and Elisens, W. J. (1995) Evidence of chloroplast capture and pollen-mediated gene flow in *Penstemon* section *Peltanthera* (Scrophulariaceae). *Systematic Botany*, **20**, 395–412.

Wright, S. (1931) Evolution in Mendelian populations. *Genetics*, **16**, 97–159.

Zink, R. M. (1994) The geography of mitochondrial DNA variation, population structure, hybridization, and species limits in the fox sparrow (*Passerella iliaca*). *Evolution*, **48**, 96–111.

Zohary, M. (1973) *Geobotanical foundations of the Middle East*, Gustav Fischer Verlag, Stuttgart.

Appendix   States of 27 polymorphisms among 16 cpDNA haplotypes observed in the *Senecio* species studied. Character states are indicated as ancestral (0) or derived (1), based on comparison with *S. malacitanus* (haplotype P). Description of character states and polymorphisms are given in Table 9.2.

cpDNA haplotype	1	2	3	4	5	6	7	8	9	10	11	12	13	14	15	16	17	18	19	20	21	22	23	24	25	26	27
A	0	0	1	0	0	0	1	1	0	0	0	0	1	0	0	0	0	0	0	0	0	1	1	0	1	0	0
B	0	0	1	0	0	0	1	1	1	1	0	0	0	1	0	0	0	0	0	0	0	1	1	0	1	0	0
C	1	0	1	0	0	0	1	0	0	0	1	0	1	0	1	0	0	0	0	0	0	1	1	0	1	0	0
D	0	1	1	0	0	0	1	0	0	0	0	0	0	0	0	0	0	0	0	0	0	1	1	1	1	0	0
E	0	0	1	0	0	0	1	0	0	1	0	0	1	0	0	1	0	0	0	0	?	?	?	?	?	?	?
F	1	0	1	0	0	0	1	0	0	0	1	0	1	0	0	0	0	0	0	0	0	1	1	0	1	0	0
G	0	0	1	0	0	0	1	1	1	0	0	0	1	1	0	0	0	0	?	?	?	?	1	0	1	0	0
H	0	0	1	0	1	0	1	1	1	0	0	1	1	0	0	0	0	0	?	?	?	?	1	0	1	0	0
I	1	0	1	0	0	0	1	0	0	0	1	0	1	0	1	0	0	0	0	1	?	?	1	0	1	0	0
J	0	0	1	0	0	0	1	0	0	0	0	0	0	0	0	0	0	0	0	0	0	1	1	0	1	0	0
K	1	0	1	1	0	0	1	0	0	0	0	1	0	1	0	0	0	0	?	0	0	?	1	0	1	0	0
L	0	0	1	0	0	1	1	1	1	0	0	1	1	0	0	0	0	0	0	0	?	?	1	0	1	0	0
M	0	0	1	0	0	0	0	0	0	0	0	0	0	0	0	0	0	0	0	0	0	1	1	0	1	0	0
N	0	0	1	0	0	0	0	0	0	0	0	0	0	0	0	1	1	0	1	0	1	0	1	0	1	0	0
O	0	0	0	0	0	0	0	0	0	0	0	0	0	0	0	1	0	1	0	0	0	0	0	0	1	1	1
P	0	0	0	0	0	0	0	0	0	0	0	0	0	0	0	0	0	0	0	0	0	0	0	0	0	0	0

The value of genomic *in situ* hybridization (GISH) in plant taxonomic and evolutionary studies

C. A. Stace and J. P. Bailey

ABSTRACT

Genomic *in situ* hybridization (GISH) is a technique that enables one to label differentially genomes or parts of genomes within cells. A broad survey of the applications of GISH is divided into four sections: chromosome disposition in interphase and mitotic cells; identification of genomes; recognition of parts of genomes (e.g. alien chromosomes, intergenomic translocations or recombinations, B-chromosomes); and meiotic studies ('what is pairing with what?'). Particular attention is paid to two topics: firstly, the lowest taxonomic levels at which GISH can discriminate (frequently this is at the species level, but can be well above that); and secondly, the reasons for the partial labelling of chromosomes when complete labelling was expected, for example the labelling of only the end regions of *Festuca rubra* chromosomes when probed with total genomic DNA from *F. rubra*. It is concluded that it is primarily the highly repetitive DNA rather than the non-repetitive transcribed and translated DNA that responds to probing by GISH. Hence best results are obtained with species that have much repetitive DNA (i.e. high C-values) which is dispersed more or less evenly along the length of the chromosomes.

10.1 Introduction

Genomic *In situ* Hybridization (GISH) involves the extraction of total DNA from one organism, labelling this in some way so that it can be detected later *in situ* by a fluorescent dye, and using this as a probe to target cells from another organism prepared using normal cytological squash methods. Those parts of the chromosomes of the target organism that are sufficiently similar to the probe material form probe–target complexes which are labelled by the fluorescent dye. Other parts not forming such a complex do not become fluorescent but can be counter-stained with a dye of contrasting colour.

Early attempts to identify individual chromosomes or genomes in a cell had to rely on morphology alone, using overall size and the position of various markers such as centromeres and secondary constrictions. The development of chromosome banding techniques around 1970 (cf. Vosa, 1976) represented a tremendous advance, since many more sites can be recognized within a genome, and often each chromosome has a unique pattern of bands.

In *Molecular systematics and plant evolution* (1999) (eds P. M. Hollingsworth, R. M. Bateman and R. J. Gornall), Taylor & Francis, London, pp. 199–210.

The first *In situ* Hybridization (ISH) studies developed in the late 1960s and 1970s (Gall and Pardue, 1969; John *et al.*, 1969) differed from GISH in that (a) the probes were of small specific regions of the genome, and (b) labelling was radio-active. They were effectively a controlled variant of chromosome banding, in which the molecular nature of the labelled bands or regions was at least potentially known. The substitution in the 1980s of fluorescence for radio-activity in Fluorescent *In situ* Hybridization (FISH) (Bauman *et al.*, 1980) was an advance over ISH partly because the technique is quicker, safer and cheaper, and also because the results can be polychromatic and have a higher resolution. GISH can be considered a further advance over FISH, because in the former whole chromosomes become labelled; this is highly advantageous in many studies (e.g. genome identification; meiotic analyses), and is applicable to cells at interphase as well as those under-going division. However, in many investigations FISH is still the preferred technique, e.g. in the use of NOR-sequences, telomere-sequences or specific repetitive sequences. In some cases, where a repetitive sequence is well dispersed throughout the whole genome, FISH using such a sequence as a probe produces a result like that of GISH, with the whole target genome being labelled. Lapitan *et al.* (1986) were able to label all *Secale* chromatin in *Triticum* addition lines by probing with a 120 bp repetitive DNA sequence from *Secale*.

GISH was originally developed for animal hybrid cell-lines (Pinkel *et al.*, 1986), and first used on plants in 1987 at the Plant Breeding Institute, Cambridge (Schwarzacher *et al.*, 1989; this paper coined the term GISH). Since then many groups have published examples of the use of the technique in solving a range of problems, and there have been good reviews on both applications (Bennett, 1995; Heslop-Harrison and Schwarzacher, 1996) and techniques (Leitch *et al.*, 1994; Anamthawat-Jónsson and Heslop-Harrison, 1995; Kenton *et al.*, 1997).

The existence of technique-orientated reviews obviates the need for us to provide one, and we shall not cover technical details except to emphasize one aspect, i.e. the fact that GISH is a *comparative*, not an *absolute*, approach. According to the experimental conditions imposed, virtually any total genomic DNA probe could be used to produce positive labelling with any target total genomic DNA, or conversely not to produce positive labelling (even with identical DNA). Since it is extremely difficult to reproduce exact experimental conditions, GISH is not a tech-nique that lends itself to probing a range of target diploid taxa with DNA from a related diploid species to establish relationships, but rather one that probes a target cell that contains two or more (or elements of two or more) different genomes. The trick is to provide the experimental conditions that allow the probe to label one of the genomes in the target cell, but not the other.

This is done by (usually both of) two methods. Firstly, the *stringency* of the experimental conditions can be varied whereby the degree of similarity of the two DNA sequences that would allow re-annealing (and therefore labelling) can be raised or lowered. Such factors as temperature, time, reagent strength and degree of DNA fragmentation can be used to vary this factor. Secondly, the technique of *blocking* can be used, whereby the labelled total genomic DNA probe is mixed with an excess of unlabelled total DNA (the block) from the non-target genome in the target cells. The block hybridizes with sequences in common between the block and probe, so that mainly genome-specific sequences in the target genome remain as sites for hybridization with the labelled probe. The degree of excess of block over probe can be a critical experimental factor. For rather distantly related probe–target pairs (e.g. *Triticum–Secale*) often no blocking at all is needed, whereas some workers have used blocks in ratios higher than 100:1.

This paper summarizes very briefly the nature of the four main lines of investigation in which GISH has been utilized, and then concentrates on two important questions that our own work has emphasized.

10.2 The use of genomic *in situ* hybridization

The uses to which GISH has been put may be placed in four major categories.

10.2.1 Chromosome disposition

The concept of chromosomes in a nucleus not being randomly intermingled, but instead occupying discrete chromosomal domains, was largely developed in the 1980s but had its roots in ideas forwarded nearly 100 years before (cf. Manuelidis, 1985; Cremer *et al.*, 1995). Such an hypothesis is ideal for investigation by GISH, and many studies have now been published (e.g. Schwarzacher *et al.*, 1989; Leitch *et al.*, 1990; Leitch *et al.*, 1991). These GISH-based studies mainly relate to sexual or somatic hybrids, or to polyploids, where two or more different genomes occur together in a cell. They can be equally informative at interphase or during mitosis.

The existence of separate chromosomal or genomic domains is clearly highly significant in several ways, e.g. control of transcription and gene expression, preferential loss of chromosomes in wide sexual or in somatic hybrids, and pairing at meiosis.

10.2.2 Genome identification

The use of GISH to label the whole of one genome but none of the other genome(s) in cells with two or more genomes, and with it the ability to 'paint' the genomes different colours, was described by Bennett (1995) as 'a dream come true'. Whatever the precise nature of the relationship between a probe and the chromatin with which it anneals and which it therefore labels, it seems to be an efficient method of genome analysis, because in most cases there is a close correlation between the genome homologies suggested by GISH and those deduced from classical biosystematic methods. This is nowhere better illustrated than in hexaploid wheat (*Triticum aestivum*), in which all three genomes can be distinguished using simultaneous probes from their putative ancestors (Mukai *et al.*, 1993). The use of different dyes produced tricoloured target chromosome spreads.

The application of GISH to genome analysis has been equally successful in four groups of cases:

(a) Artificial hybrids, e.g. *Triticum × Secale* (Schwarzacher *et al.*, 1992), *Leymus × Hordeum* (Ørgaard and Heslop-Harrison, 1994);
(b) Natural hybrids, e.g. *Festuca × Vulpia* (Bailey *et al.*, 1993), *Allium* (Hizume, 1994);

(c) Somatic hybrids, e.g. *Nicotiana* (Parokonny *et al.*, 1992);
(d) Amphidiploids, e.g. *Milium* (Bennett *et al.*, 1992), *Festuca* (Humphreys *et al.*, 1995), *Avena* (Chen and Armstrong, 1994).

As indicated by the above examples, the Poaceae, particularly the tribe Triticeae, have been investigated more than any other plant family. Many other families have also been successfully studied, e.g. Brassicaceae, Fabaceae and Solanaceae in the dicots, and Commelinaceae, Iridaceae and Liliaceae *sensu lato* in the monocots, but we are not aware of any plant work outside the angiosperms. Highly polyploid groups such as ferns, and haploids such as bryophytes, would seem to be ideally suited to GISH-based investigations. Few investigations have been carried out on cytologically 'difficult' or 'intransigent' groups, such as those with abundant tannins or resins or with very small chromosomes. In *Glycine* (with a small amount of DNA: 1C DNA value = c. 1pg) fluorescence was uneven, and was intense only in regions of repetitive DNA (Shi *et al.*, 1996), but successful experiments have been reported in *Salix*, *Betula* and *Citrus*, with even smaller amounts of DNA (1C DNA value = c. 0.4pg).

10.2.3 Recognition of parts of genomes

GISH is equally effective in detecting odd chromosomes or segments of chromosomes in foreign cells. Bennett (1995) reported that fragments as small as 50–100 Mbp should be detectable by GISH; this is about five times greater than the minimum size detectable by ISH.

Three groups of cytological situations may be summarized:

10.2.3.1 Alien chromosomes in addition and substitution lines

Schwarzacher *et al.* (1992) were able to detect chromosomes or chromosome arms of four alien genera (*Leymus*, *Thinopyrum*, *Hordeum* and *Secale*) in *Triticum* cells by probing with total genomic DNA from the alien genus.

10.2.3.2 Translocations or recombinations in cells containing two or more genomes

Parokonny *et al.* (1992) found that, in somatic hybrids between species of *Nicotiana*, intergenomic translocations in both directions were detectable.

Jiang and Gill (1993) found that the 5A chromosome in a line of *Triticum* containing *Elymus trachycaulus* chromosomes exhibited a 'zebra'-type striping which they attributed to multiple intergenomic translocations; this interpretation was supported by meiotic studies, when the zebra 5A paired with both *Triticum* and *Elymus* 5A chromosomes.

On the other hand, Zhang *et al.* (1996) found that, in partial *Triticum–Thinopyrum* amphidiploids, intergenomic translocations had occurred in some lines but not in others. Similarly, Parokonny and Kenton (1995) showed that in different cultivars of the tetraploid *Nicotiana tabacum* there were different (i.e. cultivar-specific) patterns of intergenomic translocations; the recombinant chromosomes were all in homozygous pairs.

Takahashi *et al.* (1997) examined the partially (c. 10%) fertile intergeneric hybrid *Gasteria lutzii* × *Aloe aristata*. In the F_1 the two genomes remained genetically discrete, but in backcross progeny numerous intergenomic recombinations were evident, presumably arising during F_1 meiosis. The recombinations as detected by GISH were more numerous than from predictions following standard cytological analysis of chiasmata, indicating the visual shortcomings of the latter technique.

In potato breeding, dihaploids of the tetraploid *Solanum tuberosum* are produced by the trigger of hybridization with pollen from the diploid *S. phureja*. Wilkinson *et al.* (1995) found, contrary to expectation, that such dihaploids, when probed with *S. phureja*, revealed *S. phureja* chromatin segments on three separate chromosomes, indicating that the *S. phureja* pollen is more intimately involved at the chromosome level in these crosses than had previously been believed.

Thomas *et al.* (1994) have been able to utilize GISH to localize the parental chromatin in various crosses between *Lolium* and *Festuca* section *Bovinae*, including the presence of a small segment of *Festuca pratensis* carrying a known gene in an introgressed *Lolium multiflorum* line. Similarly, Pasakinskiene *et al.* (1997) found that the position of a drought resistant gene from *F. arundinacea* in diploid recombinants of *F. arundinacea* (6x) × *L. multiflorum* (2x) could be detected by probing not only with *F. arundinacea*, but also with *F. pratensis*, a diploid progenitor of the hexaploid *F. arundinacea*.

Clearly, the application of GISH to *somatic* cells provides extremely valuable and moreover new information on pairing, crossing over and recombination in sexual hybrids, and on translocations in sexual and somatic hybrids.

10.2.3.3 B-chromosomes

GISH seems ideally suited to investigating the origin of B-chromosomes. Jamelina *et al.* (1994) probed B-containing *Crepis capillaris* with total genomic DNA from B-lacking *C. capillaris*, and found that the B-chromosomes were labelled, suggesting their origin from the A-chromosomes.

Morais-Cecilio *et al.* (1996), on the other hand, found that *Secale* B-chromosomes could be labelled only in part by total genomic DNA from B-lacking plants. But the origin of the extra B-DNA, from another outside source or via newly evolved heterochromatin, remains open to question. Probably the use of specific sequences via FISH is needed to solve such problems.

10.2.4 Meiotic studies

GISH is theoretically and in practice equally as applicable to meiotic as to interphase or mitotic problems, but relatively few studies have so far been made, probably because optimum preparations are less easily obtained. Stages from pachytene onwards are suitable for study by GISH.

Bailey *et al.* (1993), investigating the natural F_1 hybrid *Festuca rubra* (6x) × *Vulpia fasciculata* (4x), found that in most pollen mother cells of the pentaploid F_1, despite a normally low level of bivalent formation, *Festuca–Festuca*, *Vulpia–Vulpia* and *Festuca–Vulpia* bivalents were present. This confirmed their earlier results (Bailey and Stace, 1992) based on the distinction of the *Festuca* and *Vulpia* chromosomes by their very different patterns of Giemsa C-banding. The extent to which the heterogenetic (*Festuca–Vulpia*) pairing results in recombination and introgression via backcrossing to *Festuca* is not yet known.

In *Triticum* × *Secale* crosses, Benavente *et al.* (1996) found that the frequency of *Triticum–Secale* metaphase-I associations greatly exceeded levels of subsequent genetic recombination, indicating that not all of the metaphase-I associations were chiasmate. On the other hand, in the *Festuca* × *Vulpia* F$_1$, most of the diplotene associations do give the appearance of being chiasmate (see plates in Bailey and Stace, 1992 and Bailey *et al.*, 1993). Nevertheless it seems a general finding that there are more associations (both bivalents compared with univalents, and multivalents compared with bivalents) at pachytene than at metaphase I (Schwarzacher and Heslop-Harrison, 1995).

King *et al.* (1993, 1994) found that in meiosis of triploid F$_1$ hybrids between tetraploid *Triticum* and diploid *Thinopyrum* or diploid *Secale* some pairing between heterologues (*Triticum–Thinopyrum* or *Triticum–Secale*) was evident; if the pairing was chiasmate it indicates that crossing over had occurred between these genera. The number of *Triticum–Secale* bivalents was unexpectedly high (one quarter of the total).

The ability to see precisely which chromosomes in hybrids form bivalents does indeed conform with Bennett's (1995) 'dream come true'. The question of exactly what constitutes 'chromosome homology' is still to be answered, but it seems likely that GISH will be important in addressing this problem.

10.3 Taxonomic distance necessary for discrimination by GISH

Many attempts have been made over the years to define the various taxonomic ranks, especially the species, by objective simplistic criteria. Chromosome number and the ability to interbreed have been most persistently pursued in this aim, but only approximate correlations have been revealed. Most taxonomists today do not expect any given character or set of data to differentiate taxa at the same rank across a wide taxonomic spectrum (e.g. dicotyledons, or a family); indeed this unpredictability of characters has become a taxonomic principle (Stace, 1989). It is therefore not to be expected, at least by practising plant taxonomists, that GISH or any related technique will always differentiate taxa at one particular rank, and not below it, simply because the kind and level of DNA differences that allow such differentiation are not precisely correlated with any other set of data (Anamthawat-Jónsson and Heslop-Harrison, 1995). This prediction is borne out in practice. Nevertheless, some broad generalizations can be made.

In most cases investigated genera can be readily distinguished by GISH, e.g. *Leymus/Hordeum*, *Triticum/Secale*, *Festuca/Vulpia*, *Festuca/Lolium*, *Gasteria/Aloe*, *Solanum/Lycopersicon*, often without the need for DNA-blocking. On the other hand, Ørgaard and Heslop-Harrison (1994) could not differentiate between *Leymus* and *Psathyrostachys*.

In many (probably most) cases, workers have also found it possible to differentiate between the species within one genus, e.g. *Hordeum*, *Gibasis*, *Solanum*, *Milium*, *Nicotiana*, though usually careful blocking is necessary. Bennett (1995) reported, however, the inability to distinguish between *Leontodon hispidus* and *L. saxatilis* (*L. taraxacoides* nom. illegit.), two species that form natural hybrids.

In *Avena*, Chen and Armstrong (1994) and Jellen *et al.* (1994) found that diploids with the A genome (e.g. *A. strigosa*) can be distinguished from those with the C genome (e.g. *A. eriantha*, *A. pilosa*), and both of these can be used successfully to probe the genomes in cultivated hexaploid *A. sativa* (AACCDD). Unfortunately the D genome is not known in diploids or tetraploids, and in the hexaploid the A and D genomes cannot be distinguished by GISH. In contrast, all three genomes (A, B, D) can be distinguished in the hexaploid *Triticum aestivum* (Mukai *et al.*, 1993).

In *Festuca* section *Bovinae*, Humphreys *et al.* (1995) found that the genome of the diploid *F. pratensis* can be distinguished from both genomes in the related tetraploid *F. glaucescens*, and probes from these two species supported the hypothesis that they were the ancestors of the hexaploid *F. arundinacea*.

In *Festuca* section *Festuca*, we have been able to differentiate between species within the *F. rubra* and *F. ovina* aggregates respectively (Bailey and Stace, unpublished). The hybrid material used in these experiments was artificial *F. rubra* (6x) × *F. lemanii* (6x) probed with *F. rubra* and blocked with 10 : 1 excess *F. lemanii* total genomic DNA (*F. lemanii* belongs to the *F. ovina* aggregate). Under optimal experimental conditions the chromosomes of the two parents (21 of each) could be clearly distinguished. However, we could not distinguish between species within either of the two aggregates. Since at present more than 100 species are recognized in Europe alone within the *F. ovina* aggregate, and more than 20 within the *F. rubra* aggregate (Markgraf-Dannenberg, 1980), this does not suggest a very highly discriminatory role for GISH. However, these segregate species represent extreme taxonomic splitting in a very difficult taxonomic group; Hackel's (1882) classic monograph recognized only one species in each aggregate.

The fact that 'species' as currently recognized are not necessarily 'equivalent' entities in different genera adds to the unpredictability of the level at which GISH will discriminate taxa. At present we can go no further than say that it can be reasonably predicted that GISH will usually discriminate taxa at around the species level, but that sometimes it will fail at even the generic level, and probably sometimes will succeed at infraspecific levels. The optimism expressed in the last clause can be supported not only by the steady improvement in techniques, but also by the fact that many polyploid complexes at present recognized as one species (e.g. *Dactylis glomerata*) consist of a range of diploids plus variously derived polyploids, and it is to be expected that several different genomes (A, B, etc.) will be involved within some of these complexes.

10.4 Partial labelling of genomes

The normal expectation when probing with total genomic DNA is for a complete genome to be either labelled or not. Where labelling is obviously not of a whole genome it usually indicates the presence of partial genomes (e.g. alien chromosomes or their segments), or the occurrence of sexual recombination or somatic translocation. Several authors in the past have used the terms recombination and translocation too loosely, but it is in fact not possible to distinguish the two in a later generation of a sexual hybrid unless chromosome homologies have been established.

In our work with *Festuca* sect. *Festuca*, *Vulpia* and *Lolium*, however, we have regularly encountered a different phenomenon. We used the natural pentaploid hybrid *Festuca rubra* × *Vulpia fasciculata* (= X *Festulpia hubbardii*) and the artificial tetraploid hybrid (synthesized at Leicester by Ana-Teresa Romero and Pilar Catalán) *Festuca rubra* × *Lolium perenne* (no hybrid binomial). We refer to these two in shorthand by their nothogeneric names X *Festulpia* and X *Festulolium* respectively; genomically their formulae may be designated FFFVV and FFFL respectively. These two hybrids were probed with labelled total genomic DNA from each of the parents separately, in each case usually using 10 : 1 excess unlabelled blocking DNA from the other parent.

If the X *Festulpia* or X *Festulolium* hybrids were probed with the non-*Festuca* parent and blocked with the *Festuca* parental DNA, the 14 or 7 chromosomes from *Vulpia* or *Lolium* respectively were strongly labelled evenly along their length, as expected. The 21 *Festuca* chromosomes were for most of their length extremely faintly labelled, but at one or both short terminal segments they were not labelled at all. In the reverse experiment, when the hybrids were probed with *Festuca* DNA and blocked with the non-*Festuca* parental DNA, only the short terminal segments of the *Festuca* chromosomes were strongly labelled. A similar phenomenon has recently been reported in *Alstroemeria* by Kuipers *et al.* (1997), who used the expressions positive and negative GISH-banding. No such positive and negative GISH-banding has been demonstrated by Thomas *et al.* (1994, and related papers) in their work on *Festuca* section *Bovinae*, in which the Giemsa and DAPI patterns are quite different from those in *Festuca* section *Festuca*, there being no conspicuous terminal segments.

Clearly, in *F. rubra*, some regions of DNA hybridize to the probe more readily than others. These terminal segments in *F. rubra* are precisely those that form strong Giemsa C-bands (Bailey and Stace, 1992) or form DAPI-bright areas (Bailey *et al.*, 1993), and therefore can be taken as regions of AT-rich highly repetitive DNA (heterochromatin). It is now well established that it is the highly repetitive DNA that forms most of the species-specific DNA in the chromosomes, the gene-sequences being relatively uniform in related species (Anamthawat-Jónsson and Heslop-Harrison, 1995). These repetitive sequences vary in amount (lowest in taxa with low C-values) and distribution. In *Glycine* (1C = c. 1pg) GISH did not produce strong labelling along the whole length of the chromosomes, but hybridization was concentrated in interstitial regions correlating with the position of heterochromatin (Shi *et al.*, 1996).

In *Secale* (1C = c. 8–10.5pg), where the heterochromatin is dispersed along the whole chromosome length, Lapitan *et al.* (1986) were able to obtain GISH-like results (i.e. labelling of the whole genome) via FISH using a highly-repetitive DNA probe. Shields (1993) concluded that synteny (gene order) has been maintained across large chromosome sections of at least three of the five major subfamilies of grasses, and that the most rapid genome evolution occurs in the repetitive DNA. Two distinct classes of highly repetitive DNA are found: one mostly in paracentromeric and telomeric regions; and the other dispersed along the chromosome length. Presumably these two classes predominate respectively in *F. rubra* and *Secale cereale*.

Applying these data to our *Festuca* hybrids, we first formed the hypothesis that when X *Festulpia* or X *Festulolium* is probed with *Festuca* and blocked with the other parent, the non-labelling of most of the length of the *Festuca* chromosomes is caused by the blocking *Vulpia* or *Lolium* DNA, which is similar to that of *Festuca* in those areas. The terminal segments are strongly labelled because the *Vulpia* or *Lolium* DNA has no blocking effect on them. Conversely, the very faint labelling of the *Festuca* chromosomes (except at the terminal regions) when probed with *Vulpia* or *Lolium* might be due to some homology with the latters' DNA, perhaps aided by insufficient blocking. This hypothesis, however, is disproved by the fact that we obtained exactly the same patterns of hybridization when we repeated the experiments with no blocking DNA at all.

We have no reason to doubt that our probes contained ample biotinylated DNA that was homologous to all the DNA along the total length of the target chromosomes, nor that ample fluorescent dye was available. Therefore we are forced to the conclusion that it must be a physical and/or chemical difference between the DNA of euchromatin and heterochromatin that is causing the difference in labelling between those two regions. A number of possibilities offer themselves. The strongly labelled heterochromatin regions are also brightly DAPI-staining and therefore, we believe, AT-rich. Since the probe-DNA was labelled with biotinylated ATP, it is possible that more fluorescent dye is taken up by AT-rich regions than by those with approximately 50% AT. This should be tested by using biotinylated GTP instead, to see if the pattern of labelling is modified or even reversed. It could also be the case that regions of repetitive DNA hybridize more readily with the probe because matching of a sequence is possible along a considerable length of chromosome, or even because heterochromatin is in some physical way more exposed to hybridization. Another possibility is that the lower dissociation temperature of AT-rich regions leaves a longer window in which hybridization can occur during the heating-cooling cycle.

10.5 Conclusions

In experiments using GISH with or without a block, the probe total genomic DNA preferentially hybridizes with the repetitive DNA (or the fluorescent dye preferentially binds to the repetitive DNA regions). The spectacular success of GISH in probing cereal (and other grass) genomes is probably related to the facts that most grasses have relatively large C-values (i.e. large amounts of repetitive DNA), and that these repetitive sequences are dispersed along the chromosome length. Where the repetitive DNA is less abundant (e.g. *Glycine*), or highly localized (e.g. *Festuca rubra*), the significance of the results may well be less immediately obvious, but they are interpretable and then are of no less value than in the cereals. If the uneven labelling of, for example, *Festuca* section *Festuca* chromosomes is due to the use of biotinylated ATP, as discussed above, then probably biotinylated ATP–GTP mixtures should be used instead. Even a superficial knowledge of genomic structure of a taxon will greatly aid the interpretation of the results of GISH experiments.

The dispersed and localized kinds of repetitive DNA might diverge at different rates from each other, and/or in different taxa, and/or in different chromosomes. Their rates of divergence may be relatively rapid or relatively slow, perhaps in some cases similar to the rates of evolution of non-repetitive gene sequences. Since the results of GISH experiments depend upon the strongest probe–target interactions, it is not at all surprising that the taxonomic level of discrimination by GISH should vary from one plant group to another.

GISH is certainly 'a dream come true', but it is not the answer to one's ultimate dream. The latter might pertain to a technique in which low copy-number gene sequences rather than the repetitive DNA were the preferred target of total genomic DNA probes. But GISH is a powerful and revolutionary technique to add to the many other tools available to the plant taxonomist and evolutionist.

ACKNOWLEDGEMENTS

We are grateful to Dr C. Ferris and Dr I. Leitch for helpful discussion and advice, and to J. Wentworth for carrying out some of the experiments. Our GISH work on *Festuca, Vulpia* and *Lolium* was carried out with the aid of NERC Small Grant GR9/1130 'A'.

REFERENCES

Anamthawat-Jónsson, K. and Heslop-Harrison, J. S. (1995) Establishing relationships between closely related species using total genomic DNA as a probe, in *Methods in molecular biology*, **50**, (ed. J. P. Clapp), Humana Press Inc., Totowa, NJ, USA, pp. 209–225.

Bailey, J. P. and Stace, C. A. (1992) Chromosome banding and pairing behaviour in *Festuca* and *Vulpia* (Poaceae, Pooideae). *Plant Systematics and Evolution*, **182**, 21–28.

Bailey, J. P., Bennett, S. P., Bennett, M. D. and Stace, C. A. (1993) Genomic *in situ* hybridization identifies parental chromosomes in the wild grass hybrid X *Festulpia hubbardii*. *Heredity*, **71**, 413–420.

Bauman, J. G. J., Wiegant, J., Borst, P. and van Duijn, P. (1980) A new method for fluorescence microscopical localization of specific DNA sequences by *in situ* hybridization of fluorochrome-labelled RNA. *Experimental Cell Research*, **128**, 485–490.

Benavente, E., Fernandez-Calvin, B. and Orellana, J. (1996) Relationship between the levels of wheat–rye metaphase-I chromosome pairing and recombination revealed by GISH. *Chromosoma*, **105**, 92–96.

Bennett, M. D. (1995) The development and use of genomic *in situ* hybridization (GISH) as a new tool in plant biosystematics, in *Kew chromosome conference IV*, (eds P. F. Brandham and M. D. Bennett), Royal Botanic Gardens, Kew, pp. 167–183.

Bennett, S. T., Kenton, A. Y. and Bennett, M. D. (1992) Genomic *in situ* hybridization reveals the allopolyploid nature of *Milium montianum* (Gramineae). *Chromosoma*, **101**, 420–424.

Chen, Q. F. and Armstrong, K. (1994) Genomic *in situ* hybridization in *Avena sativa*. *Genome*, **37**, 607–612.

Cremer, T., Dietzel, S., Eils, R., Lichter, P. and Cremer, C. (1995) Chromosome territories, nuclear matrix filaments and inter-chromatin channels: a topological view on nuclear architecture and function, in *Kew chromosome conference IV*, (eds P. F. Brandham and M. D. Bennett), Royal Botanic Gardens, Kew, pp. 63–81.

Gall, J. G. and Pardue, M. L. (1969) Formation and detection of RNA–DNA hybrid molecules in cytological preparations. *Proceedings of the National Academy of Sciences USA*. **63**, 378–383.

Hackel, E. (1882) *Monographia Festucarum Europaearum*, Theodor Fischer, Kassel & Berlin.

Heslop-Harrison, J. S. and Schwarzacher, T. (1996) Genomic Southern and *in situ* hybridization for plant genome analysis, in *Methods of genome analysis in plants*, (ed. P. P. Jauhar), CRC, Boca Raton, pp. 163–179.

Hizume, M. (1994) Allodiploid nature of *Allium wakegi* revealed by genomic *in situ* hybridization and localization of 5S and 18S rDNAs. *Japanese Journal of Genetics*, **69**, 407–415.

Humphreys, M. W., Thomas, H. M., Morgan, W. G., Meredith, M. R., Harper, J. A., Thomas, H. *et al.* (1995) Discriminating the ancestral progenitors of hexaploid *Festuca arundinacea* using genomic *in situ* hybridization. *Heredity*, 75, 171–174.

Jamelina, M., Rejon, C. R. and Rejon, M. R. (1994) A molecular analysis of the origin of the *Crepis capillaris* B-chromosome. *Journal of Cell Science*, **107**, 703–708.

Jellen, E. N., Gill, B. S. and Cox, T. S. (1994) Genomic *in situ* hybridization differentiates between A/D-genome and C-genome chromatin and detects inter-genomic translocations in polyploid oat species (genus *Avena*). *Genome*, **37**, 613–618.

Jiang, J. M. and Gill, B. S. (1993) A zebra chromosome arising from multiple translocations involving non-homologous chromosomes. *Chromosoma*, **102**, 612–617.

John, H. A., Birnstiel, M. L. and Jones, K. W. (1969) RNA–DNA hybrids at the cytological level. *Nature*, **223**, 582–587.

Kenton, A. Y., Parokonny, A. S., Leitch, I. J. and Bennett, M. D. (1997) Cytological characterization of somatic hybrids: detection of genome origin by genomic *in situ* hybridization (GISH), in *Methods in molecular biology*, (ed. M. S. Clark), Springer Verlag, Heidelberg, pp. 486–508.

King, I. P., Purdie, K. A., Orford, S. E., Reader, S. M. and Miller, T. E. (1993) Detection of homoeologous chiasma formation in *Triticum durum* × *Thinopyrum bessarabicum* hybrids using genomic *in situ* hybridization. *Heredity*, **71**, 369–372.

King, I. P., Reader, S. M., Purdie, K. A., Orford, S. E. and Miller, T. E. (1994) A study of the effect of a homoeologous pairing promoter on chromosome pairing in wheat/rye hybrids using genomic *in situ* hybridization. *Heredity*, **72**, 318–321.

Kuipers, A. G. J., van Os, D. P. M., de Jong, J. H. and Ramanna, M. S. (1997) Molecular cytogenetics of *Alstroemeria*: identification of parental genomes in interspecific hybrids and characterization of repetitive DNA families in constitutive heterochromatin. *Chromosome Research*, **5**, 31–39.

Lapitan, N. L. V., Sears, R. G., Rayburn, A. L. and Gill, B. S. (1986) Detection of chromosome breakpoints by *in situ* hybridization with a biotin-labelled DNA probe. *Journal of Heredity*, **77**, 415–419.

Leitch, A. R., Mosgöller, W., Schwarzacher, T., Bennett, M. D. and Heslop-Harrison, J. S. (1990) Genomic *in situ* hybridization to sectioned nuclei shows chromosome domains in grass hybrids. *Journal of Cell Science*, **95**, 335–341.

Leitch, A. R., Schwarzacher, T., Mosgöller, W., Bennett, M. D. and Heslop-Harrison, J. S. (1991) Parental genomes are separated throughout the cell cycle in a plant hybrid. *Chromosoma*, **101**, 206–213.

Leitch, A. R., Schwarzacher, T., Jackson, D. and Leitch, I. J. (1994) *In situ hybridization: a practical guide*, BIOS Scientific Publishers.

Manuelidis, L. (1985) Individual interphase chromosome domains revealed by *in situ* hybridization. *Human Genetics*, **71**, 288–293.

Markgraf-Dannenberg, I. (1980) *Festuca*, in *Flora Europaea*, 5, (eds T. G. Tutin, V. H. Heywood, N. A. Burges, D. H. Moore, D. H. Valentine, S. M. Walters and D. A. Webb), Cambridge University Press, Cambridge, pp. 125–153.

Morais-Cecilio, L., Delgado, M., Jones, R. M. and Viegas, W. (1996) Painting rye B-chromosomes in wheat – interphase chromatin organization, nuclear disposition and association in plants with 2, 3 or 4 Bs. *Chromosome Research*, **4**, 195–200.

Mukai, Y., Nakahara, Y. and Yamamoto, M. (1993) Simultaneous discrimination of the three genomes in hexaploid wheat by multicolour fluorescence *in situ* hybridization using total genomic and highly repeated DNA probes. *Genome*, 36, 489–494.

Ørgaard, M. and Heslop-Harrison, J. S. (1994) Investigations of genome relationships between *Leymus, Psathyrostachys* and *Hordeum* inferred by genomic DNA–DNA *in situ* hybridization. *Annals of Botany*, **73**, 195–203.

Parokonny, A. S. and Kenton, A. Y. (1995) Comparative physical mapping and evolution of the *Nicotiana tabacum* L. karyotype, in *Kew chromosome conference IV*, (eds P. F. Brandham and M. D. Bennett), Royal Botanic Gardens, Kew, pp. 301–320.

Parokonny, A. S., Kenton, A. Y., Gleba, Y. Y. and Bennett, M. D. (1992) Genome reorganization in *Nicotiana* asymmetric somatic hybrids analysed by *in situ* hybridization. *Plant Journal*, **2**, 863–874.

Pasakinskiene, I., Anamthawat-Jónsson, K., Humphreys, M. W. and Jones, R. N. (1997) Novel diploids following chromosome elimination and somatic recombination in *Lolium multiflorum × Festuca arundinacea* hybrids. *Heredity*, **78**, 464–469.

Pinkel, D., Straume, T. and Gray, J. W. (1986) Cytogenetic analysis using quantitative, high sensitivity fluorescence hybridization. *Proceedings of the National Academy of Sciences USA*, **83**, 2934–2938.

Schwarzacher, T., Leitch, A. R., Bennett, M. D. and Heslop-Harrison, J. S. (1989) *In situ* localization of parental genomes in a wide hybrid. *Annals of Botany*, **64**, 315–324.

Schwarzacher, T. Anamthawat-Jónsson, K., Harrison, G. E., Islam, A. K. M. R., Jia, J. Z., King, I. P. *et al.* (1992) Genomic *in situ* hybridization to identify alien chromosomes and chromosome segments in wheat. *Theoretical and Applied Genetics*, **84**, 778–786.

Schwarzacher, T. and Heslop-Harrison, J. S. (1995) Molecular cytogenetic investigations of meiosis, in *Kew chromosome conference IV*, (eds P. F. Brandham and M. D. Bennett), Royal Botanic Gardens, Kew, pp. 407–416.

Shi, L., Zhu, T., Morgante, M., Rafalski, J. A. and Keim, P. (1996) Soybean chromosome painting – a strategy for somatic cytogenetics. *Journal of Heredity*, **87**, 308–313.

Shields, R. (1993) Pastoral synteny. *Nature*, **365**, 297–298.

Stace, C. A. (1989) *Plant taxonomy and biosystematics*, 2nd edn, Edward Arnold, London.

Takahashi, C., Leitch, I. J., Ryan, A., Bennett, M. D. and Brandham, P. E. (1997) The use of genomic *in situ* hybridization (GISH) to show transmission of recombinant chromosomes by a partially fertile bigeneric hybrid, *Gasteria lutzii × Aloe aristata* (Aloaceae), to its progeny. *Chromosoma*, **105**, 342–348.

Thomas, H. M., Morgan, W. G., Meredith, M. R., Humphreys, M. W., Thomas, H. and Leggett, J. M. (1994) Identification of parental and recombined chromosomes in hybrid derivatives of *Lolium multiflorum × Festuca pratensis* by genomic *in situ* hybridization. *Theoretical and Applied Genetics*, **88**, 909–913.

Vosa, C. G. (1976) The use of Giemsa and other staining techniques in karyotype analysis, in *Commentaries in plant science*, (ed. H. Smith), Pergamon Press, Oxford, pp. 183–197.

Wilkinson, M. J., Bennett, S. T., Clulow, S. A., Allainguillaume, J., Harding, K. and Bennett, M. D. (1995) Evidence for somatic translocation during potato dihaploid induction. *Heredity*, **74**, 146–151.

Zhang, X. Y., Dong, Y. S. and Wang, R. R. C. (1996) Characterization of genes and chromosomes in partial amphiploids of the hybrid *Triticum aestivum × Thinopyrum ponticum* by *in situ* hybridization, isozyme analysis and RAPD. *Genome*, **39**, 1062–1071.

Chapter 11

RAPDs in systematics – a useful methodology?

S. A. Harris

ABSTRACT

The simplicity of the technique and the number of markers that can be generated has encouraged the application of randomly amplified polymorphic DNA (RAPD) markers to many types of biodiversity-associated problems since 1990. This paper reviews the basis of RAPD polymorphism, the advantages and disadvantages of the methodology and the importance of data analysis methodologies. It is concluded that the early promise of RAPDs as a quick, reliable and easily utilised DNA-based technology for the investigation of a wide-range of systematic problems has not been fulfilled.

11.1 What are RAPDs?

Randomly amplified polymorphic DNAs (RAPDs) are one of a family of techniques that produce arbitrary fragment length polymorphisms and are collectively described as multiple arbitrary amplicon profiling (MAAP) (Caetano-Anollés, 1994). Other methodologies in this family include those using an arbitrarily primed polymerase chain reaction (AP–PCR) (Welsh and McClelland, 1990) and DNA amplification fingerprinting (DAF) (Caetano-Anollés *et al.*, 1991). The RAPDs technique utilises single, arbitrary, decamer DNA oligonucleotide primers to amplify regions of the genome using PCR (Welsh and McClelland, 1990; Williams *et al.*, 1990; Hadrys *et al.*, 1992; Newbury and Ford-Lloyd, 1993; Williams *et al.*, 1993). Priming sites are thought to be randomly distributed throughout a genome and polymorphism in these regions results in differing amplification products. The methodology is simple and has been widely used for the assessment of 'genetic' diversity, investigation of hybridisation and introgression events, examining species relationships and investigating population structure and processes.

11.2 What is the basis of the RAPDs technique?

Despite the widespread use of RAPDs there have been few attempts to describe those molecular mechanisms that underlie the RAPDs phenomenon (Caetano-Anollés *et al.*, 1992; Wagner *et al.*, 1994) and that may influence the value of the data in systematics. Venugopal *et al.* (1993), in apparently the only empirical study of the probable mechanism by which a single primer generates DNA polymorphism,

In *Molecular systematics and plant evolution* (1999) (eds P. M. Hollingsworth, R. M. Bateman and R. J. Gornall), Taylor & Francis, London, pp. 211–228.

suggest that there are a number of sites in the genome flanked by perfect or imperfect invert repeats which permit multiple annealing of the primer to occur (Williams *et al.*, 1993). Primer annealing sites are scattered throughout the nuclear and cytoplasmic genomes (Williams *et al.*, 1993; Thormann *et al.*, 1994; Aagaard *et al.*, 1995), in all classes of DNA from single-copy DNA to multiple-copy DNA (Williams *et al.*, 1993), and in coding and non-coding regions (Caetano-Anollés, 1993). Furthermore, Caetano-Anollés and colleagues (Caetano-Anollés *et al.*, 1991; Caetano-Anollés *et al.*, 1992; Caetano-Anollés, 1993) have proposed a complex reaction mechanism to explain the interactions between the variously primed DNA molecules using a single arbitrary primer, in which product amplification depends on two levels of genomic 'screening'. During the first screening level there is the selection of primer annealing sites that is determined by the primer sequence and the reaction environment, whilst during the second level there is differential amplification of the initial amplification products, which depends on the factors such as DNA concentration and template sequence (Caetano-Anollés, 1993). Such complex, stochastic and dynamic equilibria make RAPD reactions sensitive to both the experimental conditions and the sequences that are initially amplified.

RAPD polymorphism between DNAs appears to be the result of a range of processes, including nucleotide substitutions that create or abolish primer sites, formation of secondary structures between priming sites (Bowditch *et al.*, 1993) and insertion, deletion or inversion of either priming sites or segments between priming sites (Williams *et al.*, 1993). Furthermore, the effect of a nucleotide substitution appears to depend on its position in the primer recognition region (Caetano-Anollés *et al.*, 1992), especially in the 3′ region of the primer. In addition to polymorphism in product size distribution within a RAPD profile there is also polymorphism with respect to product intensity (Williams *et al.*, 1990; Caetano-Anollés *et al.*, 1991; Demeke *et al.*, 1992; Williams *et al.*, 1993). It has been suggested that intensity polymorphism is the result of product copy number differences, competition between PCR products, heterozygosity, comigration or partial mismatching of primer sites (Adams and Demeke, 1993; Venugopal *et al.*, 1993; Williams *et al.*, 1993; Lorenz *et al.*, 1994; Dowling *et al.*, 1996).

11.3 How have RAPDs been used?

The simplicity of the technique and the speed of data generation has attracted many researchers, particularly those interested in either genetic 'fingerprinting' or the patterns and levels of 'genetic' diversity (e.g. Chalmers *et al.*, 1992; Kazan *et al.*, 1993; Koller *et al.*, 1993; Megnegneau *et al.*, 1993; Russell *et al.*, 1993; Vierling and Nguyen, 1993; Demeke and Adams, 1994; Sharma *et al.*, 1995; Wang *et al.*, 1996; Wolff and Morgan-Richards, 1999 – this volume). An incomplete survey of the plant-based literature (excluding those investigations primarily concerned with genome mapping and the study of quantitative trait loci) between 1992 and 1996 showed a gradual increase in the number of papers published using RAPD technology. In addition to studies of 'genetic' diversity there have been increasing numbers of papers concerned with population genetics (e.g. Huff *et al.*, 1993; Nesbitt *et al.*, 1995; Crochemore *et al.*, 1996), phylogenetics (e.g. Demeke *et al.*, 1992; Adams and Demeke, 1993; Boehm *et al.*, 1993; van Buren *et al.*, 1994;

Stewart and Porter, 1995) and hybridisation/introgression (e.g. Arnold, 1993; Sale *et al.*, 1996; Sedgley *et al.*, 1996; Comes and Abbott, 1999 – this volume).

11.4 Advantages of RAPDs

An ideal, maximally informative, systematic marker would be expected to show: (i) variation at a suitable taxonomic rank; (ii) no environmental or developmental influences; (iii) simple codominant inheritance; (iv) independence of other markers; (v) reliability and reproducibility; (vi) a random scattering across the genome; and (vii) a state identity or similarity attributable to common ancestry. Thus an ideal marker allows the possibility of unambiguously characterising taxa.

RAPDs have considerable appeal for surveys of genomic variation and the identification of DNA-based characters for systematics since they are relatively inexpensive, randomly sample a potentially large number of loci and sequence information is not necessary for primer design (Hadrys *et al.*, 1992; Huff *et al.*, 1993; Williams *et al.*, 1993; Weising *et al.*, 1995), thus the technology can be easily applied to any taxon (Avise, 1994). Unfortunately, the technical simplicity of the RAPDs procedure has obscured the difficulties of understanding product banding patterns which have led to criticisms of this approach (Hillis, 1994; Karp and Edwards, 1997).

11.5 Disadvantages of RAPDs

The disadvantages of RAPDs may conveniently be summarised as practical problems associated with technology and intellectual problems associated with the RAPD phenomenon, some of which are discussed at more length below.

11.5.1 Reproducibility

Reproducibility is an acknowledged problem in RAPD analyses and strict procedural standardisation is required to obtain reproducible results (e.g. Williams *et al.*, 1993; Weising *et al.*, 1995; Jones *et al.*, 1997). Factors that influence RAPD reproducibility include: (i) primer and primer concentration; (ii) *Taq* polymerase source; (iii) magnesium ion concentration; (iv) template concentration; (v) thermocycler; (vi) temperature profile (Devos and Gale, 1992; Williams *et al.*, 1993; Weising *et al.*, 1995; Staub *et al.*, 1996). However, in strictly controlled situations within a single laboratory RAPD reproducibility problems can be overcome, although problems still exist with respect to transferring markers between laboratories, particularly if different gel systems or thermocyclers are routinely in use. Staub *et al.* (1996) suggested that for reproducible PCR results it may be necessary to optimise reactions for specific lots of PCR reagents. As with all PCR-based analyses DNA contamination is a continual problem, although this can be overcome by careful experimental procedures (Williams *et al.*, 1993).

11.5.2 Primer structure

There are 1,048,576 possible combinations of ten bases for RAPD primers. However, the majority of RAPD primers, rather than having random nucleotide contents, are constrained to have 60–70% GC (24,576 possible combinations) (Bowditch *et al.*, 1993; Williams *et al.*, 1993). It is known that GC content is not evenly distributed in plant genomes (Li and Graur, 1991), thus the RAPD technique may preferentially screen GC-rich regions. Clark and Lanigan (1993) highlighted the problem of biased GC-content in RAPD primers and argued that the high degree of polymorphism at CpG sites (methylation regions of eukaryotic genomes) may result in overestimates of nucleotide diversities of very similar taxa. The high GC content of RAPD primers appears to be necessary for successful low temperature annealing that allows a degree of imperfect priming (Welsh and McClelland, 1994), and is consistent with the observation that the number of amplified PCR products is positively correlated with the GC content of the RAPD primer (Caetano-Anollés *et al.*, 1991; Bucci and Menozzi, 1995). Mismatch offered by GC-rich RAPD primers means that there is the possibility that different primers bind at the same or overlapping priming sites and that nested priming may occur. Thus products from different primers may be either identical or inter-dependent.

11.5.3 Dominance

Direct analysis of ethidium bromide-stained gels in RAPD analysis makes the implicit assumption that RAPD markers are dominant, that is markers are scored as either present (+) or absent (−), and therefore dominant homozygotes (++) and heterozygotes (+−) are indistinguishable from each other. Experimental data from Williams *et al.* (1993) and Fritsch and Rieseberg (1992) showed that at least 95% of RAPD products behave as dominant markers, whilst Echt *et al.* (1992) found no codominant RAPD products. Unfortunately, in the majority of systematic studies it is rarely possible to perform extensive crossing programmes to test for marker segregation patterns. However, analysis of half-sib progeny arrays may provide useful information about segregation patterns, if the maternal genotype is known. Dominance of RAPD markers means that the estimation of genetic diversity and partitioning can only be made indirectly.

11.5.4 Product competition

Stochasticity of the RAPD reaction means that competition between products is a serious concern (Williams *et al.*, 1993). Such competition may be expected to occur at many different levels (between sites within a genome, between genomes and between genotypes) (Heun and Helentjaris, 1993) and may be enhanced in bulked segregant analyses (Michelmore *et al.*, 1991; Yu and Pauls, 1993). Halldén *et al.* (1996) have shown that competition is a serious source of error in genotyping *Brassica* and that it leads to an underestimate of genetic relatedness. Welsh and McClelland (1994) suggested that this effect could be overcome by using primers that generated more complex patterns, since there appears to be an inverse correlation between competition effects and product pattern complexity. However, repeating RAPD amplifications will not eliminate artefact bands if competition is a major factor in their occurrence.

11.5.5 Homology

Homology between RAPD products is most frequently based on the assumption that products that migrate to the same position on a gel are identical by descent, and is similar to the scoring of restriction fragment data (Bremer, 1991). Products of the same size may be non-homologous due to: (i) limitations in gel resolution; and (ii) size convergence, due to deletions and insertions between primer sites and primer site loss/gain (Williams *et al.*, 1993; Smith *et al.*, 1994). Furthermore, primer site distribution may result in products that display partial sequence similarity and hence are not independent (Smith *et al.*, 1994). Such problems are presumably correlated with genetic distance, becoming more significant for comparisons between distantly related species than within species or between closely related species. Empirical data for such concerns have been found by studies in the bacterium *Xanthomonas campestris*, where products with partial sequence similarity were amplified and non-homologous loci were indistinguishable by size (Smith *et al.*, 1994). Rieseberg (1996) showed that 91% of 220 comigrating products were homologous (as determined by either Southern blot or restriction enzyme analyses) among three species of *Helianthus*, indicating that product size may be a good predictor of homology, although a study of six *Brassica* species and *Raphanus sativa* showed that 20% of 15 RAPD products did not hybridise to all bands of the same mobility (Thormann *et al.*, 1994). Furthermore, in *Helianthus* approximately 13% of the homologous products appear to be paralogous, rather than orthologous, when compared to linkage mapping data (Rieseberg, 1996).

11.5.6 Allelic variation

Direct visualisation of RAPD products assumes that there are only two alleles (+ or –) per locus. Two models have been applied for understanding the systematic consequences of allelic variation at a locus, the 'allele-as-character' model (independent allele model) and the 'locus-as-character' model (Buth, 1984).

The 'allele-as-character' model treats individual alleles and their presence or absence as binary characters. Unfortunately alleles in this model are not independent since they must sum to one; if one allele decreases in frequency then the other must increase. Thus the 'loss' of a RAPD product is not an independent event but must involve either the occurrence of a new allele or the change in frequency of an existing allele. Similar RAPD phenotypes may therefore mask different transformation events.

The failure of the 'allele-as-character' model has been recognised and corrections have been tried using a 'locus-as-character' model, where different allelic combinations are used as the character's states (Buth, 1984). This model requires positive allele detection, although evidence exists that RAPDs may be multiallelic (Haymer, 1994), whilst population genetic models assume that RAPD loci are diallelic (Lynch and Milligan, 1994), and allele 'loss' may be the result of different events. Thus there is no simple way of identifying alleles or their combinations at different loci, meaning that the 'allele-as-character' model is not suitable for the analysis of RAPD data.

11.5.7 Genome sampling

Theoretically the RAPD methodology randomly samples all of the plant genome, whether nuclear, mitochondrial or chloroplast and whether highly repetitive, moderately repetitive or single-copy (Williams *et al.*, 1993). Unfortunately there is relatively little information regarding the distribution of such genomes and sequences within and between RAPD profiles (Smith *et al.*, 1994). Lorenz *et al.* (1994) and Thormann *et al.* (1994) showed that both the mitochondrial and chloroplast genomes made a contribution to the RAPD phenotypes of *Beta vulgaris* and *Brassica* species respectively, the contribution of the mitochondrial genome being greater than that of the chloroplast genome. Similarly, Aagaard *et al.* (1995) found that 45% of the scored RAPD products were of mitochondrial origin in *Pseudotsuga menziesii*. Since the population genetics of organelle genomes differ from those of the nuclear genome the distribution of organelle and nuclear markers within a RAPD profile may have important consequences for data interpretation (McCauley, 1995; Ennos *et al.*, 1999 – this volume).

11.5.8 Non-independence of loci

Non-independence between RAPD products may arise from sources that include RAPD products being associated with repetitive sequences in the genome, the unknown allelic relationships between products, heteroduplex formation and non-genetic artefacts (Hatcher *et al.*, 1992; Ellsworth *et al.*, 1993; Ayliffe *et al.*, 1994; Novy *et al.*, 1994; Novy and Vorsa, 1996). Furthermore, competition between products and product comigration may lead to non-independence of loci (Heun and Helentjaris, 1993; Williams *et al.*, 1993). In the absence of detailed genetic analyses, confirmation of independence between RAPD products cannot be made (Smith *et al.*, 1994; Dowling *et al.*, 1996).

11.6 Methods of analysing RAPD data

Product scoring is the first step in RAPD data analysis and requires a choice to be made between either scoring product presence/absence or accounting for product intensity (e.g. Demeke *et al.*, 1992; Adams and Demeke, 1993; Huff *et al.*, 1993; Nesbitt *et al.*, 1995). RAPD data is most often scored on the basis of product presence/absence, as a result of the difficulty of identifying the bases of intensity differences. Details of how RAPD gels are scored are, however, rarely given, although this is an important issue because of the reproducibility problems associated with the technique. Staub *et al.* (1996) have suggested that RAPD profiles be scored on a conservative basis: (i) only products in a limited size range are scored; (ii) only products of high intensity are scored since these generally show greater reproducibility compared to lower intensity products; (iii) products are scored within a region delimited by monomorphic large and small products, so that the quality of a RAPD amplification may be determined. Such procedures, whilst they decrease the number of products scored, are likely to increase confidence in any RAPD product scored. Re-extraction of DNA, repetition of RAPD amplifications and the inclusion of suitable controls to check RAPD reproducibility and the absence of contamination further enhances confidence in the data.

Gel resolution is usually low on standard agarose gels, where different sized products may migrate to apparently the same gel position. In contrast, much greater resolution of RAPD markers may be obtained using a combination of silver staining and either polyacrylamide or cellulose acetate gels (Dowling *et al.*, 1996; Valentini *et al.*, 1996). Limitations of RAPD data have been recognised and detailed analyses of RAPD products through Southern blotting and restriction digestion experiments, and analyses of genomic origin and DNA class have been suggested (Dowling *et al.*, 1996). Other workers have proposed the use of DNA pooling and RAPD product selection strategies to overcome intrataxon variation (Furman *et al.*, 1997).

Once RAPD data have been scored there are three primary methods of analysing the data, depending on the design of the experiment: (i) similarity measures (e.g. Abo-elwafa *et al.*, 1995); (ii) character measures (e.g. van Buren *et al.*, 1994); (iii) frequency measures (e.g. Nesbitt *et al.*, 1995).

11.6.1 Similarity measures

The most widely used approach to treating RAPD data has been the calculation of a similarity measure using one of three techniques: (i) the simple matching coefficient (SMC; Sneath and Sokal, 1973), which measures the proportion of shared product presences and absences between two RAPD profiles; (ii) the Jaccard's coefficient (J; Jaccard, 1908), which measures the proportion of shared product presences; (iii) the Nei and Li coefficient (NL; Nei and Li, 1979) which measures the probability of a product amplified in one sample also being amplified in another sample. An advantage of these similarity measures, compared to many of the others available (Birks, 1987), is that their statistical properties are well understood: SMC (Goodall, 1967); J (Real and Vargas, 1996) and NL (Nei and Li, 1979).

SMC, J and NL are based on different assumptions, which may not be met by RAPD data. In the case of SMC all shared bands (both presences and absences) are taken into account, even though the reason why RAPD products are absent may be unknown, an assumption that is not made by either J or NL. NL is a measure that may be interpreted biologically (the expected proportion of products shared because they are inherited from a common ancestor), but there are problems in understanding homology between RAPD products. These differences have led to discussions over which is the most appropriate measure to use for RAPD data. Some authors choose to use RAPD similarity measures that exclude shared absences (i.e. J or NL) (Kazan *et al.*, 1993), whilst many authors choose SMC (e.g. Lifante and Aguinagalde, 1996). In comparisons between the different measures some authors have found that different similarity measures give essentially the same result (e.g. Stiles *et al.*, 1993; Asemota *et al.*, 1996), which is not, perhaps, surprising given that the three measures are monotonic.

When considering RAPD data it is important to understand how different similarity coefficients are affected by the types of error that are associated with the RAPD technique. Two major groups of error may be identified: (i) false positive (a product that is present but should be absent); (ii) false negative (a product that is absent but should be present). The consequences of these two types of error on estimates of SMC, J and NL have been explored by Lamboy (1994a, 1994b) with respect to experimental artefacts. However, such problems still occur with RAPD data in the absence of experimental artefacts. For example, false positives are scored in instances of mistaken product homology, whilst false negatives are scored when products are present below the limit of gel resolution (Smith *et al.*, 1994).

Since SMC, J and NL are monotonic the effects of the proportion of false positives (f_p) and the proportion of false negatives (f_n) in a RAPD data set are the same for each measure. That is, if the proportion of shared presences (p) is greater than 0.5 then f_p results in a greater percentage bias (percentage difference between the 'true' and estimated values) than f_n. However, if p < 0.5 then f_n results in a greater percentage bias than f_p (Lamboy, 1994b). However, differences do exist between the measures with respect to bias distribution. Using Lamboy's (1994b) equations, and setting the proportion of shared bands (s; + or −) and p to 0.01, 0.20, 0.40, 0.60, 0.80 and 0.99 and f_p and f_n to vary between 0.00 and 0.10 in 0.01 steps, the percentage bias for SMC and NL is always less than or equal to that for J. Furthermore, < 10% bias was found on approximately 52% of occasions for SMC and NL, in contrast to J where < 10% bias was only found approximately 39% of the time. As the degree of acceptable bias is reduced then the differential between SMC and NL or J becomes less (at < 1% bias approximately 5.5% or 5% respectively). These data would tend to argue that NL is the most acceptable measure for similarity measures in RAPD analyses since it has biological meaning, only shared positive data are taken into account, and the percentage bias is less compared to the other measure, J.

Unfortunately for any one RAPD reaction the relative values of f_p and f_n cannot be predicted for mistaken homology, although Lamboy (1994a) has proposed corrections for similarity measures based on the occurrence of PCR artefacts in RAPD replications. Even the use of restriction enzyme analysis of specific RAPD products and Southern hybridisation (Dowling *et al.*, 1996; Rieseberg, 1996) cannot, as Stothard (1997) pointed out, be used to prove homology since products that share sequence identity may be paralogues rather than orthologues. Furthermore, Southern hybridisation will only detect sequences of a similarity determined by the washing conditions in the analysis (Maniatis, 1982).

The presence of errors within RAPD data (from whatever cause) can have dramatic effects on the resulting similarity matrices and hence clustering procedures (e.g. Lamboy, 1994a). Clearly without an estimate of the error within RAPD data it is not feasible to judge the quality of any clustering diagrams that result, since error effects are dependent on the proportion of shared bands in the comparison. The properties of clustering diagrams have been explored (Nei, 1987), although most workers choose to use either unweighted pair–group mean analysis (UPGMA) (Sneath and Sokal, 1973) or neighbour-joining (NJ) (Saitou and Nei, 1987) dendrograms. Furthermore, clustering diagrams based on RAPD data often have long terminal branches and short internode distances (e.g. Lifante and Aguinagalde, 1996; Millán *et al.*, 1996) and the identification of groups is based on subjective judgement. Such issues have traditionally been raised regarding numerical procedures in systematics and have been tackled by some workers through the estimation of errors associated with nodes (Ritland, 1989). Rieseberg (1996) suggested that split decomposition analysis (Bandelt and Dress, 1992) may be appropriate to measure random noise within such distance dendrograms, and recommended the use of bootstrap procedures (Felsenstein, 1985) to assess internal consistency within the data.

11.6.2 Character measures

The view that similarity measures reduce the information content of data has led some workers to explore RAPD products as character data (e.g. van Buren *et al.*, 1994; Stewart and Porter, 1995; Hoey *et al.*, 1996). At first glance RAPD data would appear to be effective character data since they are easily scored as character (product) presence/absence. For RAPD products to be utilised as character data they must be variable, independent and homologous (Swofford and Olsen, 1990). RAPD characters are variable, but it is clear that such characters may not conform to either the independence or homology criteria (Smith *et al.*, 1994; Backeljau *et al.*, 1995; van de Zande and Bijlsma, 1995; Rieseberg, 1996). Some of the problems associated with product homology and independence have already been considered.

A consideration raised by Backeljau *et al.* (1995) was the choice of an appropriate parsimony model, which is similar to the debate over appropriate parsimony models for RFLP data (e.g. deBry and Slade, 1985; Swofford *et al.*, 1996). Some workers (e.g. van Buren *et al.*, 1994) utilising the parsimony approach have used either Wagner or Fitch parsimony, although others (Stewart and Porter, 1995) have recognised that free reversibility of character states does not apply to RAPD products and have applied Dollo parsimony. If a ten base sequence is one base away from being a priming site, and a single base mismatch precludes amplification, then only one of 60 possible substitutions will change a non-priming site to a priming site (i.e. − to +), but any substitution will change a priming site to a non-priming site (+ to −). Backeljau *et al.* (1995) have argued that Dollo parsimony is not appropriate for RAPD data since RAPD polymorphisms do not depend solely on substitutions in the priming site (Williams *et al.*, 1993). Furthermore, the appropriateness of Dollo parsimony may be affected by whether product 'loss' results in a band being produced in another position on the RAPD profile, i.e. scored as separate loci (Holsinger and Jansen, 1993) and by the rates of nucleotide evolution and transversion/transition ratios in different parts of the genome (Gojobori *et al.*, 1982; Li and Graur, 1991). Such considerations led Backeljau *et al.* (1995) to suggest that weighted parsimony may overcome some of the problems, but in the absence of a detailed understanding of the RAPD reaction, there appears to be no obvious way that the necessary transformations can be applied and hence it is inappropriate to consider RAPD products as character data in parsimony analyses.

11.6.3 Frequency measures

RAPDs have been widely used for the assessment of 'genetic' diversity within and between either species, cultivars or natural populations, and many of these studies have concentrated on the use of single accessions (e.g. Chalmers *et al.*, 1994) rather than population level samples. Increasingly, however, RAPD analyses have utilised frequency of RAPD products (e.g. Huff *et al.*, 1993; Nesbitt *et al.*, 1995), and from these data levels and patterns of 'genetic' diversity are calculated. However, since one does not understand the genetic basis of RAPD markers in most studies based on natural populations then phenotypic rather than genotypic diversity is being measured and RAPD profiles should be treated as DNA level phenotypes. Such an approach has been suggested by Smith *et al.* (1994), where it was proposed that some of the problems of RAPDs may be overcome by treating RAPD profiles as haplotypes.

The majority of researchers consider individual RAPD products as independent markers with two states (+ and –) and have calculated diversity using; (i) similarity measures (e.g. Chalmers *et al.*, 1994; Wachira *et al.*, 1995, 1997); (ii) Shannon's measure (e.g. Russell *et al.*, 1993; Wolff and Morgan-Richards, 1999 – this volume); (iii) analysis of molecular variance (AMOVA; e.g. Huff *et al.*, 1993; Nesbitt *et al.*, 1995; Wolff and Morgan-Richards, 1999 – this volume). Shannon's measure and AMOVA both apportion diversity of RAPD markers within and between populations and make no assumptions regarding Hardy–Weinberg equilibrium. In contrast, some researchers have chosen to assume that their study populations are in Hardy–Weinberg equilibrium and hence calculate gene diversities and genetic differentiation parameters using standard techniques.

The utilisation of RAPD markers for studies of diversity has been investigated theoretically by Clark and Lanigan (1993) and Lynch and Milligan (1994). Clark and Lanigan (1993) proposed nine criteria that must be satisfied before RAPDs can be effectively used to estimate 'genetic' diversity between closely related taxa: (i) primer selection must not be biased to those revealing most polymorphism; (ii) all polymorphic and monomorphic bands must be carefully scored; (iii) polymorphic bands must be shown to behave as Mendelian factors; (iv) allelism of bands must be ascertained by either Southern analysis or segregation analysis; (v) fragment homology must be demonstrated; (vi) for diploids, population samples must be examined to determine product frequencies; (vii) true nucleotide sequence divergence should not exceed 10%; (viii) single nucleotide substitutions are assumed to result in a loss of amplification product; (ix) insertion/deletion variation that results in product presence/absence is assumed to be rare. It is clear that the majority of these conditions are not fulfilled in RAPD studies.

Dowling *et al.* (1996) have detailed a scheme whereby RAPD markers may be effectively used for population genetic and systematic studies. The approach combines half-sib and/or full-sib progeny analyses with Southern blot and restriction enzyme investigations of individual RAPD products. Unfortunately, in such a scheme the main advantages of RAPDs (speed and cost) are undermined and it may be appropriate to consider the use of other methodologies where time is used for reliable marker development, rather than assessing the quality of suspect data.

Since different RAPD studies use different sampling strategies, different primers identified by different criteria, different reaction conditions and different ways of scoring products, comparisons between studies are impossible, unless they are conducted by the same research group using identical strategies. Hamrick *et al.* (1979) have pointed out the difficulties of making comparisons between allozyme studies and such problems are magnified in RAPD studies. Such difficulties therefore limit the value of RAPD markers for understanding evolutionary processes unless additional information about RAPD marker characteristics is obtained.

11.7 Conclusions

The early promise of RAPDs as a quick, reliable and easily utilised DNA-based technology for the investigation of a wide-range of systematic problems has not been fulfilled. The majority of problems associated with RAPDs may be summarised as arising from the problems associated with reproducibility and product homology.

If these can be effectively overcome then RAPDs may become an effective and applicable tool for a wide range of systematic problems, including taxon delimitation, hybridisation, understanding patterns of genetic variation and phylogenetics. Unfortunately, the time and resources needed to overcome the first problem and the difficulty of understanding the basis of the second, render the technique inappropriate for many systematic studies.

Product reproducibility is determined by a variety of factors and is influenced by conditions within individual laboratories. Changes that result from the source of *Taq* polymerase and from thermal cycler design make the repetition of experiments between laboratories problematic, and raises questions about the future value of any markers that are identified using RAPDs. This problem is perhaps not such a concern if all that is required of a marker is the pattern of variation. If, however, markers are being used for species delimitation or the identification of taxon-specific markers for introgression and hybridisation studies then this becomes an important issue.

The homology problem in RAPDs may be solved in some cases by using improved systems of product resolution and gaining a detailed understanding of each of the products through genetic analysis and a combination of Southern blot and restriction enzyme studies. Such approaches have resulted in the generation of sequence characterised amplified regions (SCARs) (Paran and Michelmore, 1993; Fuchs *et al.*, 1996). Unfortunately, the differentiation of orthology from paralogy is not possible using these approaches and ultimately sequence analysis of individual RAPD products may not clarify this issue (Stothard, 1997).

Such criticisms are not to say that RAPDs are useless for systematic research; valuable data about patterns of variation within and between plant taxa have been obtained (Wolff and Morgan-Richards, 1999 – this volume). Furthermore, pragmatism often dictates that this is the only type of methodology available. The greatest attraction of the method is that it generates DNA data that are, theoretically at least, randomly scattered across the genome. Such markers are attractive for studies that involve differentiation of similar species (e.g. Chalmers *et al.*, 1992; Zamora *et al.*, 1996), hybridisation (e.g. Sale *et al.*, 1996; Sedgley *et al.*, 1996) and identification of patterns of variation (e.g. Hormaza *et al.*, 1994; Moreau *et al.*, 1994; Vasconcelos *et al.*, 1996). Rieseberg (1996) suggested that RAPDs may be useful for investigations within species or between closely related species and Dowling *et al.* (1996) have provided a rigorous framework within which RAPD studies may be conducted and the data interpreted.

Investigations using RAPDs cannot be undertaken in a biological vacuum and if such markers are to be utilised then they must be based on an understanding of the biology of the organism under study. Molecular panaceas for systematics do not exist; there is little value in using either RAPDs or organisms as black boxes. Some questions cannot be asked of RAPD data and some conclusions cannot be drawn from RAPD data, just as with any other technology.

Given the limitations that exist with RAPD analysis, do such criticisms aid in understanding the characteristics of other types of molecular data for systematics? Over the past few years increasing numbers of molecular techniques have become available to researchers, for example single-stranded conformation polymorphism (SSCP) (Hayashi, 1992; Jordan *et al.*, 1998), amplified fragment length polymorphism (AFLP) (Vos *et al.*, 1995; Matthes *et al.*, 1998) and microsatellites (SSR)

(Beckman and Soller, 1990; Morgante *et al.*, 1998). SSCP is known to be technically unreliable and the interpretation of the data is problematic (Dowling *et al.*, 1996). Such considerations indicate that it may not be an effective system to use for the majority of systematic questions. AFLPs, whilst much more reliable and reproducible than RAPDs, suffer from similar problems of data interpretation to RAPDs, for example marker origin is unknown, locus/allele designations are unclear and homology statements regarding bands are uncertain. Thus the analysis of AFLP data may be as problematic as RAPD data. In the case of SSR data, whilst they are a potentially valuable source of information for population genetic studies (Jarne and Lagode, 1996), SSRs tend to be restricted to a few marker loci in a small number of species and, although codominance is displayed, the homology of individual SSRs is based on product size and therefore suffers from the same difficulties as RAPDs if used in phylogenetic studies (Jarne and Lagode, 1996). Furthermore, the repeated nature of SSR loci and their high mutation rates mean that it may not be possible to confirm that bands of identical size in two taxa or widely divergent populations are evolutionarily homologous, since different mutation events in the SSR may produce products of similar size (Jarne and Lagode, 1996; Provan *et al.*, 1999 – this volume).

ACKNOWLEDGEMENTS

My thanks to Sarah Rendell, Julian Robinson and an anonymous reviewer for comments on this manuscript.

REFERENCES

Aagaard, J. E., Vollmer, S. S., Sorensen, F. C. and Strauss, S. H. (1995) Mitochondrial DNA products among RAPD profiles are frequent and strongly differentiated between races of Douglas-fir. *Molecular Ecology*, **4**, 441–446.

Abo-elwafa, A., Murai, K. and Shimada, T. (1995) Intra- and inter-specific variations in *Lens* revealed by RAPD markers. *Theoretical and Applied Genetics*, **90**, 335–340.

Adams, R. P. and Demeke, T. (1993) Systematic relationships in *Juniperus* based on random amplified polymorphic DNAs (RAPDs). *Taxon*, **42**, 553–571.

Arnold, M. L. (1993) *Iris nelsonii* (Iridaceae): origin and genetic composition of a homoploid hybrid species. *American Journal of Botany*, **80**, 577–583.

Asemota, H. N., Ramser, J., Lopez-Peralta, C., Weising, K. and Kahl, G. (1996) Genetic variation and cultivar identification of Jamaican yam germplasm by random amplified polymorphic DNA analysis. *Euphytica*, **92**, 341–351.

Avise, J. C. (1994) *Molecular markers, natural history and evolution*, Chapman & Hall, London.

Ayliffe, M. A., Lawrence, G. J., Ellis, J. G. and Pryor, A. J. (1994) Heteroduplex molecules formed between allelic sequences cause nonparental RAPD bands. *Nucleic Acids Research*, **22**, 1632–1636.

Backeljau, T., de Bruyn, L., de Wolf, H., Jordaens, K., van Dongen, S., Verhagen, R. *et al.* (1995) Random amplified polymorphic DNA (RAPD) and parsimony methods. *Cladistics*, **11**, 119–130.

Bandelt, H. J. and Dress, A. W. M. (1992) Split decomposition: a new and useful approach to phylogenetic analysis of distance data. *Molecular Phylogenetics and Evolution*, **1**, 242–252.

Beckman, J. S. and Soller, M. (1990) Toward a unified approach to gene mapping of eukaryotes based on sequence tagged microsatellite sites. *Bio/Technology*, **8**, 930–932.

Birks, H. J. B. (1987) Recent methodological developments in quantitative descriptive biogeography. *Annales Zoologici Fennici*, **24**, 165–178.

Boehm, M., Loew, R., Haag-Kerwer, A., Luettge, U. and Rausch, T. (1993) Evaluation of comparative DNA amplification fingerprinting for rapid species identification within the genus *Clusia*. *Botanica Acta*, **106**, 448–453.

Bowditch, B. M., Albright, D. G., Williams, J. G. K. and Braun, M. J. (1993) Use of amplified polymorphic DNA markers in comparative genome studies. *Methods in Enzymology*, **224**, 294–309.

Bremer, B. (1991) Restriction data from chloroplast DNA for phylogenetic reconstruction: Is there only one accurate way of scoring? *Plant Systematics and Evolution*, **175**, 39–54.

Bucci, G. and Menozzi, P. (1995) Segregation analysis of random amplified polymorphic DNA (RAPD) markers in *Picea abies* Karst. *Molecular Ecology*, **2**, 227–232.

Buth, D. G. (1984) The applications of electrophoretic data in systematic studies. *Annual Review of Ecology and Systematics*, **15**, 501–522.

Caetano-Anollés, G. (1993) Amplifying DNA with arbitrary oligonucleotide primers. *PCR Methods and Applications*, **3**, 85–94.

Caetano-Anollés, G. (1994) MAAP: a versatile and universal tool for genome analysis. *Plant Molecular Biology*, **25**, 1011–1026.

Caetano-Anollés, G., Bassam, B. and Gresshoff, P. M. (1991) DNA amplification fingerprinting using very short arbitrary primers. *Bio/technology*, **9**, 553–557.

Caetano-Anollés, G., Bassam, B. and Gresshoff, P. M. (1992) Primer–template interactions during DNA amplification fingerprinting with single arbitrary oligonucleotides. *Molecular and General Genetics*, **235**, 157–165.

Chalmers, K. J., Waugh, R., Sprent, J. I., Simons, A. J. and Powell, W. (1992) Detection of genetic variation between and within populations of *Gliricidia sepium* and *G. maculata* using RAPD markers. *Heredity*, **69**, 465–472.

Chalmers, K. J., Newton, A. C., Waugh, R., Wilson, J. and Powell, W. (1994) Evaluation of the extent of genetic variation in mahoganies (Meliaceae) using RAPD markers. *Theoretical and Applied Genetics*, **89**, 504–508.

Clark, A. G. and Lanigan, C. M. S. (1993) Prospects for estimating nucleotide divergence with RAPDs. *Molecular Biology and Evolution*, **10**, 1096–1111.

Comes, H. P. and Abbott, R. J. (1999) Reticulate evolution in the Mediterranean species complex of *Senecio* sect. *Senecio*: uniting phylogenetic and population-level approaches, in *Molecular systematics and plant evolution*, (eds P. M. Hollingsworth, R. M. Bateman and R. J. Gornall), Taylor & Francis, London, pp. 171–198.

Crochemore, M. L., Huyghe, C., Kerlan, M. C., Durand, F. and Julier, B. (1996) Partitioning and distribution of RAPD variation in a set of populations of the *Medicago sativa* complex. *Agronomie (Paris)*, **16**, 421–432.

DeBry, R. W. and Slade, N. A. (1985) Cladistic analysis of restriction endonuclease cleavage maps within a maximum-likelihood framework. *Systematic Zoology*, **34**, 21–34.

Demeke, T. and Adams, R. P. (1994) The use of RAPDs to determine germplasm collection strategies in the African species *Phytolacca dodecandra* (Phytolaccaceae), in *Conservation of plant genes II: Utilization of ancient and modern DNA*, (eds R. P. Adams, J. S. Miller, E. M. Golenberg and J. E. Adams), Missouri Botanical Garden, Missouri, pp. 131–140.

Demeke, T., Adams, R. P. and Chibbar, R. (1992) Potential taxonomic use of random amplified polymorphic DNA (RAPD): a case study in *Brassica*. *Theoretical and Applied Genetics*, **84**, 990–994.

Devos, K. M. and Gale, M. D. (1992) The use of random amplified polymorphic DNA markers in wheat. *Theoretical and Applied Genetics*, **84**, 567–572.

Dowling, T. E., Moritz, C., Palmer, J. D. and Rieseberg, L. H. (1996) Nucleic acids III: Analysis of fragments and restriction sites, in *Molecular systematics*, (eds D. M. Hillis, C. Moritz and B. K. Mable), Sinauer Associates, Sunderland, Mass., pp. 249–320.

Echt, C. S., Erdahl, L. A. and McCoy, T. J. (1992) Genetic segregation of random polymorphic DNA in diploid cultivated alfalfa. *Genome*, **35**, 64–67.

Ellsworth, D. L., Rittenhouse, K. D. and Honeycutt, R. L. (1993) Artifactual variation in randomly amplified polymorphic DNA banding patterns. *Biotechniques*, **14**, 214–217.

Ennos, R. A., Sinclair, W. T., Hu, X-S. and Langdon, A. (1999) Using organelle markers to elucidate the history, ecology and evolution of plant populations, in *Molecular systematics and plant evolution*, (eds P. M. Hollingsworth, R. M. Bateman and R. J. Gornall), Taylor & Francis, London, pp. 1–19.

Felsenstein, J. (1985) Confidence limits on phylogenies: an approach using the bootstrap. *Evolution*, **39**, 783–791.

Fritsch, P. and Rieseberg, L. H. (1992) High outcrossing rates maintain male and hermaphrodite individuals in populations of the flowering plant *Datisca glomerata*. *Nature*, **359**, 633–636.

Fuchs, H., Anastassiadis, C., Schlee, P., Stein, H. and Rottmann, O. (1996) Detection of a G/T base substitution in two bleak (*Alburnus alburnus*) populations by RAPD. *Journal of Applied Ichthyology*, **12**, 95–97.

Furman, B. J., Grattapaglia, D., Dvorak, W. S. and O'Malley, D. M. (1997) Analysis of genetic relationships of Central American and Mexican pines using RAPD markers that distinguish species. *Molecular Ecology*, **6**, 321–331.

Gojobori, T., Li, W. -H. and Graur, D. (1982) Patterns of nucleotide substitution in pseudogenes and functional genes. *Journal of Molecular Evolution*, **18**, 360–369.

Goodall, D. (1967) The distribution of the matching coefficient. *Biometrics*, **23**, 647–656.

Hadrys, H., Balick, M. and Schierwater, B. (1992) Applications of random amplified polymorphic DNA (RAPD) in molecular ecology. *Molecular Ecology*, **1**, 55–63.

Halldén, C., Hansen, M., Nilsson, N. O., Hjerdin, A. and Sall, T. (1996) Competition as a source of errors in RAPD analysis. *Theoretical and Applied Genetics*, **93**, 1185–1192.

Hamrick, J. L., Linhart, Y. B. and Mitton, J. B. (1979) Relationships between life history characteristics and electrophorectically detectable genetic variation in plants. *Annual Review of Ecology and Systematics*, **10**, 173–200.

Hatcher, S. L. S., Lambert, Q. T., Teplitz, R. L. and Carlson, J. R. (1992) Heteroduplex formation: a potential source of genotyping error from PCR products. *Prenatal Diagnosis*, **13**, 171–177.

Hayashi, K. (1992) PCR–SSCP: A method for detection of mutations. *Genetic Analysis: Techniques and Applications*, **9**, 73–79.

Haymer, D. S. (1994) Random amplified polymorphic DNAs and microsatellites: what are they, and can they tell us anything we don't already know? *Annals of the Entomological Society of America*, **87**, 717–722.

Heun, M. and Helentjaris, T. (1993) Inheritance of RAPDs in F1 hybrids of corn. *Theoretical and Applied Genetics*, **85**, 961–968.

Hillis, D. M. (1994) Homology in molecular biology, in *Homology. The hierarchical basis of comparative biology*, (ed. B. K. Hall), Academic Press, London, pp 339–368.

Hoey, B. K., Crowe, K. R., Jones, V. M. and Polans, N. O. (1996) A phylogenetic analysis of *Pisum* based on morphological characters, and allozyme and RAPD markers. *Theoretical and Applied Genetics*, **92**, 92–100.

Holsinger, K. E. and Jansen, R. K. (1993) Phylogenetic analysis of restriction site data. *Methods in Enzymology*, **224**, 439–455.

Hormaza, J. I., Dollo, L. and Polito, V. S. (1994) Determination of relatedness and geographical movements of *Pistacia vera* (Pistachio; Anacardiaceae) germplasm by RAPD analysis. *Economic Botany*, **48**, 349–358.

Huff, D. R., Peakall, R. and Smouse, P. E. (1993) RAPD variation within and among natural populations of outcrossing buffalo grass [*Buchloë dactyloides* (Nutt.) Engelm.]. *Theoretical and Applied Genetics*, **86**, 927–934.

Jaccard, P. (1908) Nouvelles recherches sur la distribution florale. *Bulletin de la Société Vaudense des Sciences Naturelles*, **44**, 223–270.

Jarne, P. and Lagode, P. J. L. (1996) Microsatellites, from molecules to populations and back. *Trends in Ecology and Evolution*, **11**, 424–429.

Jones, C. J., Edwards, K. J., Castaglione, S., Winfield, M. O., Sala, F., van der Wiel, C. et al. (1997) Reproducibility testing of RAPD, AFLP and SSR markers in plants by a network of European laboratories. *Molecular Breeding*, **3**, 381–390.

Jordan, W. C., Foley, K. and Bruford, M. W. (1998) Single-stranded conformation polymorphism (SSCP) analysis, in *Molecular tools for screening biodiversity*, (eds A. Karp, P. G. Isaac and D. S. Ingram), Chapman & Hall, London, pp. 152–156.

Karp, A. and Edwards, K. J. (1997) Molecular techniques in the analysis of the extent and distribution of genetic diversity, in *Molecular genetic techniques for plant genetic resources*, (eds W. G. Ayad, T. Hodgkin, A. Jaradat et al.), IPGRI, Rome, pp. 11–22.

Kazan, K., Manners, J. M. and Cameron, D. F. (1993) Genetic variation in agronomically important species of *Stylosanthes* determined using random amplified polymorphic DNA markers. *Theoretical and Applied Genetics*, **85**, 882–888.

Koller, B., Lehmann, A., McDermott, J. M. and Gessler, C. (1993) Identification of apple cultivars using RAPD markers. *Theoretical and Applied Genetics*, **85**, 901–904.

Lamboy, W. (1994a) Computing genetic similarity coefficients from RAPD data: correcting for the effects of PCR artefacts caused by variation in experimental conditions. *PCR Methods and Applications*, **4**, 38–43.

Lamboy, W. (1994b) Computing genetic similarity coefficients from RAPD data: the effects of PCR artefacts. *PCR Methods and Applications*, **4**, 31–37.

Li, W. -H. and Graur, D. (1991) *Fundamentals of molecular evolution*, Sinauer Associates, Sunderland, Mass.

Lifante, Z. D. and Aguinagalde, I. (1996) The use of random amplified polymorphic DNA (RAPD) markers for the study of taxonomical relationships among species of *Asphodelus* sect. *Verinea* (Asphodelaceae). *American Journal of Botany*, **83**, 949–953.

Lorenz, M., Weihe, A. and Börner, T. (1994) DNA fragments of organellar origin in random amplified polymorphic DNA (RAPD) patterns of sugar beet (*Beta vulgaris* L.). *Theoretical and Applied Genetics*, **88**, 775–779.

Lynch, M. and Milligan, B. G. (1994) Analysis of population genetic structure with RAPD markers. *Molecular Ecology*, **3**, 91–99.

Maniatis, T., Fritsch, E. F. and Sambrook, J. (1982) *Molecular cloning: a laboratory manual*, Cold Spring Harbor Laboratory, Cold Spring Harbor.

Matthes, M. C., Daly, A. and Edwards, K. J. (1998) Amplified fragment length polymorphism (AFLP), in *Molecular tools for screening biodiversity*, (eds A. Karp, P. G. Isaac and D. S. Ingram), Chapman & Hall, London, pp. 183–190.

McCauley, D. E. (1995) The use of chloroplast DNA polymorphism in studies of gene flow in plants. *Trends in Ecology and Evolution*, **10**, 198–202.

Megnegneau, B., Debets, F. and Hoekstra, R. F. (1993) Genetic variability and relatedness in the complex group of black Aspergilli based on random amplification of polymorphic DNA. *Current Genetics*, **23**, 323–329.

Michelmore, R. W., Paran, I. and Kesseli, R. V. (1991) Identification of markers linked to disease-resistance genes by bulked segregant analysis: a rapid method to detect markers in specific genomic regions by using segregating populations. *Proceedings of the National Academy of Sciences of the USA*, **88**, 9828–9832.

Millán, T., Osuna, F., Cobos, S., Torres, A. M. and Cubero, J. I. (1996) Using RAPDs to study phylogenetic relationships in *Rosa*. *Theoretical and Applied Genetics*, **92**, 273–277.

Moreau, F., Kleinschmit, J. and Kremer, A. (1994) Molecular differentiation between *Q. petraea* and *Q. robur* assessed by random amplified DNA fragments. *Forest Genetics*, **1**, 51–64.

Morgante, M., Pfeiffer, A., Jurman, I., Paglia, G. and Olivieri, A. M. (1998) Isolation of microsatellite markers in plants, in *Molecular tools for screening biodiversity*, (eds A. Karp, P. G. Isaac and D. S. Ingram), Chapman & Hall, London, pp. 288–296.

Nei, M. (1987) *Molecular evolutionary genetics*, Columbia University Press, New York.

Nei, M. and Li, W. (1979) Mathematical model for studying genetic variation in terms of restriction endonucleases. *Proceedings of the National Academy of Sciences of the USA*, **76**, 5269–5273.

Nesbitt, K. A., Potts, B. M., Vaillencourt, R. E., West, A. K. and Reid, J. B. (1995) Partitioning and distribution of RAPD variation in a forest tree species, *Eucalyptus globulus* (Myrtaceae). *Heredity*, **74**, 628–637.

Newbury, H. J. and Ford-Lloyd, B. V. (1993) The use of RAPD for assessing variation in plants. *Plant Growth Regulation*, **12**, 43–51.

Novy, R. G. and Vorsa, N. (1996) Evidence for RAPD heteroduplex formation in cranberry: implications for pedigree and genetic-relatedness studies and a source of co-dominant RAPD markers. *Theoretical and Applied Genetics*, **92**, 840–849.

Novy, R. G., Kobak, C., Goffreda, J. and Vorsa, N. (1994) RAPDs identify varietal misclassifications and regional divergence in cranberry [*Vaccinium macrocarpon* (Ait.) Pursh.]. *Theoretical and Applied Genetics*, **88**, 1004–1010.

Paran, I. and Michelmore, R. W. (1993) Development of reliable PCR based markers linked to downy mildew resistance genes in lettuce. *Theoretical and Applied Genetics*, **85**, 989–993.

Provan, J., Soranzo, N., Wilson, N. J., McNicol, J. W., Morgante, M. and Powell, W. (1999) The use of uniparentally inherited simple sequence repeat markers in plant population studies and systematics, in *Molecular systematics and plant evolution*, (eds P. M. Hollingsworth, R. M. Bateman and R. J. Gornall), Taylor & Francis, London, pp. 35–50.

Real, R. and Vargas, J. M. (1996) The probabilistic basis of Jaccard's index of similarity. *Systematic Biology*, **45**, 380–385.

Rieseberg, L. H. (1996) Homology among RAPD fragments in interspecific comparisons. *Molecular Ecology*, **5**, 99–105.

Ritland, K. (1989) Genetic differentiation, diversity, and inbreeding in the mountain monkeyflower (*Mimulus caespitosus*) of the Washington Cascades. *Canadian Journal of Botany*, **67**, 2017–2024.

Russell, J. R., Hosein, F., Johnson, E., Waugh, R. and Powell, W. (1993) Genetic differentiation of cocoa (*Theobroma cacao* L.) populations revealed by RAPD analysis. *Molecular Ecology*, **2**, 89–97.

Saitou, N. and Nei, M. (1987) The neighbor-joining method: a new method for reconstructing phylogenetic trees. *Molecular Biology and Evolution*, **4**, 406–425.

Sale, M. M., Potts, B. M., West, A. K. and Reid, J. B. (1996) Molecular differentiation within and between *Eucalyptus risdonii, E. amygdalina* and their hybrids using RAPD markers. *Australian Journal of Botany*, **44**, 559–569.

Sedgley, M., Wirthensohn, M. G. and Delaporte, K. L. (1996) Interspecific hybridisation between *Banksia hookeriana* Meisn. and *Banksia prionotes* Lindl. (Proteaceae). *International Journal of Plant Sciences*, **157**, 638–643.

Sharma, S. K., Dawson, I. K. and Waugh, R. (1995) Relationships among cultivated and wild lentils revealed by RAPD analysis. *Theoretical and Applied Genetics*, **91**, 647–654.

Smith, J. J., Scott-Craig, J. S., Leadbetter, J. R., Bush, G. L., Roberts, D. L. and Fulbright, D. W. (1994) Characterization of random amplified polymorphic DNA (RAPD) products from *Xanthomonas campestris* and some comments on the use of RAPD products in phylogenetic analysis. *Molecular Phylogenetics and Evolution*, **3**, 135–145.

Sneath, P. H. A. and Sokal, R. R. (1973) *Numerical taxonomy*, Freeman, San Francisco.

Staub, J., Bacher, J. and Poetter, K. (1996) Sources of potential errors in the application of random amplified polymorphic DNAs in cucumber. *Hortscience*, **31**, 262–266.

Stewart, C. N. and Porter, D. M. (1995) RAPD profiling in biological conservation: an application to estimating clonal variation in rare and endangered *Iliamna* in Virginia. *Biological Conservation*, **74**, 135–142.

Stiles, J. I., Lemme, C., Sondur, S., Morshidi, M. B. and Manshardt, R. (1993) Using randomly amplified polymorphic DNA for evaluating genetic relationships among papaya cultivars. *Theoretical and Applied Genetics*, **85**, 697–701.

Stothard, J. R. (1997) Phylogenetic inference with RAPDs: some observations involving computer simulation with viral genomes. *Journal of Heredity*, **88**, 222–228.

Swofford, D. L. and Olsen, G. J. (1990) Phylogeny construction, in *Molecular systematics*, (eds D. M. Hillis and C. Moritz), Sinauer Associates, Sunderland, Massachusetts, pp. 411–501.

Swofford, D. L., Olsen, G. J., Waddell, P. J. and Hillis, D. M. (1996) Phylogenetic inference, in *Molecular systematics*, (eds D. M. Hillis, C. Moritz and B. K. Mable), Sinauer Associates, Sunderland, Massachusetts, pp. 407–514.

Thormann, C. E., Ferreira, M. E., Camargo, L. E. A., Tivang, J. G. and Osborn, T. C. (1994) Comparison of RFLP and RAPD markers to estimating genetic relationships within and among cruciferous species. *Theoretical and Applied Genetics*, **88**, 973–980.

Valentini, A., Timperio, A. M., Cappuccio, I. and Zolla, L. (1996) Random amplified polymorphic DNA (RAPD) interpretation requires a sensitive method for the detection of amplified DNA. *Electrophoresis*, **17**, 1553–1554.

van Buren, R., Harper, K. T., Andersen, W. R., Stanton, D. J., Seyoum, S. and England, J. L. (1994) Evaluating the relationship of autumn buttercup (*Ranunculus acriformis* var. *aestivalis*) to some close congeners using random amplified polymorphic DNA. *American Journal of Botany*, **81**, 514–519.

van de Zande, L. and Bijlsma, R. (1995) Limitations of the RAPD technique in phylogeny reconstruction in *Drosophila. Journal of Evolutionary Biology*, **8**, 645–656.

Vasconcelos, M. J. V., de Barros, E. G., Moreira, M. A. and Vieira, C. (1996) Genetic diversity of the common bean *Phaseolus vulgaris* L. determined by DNA-based molecular markers. *Brazilian Journal of Genetics*, **19**, 447–451.

Venugopal, G., Mohapatra, S., Salo, D. and Mohapatra, S. (1993) Multiple mismatch annealing: basis for random amplified polymorphic DNA fingerprinting. *Biochemical and Biophysical Research Communications*, **197**, 1382–1387.

Vierling, R. A. and Nguyen, H. T. (1993) Use of RAPD markers to determine the genetic diversity of diploid wheat genotypes. *Theoretical and Applied Genetics*, **84**, 835–838.

Vos, P., Hogers, R., Bleeker, M., Reijans, M., Vandelee, T., Hornes, M. *et al.* (1995) AFLP: a new technique for DNA fingerprinting. *Nucleic Acids Research*, **23**, 4407–4414.

Wachira, F. N., Powell, W. and Waugh, R. (1997) An assessment of genetic diversity among *Camellia sinensis* L. (cultivated tea) and its wild relatives based on randomly amplified polymorphic DNA and organelle-specific STS. *Heredity*, **78**, 603–611.

Wachira, F. N., Waugh, R., Hackett, C. A. and Powell, W. (1995) Detection of genetic diversity in tea (*Camellia sinensis*) using RAPD markers. *Genome*, **38**, 201–210.

Wagner, A., Blackstone, N., Cartwright, P., Dick, M., Misof, B., Snow, P. *et al.* (1994) Surveys of gene families using polymerase chain reaction: PCR selection and PCR drift. *Systematic Biology*, **43**, 250–261.

Wang, C. T., Wang, W. Y., Chiang, C. H., Wang, Y. N. and Lin, T. P. (1996) Low genetic variation in *Amentotaxus formosana* Li revealed by isozyme analysis and random amplified polymorphic DNA markers. *Heredity*, **77**, 388–395.

Weising, K., Nybom, H., Wolff, K. and Meyer, W. (1995) *DNA fingerprinting in plants and fungi*, CRC Press, London.

Welsh, J. and McClelland, M. (1990) Fingerprinting genomes using PCR with arbitrary primers. *Nucleic Acids Research*, **18**, 7213–7218.

Welsh, J. and McClelland, M. (1994) Fingerprinting using arbitrary primed PCR: application to genetic mapping, population biology, epidemiology, and detection of differentially expressed RNAs, in *The polymerase chain reaction*, (eds K. B. Mullis, F. Ferré and R. A. Gibbs), Birkhäuser, Boston, pp. 295–303.

Williams, J. G. K., Hanafey, M. K., Rafalski, J. A. and Tingey, S. V. (1993) Genetic analysis using random amplified polymorphic DNA markers. *Methods in Enzymology*, **218**, 704–740.

Williams, J. G. K., Kubelik, A. R., Livak, K. J., Rafalski, J. A. and Tingey, S. V. (1990) DNA polymorphisms amplified by arbitrary primers are useful as genetic markers. *Nucleic Acids Research*, **18**, 6531–6535.

Wolff, K. and Morgan-Richards, M. (1999) The use of RAPD data in the analysis of population genetic structure: case-studies of *Alkanna* (Boraginaceae) and *Plantago* (Plantaginaceae), in *Molecular systematics and plant evolution*, (eds P. M. Hollingsworth, R. M. Bateman and R. J. Gornall), Taylor & Francis, London, pp. 51–73.

Yu, K. and Pauls, K. P. (1993) Rapid estimation of genetic relatedness among heterogeneous populations of alfalfa by random amplification of bulked genomic DNA samples. *Theoretical and Applied Genetics*, **86**, 788–794.

Zamora, R., Jamilena, M., Ruiz Rejón, M. and Blanca, G. (1996) Two new species of the carnivorous genus *Pinguicula*, (Lentibulariaceae) from Mediterranean habitats. *Plant Systematics and Evolution*, **200**, 41–60.

Chapter 12

Nuclear protein-coding genes in phylogeny reconstruction and homology assessment: some examples from Leguminosae

J. J. Doyle and J. L. Doyle

ABSTRACT

Nuclear, protein-coding genes present both advantages and disadvantages for plant systematists. Analyses of gene expression in a phylogenetic context may provide criteria for assessing homologies of structural or functional phenotypic characters. Recombination introduces technical and theoretical problems, but permits different nuclear genes to serve as independent historical markers, a feature that may be particularly critical in studies of closely related species. Because gene duplication has been and continues to be a major force in plant genome evolution, identifying orthologous sequences is an important concern. Concerted evolution can alleviate this problem, but varies in its effectiveness as a homogenizing agent not only across different gene families but also at different taxonomic levels and even from taxon to taxon in the same gene family. Among genes we are engaged in studying, the nuclear copy of the normally mitochondrially-encoded cytochrome oxidase subunit II (*cox2*) is a true single copy gene in some legumes, and has proven useful in studies of phylogeny within the papilionoid tribe Phaseoleae and allies. In studies of *Glycine*, introns of one member of the complex histone H3 gene family are more variable than nrDNA ITS sequences and at the same time more congruent with genomic relationships than is cpDNA. Genes encoding leghemoglobin vary in their degree of concerted evolution both across the Papilionoideae and within Phaseoleae, and represent a cautionary note about assumptions concerning this phenomenon. Glutamine synthetase genes have been used both for studies at lower taxonomic levels and also to address questions concerning the homologies of nodules in different legume subfamilies. The gene family seems to be much more complex than it appeared from studies of expressed sequences in pea or *Phaseolus*.

12.1 Introduction

The three genomes of plants contain a wealth of sequences with potential for use in systematics. Until the widespread use of the polymerase chain reaction (PCR) made DNA sequencing accessible to systematists, the chloroplast genome was by far the most widely used source of characters for molecular phylogenetic analysis in flowering plants. With the advent of PCR, chloroplast sequencing studies initially concentrated on relatively conservative genes such as *rbcL*, and addressed questions

In *Molecular systematics and plant evolution* (1999) (eds P. M. Hollingsworth, R. M. Bateman and R. J. Gornall), Taylor & Francis, London, pp. 229–254.

at higher taxonomic levels (e.g. Chase *et al.*, 1993). Individual chloroplast sequences have also been developed for use intra- and inter-specific studies (e.g. Taberlet *et al.*, 1991; Dumolin-Lapegue *et al.*, 1997; Oxelman *et al.*, 1997; see Soltis and Soltis, 1998), but since sequencing replaced restriction mapping as the principal molecular systematic methodology, the primary sequence used at lower taxonomic levels has been the internal transcribed spacer region of nuclear 18S-26S ribosomal RNA gene family (nrDNA ITS: e.g. Baldwin *et al.*, 1995; Hershkovitz *et al.*, 1999 – this volume). At the same time, the 18S and, to a lesser degree, the 26S coding regions of that locus provide useful characters for studies at higher taxonomic levels (reviewed in Soltis and Soltis, 1998). Even the mitochondrial genome, long neglected because of its structural complexity and overall very slow rate of substitution (Palmer, 1992; Laroche *et al.*, 1997) has begun to be used at higher taxonomic levels (e.g. Malek *et al.*, 1996; Davis *et al.*, 1998).

Despite the availability of these diverse molecular tools, additional sources of characters for molecular systematic studies are needed for resolving relationships at both ends of the phylogenetic spectrum. Studies at higher taxonomic levels seek to reconstruct ancient divergences, where many original synapomorphies are likely to have been obscured by superimposed substitutions ('multiple hits'); also, many divergences involve relatively rapid radiations, where few characters were available to support branching patterns in the first place. At the opposite end of the spectrum, genes of recently diverged species will usually differ by few characters, so any single gene may not provide sufficient resolution. Even when single genes do provide a suitable hypothesis, independent estimates of phylogenetic pattern are also required, particularly for studies involving lower taxonomic categories. The nuclear genome is the most likely source of many of the sequences that will be developed for phylogeny reconstruction in years to come. It is by far the largest plant genome, with a correspondingly broad range of evolutionary processes and rates. Moreover, unlike the single linkage groups of organellar genomes, the nuclear genome can provide numerous independent historical estimates because of recombination and independent assortment of chromosomes.

In addition, the vast majority of genes that underlie morphological, anatomical, and chemical phenotypes are encoded by the nuclear genome. These phenotypes are the sources of the characters that have traditionally been used in plant classification and phylogeny and play important functional roles, so obtaining an understanding of their molecular bases is of obvious relevance for both phylogenists and biosystematists. Rapid advances in molecular developmental genetics are providing systematists with opportunities to address questions of homology at a level hitherto unavailable (e.g. Doyle, 1994a, b; Albert *et al.*, 1998; Gustafsson and Albert, 1999 – this volume).

In this chapter, the systematic utility of nuclear genes encoding proteins (as opposed to structural RNAs) is addressed. After discussing some general concerns about the use of nuclear genes, some examples are given of genes currently under study in our laboratory, most of which involve Leguminosae. One of these, cytosolic glutamine synthetase, is described as an example of how gene sequences and expression patterns could be used to study homology of morphological features. Finally, we discuss the use of intron-containing genes in studies of congeneric species and closely related genera, with emphasis on a histone locus in *Glycine*.

12.2 Some problems with nuclear genes in systematic studies

The difficulties of using nuclear low-copy genes are readily illustrated by contrast with cpDNA. Most current systematic studies rely on PCR to amplify a gene of interest for sequencing, which for several reasons is a relatively simple matter for cpDNA. Numerous copies of the chloroplast genome are usually present in each of the many chloroplasts found in a typical leaf cell. Thus chloroplast genes occur in high copy number in the cell relative to any given nuclear low-copy gene, facilitating amplification (similarly, amplification of highly repeated nrDNA sequences is also expected to be easier than amplification of a low-copy nuclear gene).

The chloroplast genome is small and compact, is saturated with genes and has short intergenic regions, rarely contains repeated sequences (other than the two identical inverted repeat [IR] regions, whose genes are effectively single copy because of copy correction), and is relatively stable in gene content and order (reviewed by Palmer, 1991). Since the 1980s, when the complete chloroplast sequences became available from a non-vascular plant (*Marchantia*: Ohyama *et al.*, 1986) a dicotyledonous angiosperm (*Nicotiana*: Shinozaki *et al.*, 1986) and a monocot (*Oryza*: Hiratsuka *et al.*, 1989), it has been relatively simple to design oligonucleotide primers for PCR amplification and sequencing of particular genes or intergenic regions that can be used over a wide range of land plant taxa. In contrast, the nuclear genome is huge and highly variable in size (Bennett and Leitch, 1997), and its genes are usually widely separated from one another by large regions of noncoding DNA and scattered among multiple chromosomes. Repeats, often belonging to families of retrotransposons, can make up half or more of the genome (e.g. Sanmiguel *et al.*, 1996). Many protein coding gene sequences are available from the GenBank/EMBL databases, but because nuclear genes overall have a higher rate of evolution than do chloroplast genes (Wolfe *et al.*, 1989; Gaut, 1998), it is in general more difficult to construct universal primers for nuclear genes.

As opposed to the majority of chloroplast genes, many nuclear genes contain introns that usually show considerably higher levels of variation than do surrounding coding regions. From the standpoint of reconstructing phylogenies of higher taxa, introns are a nuisance. They must often be sequenced along with exons, but are usually too divergent among higher taxa to provide any relevant phylogenetic information. The number of useful characters per kilobase of sequence is thus dramatically reduced.

In diploid sexual plant species, heterozygosity is expected at many nuclear loci. Heterozygosity presents technical problems, because the two (or more, in polyploids) alleles will be amplified together in heterozygotic individuals. Heterogeneity within individuals can be useful in analyzing cases of hybridisation (for an example of the use of such patterns in nrDNA ITS, see Sang *et al.*, 1995), but it can also be very difficult to interpret. If alleles differ in sequence but not in length, direct sequencing of PCR products from a heterozygote should reveal the polymorphic nature of the amplified product, with the variant nucleotides at a particular position occurring together as double bands on gels or superimposed peaks on sequence printouts. If alleles differing in length are sequenced together, sequences downstream from insertion/deletion (indel) sites will be completely uninterpretable due to the offsetting of homologous positions, resulting in many superimposed nucleotides.

The molecular systematics research programme involves inferring species trees from gene trees, an inference that requires substitution of taxon names for the actual OTUs used in the molecular phylogenetic analysis (discussed at length by Doyle, 1997). This substitution is usually straightforward for cpDNA, where there is a single haplotype per individual except in rare cases of heteroplasmy. For nuclear genes, however, a heterozygous individual organism and the taxon it represents cannot be represented adequately by any single allele, so no unique substitution of taxon for gene can be made (Doyle, 1995).

The presence of multiple alleles at a locus also contributes to yet another problem. Where alleles with different histories are brought together sexually in the genome of a diploid individual, the potential exists for allelic recombination. Like taxic hybridisation, recombination is a form of reticulation and as such does not conform to the expectation that the taxa being studied have a singular history (Nixon and Wheeler, 1992; Doyle, 1996, 1997). Again like taxic hybridisation (e.g. McDade, 1995), recombination can have severe consequences for the reconstruction of allele phylogenies (Doyle, 1997). Although intramolecular recombination occurs in cpDNA (leading, for example, to the presence of isomeric forms of the molecule by 'flip-flop' recombination (Palmer, 1991)), effective recombination of angiosperm chloroplast genomes with different histories is unknown in nature. Thus, the chloroplast genome behaves historically as a single gene; it is a coalescent gene or 'c-gene' *sensu* Doyle (1995). This is not expected to be true of many nuclear genes, where the historically independent unit can be as small as a single nucleotide (Hudson, 1990).

Many plant protein-coding nuclear genes belong to multigene families that range in size from two to many hundreds of members, and evolution in such families is poorly understood (Clegg *et al.*, 1997). Gene duplication complicates the conventional definition of homology (similarity due to common descent): sequences related by gene duplication (e.g. genes belonging to different subfamilies of a multigene family, such as α vs. β hemoglobins) are paralogous, whereas genes related by taxic cladogenesis (e.g. genes of the same gene subfamily in different species, such as chimp vs. gorilla β hemoglobins) are orthologous (Fitch, 1970). The inclusion of paralogous sequences in a phylogenetic analysis can lead to incorrect inferences of taxic relationships (e.g. Doyle, 1992; Doyle and Davis, 1998). In contrast, chloroplast genes are with few known exceptions actually or (in the IR) effectively single copy, which confines paralogy problems to those sequences that have been transferred from the chloroplast to other genomes (e.g. Olmstead and Palmer, 1994).

The sampling problem for duplicated genes can be mitigated by concerted evolution (Zimmer *et al.*, 1980). The processes responsible for concerted evolution (gene conversion, unequal crossing over) may completely homogenise the members of a gene family such that only multiple copies of a single species-specific sequence are observed (Doyle and Davis, 1998). This is the typical expectation for nrDNA, which is often but not always realised in practice (Suh *et al.*, 1993; Buckler *et al.*, 1997). Plant nuclear protein-coding genes also are known to evolve concertedly, though to varying degrees. Divergent paralogous clades are well known for mammalian globin genes, yet it was in a study of globins that the term 'concerted evolution' was first coined (Zimmer *et al.*, 1980). Similarly, paralogous actin gene clades have been maintained over long evolutionary periods, but examples of gene conversion have been hypothesised among some members of this gene family in maize (Moniz de Sá and Drouin, 1996). In contrast, the phylogenetic tree of the small *rbcS* gene family shows taxon-specific clades of genes; in Solanaceae, for example, such groupings clearly include paralogous copies whose coding sequences (but not intron number) have been homogenised (Meagher *et al.*, 1989). Whereas complete homogenisation by concerted evolution eliminates sampling problems associated with paralogy, simulation studies have shown that incomplete levels of concerted evolution can produce highly misleading phylogenies (Sanderson and Doyle, 1992). This occurs because there are two competing phylogenetic signals: (1) that due to divergence, which produces synapomorphies linking orthologous sequences across taxa, and (2) that due to concerted evolution, where characters spread by recombination among paralogous sequences within a species become synapomorphies, uniting those paralogues in species-specific clades (Doyle and Davis, 1998).

12.3 'Single copy' nuclear genes versus gene families

Given the problems associated with multigene families, it would seem reasonable to use genes that are truly single copy in the nucleus. This, however, turns out to be more easily said than done, because it appears that most plant genes belong to gene families. 'Single copy' is in any case somewhat misleading, because gene duplication has been a dominant mechanism in the evolution of all genes, and therefore even what appear to be single copy genes may be related to some other gene, albeit distantly. For the systematist, the practical question is whether or not the closest paralogues of a locus under investigation are sufficiently divergent so as not to confuse inferences of taxic history from the gene tree. This requirement can be met even when it is clear that a gene belongs to a gene family, as in the case of the histone H3 gene family (Doyle *et al.*, 1996; discussed below), or the phytochrome gene family (Mathews *et al.*, 1995; Lavin *et al.*, 1998).

One gene that appears to be single copy (at least in angiosperms) is a locus involved in the control of flowering, known as *leafy* in *Arabidopsis*, *unifoliolata* in *Pisum*, *nfl* in *Nicotiana*, and *floricaula* in *Antirrhinum* (e.g. Hofer *et al.*, 1997). Exon sequences and intron positions of this gene are highly conserved over seed plant evolution (Frohlich and Meyerowitz, 1997), but the gene has not yet been used in comprehensive phylogenetic studies. Other genes that appear to be single copy in at least some taxa and are being developed for phylogenetic studies include *waxy* (Mason-Gamer and Kellogg, 1996), RNA polymerase II (Denton *et al.*, 1998), alcohol dehydrogenase (ADH: Sang *et al.*, 1997), and arginine decarboxylase (Galloway *et al.*, 1997); however, the last three genes are duplicated in at least some diploid taxa, including a recent ADH duplication in one section of *Paeonia* (Sang *et al.*, 1997).

In the following sections, we describe genes that we are investigating for their potential in our studies of phylogeny and evolution of legume genera and tribes. One (nuclear *cox2*) is a true single copy gene in the nucleus whereas the other (symbiotic leghemoglobin) belongs to a gene family. Another gene family, that encoding cytosolic glutamine synthetase, is discussed under the heading of genes used for molecular tests of morphological homology.

12.3.1 Nuclear cox2: a relatively recent transfer to the nucleus

The evolution of organelle genomes has involved the massive loss of genes from the genomes of the originally free-living eubacteria that are hypothesised to have evolved endosymbiotically into chloroplasts and mitochondria. Many of these genes apparently were transferred to the nucleus, a process that although presumably most prevalent early in the evolution of these genomes has continued even in modern taxa. One such mitochondrial gene, cytochrome oxidase subunit II (*cox2*) appears to have been transferred as an edited RNA copy to the nucleus some 60–200 million years ago, and to have been lost subsequently from the mitochondrial genome in some lineages of modern legumes (Nugent and Palmer, 1991; J. J. Doyle, J. L. Doyle, K. Song, K. Adams and J. D. Palmer, unpublished). The nuclear gene has acquired a transit peptide sequence and a typical nuclear spliceosomal intron, and appears to be a true single copy gene in the nuclei of most species studied. Amplification primers designed specifically for the nuclear copy were used to amplify nu-*cox2* from diverse legumes in the papilionoid tribe Phaseoleae and its presumed allies.

Phylogenetic analysis of this gene (J. J. Doyle, J. L. Doyle, K. Song, K. Adams and J. D. Palmer, unpublished) suggests that it has potential for studies at taxonomic levels ranging from tribe to species. The overall topology is consistent with results from *rbcL* analyses for Phaseoleae and allies (Doyle *et al.*, 1997). In particular, some initially surprising results from the cpDNA studies were corroborated by these analyses, such as the nesting of taxa from the morphologically quite distinct tribes Desmodieae and Psoraleeae within the large and diverse Phaseoleae (Fig. 12.1). At the lower end of the taxonomic scale, the more variable regions of this gene (notably the intron) provide sufficient characters to resolve relationships among species of *Phaseolus*, and produced a topology that was largely congruent with previous cpDNA restriction site studies (Delgado-Salinas *et al.*, 1993). It is unclear how useful this gene will be outside Leguminosae, because of a lack of sampling for the nuclear copy in other families.

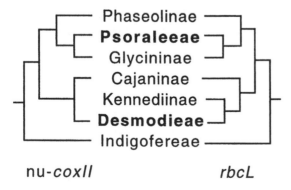

nu-*coxII*   *rbcL*

Figure 12.1 Congruence between topologies based on the chloroplast gene *rbcL* (Doyle *et al.*, 1997, and unpublished) and the nuclear copy of *cox2* (K. Song, K. Adams, J.D. Palmer and J.J. Doyle, in prep.). Taxa in boldface are tribes, while the remaining taxa are all members of the tribe Phaseoleae. Members of subtribe Kennediinae have not yet been sampled for *cox2*.

12.3.2 Leghemoglobin in Glycine subgenus Glycine: unexpected maintenance of paralogous clades

Leghemoglobins (Lb), as the name implies, are globins produced by legumes. These oxygen-binding proteins are most prevalent in the nodules of leguminous plants that participate in symbiosis with nitrogen-fixing soil bacteria ('rhizobia' in a broad, non-phylogenetic sense (Young, 1996)). Their role in the nodule is thought to involve binding and release of oxygen at a partial pressure that allows bacteroid growth without poisoning the oxygen-sensitive bacterial nitrogenase. Although these globins were once thought to be a prime example of horizontal gene transfer from animals, it was shown some time ago that globin genes occur in the genomes of non-nodulating legumes and in plants other than legumes, so that there is no need to postulate anything but vertical evolution (Appleby et al., 1990). The symbiotic leghemoglobins are encoded by a small gene family (e.g. Lee and Verma, 1984). Recently the identification of a non-symbiotic leghemoglobin gene from soybean (*Glycine max*) has shown that the gene family is more complex than originally believed (Andersson et al., 1996). The role of plant globins beyond nodulation remains a matter of speculation.

Phylogenetic analyses of the symbiotic leghemoglobins show a complex pattern in which some but not all taxa appear to evolve concertedly (Fig. 12.2a; Doyle, 1994b; Frühling et al., 1997). Thus, for example, in such distantly related papilionoid legumes as *Glycine* and *Lupinus*, the Lb genes of a taxon all group together in a pattern suggestive of concerted evolution. In contrast, sequences from *Medicago* and *Vicia* do not each form single taxon-specific clades, but instead show a pattern most consistent with maintenance of paralogous relationships. Intriguingly, a very similar pattern is observed in the gene family encoding the 7S seed storage proteins. In that case again, members of the temperate herbaceous group of legumes (e.g. *Pisum, Medicago, Vicia, Cicer*) appear to have preserved paralogous relationships whereas sequences from *Glycine* and its allies (e.g. *Vigna*) evolve concertedly, retaining no evidence of paralogy/orthology relationships (Doyle, 1994b; Saenz de Miera and Perez de la Vega, 1998).

Using amplification primers designed against both *Phaseolus vulgaris* and *Glycine max* (Skroch et al., 1993), we have cloned Lb genes from several species of *Glycine* subgenus *Glycine*, the perennial relatives of the cultivated soybean. Because *Glycine max* sequences grouped together in the Lb tree (Fig. 12.2a), we anticipated that Lb sequences from *Glycine* species would conform to the classic concerted evolutionary pattern, with species-specific clusters of genes rather than several paralogous clades (Fig. 12.2b). Instead, strongly supported clades including putatively orthologous genes from both subgenera are observed (Fig. 12.2c), and it therefore appears that concerted evolution at the symbiotic Lb loci within *Glycine* is either absent or is too weak to homogenise the paralogous sequences and produce a classic concerted evolution pattern (Doyle et al., 1998). This result probably could have been anticipated to some degree. Unlike nrDNA sequences, whose sometimes thousands of copies are mostly totally homogenised even when they occur at several different chromosomal locations, the various *G. max* leghemoglobin sequences clearly differ from one another. Thus, although concerted evolutionary forces appear to be sufficiently strong to produce a *Glycine*-specific gene clade relative to other genera, these forces have not resulted in complete homogenisation.

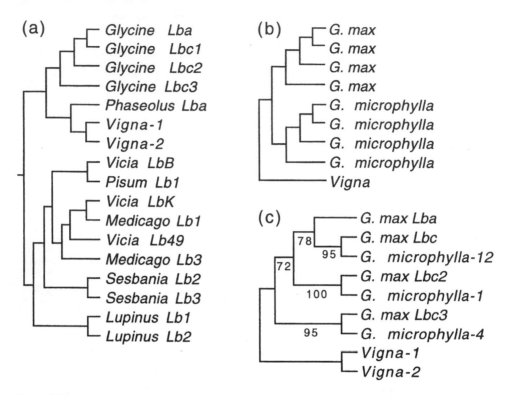

Figure 12.2 Leghemoglobin evolution in *Glycine*. (a) Relationships of Lb genes in legumes, abstracted from trees in Doyle (1994b) and Frühling *et al.* (1997). In *Glycine, Vigna, Lupinus*, and *Sesbania*, the multiple Lb sequences group together, whereas relationships are more complex among Lb genes of other genera, suggesting that concerted evolution has operated to varying degrees in different taxa. (b) Expected relationships of Lb genes sampled from a member of subgenus *Glycine* (*G. microphylla*) relative to the Lb genes of *G. max*. If concerted evolution is homogenizing Lb loci in *G. max* relative to other genera (Fig. 2a) then it seems likely that a similar pattern of concerted evolution would be observed between species in the genus, resulting in species-specific gene clades. (c) The single most parsimonious Lb gene tree for published sequences of *G. max* and several clones isolated from *G. microphylla* (Doyle *et al.*, 1998). The tree was obtained by a branch and bound parsimony search, and has a length of 554 steps, a consistency index of 0.91 (0.84 without autapomorphies), and a retention index of 0.83. Bootstrap replicate values (numbers along internal branches) strongly support the existence of clades of orthologous sequences in the two *Glycine* species. There is no evidence of concerted evolution.

Most recently we have explored the Lb gene families of genera more closely related to *Glycine* than are *Vigna* or *Phaseolus*, to ascertain whether their genes group with putative orthologues from *Glycine*. Thus far the pattern is quite complex, with no definitive cases of orthology yet observed (J.J. Doyle and G. Anderson, unpublished). An alternative explanation for the clustering of *Glycine* Lb genes relative to those of other genera is that these genes are the products of a series of gene duplications in the most recent common ancestor of the two *Glycine* subgenera. As noted for gene families in general by Li (1997), such a hypothesis also requires the loss of paralogous sequences in each of several taxa prior to these more recent duplications. This may seem implausible, or at least unparsimonious, but cannot be ruled out given our general ignorance of the forces shaping gene family evolution (Clegg *et al.*, 1997).

12.4 Molecular criteria for homology determination: nodulation and nodulin gene evolution

By identifying genes that influence morphological features and describing their expression patterns, molecular biology provides a source of new criteria for hypothesizing homologies of phenotypic features (e.g. Doyle, 1994a, 1997; Abouheif, 1997; Abouheif *et al.*, 1997). One morphological feature of Leguminosae about which homology questions exist is the legume nodule (Doyle, 1994b; Hirsch and Larue, 1997). Nodulation is a hallmark of Leguminosae, but is not present uniformly throughout the family, being prevalent only in two of the three subfamilies (Papilionoideae and Mimosoideae). Nodulation in the third subfamily, the paraphyletic Caesalpinioideae, has been confirmed only in the genus *Chamaecrista* and in genera that are part of the mimosoid lineage (de Faria *et al.*, 1989). A parsimonious optimisation of nodulation on phylogenetic hypotheses for the family (Doyle *et al.*, 1997) suggests that the syndrome has arisen multiple times in the family: at least once in Papilionoideae, once in the caesalpinioid ancestors of Mimosoideae, and once in *Chamaecrista*. However, nodulation is a complex syndrome, and a hypothesis of independent origins must posit parallel evolution of the nodule, which is a novel organ of unknown homology (Hirsch and Larue, 1997), and the symbiosome, a novel organelle that includes plant peribacteroid membranes and the enclosed symbiont bacteroids (Verma *et al.*, 1992). This complexity suggests that a single origin and several independent losses of the trait may be more likely than multiple origins. Testing these competing hypotheses involves determining whether nodules in distantly related members of the Leguminosae are homologous. Positional, structural and functional similarities among nodules of all legumes appear to support homology.

The nodulation syndrome is also characterised by the presence of 'nodulins', a diverse group of proteins originally thought to be present only in the nodule (Legocki and Verma, 1980). However it is now known that many nodulin genes are expressed in other organs than nodules, though often at low levels. It is likely that nodulins are not novel proteins, therefore, but products of genes that have been recruited from existing functions and are now most strongly expressed in nodules (Nap and Bisseling, 1990).

Most nodulins are encoded by members of gene families that also encode genes expressed in other tissues. Doyle (1994b) proposed tests of nodule homology using the expression patterns and orthology relationships of such gene families. The test itself is simple: for example, if a particular nodulin is found to be encoded by one member of a gene family in a mimosoid, but by a paralogous member of the same gene family in a papilionoid, the simplest explanation would be independent recruitment of this nodulin in these two legume subfamilies. The occurrence of such separate recruitments would constitute evidence that nodules in Mimosoideae are not homologous with those in Papilionoideae, which in turn would support two parallel, independent origins of nodulation. If the genes encoding this nodulin are orthologous, the test would be inconclusive, because the observed pattern could be due either to a single recruitment by a common nodulating ancestor or to chance parallel recruitment of the same gene in non-homologous nodules. Because many factors could cause any single gene family to give a misleading answer (e.g. expression shifts could cause paralogous genes to be expressed in homologous nodules), the situation is much like vicariance biogeography, where multiple independent phylogenies must be obtained before a general area cladogram can be constructed.

Applying the test is not a simple matter. Few nodulating species have been investigated at the molecular level, even in Papilionoideae. Furthermore, the structural, developmental and biochemical diversity of nodules in that subfamily is also reflected at the molecular level, and in many cases no homologues have been described in one species for genes that encode well-characterised and important nodulin proteins in another (Doyle, 1994b). The situation is still worse in the other two subfamilies, where no nodulin genes have yet been characterised. Furthermore, not all gene families can be used for such studies. Gene families that show extensive concerted evolution, such as leghemoglobin, do not preserve the paralogous relationships that are required for this test.

12.4.1 Glutamine synthetase: a complex gene family

One gene family that has at least been studied in a diversity of papilionoid species and in some non-legumes is glutamine synthetase (GS), a key enzyme in nitrogen metabolism that is expressed in many tissues, including the nodule. GS isozymes are encoded by a nuclear gene family marked by an ancient duplication between the usually single copy gene encoding a chloroplast-expressed isozyme and the small multigene family encoding isozymes expressed in the cytosol (Doyle, 1991; Pesole *et al.*, 1991, 1995; Biesiadka and Legocki, 1997). The cytosolic subfamily includes nodule-enhanced isozymes in several legume species (e.g. Boron and Legocki, 1993; Roche *et al.*, 1993; Temple *et al.*, 1995; Guan *et al.*, 1996).

Topologies of our parsimony analyses of published GS sequences agree in many respects with a neighbour-joining tree that includes many of the same sequences (Biesiadka and Legocki, 1997), with differences mainly involving weakly-supported clades, of which there are several. Sampling is very limited; even for legumes, one or more sequences are available for only seven papilionoid genera out of the approximately 650 genera of the family. These topologies suggest a complex history of duplications and extinctions, some quite recent. Both analyses identify several groups of putatively orthologous legume sequences, among which relationships are poorly supported. A separate analysis of legume and other sequences that group together in the larger studies identifies three major clades of legume sequences, although only one of these is well-supported (Fig. 12.3).

We have designed PCR primers for cytosolic GS, which we have used to amplify genomic sequences from representatives of several additional legume genera, notably the nodulating caesalpinioid genus *Chamaecrista* and its non-nodulating putative sister genus, *Senna*, both formerly part of *Cassia* (Irwin and Barneby, 1981). We have identified sequences belonging to at least two of the three legume clades (Doyle and Doyle, 1997). As we began this project, only one genomic sequence

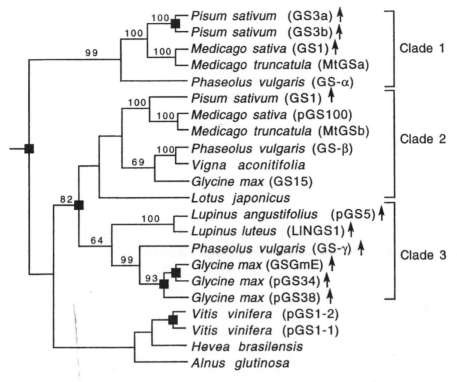

Figure 12.3 Phylogeny of cytosolic glutamine synthetase genes in legumes. Parsimony analysis of complete sequences from GenBank identified two most parsimonious trees (length = 3216; consistency index = 0.35; retention index = 0.58), the strict consensus of which is shown here. Black squares represent hypothesised gene duplication events. Bootstrap values are shown above branches. Three clades of legume sequences are indicated.

was known, that of a *Medicago* gene (Tischer *et al.*, 1986); this sequence contains 11 introns that, combined with the *ca* 1 kb of coding sequence, produce a total sequence length of nearly 5 kb. We found that intron positions were conserved among all of the sequences analyzed, with the exception of a *Senna* clone that lacked one intron. This intron occurs at a homologous position even in the chloroplast-expressed GS paralogue, and its loss could represent an interesting taxonomic character. Not unexpectedly for ancient paralogues, we found that introns varied considerably in size and sequence among the genes we isolated. They were difficult to align, and were therefore removed from analyses. In contrast, there was virtually no length variation among exon sequences and they were simple to align among our sequences and with the various complementary DNA (cDNA) sequences from GenBank/EMBL.

The existence of several paralogous GS clades provides one of the prerequisites for the test of nodule homologies using nodulin genes. The second requirement is the determination of expression patterns of GS genes in *Chamaecrista* for comparison with those in Papilionoideae. In *Chamaecrista*, analysis of cDNAs cloned from reverse-transcript PCR (RT–PCR) experiments show that the same gene is transcribed in nodules as in hypocotyls (Doyle and Doyle, unpublished), and this gene appears to be a member of Clade 3 (Doyle and Doyle, 1997; Fig. 12.3). However, expression patterns of GS among Papilionoideae are more complex than was originally believed, and although Clade 3 includes genes with enhanced expression in nodules, so do at least some genes in the two other clades. Thus GS does not appear to provide the information necessary to test nodule homologies, as we had hoped would be possible. Other gene families that include nodule-expressed members and which are well understood phylogenetically may prove to be better candidates for this purpose, such as the MADS-box transcription factor family (e.g. Heard and Dunn, 1995; Heard et al., 1997). Nevertheless, determining the expression patterns of GS genes is still of interest. Our RT–PCR results indicate differences in cytosolic GS gene expression between *Chamaecrista* and *Senna*; an evolutionary shift in gene expression could prove to be interesting from a phylogenetic perspective.

12.5 Studies of closely related species: the need for additional sources of characters

It is at lower taxonomic levels that the gene tree versus species tree problem is likely to be most severe. When more than one chloroplast gene is sampled in phylogenetic studies at higher taxonomic levels, it is primarily to increase the phylogenetic signal so as to infer more accurately the single historical pattern underlying all of the linked genes of this genome. Although the availability of numerous linked genes is clearly advantageous for obtaining increased resolution, the fact that only a single history *does* exist for all of these genes is a liability if the history of the chloroplast genomes differs from the histories of the taxa bearing those genomes. Major causes of this gene-species tree problem are hybridisation and lineage sorting (Neigel and Avise, 1986; Nei, 1987; Doyle, 1992; Wendel and Doyle, 1998; Comes and Abbott, 1999 – this volume). The timescale over which these problems can persist is unknown, but they are assumed to be most severe at lower taxonomic levels, where gene flow may only recently have ceased among closely related species and time may not have sufficed for allele fixation. The problem may be particularly severe for organellar genes (e.g. Rieseberg and Soltis, 1991; Hoelzer, 1997), although the argument has been made that, because of the smaller effective population size of uniparentally inherited organellar genomes, lineage sorting is actually more of a problem for nuclear genes (Moore, 1995, 1997).

Regardless of whether any particular class of gene is more or less likely than others to capture taxic history faithfully, it seems advisable to seek corroborating data from more than one gene. But not just any gene will do if the concern is incongruent histories – it must be a gene in the coalescent sense, one that is historically independent of the gene for which a tree already has been obtained. Nuclear genes in sexually reproducing species potentially are a rich source of such independent phylogenetic markers, because recombination (including both independent assortment of chromosomes and crossing over) breaks up linkage groups or even genes (in the standard, 'molecular' sense) into multiple coalescent genes. Nuclear genes therefore provide a check on each other and on the single historical hypothesis provided by an organellar genome. Recombination can be a liability, but is also a potential benefit (e.g. Avise, 1989; Hudson, 1990).

12.5.1 Nuclear genes at lower taxonomic levels: introns as sources of characters

As noted for GS, nuclear gene introns are a nuisance if genes have diverged substantially from one another, as is expected to be the case at higher taxonomic levels or when ancient paralogues are studied. This is because introns can add considerably to the amount of sequence that must be obtained but provide few if any characters of phylogenetic utility, due to substitutional saturation and the difficulty of obtaining accurate alignments. At lower taxonomic levels, however, introns represent a potential source of useful characters. Noncoding sequences of a protein-coding gene are expected to evolve at about the same rate as the most variable nucleotide positions in a coding sequence, four-fold degenerate codon sites (e.g. Li, 1997). However, unlike such positions, which are scattered among much more highly conserved sites, noncoding sequences occur as blocks of contiguous nucleotides. Of the three genic noncoding sequence classes (5'- and 3'-untranslated regions and introns), introns have the advantage of being embedded within more highly conserved coding regions, permitting the construction of PCR primers that flank the region of interest. This is not the case for the other noncoding genomic regions, for which primers at only one end can be constructed readily. Introns are thus technically comparable to the nrDNA ITS, which is flanked by the highly conserved 18S and 26S rRNA genes that are a ready source of universal primers. Strand *et al.* (1997) describe the sort of common sense rules for primer design in nuclear protein-coding genes that we have used in our various studies.

Nuclear gene intron sequences are not without their problems, however, even at lower taxonomic levels. The purely technical issues associated with low-abundance sequences have already been mentioned. In addition, compared with genes of the relatively conservative chloroplast genome, or the highly conserved ribosomal genes flanking the nrDNA ITS, the greater variability of nuclear gene coding sequences can make the design of 'universal' amplification and sequencing primers a significant challenge. When primers designed for one group do not initially work on a new set of taxa, one obvious strategy is to lower the stringency of primer annealing. However, if the gene of interest belongs to a multigene family, as is often the case for nuclear protein-coding genes, this may result in the amplification of undesired paralogues. On the other hand, primers that are more taxonomically universal can amplify paralogous copies even at standard stringencies. The complex interactions between PCR primers and the target sequences of diverse members of a multigene family (Wagner *et al.*, 1994) make it difficult to be certain that the desired member will be amplified. Recombination among PCR products derived from paralogous sequences is also a potential problem, as it also is for alleles in heterozygous individuals (Bradley and Hillis, 1997).

Paralogy may be a problem even for seemingly well-characterised genes thought to be single copy. Little is known about most gene families, and it is often difficult to be sure that what is single copy in one group will not exist as multiple paralogous copies in another. This is particularly likely due to the prevalence of polyploidy and subsequent diploidisation in plants, but gene duplications may also be very recent in diploid species, as in cytosolic GS of *Pisum* (Walker *et al.*, 1995).

There is also no guarantee that an intron present in one group of taxa will be present in others. As noted for GS, one *Senna* gene seems to have lost an otherwise well-conserved intron, and intron loss is known to occur in other gene families, such as actins (Drouin and Moniz de Sá, 1997). Even if present, introns can vary greatly in size. The *leafy* gene contains introns that are present in a wide range of seed plants but range in length from under 100 bp to several kilobases (Frohlich and Meyerowitz, 1997). Individual small introns are too short to provide sufficient useful characters, whereas very long introns present technical challenges to conventional PCR methods.

Introns may present problems even if all of the above difficulties are overcome. Plant introns are often highly A-T rich (particularly T-rich: Ko *et al.* 1998). As has been noted for the mitochondrial DNA control region (Zhang and Hewitt, 1997), compositional biases can make sequences less useful for phylogenetic studies by reducing the number of possible character states. Furthermore, noncoding sequences are free of the constraints that often prohibit length mutation in exons. Length variation in introns is often due to different numbers of simple sequence repeats (in many cases mononucleotide runs), which complicates primary assessment of positional homology. Because alignment remains a very difficult area of phylogenetic analysis (e.g. Wheeler, 1997; Doyle and Davis, 1998), length variation can be a major concern.

Despite these problems, it is difficult to see what better alternatives exist among nuclear genes. If, as expected, introns evolve at close to the neutral evolutionary rate, the only sequences expected to accumulate more variability at the substitutional level are genes subject to diversifying selection. The best plant examples of such genes are incompatibility loci, but the high degree of polymorphism within species, the retention of alleles across species boundaries (e.g. Richman *et al.*, 1996), and the potential loss of entire loci in self-compatible taxa (Conner *et al.*, 1998) make such genes poor choices for phylogenetic studies. Highly variable amplification patterns can be obtained using such methods as random amplified polymorphic DNA (RAPD: Williams *et al.*, 1990) and amplified fragment length polymorphism (AFLP: Vos *et al.*, 1995) markers, but quite apart from technical problems, homology determination for such anonymous markers is difficult, complicating their use in phylogenetic studies (e.g. Backeljau *et al.*, 1995; Harris, 1999 – this volume). Microsatellite loci can provide considerable variation even within species, but slipped strand mispairing, the mutational process responsible for their size variation (Levinson and Gutman, 1987), also makes homology determination difficult (see Provan *et al.*, 1999 – this volume). This has suggested to at least some workers that microsatellites are poor candidates for phylogeny reconstruction (e.g. Garza *et al.*, 1995). In any case, microsatellite variation is not amenable to constructing gene trees for a single locus. Introns may therefore represent the only viable alternatives to ribosomal gene spacers and cpDNA for constructing gene trees in closely related species.

12.5.2 Intron-containing nuclear protein-coding genes in use or under development, primarily in Glycine (Leguminosae)

We are testing several nuclear genes for utility in phylogeny reconstruction in several families, but particularly in *Glycine* subgenus *Glycine*, where we have built on our previous phylogenetic studies (e.g., Doyle *et al.*, 1990a). *Glycine* is genetically a diploid, but is considered to be a paleopolyploid at $2n = 4x = 40$. It retains some evidence of a hypothesised polyploid origin both at individual genetic loci such as Lb (Lee and Verma, 1984) and in its overall linkage map (Zhu et al., 1994; Shoemaker et al., 1996). The genus is best known for the cultivated soybean, G. max, which together with its wild progenitor form the north Asian annual subgenus Soja. The other subgenus, Glycine, comprises approximately 15 diploid perennial species, all native to Australia, as well as several polyploid complexes, some of which have colonised Pacific islands.

Introns of some of the genes described above should be useful for inferring phylogenies within Glycine. We have recently begun studies of nu-cox2 in the subgenus, having developed primers to amplify a smaller region that contains the ca 600 bp intron. In the case of leghemoglobin, the finding that paralogous clades are maintained among Glycine species suggests the possibility of constructing paralogue-specific primers for each clade. Each paralogue, with its three introns, could in theory provide an estimate of species phylogeny. However, in G. max at least some of the Lb genes are linked (Lee and Verma, 1984), so different gene trees might not be independent. Paralogous loci of the cytosolic subfamily of glutamine synthetase, with their 11 introns flanked by short exons of relatively conserved sequence, could also be used. The complexity of this subfamily, however, makes it a rather unsatisfactory subject for phylogenetic studies.

Better prospects exist for the chloroplast-expressed but nuclear-encoded GS gene (ncpGS), which is single copy in most plants. Our work has shown that intron positions are highly conserved between the chloroplast and cytosolic GS genes. We have developed primers specific for ncpGS genes, and have used various primer combinations to obtain ncpGS intron sequences for several genera. This locus has been very useful in diploid members of the Oxalis tuberosa complex (Oxalidaceae), where we have found levels of variation that are in general greater than those observed for the nrDNA ITS (Emshwiller and Doyle, in press). Multiple loci occur in the octoploid Oxalis tuberosa, and these putatively homoeologous sequences, along with ITS data (Emshwhiller and Doyle, 1998), are being used to trace genomic contributions to this minor domesticate. The situation appears to be more complex in Glycine, where we have found two loci in $2n = 40$ species. Our initial assumption was that these were homoeologous loci, present due to the paleopolyploid condition of the genus. However, we have found that at least some diploid genera related to Glycine also possess more than one gene, suggesting the possibility of duplications and gene losses (J.J. Doyle, J.L. Doyle, and C. Harbison, unpublished).

Leafy has already been mentioned as a single copy gene with some promise for higher level taxonomic studies (see above); very degenerate primers provided by Michael Frohlich (U. of Michigan) amplify the highly variable intron described above. These primers have been used to amplify a single product in genera of Celastrales and Asclepiadaceae, as well as in Glycine. In Celastrales, however, intron sizes were too small (ca 100 bp) to warrant the use of this locus for studies of closely related species (M. Simmons, pers. comm., 1997). In Dischidia and allies (Asclepiadaceae) the intron is ca 740 bp and is being actively studied (T. Livshultz, pers. comm., 1998). In Glycine species the intron is several hundred bp long, and the sequences thus far obtained show considerable interspecific variation, but as with ncpGS there may be more than one paralogous gene present (J.L. Doyle and

J.J. Doyle, unpublished). Two additional genes being studied in other families may also prove useful in our *Glycine* work. The malate synthase gene contains a large (*ca* 1 kb) intron that is more variable in members of the arecoid palm subtribe Oncospermatinae than are noncoding chloroplast gene regions (Carl Lewis, pers. comm., 1998). The gene *pistillata*, a member of the MADS-box family (Goto and Meyerowitz, 1994; Münster *et al.*, 1997), contains a large (> 1 kb) intron that is being used in studies of Brassicaceae, where it provides more phylogenetically useful characters than either nrDNA ITS or chloroplast noncoding sequences (Bailey and Doyle, in press). A primer set constructed against conserved sequences flanking this *pistillata* intron amplifies a single product in taxa from other families, including legumes (Bailey and Doyle, in press).

12.5.2.1 Histone H3-D in Glycine subgenus Glycine

One member of the large gene family encoding the core histone H3 (Old and Woodland, 1984) is already being used extensively in our phylogenetic studies of *Glycine*. Amplifications of *G. max* genomic DNA with a pair of PCR amplification primers 'universal' for plant histone H3 genes produced a complex set of products whose predominant class lacked introns (Kanazin *et al.*, 1996). Larger amplification products, when cloned, were found to possess from 1–3 introns in different angiosperm families; two loci in *Glycine*, each with three introns, were called H3-B and H3-D (Kanazin *et al.*, 1996). New primers that were specific for the H3-B and H3-D loci were developed from sequences of the cloned *G. max* genes. This was done by designing primers that span the intron/exon junctions, and so are incapable of amplifying genes that lack introns. Three such primers were constructed, a common 3′ primer and two paralogue-specific 5′ primers, making it possible to amplify H3-B and H3-D genes separately. Duplicate H3-B genes were found in *Glycine* species and some allied genera in Phaseoleae subtribe Glycininae, and H3-B gene trees proved too complex to infer species relationships (Doyle *et al.*, 1996). Reasons for this complexity were not readily apparent, but some degree of concerted evolution could be obscuring orthology relationships.

In contrast, H3-D appears to be a single copy locus in *Glycine* and most of its allies. Phylogenetic analyses of *Glycine* H3-D genes resulted in a topologically identical single tree (but with varying clade support) over the range of slightly different alignments used (Doyle *et al.*, 1996). The H3-D topology for *Glycine* subgenus *Glycine* shared many similarities with trees based on cpDNA restriction site studies (Doyle *et al.*, 1990a) or nrDNA ITS sequences (Nickrent and Doyle, 1995; Kollipara *et al.*, 1997). Although particular cases of incongruence were also observed, the most important general observation concerns levels of variation among the three sources of data. The histone H3-D sequences, though relatively short (*ca* 500 bp, of which *ca* 300 bp is intron: Fig. 12.4), contain more variation than was observed in the nrDNA ITS. H3-D and cpDNA restriction site data were complementary, with cpDNA providing considerable resolution among groups of species, but much less resolution within species groups, whereas the opposite was true of H3-D sequences (Doyle *et al.*, 1996). Some of this difference could be due to biologically meaningful incongruence, if, as now seems likely, cpDNA introgression has occurred.

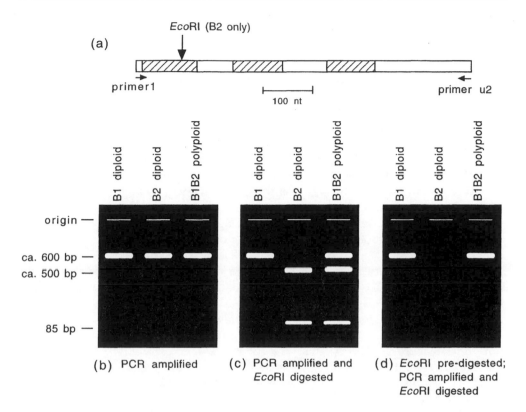

Figure 12.4 Strategy used for amplifying homoeologous histone H3 genes in *Glycine* polyploids. (a) Map of region amplified using a primer specific for histone H3-D (primer 1) and a universal histone H3 primer (primer u2). Exons are white, introns are hatched. The position of the single *Eco*R-I restriction site found in alleles at the homoeologous B2 locus is indicated with a vertical arrow. (b) PCR amplification of genomic DNA from two diploids (B1 and B2) and their allopolyploid (B1B2) using these primers produces a single band of approximately 600 bp. (c) Digestion of amplification products with *Eco*R-I reveals differences between B1 and B2 products, and shows that the allopolyploid contains both of these products. In (d), genomic DNA from each sample was digested with *Eco*R-I prior to PCR amplification with histone primers. This has no effect on the B1 allele, which has no *Eco*R-I site; this product is also obtained from the B1 homoeologue in the allopolyploid. Amplification of the B2 allele and of the B2 homoeologue in the allopolyploid is prevented by cleavage of the target fragment. The B1 homoeologue in the allopolyploid can now be sequenced directly.

We are using the ability of H3-D genes to resolve relationships among alleles from closely related species in two ongoing studies involving groups within subgenus *Glycine* for which parallel cpDNA data exist: (1) the diploid B-genome species (Doyle *et al.*, 1990c), and (2) the *G. tabacina* polyploid complex (Doyle *et al.*, 1990b, d). Little variation was found within species groups in our initial survey of cpDNA variation in subgenus *Glycine*, despite the fact that 29 restriction enzymes were used (Doyle *et al.*, 1990a). A subsequent study sampled 72 accessions of the B-genome species group, using additional, frequent-cutting enzymes to identify 27 different chloroplast haplotypes, whose distribution in many cases disagreed with classifications based on morphological characters (Doyle *et al.*, 1990c). We have now amplified and directly sequenced H3-D from over 40 of these same B-genome accessions (Doyle *et al.*, 1999). Over 20 alleles have been identified, with sufficient substitutional variation to construct an allele tree; heterozygosity was rare, which is not surprising in this group of mostly autogamous species. Resolution provided by these relatively short sequences is roughly comparable to that obtained with extensive cpDNA restriction site mapping of the same accessions, but histone H3-D and cpDNA data are incongruent. For example, *G. microphylla*, which was polymorphic for several divergent cpDNA haplotypes, has only a single histone H3-D allele. The agreement between morphology and histone H3-D variation suggests that histone H3-D, unlike cpDNA, is tracking cladogenesis in this species group. Chloroplast haplotypes that transgress species boundaries are estimated to be younger than histone H3-D alleles, and thus appear to post-date speciation. This suggests that hybridisation, rather than sorting of ancient ancestral polymorphisms, explains the observed cpDNA incongruence (Doyle *et al.*, 1999).

Two major classes of polyploids exist within *G. tabacina*, one of which is a wide allopolyploid involving the A and B genomes (AAB_2B_2) whereas the second is somewhat narrower genetically, combining various B genomes (BBB_2B_2: Doyle *et al.*, 1990b, d). Individuals of both classes show evidence of fixed heterozygosity for nrDNA, although to varying degrees in different accessions, and always with one repeat more abundant than the other (Doyle *et al.*, 1990b). Polyploid individuals should possess a single copy of each of the two homoeologues at a locus such as histone H3-D. However, we have found that obtaining sequences of both loci can be difficult. Even though the locus amplifies well in both diploid progenitors, only one homoeologue may amplify in a particular polyploid accession. This seems to be due at least in part to stochastic effects ('PCR drift' of Wagner *et al.*, 1994), because subsequent replicates of the same experiment often amplify both loci. However, even in such cases there is often preferential amplification of one locus, suggesting that one template is competing more efficiently for primers ('PCR selection').

Such unpredictable results have led us to develop methods for amplifying only one homoeologous locus from fixed heterozygotes. In the case of the BBB_2B_2 polyploid, we were primarily interested in surveying alleles from the BB diploid genome donors. Our studies of the diploid B-genome group revealed that B_2 alleles all differed from B alleles by the presence of an *Eco*R-I restriction site. Digestion of genomic DNA from a BBB_2B_2 polyploid prior to amplification resulted in the elimination of the B_2 product, enabling us to sequence the B allele directly (Fig. 12.4). Using this strategy we have now directly sequenced B alleles from over 40 polyploid accessions. A comparison of allele distributions and topologies is providing striking contrasts with cpDNA haplotype patterns. Our data suggest a young polyploid that has rapidly expanded its range both within Australia and onto Pacific islands and shows evidence of gene flow among its populations (J.J. Doyle, J.L. Doyle and A.H.D. Brown, unpublished data).

Our primers also have been used to amplify histone H3-D genes from other taxa. In members of the *Dichrostachys* group of mimosoid legumes, initial utilisation of the two H3-D intron-spanning primers produced weak amplifications; this was not surprising, because the intron sequences that are part of the primers are expected to be variable, and had been made specifically from *Glycine* sequence information with no attempt at universality. Stronger amplifications that still produced single products were obtained using the 5' H3-D primer and a general plant histone H3 primer from the 3' end of the gene (Fig. 12.4; Kanazin *et al.*, 1996). Useful phylogenetic data were obtained this way, but it is possible that some paralogous sequences were amplified (Livshultz and Luckow, 1997). Similar paralogy problems have been reported in *Medicago* (M. Wojciechowski, pers. comm., 1997), an allotetraploid. These primers appear to be working well, however, in *Gastrolobium*, another papilionoid legume (M. Crisp, pers. comm., 1998).

Probably the best way to use the histone H3 family outside of *Glycine* or its close allies is to repeat the steps taken to produce primers specific for the H3-D locus in *Glycine* (Doyle *et al.*, 1996; Kanazin *et al.*, 1996). This approach involves: (1) beginning with the Kanazin *et al.* (1996) universal angiosperm primers; (2) cloning and sequencing large products predicted to include one or more introns; (3) developing intron-specific primers from the resulting sequences; and (4) testing the products of such primers for their utility in inferring taxic relationships.

12.6 Conclusions

Low-copy nuclear genes pose numerous difficulties for systematists compared with sequences from the chloroplast genome. Most significant are the problems associated with the presence of multiple, divergent copies within individual organisms. Whether these copies are alleles at a single locus or multiple paralogous loci, such heterogeneity introduces technical, analytical and theoretical complications. Special consideration should be given to sampling, definition of taxa for data combination and the possibility of non-hierarchical relationships caused by recombination. Despite these difficulties, nuclear low-copy sequences are beginning to find use in phylogeny reconstruction, due to their sheer number and the fact that they represent many potentially independent estimates of taxic history.

ACKNOWLEDGEMENTS

We thank Donovan Bailey, Eve Emshwiller, Carl Lewis, Richard Bateman, and an anonymous reviewer for helpful comments on an earlier draft of the manuscript. We are grateful to Donovan Bailey, Eve Emshwiller, Carl Lewis, Tanya Livshultz, and Mark Simmons for sharing unpublished results with us. Studies described here were supported by various grants from the US National Science Foundation, most recently DEB-9420215 and DEB-9614984.

REFERENCES

Abouheif, E. (1997) Developmental genetics and homology: a hierarchical approach. *Trends in Ecology and Evolution*, **12**, 405–8.
Abouheif, E., Akam, M., Dickinson, W. J., Holland, P. W. H., Meyer, A., Patel, N. H. *et al.* (1997) Homology and developmental genes. *Trends in Genetics*, **13**, 432–3.

Albert, V. A., Gustafsson, M. H. G. and Di Laurenzio, L. (1998) Ontogenetic systematics, molecular developmental genetics, and the angiosperm petal, in *Molecular systematics of plants II: DNA sequencing*, (eds D. E. Soltis, P. S. Soltis and J. J. Doyle), Chapman & Hall, New York, pp. 349–74.

Andersson, C. R., Jensen, E. O., Llewellyn, D. J., Dennis, E. S. and Peacock, W. J. (1996) A new hemoglobin gene from soybean: a role for hemoglobin in all plants. *Proceedings of the National Academy of Sciences USA*, **93**, 5682–7.

Appleby, C. A., Dennis, E. S. and Peacock, W. J. (1990) A primaeval origin for plant and animal hemoglobins. *Australian Systematic Botany*, **3**, 81–90.

Avise, J. C. (1989) Gene trees and organismal history: a phylogenetic approach to population biology. *Evolution*, **43**, 1192–1208.

Backeljau, T., De Bruyn, L., De Wolf, H., Jordaens, K., Van Dongen, S., Verhagen, R. *et al.* (1995) Random amplified polymorphic DNA (RAPD) and parsimony methods. *Cladistics*, **11**, 119–30.

Bailey, C. D. and Doyle, J. J. Potential phylogenetic utility of the low-copy nuclear gene pistillata in dicotyledonous plants: comparison to nrDNA ITS and *trnL* intron in *Sphaerocardamum* and other Brassicaceae. *Molecular Phylogenetics and Evolution*, in press.

Baldwin, B. G., Sanderson, M. J., Porter, J. M., Wojciechowski, M. F., Campbell, C. S. and Donoghue, M. J. (1995) The ITS region of nuclear ribosomal DNA: a valuable source of evidence on angiosperm phylogeny. *Annals of the Missouri Botanical Garden*, **82**, 247–77.

Bennett, M. D. and Leitch, I. J. (1997) Nuclear DNA amounts in angiosperms – 583 new estimates. *Annals of Botany, (London)*, **80**, 169–96.

Biesiadka, J. and Legocki, A. B. (1997) Evolution of the glutamine synthetase gene in plants. *Plant Science*, **128**, 51–8.

Boron, L. J. and Legocki, A. B. (1993) Cloning and characterization of a nodule-enhanced glutamine synthetase-encoding gene from *Lupinus luteus*. *Gene*, **13**, 95–102.

Bradley, R. D. and Hillis, D. M. (1997) Recombinant DNA sequences generated by PCR amplification. *Molecular Biology and Evolution*, **14**, 592–3.

Buckler, E. S. IV, Ippolito, A. and Holtsford, T. P. (1997) The evolution of ribosomal DNA: divergent paralogues and phylogenetic implications. *Genetics*, **145**, 821–32.

Chase, M. W., Soltis, D. E., Olmstead, R. G., Morgan, D., Les, D. H., Mishler, B.D. *et al.* (1993) Phylogenetics of seed plants: an analysis of nucleotide sequences from the plastid gene *rbcL*. *Annals of the Missouri Botanical Garden*, **80**, 528–80.

Clegg, M. T., Cummings, M. P. and Durbin, M. L. (1997) The evolution of plant nuclear genes. *Proceedings of the National Academy of Sciences USA*, **94**, 7791–8.

Comes, H. P. and Abbott, R. J. (1999) Reticulate evolution in the Mediterranean species complex of *Senecio* sect. *Senecio*: Uniting phylogenetic and population-level approaches, in, *Molecular systematics and plant evolution*, (eds P. M. Hollingsworth, R. M. Bateman and R. J. Gornall), Taylor & Francis, London, pp. 171–98.

Conner, J. A., Conner, C., Nasrallah, M. E. and Nasrallah, J. B. (1998) Comparative mapping of the *Brassica S* locus region and its homeolog in arabidopsis: its implications for the evolution of mating systems in Brassicaceae. *Plant Cell*, **10**, 801–12.

Davis, J. I., Simmons, M. P., Stevenson, D. W. and Wendel, J. F. (1998) Data decisiveness, data quality and incongruence in phylogenetic analysis: an example from the monocotyledons using mitochondrial *atpA* sequences. *Systematic Biology*, **47**, 282–310.

de Faria, S. M., Lewis, G. P., Sprent, J. I. and Sutherland, J. M. (1989) Occurrence of nodulation in the Leguminosae. *New Phytologist*, **111**, 607–19.

Delgado-Salinas, A., Bruneau, A. and Doyle, J. J. (1993) Chloroplast DNA phylogenetic studies in New World Phaseolinae (Leguminosae: Papilionoideae: Phaseoleae). *Systematic Botany*, **18**, 6–17.

Denton, A. L., McConaughy, B. L. and Hall, B. D. (1998) Usefulness of RNA polymerase II coding sequences for estimation of green plant phylogeny. *Molecular Biology and Evolution*, **15**, 1082–5.

Doyle, J. J. (1991) Evolution of higher plant glutamine synthetase genes: tissue specificity as a criterion for predicting orthology. *Molecular Biology and Evolution*, **8**, 366–77.

Doyle, J. J. (1992) Gene trees and species trees: molecular systematics as one-character taxonomy. *Systematic Botany*, **17**, 144–63.

Doyle, J. J. (1994a) Evolution of a plant homeotic multigene family: toward connecting molecular systematics and molecular developmental genetics. *Systematic Biology*, **43**, 307–28.

Doyle, J. J. (1994b) Phylogeny of the legume family: an approach to understanding the origins of nodulation. *Annual Review of Ecology and Systematics*, **25**, 325–49.

Doyle, J. J. (1995) The irrelevance of allele tree topologies for species delimitation, and a non-topological alternative. *Systematic Botany*, **20**, 574–88.

Doyle, J. J. (1996) Homoplasy connections and disconnections: genes and species, molecules and morphology, in *Homoplasy and the evolutionary process*, (eds M. J. Sanderson and L. Hufford), Academic Press, New York, pp. 37–66.

Doyle, J. J. (1997) Trees within trees: genes and species, molecules and morphology. *Systematic Biology*, **4**, 537–53.

Doyle, J. J. and Davis, J. I. (1998) Homology in molecular phylogenetics: a parsimony perspective, in *Molecular systematics of plants II: DNA sequencing*, (eds D. E. Soltis, P. S. Soltis and J. J. Doyle), Chapman & Hall, New York, pp. 101–31.

Doyle, J. J. and Doyle, J. L. (1997) Phylogenetic perspectives on the origins and evolution of nodulation in the legumes and allies. *NATO ASI Series*, **39**, 307–12.

Doyle, J. J., Doyle, J. L. and Brown, A. H. D. (1990a) A chloroplast DNA phylogeny of the wild perennial relatives of soybean (*Glycine* subgenus *Glycine*): congruence with morphological and crossing groups. *Evolution*, **44**, 371–89.

Doyle, J. J., Doyle, J. L. and Brown, A. H. D. (1990b) Analysis of a polyploid complex in *Glycine* with chloroplast and nuclear DNA. *Australian Journal of Systematic Botany*, **3**, 125–36.

Doyle, J. J., Doyle, J. L. and Brown, A. H. D. (1990c) Chloroplast DNA polymorphism and phylogeny in the B genome of *Glycine* subgenus *Glycine* (Leguminosae). *American Journal of Botany*, **77**, 772–82.

Doyle, J. J., Doyle, J. L. and Brown, A. H. D. (1999) Incongruence in the diploid B-genome species complex of *Glycine* (Leguminosae) revisited: histone H3-D alleles vs. chloroplast haplotypes. *Molecular Biology and Evolution*, **16**, 354–62.

Doyle, J. J., Kanazin, V. and Shoemaker, R. C. (1996) Phylogenetic utility of histone H3 intron sequences in the perennial relatives of soybean (*Glycine*: Leguminosae). *Molecular Phylogenetics and Evolution*, **6**, 438–47.

Doyle, J. J., Doyle, J. L., Brown, A. H. D. and Grace, J. P. (1990d) Multiple origins of polyploids in the *Glycine tabacina* complex inferred from chloroplast DNA polymorphism. *Proceedings of the National Academy of Sciences USA*, **87**, 714–7.

Doyle, J. J., Doyle, J. L., Ho, J. C. and Nesbitt, T. C. (1998) Patterns of paralogy and concerted evolution in the leghemoglobin gene family of *Glycine* (Leguminosae). *American Journal of Botany*, **85**, s64–5 (abstract).

Doyle, J. J., Doyle, J. L., Ballenger, J. A., Dickson, E. E., Kajita, T. and Ohashi, H. (1997) A phylogeny of the chloroplast gene *rbcL* in the Leguminosae: taxonomic correlations and insights into the evolution of nodulation. *American Journal of Botany*, **84**, 541–54.

Drouin, G. and Moniz de Sá, M. (1997) Loss of introns in the pollen-specific actin gene subfamily members of potato and tomato. *Journal of Molecular Evolution*, **45**, 509–13.

Dumolin-Lapegue, S., Pemonge, M. -H. and Petit, R. J. (1997) An enlarged set of consensus primers for the study of organelle DNA in plants. *Molecular Ecology*, **6**, 393–7.

Emshwhiller, E. and Doyle, J. J. Chloroplast-expressed glutamine synthetase (ncpGS): potential utility for phylogenetic studies with an example from *Oxalis* (Oxalidaceae). *Molecular Phylogenetics and Evolution*, in press.

Emshwiller, E. and Doyle, J. J. (1998) Origins of domestication and polyploidy in oca (*Oxalis tuberosa*: Oxalidaceae): nrDNA ITS data. *American Journal of Botany*, **85**, 975–85.

Fitch, W. M. (1970) Distinguishing homologous from analogous proteins. *Systematic Zoology*, **19**, 99–113.

Frohlich, M. W. and Meyerowitz, E. M. (1997) The search for flower homeotic gene homologs in basal angiosperms and Gnetales: a potential new source of data on the evolutionary origin of flowers. *International Journal of Plant Science*, **158**, S131–42.

Frühling, M., Roussel, H., Gianinazzi-Person, V., Pühler, A. and Perlick, A. M. (1997) The *Vicia faba* leghemoglobin gene VfLb29 is induced in root nodules and in roots colonized by the arbuscular mycorrhizal fungus *Glomus fasciculatum*. *Molecular Plant-Microbe Interactions*, **10**, 124–31.

Galloway, G. L., Malmberg, R. L. and Price, R. A. (1997) Phylogenetic utility of the nuclear gene arginine decarboxylase, with an example from the Brassicaceae. *American Journal of Botany*, **84**, s196 (abstract).

Garza, J. C., Slatkin, M. and Freimer, N. B. (1995) Microsatellite allele frequencies in humans and chimpanzees, with implications for constraints on allele size. *Molecular Biology and Evolution*, **12**, 594–603.

Gaut, B. S. (1998) Molecular clocks and nucleotide substitution rates in higher plants. *Evolutionary Biology*, **30**, 93–120.

Goto, K. and Meyerowitz, E. M. (1994) Function and regulation of the *Arabidopsis* floral homeotic gene *pistillata*. *Genes and Development*, **8**, 548–60.

Guan, C. G., Ribeiro, A., Akkermans, A. D. L., Jing, Y. X., Van Kammen, A., Bisseling, T. and Pawlowski, K. (1996) Nitrogen metabolism in actinorhizal nodules of *Alnus glutinosa*: expression of glutamine synthetase and acetylornithine transaminase. *Plant Molecular Biology*, **32**, 1177–84.

Gustafsson, M. H. G. and Albert, V. A. (1999) Inferior ovaries and angiosperm diversification, in *Molecular systematics and plant evolution*, (eds P. M. Hollingsworth, R. M. Bateman and R. J. Gornall), Taylor & Francis, London, pp. 403–31.

Harris, S. A. (1999) RAPDs in systematics – a useful methodology, in *Molecular systematics and plant evolution*, (eds P. M. Hollingsworth, R. M. Bateman and R. J. Gornall), Taylor & Francis, London, pp. 211–28.

Heard, J. and Dunn, K. (1995) Symbiotic induction of a MADS-box gene during development of alfalfa root nodules. *Proceedings of the National Academy of Sciences USA*, **92**, 5273–77.

Heard, J., Caspi, M. and Dunn, K. (1997) Evolutionary diversity of symbiotically induced nodule MADS box genes: characterization of *nmhC5*, a member of a novel subfamily. *Molecular Plant–Microbe Interactions*, **10**, 665–76.

Hershkovitz, M. A., Zimmer, E. A. and Hahn, W. J. (1999) Ribosomal DNA sequences and angiosperm systematics, in *Molecular systematics and plant evolution*, (eds P. M. Hollingsworth, R. M. Bateman and R. J. Gornall), Taylor & Francis, London, pp. 268–326.

Hiratsuka, J., Shimada, H., Whittier, R., Ishibashi, T., Sakamoto, M., Mori, M. *et al.* (1989) The complete sequence of the rice (*Oryza sativa*) chloroplast genome: intermolecular recombination between distinct transfer RNA genes accounts for a major plastid DNA inversion during the evolution of the cereals. *Molecular and General Genetics*, **217**, 185–94.

Hirsch, A. M. and Larue, T. A. (1997) Is the legume nodule a modified root or stem or an organ *sui generis*? *Critical Reviews in Plant Sciences*, **16**, 361–92.

Hoelzer, G. A. (1997) Inferring phylogenies from mtDNA variation: mitochondrial gene-trees versus nuclear gene-trees revisited. *Evolution*, **51**, 622–6.

Hofer, J., Turner, L., Hellens, R., Ambrose, M., Matthews, P., Michael, A. and Ellis, N. (1997) *Unifoliata* regulates leaf and flower morphogenesis in pea. *Current Biology*, **7**, 581–7.

Hudson, R. R. (1990) Gene genealogies and the coalescent process. *Oxford Surveys in Evolutionary Biology*, **7**, 1–44.

Irwin, H. S. and Barneby, R. C., (1981) Cassieae, in *Advances in Legume systematics*, part 1, (eds R. M. Polhill and P. H. Raven), Royal Botanic Gardens Kew, London, pp. 97–106.

Kanazin, V., Blake, T. and Shoemaker, R. C. (1996) Organization of the histone H3 genes in soybean, barley, and wheat. *Molecular and General Genetics*, **250**, 137–47.

Ko, C. H., Brendel, V., Taylor, R. D. and Walbot, V. (1998) U-richness is a defining feature of plant introns and may function as an intron recognition signal in maize. *Plant Molecular Biology*, **36**, 573–583.

Kollipara, K. P., Singh, R. J. and Hymowitz, T. (1997) Phylogenetic and genomic relationships in the genus *Glycine* Willd. based on sequences from the ITS region of nuclear rDNA. *Genome*, **40**, 57–68.

Laroche, J., Li, P., Maggia, L. and Bousquet, J. (1997) Molecular evolution of angiosperm mitochondrial introns and exons. *Proceedings of the National Academy of Sciences USA*, **94**, 5722–7.

Lavin, M., Eshbaugh, E., Hu, J. -M., Mathews, S. and Sharrock, R. A. (1998) Monophyletic subgroups of the tribe *Millettieae* (Leguminosae) as revealed by phytochrome nucleotide sequence data. *American Journal of Botany*, **85**, 412–33.

Lee, J. S. and Verma, D. P. S. (1984) Structure and chromosomal arrangement of leghemoglobin genes in kidney bean suggest divergence in soybean leghemoglobin gene loci following tetraploidization. *EMBO Journal*, **3**, 2745–52.

Legocki, R. P. and Verma, D. P. S (1980) Identification of 'nodule-specific' host proteins (nodulins) involved in the development of *Rhizobium*-legume symbiosis. *Cell*, **20**, 153–63.

Levinson, G. and Gutman, G. A. (1987) Slipped-strand mispairing: a major mechanism for DNA sequence evolution. *Molecular Biology and Evolution*, **4**, 203–21.

Li, W. -H. (1997) *Fundamentals of molecular evolution*. Sinauer, Sunderland, Massachusetts.

Livshultz, T. and Luckow, M. (1997) Phylogenetic relationships among the primarily Madagascan genera *Dichrostachys*, *Gagnebina*, and *Alantsilodendron* (Leguminosae: Mimosoideae). *American Journal of Botany*, **84**, s213 (abstract).

Malek, O., Laettig, K., Hiesel, R., Brennicke, A. and Knoop, V. (1996) RNA editing in bryophytes and a molecular phylogeny of land plants. *EMBO Journal*, **15**, 1403–11.

Mason-Gamer, R. J. and Kellogg, E. A. (1996) Potential utility of the nuclear gene *waxy* for plant phylogenetic analyses. *American Journal of Botany*, **83**, s178 (abstract).

Mathews, S., Lavin, M. and Sharrock, R. A. (1995) Evolution of the phytochrome gene family and its utility for phylogenetic analyses in angiosperms. *Annals of the Missouri Botanical Garden*, **82**, 296–321.

McDade, L. A. (1995) Hybridization and phylogenetics, in *Experimental and molecular approaches to plant biosystematics*, (eds P. C. Hoch, A. G. Stevenson and B. A. Schaal), Missouri Botanical Garden, St. Louis, pp. 305–31.

Meagher, R. B., Berry-Lowe, S. and Rice, K. (1989) Molecular evolution of the small subunit of ribulose bisphosphate carboxylase: nucleotide substitution and gene conversion. *Genetics*, **123**, 845–63.

Moniz de Sá, M. and Drouin, G. (1996) Phylogeny and substitution rates of angiosperm actin genes. *Molecular Biology and Evolution*, **13**, 1198–212.

Moore, W. S. (1995) Inferring phylogenies from mtDNA variation: mitochondrial-gene trees versus nuclear-gene trees. *Evolution*, **49**, 718–26.

Moore, W. S. (1997) Mitochondrial-gene trees versus nuclear-gene trees, a reply to Hoelzer. *Evolution*, 51, 627–9.

Münster, T., Pahnke, J., Di Rosa, A., Kim, J. T., Martin, W., Saedler, H. and Theissen, G. (1997) Floral homeotic genes were recruited from homologous MADS-box genes preexisting in the common ancestor of ferns and seed plants. *Proceedings of the National Academy of Sciences USA*, 94, 2415–20.

Nap, J. P. and Bisseling, T. (1990) The roots of nodulins. *Physiologia Plantarum*, 79, 407–14.

Nei, M. (1987) *Molecular evolutionary genetics*. Columbia University Press, New York.

Neigel, J. E. and Avise, J. C. (1986) Phylogenetic relationships of mitochondrial DNA under various demographic models of speciation, in *Evolutionary processes and theory*, (eds. S. Karlin and E. Nevo), Academic Press, New York, pp. 515–34.

Nickrent, D. L. and Doyle, J. J. (1995) A molecular phylogeny of diploid *Glycine* (Fabaceae) based upon nuclear ribosomal ITS sequences. *American Journal of Botany*, 82, s153 (abstract).

Nixon, K. and Wheeler, Q. D. (1992) Extinction and the origin of species, in *Extinction and phylogeny*, (eds M. J. Novacek and Q. D. Wheeler), Columbia University Press, New York, pp. 119–43.

Nugent, J. M. and Palmer, J. D. (1991) RNA-mediated transfer of the gene *coxII* from the mitochondrion to the nucleus during flowering plant evolution. *Cell*, 66, 473–82.

Ohyama, K., Fukuzawa, H., Kohchi, T., Shirai, H., Sano, T., Sano, S. *et al.* (1986) Chloroplast gene organization deduced from complete sequence of liverwort *Marchantia polymorpha* chloroplast DNA. *Nature*, 322, 572–4.

Old, R. W. and Woodland, H. R. (1984) Histone genes: not so simple after all. *Cell*, 38, 624–6.

Olmstead, R. G. and Palmer, J. D. (1994) Chloroplast DNA systematics: a review of methods and data analysis. *American Journal of Botany*, 81, 1205–24.

Oxelman, B., Lidén, M. and Berglund, D. (1997) Chloroplast *rps*16 intron phylogeny of the tribe Sileneae (Caryophyllaceae). *Plant Systematics and Evolution*, 206, 393–410.

Palmer, J. D. (1991) Plastid chromosomes: structure and evolution, in *The molecular biology of plastids*, (eds L. Bogorad and I. K. Vasil), Academic Press, San Diego, pp. 5–53.

Palmer, J. D. (1992) Mitochondrial DNA in plant systematics: applications and limitations. in *Molecular systematics of plants*, (eds P. S. Soltis, D. E. Soltis and J. J. Doyle), Chapman & Hall, New York, pp. 36–49.

Pesole, G., Bozzetti, M.P., Lanave, C., Preparata, G. and Saccone, C. (1991) Glutamine synthetase gene evolution: a good molecular clock. *Proceedings of the National Academy of Sciences USA*, 88, 522–6.

Pesole, G., Gissi, C., Lanave, C. and Saccone, C. (1995) Glutamine synthetase gene evolution in bacteria. *Molecular Biology and Evolution*, 12, 189–97.

Provan, J., Soranzo, N., Wilson, N. J., McNicol, J. W., Morgante, M. and Powell, W. (1999) The use of uniparentally inherited simple sequence repeat markers in plant population studies and systematics, in *Molecular systematics and plant evolution*, (eds P. M. Hollingsworth, R. M. Bateman and R. J. Gornall), Taylor & Francis, London, pp. 35–50.

Richman, A. D., Uyenoyama, M. K. and Kohn, J. R. (1996) Allelic diversity and gene genealogy at the self-incompatibility locus in the Solanaceae. *Science*, 273, 1212–16.

Rieseberg, L. H. and Soltis, D. E. (1991) Phylogenetic consequences of cytoplasmic gene flow in plants. *Evolutionary Trends in Plants*, 5, 65–84.

Roche, D., Temple, S. J. and Sengupta-Gopalan, C. (1993) Two classes of differentially regulated glutamine synthetase genes are expressed in the soybean nodule; a nodule-specific class and a constitutively expressed class. *Plant Molecular Biology*, 22, 971–83.

Saenz de Miera, L. E. and Perez de la Vega, M. (1998) A comparative study of vicilin genes in *Lens*: negative evidence of concerted evolution. *Molecular Biology and Evolution*, 15, 303–11.

Sanderson, M. J. and Doyle, J. J. (1992) Reconstruction of organismal phylogenies from multigene families: paralogy, concerted evolution, and homoplasy. *Systematic Biology*, **41**, 4–17.

Sang, T., Crawford, D. J. and Stuessy, T. F. (1995) Documentation of reticulate evolution in peonies (*Paeonia*) using internal transcribed spacer sequences of nuclear ribosomal DNA: implications for biogeography and concerted evolution. *Proceedings of the National Academy of Sciences USA*, **92**, 6813–17.

Sang, T., Donoghue, M. J. and Zhang, D. (1997) Evolution of alcohol dehydrogenase genes in peonies *(Paeonia)*: phylogenetic relationships of putative nonhybrid species. *Molecular Biology and Evolution*, **14**, 994–1007.

Sanmiguel, P., Tikhonov, A., Jin, Y. K., Motchoulskaia, N., Zakharov, D., Melake-Berhan, A. *et al.* (1996) Nested retrotransposons in the intergenic regions of the maize genome. *Science*, **274**, 765–8.

Shinozaki, K., Ohme, M., Tanaka, M., Wakasugi, T., Hayashida, N., Matsubayashi, T. *et al.* (1986) The complete nucleotide sequence of the tobacco chloroplast genome: its gene organization and expression. *EMBO Journal*, **5**, 2043–50.

Shoemaker, R. C., Polzin, K., Labate, J., Specht, J., Brummer, E. C., Olson, T. *et al.* (1996) Genome duplication in soybean (*Glycine* subgenus *soja*). *Genetics*, **144**, 329–38.

Skroch, P. W., Dobert, R. C., Triplett, E. W. and Nienhuis, J. (1993) Polymorphism of the leghemoglobin gene in *Phaseolus* demonstrated by polymerase chain reaction amplification. *Euphytica*, **69**, 177–83.

Soltis, D. E. and Soltis, P. S. (1998) Choosing an approach and an appropriate gene for phylogenetic analysis, in *Molecular systematics of plants II: DNA sequencing*, (eds D. E. Soltis, P. S. Soltis and J. J. Doyle), Chapman & Hall, New York, pp. 1–42.

Strand, A. E., Leebens-Mack, J. and Milligan, B. G. (1997) Nuclear DNA-based markers for plant evolutionary biology. *Molecular Ecology*, **6**, 113–18.

Suh, Y., Thien, L. B., Reeve, H. E. and Zimmer, E. A. (1993) Molecular evolution and phylogenetic implications of internal transcribed spacer sequences of ribosomal DNA in Winteraceae. *American Journal of Botany*, **80**, 1042–55.

Taberlet, P., Gielly, L., Pautou, G. and Bouvet, J. (1991) Universal primers for amplification of three non-coding regions of chloroplast DNA. *Plant Molecular Biology*, **17**, 1105–10.

Temple, S. J., Heard, J., Ganter, G., Dunn, K. and Sengupta-Gopalan, C. (1995) Characterization of a nodule-enhanced glutamine synthetase from alfalfa: nucleotide sequence, *in situ* localization, and transcript analysis. *Molecular Plant–Microbe Interactions*, **8**, 218–27.

Tischer, E., Dassarma, S. and Goodman, H. M. (1986) Nucleotide sequence of an alfalfa (*Medicago sativa*) glutamine synthetase gene. *Molecular and General Genetics*, **203**, 221–9.

Verma, D. P. S., Hu, C. A and Zhang, M. (1992) Root nodule development: origin, function and regulation of nodulin genes. *Physiologia Plantarum*, **85**, 253–65.

Vos, P., Hogers, R., Bleeker, M., Reijans, M., van de Lee, T., Hornes, M. *et al.* (1995) AFLP: a new technique for DNA fingerprinting. *Nucleic Acids Research*, **23**, 4407–14.

Wagner, A., Blackstone, N., Cartwright, P., Dick, M., Misof, B., Snow, P. *et al.* (1994) Surveys of gene families using polymerase chain reaction: PCR selection and PCR drift. *Systematic Biology*, **43**, 250–61.

Walker, E. L., Weeden, N. F., Taylor, C. B., Green, P. and Coruzzi, G. M. (1995) Molecular evaluation of duplicate copies of genes encoding cytosolic glutamine synthetase in *Pisum sativum*. *Plant Molecular Biology*, **29**, 1111–25.

Wendel, J. F. and Doyle, J. J. (1998) Phylogenetic incongruence: window into genome history and molecular evolution in *Molecular systematics of plants II: DNA sequencing*, (eds D. E. Soltis, P. S. Soltis and J. J. Doyle), Chapman & Hall, New York, pp. 265–96.

Wheeler, W. C. (1997) Optimization alignment: the end of multiple sequence alignment in phylogenetics? *Cladistics*, **12**, 1–9.

Williams, J. G. K., Kubelik, A. R. Livak, K. J., Rafalski, J. A. and Tingey, S. V. (1990) DNA polymorphisms amplified by arbitrary primers are useful as genetic markers. *Nucleic Acids Research*, **18**, 6531–5.

Wolfe, K. H., Sharp, P. M. and Li, W-H. (1989) Rates of synonymous substitution in plant nuclear genes. *Journal of Molecular Evolution*, **29**, 208–11.

Young, J. P. W. (1996) Phylogeny and taxonomy of rhizobia. *Plant and Soil*, **186**, 45–52.

Zimmer, E. A., Martin, S. L., Beverley, Y. W., Kan, S. M. and Wilson, A. (1980) Rapid duplication and loss of genes coding for the a chain of hemoglobin. *Proceedings of the National Academy of Sciences USA*, **77**, 2158–62.

Zhang, D. X. and Hewitt, G. M. (1997) Insect mitochondrial control region: a review of its structure, evolution, and usefulness in evolutionary studies. *Biochemical Systematics and Ecology*, **25**, 99–120.

Zhu, T., Schupp, J. M., Oliphant, A. and Keim, P. I. (1994) Hypomethylated sequences: characterization of the duplicate soybean genome. *Molecular and General Genetics*, **244**, 638–45.

Chapter 13

Spectral analysis – a brief introduction

M. A. Charleston and R. D. M. Page

ABSTRACT

Most phylogenetic methods emphasise tree-building – distilling a data matrix into one (ideally) or more trees. These techniques will construct a tree whether the data is tree-like or entirely random, and hence often give little insight into the structure of the data. Spectral analysis is a tool that enables more thorough exploration of the data. Central to the method is the concept of a 'split', which is simply a bipartition of a set of taxa (or sequences) into two sets. For a data set for n species there are $2^{(n-1)}$ possible splits, but a tree for n species may contain at most $2n - 3$ splits. Unless the data is perfectly clean, there will be support for more splits than can be accommodated in a single tree. Spectral analysis provides a way to visualise the support and conflict for each possible split, making it easier to see how well the data supports a particular tree. To perform spectral analysis, the computer program *Spectrum* (for both Macintosh and Windows computers) is available free on the internet: (http://taxonomy.zoology.gla.ac.uk/~mac/spectrum/spectrum.html).

13.1 Introduction

Systematics is devoted to finding evolutionary relationships, usually expressed in trees, but not all phylogenetic data is tree-like: for reasons such as hybridisation or horizontal transfer of genetic material, or conflicting information due to the stochastic nature of molecular evolution, there may be signals in a given data set which cannot correspond to the same underlying tree. Rather than select the largest subset of the data which can best be represented in a single figure such as a tree, we discuss spectral analysis (Hendy and Penny, 1993) as an intuitive method of representing all of the signals in a single vector. This *spectrum* can readily be inspected to show how 'tree-like' the original data set is, before we commit ourselves to a single tree or set of trees as representatives of the information in the data.

Spectral analysis also offers a consistent method for the correction of so-called 'multiple hits' – instances of homoplasy which are introduced by stochastic effects such as parallel mutations in separate lineages, reversals and multiple invisible changes of character state along a lineage. All of these possibilities can be accommodated given a simple enough Markov model of character evolution, but unless the underlying tree is known in advance, the 'corrections' have to operate on *dependent* quantities (say, distances between extant taxa), but as if they were *independent*.

In *Molecular systematics and plant evolution* (1999) (eds P. M. Hollingsworth, R. M. Bateman and R. J. Gornall), Taylor & Francis, London, pp. 255–267.

The *Hadamard conjugation* (described later) allows us to convert these dependent quantities into an equivalent set of independent ones, upon which the 'corrections' mentioned above can operate in a truly consistent manner.

We also describe another technique deriving naturally from spectral analysis – that of the *nearest neighbourhood* of a tree. This method of investigating phylogenetic signal, now with a specific tree in mind, is simple to implement and interpret, and is part of the suite of tools which are becoming more available, and being increasingly recognised as important, to the modern systematist.

13.2 Spectral analysis – what it is and what it isn't

Spectral analysis is *not* a tree-building method. It is a tool for investigating phylogenetic signal in a data set, that is, information which the data is providing, without resorting to a tree.

When we consider a very simple phylogenetic data set (see, for example, Table 13.1 below), it is often very easy to see what is the underlying structure of that data set. Consider each of the characters in turn: the first 4 characters (columns 1–4) merely separate the 5 taxa: they are *singleton* splits, which correspond to terminal branches in phylogenetic trees. The next two characters (columns 5 and 6) are both uniting taxa **A** and **B**, suggesting that these two are sister taxa with respect to the complete taxon set. Site 7 suggests that taxa **B** and **C** are sister taxa, but this is outweighed by the two characters putting **A** with **B**. Sites 8 and 9 are synapomorphies for the triplet **A**, **B** and **C**, and the last character is a singleton again (there being only one taxon with state 0), conveying only the information that **E** is not the same as any of the other taxa.

Note that if we were considering the tree in Fig. 13.1 as rooted (as shown) with evolution proceeding from the root toward the tips, then characters 1–4 would correspond to autapomorphies, characters 5–6 and 8–10 would be synapomorphies, and character 7 would be homoplastic.

However, when we consider more complex data sets, the pattern is not nearly so clear. With n taxa there are $(2n-5)(2n-7)\ldots(3)(1) = (2n-5)!!$ possible unrooted binary trees, and $(2n-3)(2n-5)\ldots(3)(1) = (2n-3)!!$ rooted binary trees (the double exclamation mark is read 'double factorial'). This number gets extremely big, very quickly: it grows exponentially in n – with only 26 taxa there are around 8.2×10^{32} trees from which to choose (there are 'only' about 10^{20} stars in the universe).

Table 13.1   Simple phylogenetic data set.

| Taxa | Characters | | | | | | | | | |
	1	*2*	*3*	*4*	*5*	*6*	*7*	*8*	*9*	*10*
A	1	0	0	0	1	1	0	1	1	1
B	0	1	0	0	1	1	1	1	1	1
C	0	0	1	0	0	0	1	1	1	1
D	0	0	0	1	0	0	0	0	0	1
E	0	0	0	0	0	0	0	0	0	0

Even if we are able to find the 'best' tree to represent the data, a rooted binary tree only contains $(2n - 2)$ branches, of which n must be in every tree since they correspond to autapomorphies of the n taxa themselves, leaving just $(n - 2)$ individual pieces of information about the phylogenetic relationships among the taxa of interest.

The basic routes of spectral analysis are shown in Fig. 13.2. We may begin with a set of aligned characters (with two or more states), or a set of distances, and calculate the observed bipartition spectrum from these. The Hadamard log conjugation is an optional adjustment for the problem of multiple changes of character state on the same lineage ('multiple hits') or on parallel lineages (homoplasy and 'long branch attraction'). Such a spectrum is referred to as a conjugated spectrum.

13.2.1 Calculating a spectrum

Even in complex data sets it is possible to simplify things, by considering what exactly each character is suggesting about the data and amalgamating these pieces of information into a spectrum.

Consider again the simple data set in Table 13.1. Each varying character corresponds to a *split* of the taxon set into two non-overlapping subsets whose union is the whole set – in other words, a *bipartition*. Correspondingly, each branch in a tree induces a split. To a large degree, phylogenetic methods attempt to find the tree whose splits best match the splits in the data, and to take that tree as the best estimation of the actual phylogenetic relationships among the taxa. The uniform characters correspond to the trivial, or *empty* split, which can be thought of as comprising the empty set $\varnothing$ and the complete taxon set, since such characters cannot provide separating information on the taxa.

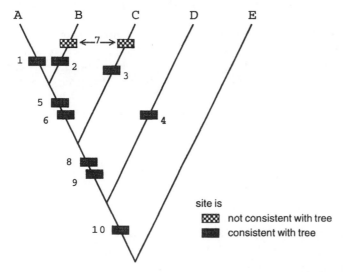

Figure 13.1 The simplest tree which could correspond to the character data in Table 13.1. There are two homoplasious character state changes indicated by character 7; all other characters are compatible with this tree. Deriving this tree is very easy for such a simple data set; for more complex data sets the patterns are less clear.

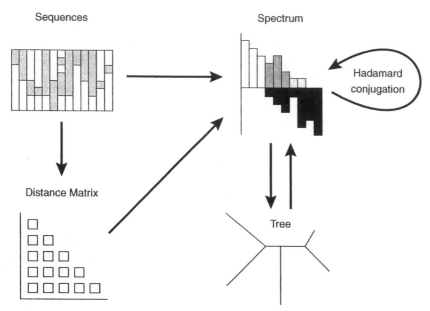

Figure 13.2 Spectral analysis allows us to get spectral information from sequences of aligned (homologous) characters or from distances, or even from trees themselves. Once obtained, the spectrum can be operated on with the **Hadamard conjugation** (see later) to adjust for multiple changes of character state.

The derivation of an *observed spectrum* is very simple for two-state (binary) data; for four-state data the process is more involved, and we shall deal with that later. For the moment, suppose we have a set of phylogenetic data such as are shown in Table 13.1.

We use a scheme devised by Hendy (1991), which assigns a unique binary number to each of the $2^{(n-1)}$ possible splits for a set of n taxa. First we must label the taxa from 1 to n. Next, we assign a '0' to each taxon which has the same state as the n-th at that character, and a '1' to all the other taxa. Note that this process effectively re-codes the character.

We then write the character states in reverse order, so if taxon i has state d_i we write $d_n d_{n-1} \ldots d_2 d_1$. We then treat the resulting string of 1's and 0's as a binary number, which is the label for that split.

Succinctly, if for a specific character d_i is the state of taxon i, the split number e is

$$e = \sum_{i=1}^{n-1} \{2^{i-1} \text{ if } d_i = 1\}$$

thus the split number 0 corresponds to a uniform character, and the 'singleton' characters have split numbers 1, 2, 4, 8, $\ldots$, $2^{(n-2)}$, $2^{(n-1)} - 1$ (the last term is for the case when all the taxa have a different character state from the n-th).

For example, the character states (x,y,x,x,y) give rise to the taxa (1 through 5) being assigned $d_1 = 1$, $d_2 = 0$, $d_3 = 1$, $d_4 = 1$, and $d_5 = 0$, corresponding to the binary number $01101 = 13$.

The data in Table 13.1 correspond to the following splits:

character	binary expression	split number
1	00001	1
2	00010	2
3	00100	4
4	01000	8
5,6	00011	3
7	00110	6
8,9	00111	7
10	01111	15

Thus the observed spectrum would look something like that shown in Fig. 13.3. Note that the 'total conflict' with a given split s is the sum of the amounts of support of all those splits which are *incompatible* with s, i.e., cannot fit on the same tree as s.

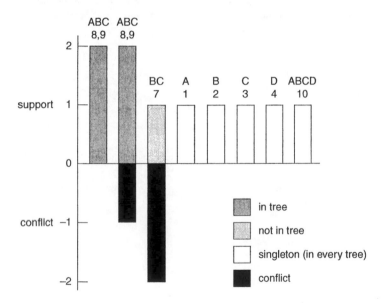

Figure 13.3 This is the **observed** spectrum for the data set shown in Table 13.1. Above the x-axis are shown the **support** values for each split (in this case corresponding to one or two characters for each split) and below the axis are shown the **conflict** values, i.e., the total amount of support for splits which contradict each given split.

13.2.2 Detection of 'good' and 'poor' signal

The spectrum immediately shows how 'tree-like' a phylogenetic data set is (see Fig. 13.4). Spectra are often ranked in descending order of support for the splits (since at this stage the split number is not really important), which makes inspection easier. 'Good' phylogenetic signal is characterised by having few well-supported splits with low conflict values, with small support (and possibly large conflict) for

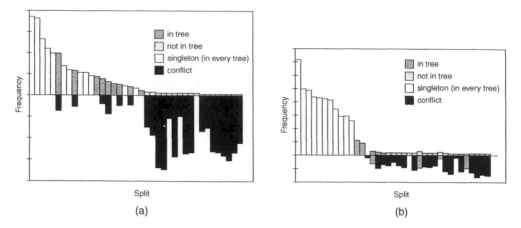

Figure 13.4 In the spectrum on the left (a) we see relatively strong phylogenetic signal: most of the largest splits will fit on the same tree, and the conflict with these splits in the rest of the data is not high. On the other hand, spectrum (b) shows much weaker phylogenetic signal: the relative support for internal branches (as opposed to the singleton splits corresponding to terminal branches) is much lower, and many of these have substantial conflict values.

the other splits. The well-supported splits are also compatible, i.e., they will all correspond to splits in the same tree. 'Weaker' phylogenetic signal is characterised by having fewer well-supported internal splits (the splits which correspond to terminal branches must be in every tree so their support is of less interest at this stage). Internal splits tend to have high conflict values.

It is also known (Lento *et al.*, 1995) that there is a high correlation between the bootstrap support (Felsenstein, 1985; Efron *et al.*, 1996) for a given split and the ratio of support to conflict for that split, which again makes the spectrum a useful tool for determining the robustness of data.

13.2.3 Spectra from different sources

When dealing with four character states, as in DNA data, there are several ways to proceed: we can treat four character states as two, mapping, for example, the four nucleotides A, C, G and T onto the purines (A, G) and pyrimidines (C, T). Alternatively we may create a kind of 'average' spectrum taken over all the seven non-trivial mappings from four onto two states (shown in Table 13.2).

If we are lucky and only encounter two character states at any given character, then the procedure is again straightforward – we simply code each character as two character states, while allowing these states to vary across characters (see Fig. 13.5). For example if a character for five taxa had nucleotides GAGAA, we could code Gs as 1s and As as 0s, giving split number 00101 = 5.

If we are not so lucky and cannot reduce the information content of each character in this way at all, we can create a *quadripartition spectrum*, which is computationally much more intense and shall not be discussed further here. The interested reader is referred to Hendy and Penny (1993).

Table 13.2 The seven non-trivial mappings from four to two character states.

{A,C,G,T} →	{A,G},{C,T}
	{A,C},{G,T}
	{A,T},{C,G}
	{A},{C,G,T}
	{C},{A,G,T}
	{G},{A,C,T}
	{T},{A,C,G}

It is also possible to derive a spectrum from a distance matrix: this *distance spectrum* is the spectrum which would generate the observed set of distances, and it relies on a special matrix, the *Hadamard* matrix (Hendy, 1991; Hendy and Charleston, 1993) to convert between a set of *path lengths* which give rise to distances, and a set of *branch lengths* on a tree. Note that we do not actually have to know the tree to do this conversion – this is a special property of the Hadamard.

It is known that the variance of branch lengths inferred from a distance matrix is reduced, in comparison with branch lengths inferred directly from character data (Charleston, 1994; Waddell *et al.*, 1994). This rather startling result means that the spectrum can in a sense be 'cleaned up' by converting the two- or four-state data matrix into a distance matrix, and then convert back to a spectrum. This facility is provided in the Spectrum program (Charleston, 1998).

A	G	T	C	A	T	C	A	T	C	C
B	A	T	T	A	C	C	A	T	T	C
C	G	T	T	G	T	T	A	T	T	A
D	A	C	C	A	C	T	C	C	C	A
E	A	C	C	G	C	C	C	C	C	A
	↓	↓	↓	↓	↓	↓	↓	↓	↓	↓
A	1	1	0	1	1	0	1	1	0	1
B	0	1	1	1	0	0	1	1	1	1
C	1	1	1	0	1	1	1	1	1	0
D	0	0	0	1	0	1	0	0	0	0
E	0	0	0	0	0	0	0	0	0	0
split	5	7	6	11	5	12	7	7	6	3

Figure 13.5 Mapping from four to two states

13.3 The Hadamard – matrix and conjugation

A Hadamard matrix is defined as a square matrix of 1s and –1s whose columns are orthogonal. Here, we restrict ourselves further and consider only Hadamard matrices of the following form:

$$H_2 = \begin{bmatrix} 1 & 1 \\ 1 & -1 \end{bmatrix};$$

$$H_3 = \begin{bmatrix} 1 & 1 & 1 & 1 \\ 1 & -1 & 1 & -1 \\ 1 & 1 & -1 & -1 \\ 1 & -1 & -1 & 1 \end{bmatrix} = \begin{bmatrix} H_2 & H_2 \\ H_2 & -H_2 \end{bmatrix};$$

etc.

In general the $(i+1)$-th Hadamard matrix (as we are considering them here) is given by the recursive formula

$$H_{i+1} = \begin{bmatrix} H_i & H_i \\ H_i & -H_i \end{bmatrix} = H_2 \otimes H_i$$

where $\otimes$ refers to the *Kronecker product* operator.

The Hadamard matrix for n taxa is of size $2^{(n-1)} \times 2^{(n-1)}$, which gets large very quickly, and it is this complexity which restricts the operation of the Hadamard calculations to relatively small n, around 20 taxa for two-state (bipartition) spectra and around 10 taxa for four-state (quadripartition) spectra.

The Hadamard matrix has a very useful property: it allows us to convert between sets of path lengths in a tree and branch lengths, *without knowing what the tree is*. This makes the Hadamard matrix a very powerful tool if we want to treat data in a consistent fashion, as explained in the following section.

13.3.1 Paths and branches

In a tree, say the unrooted binary tree shown in Fig. 13.6, there is just one path between each pair of *tips* (extant taxa), but these paths may overlap. If for example

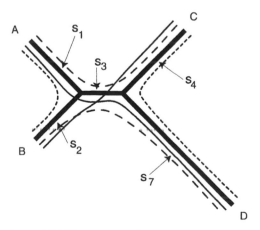

Figure 13.6 This tree on four taxa (A, B, C, D) shows how some paths intersect others; the path A–C and the path B–D both contain the internal branch of this tree.

we wished to correct for 'multiple hits' in a data set, and operated on the distances between taxa with some correction function, then we would have to sometimes treat some parts of the underlying tree differently, for each pair of taxa.

In the figure, the paths which correspond to the same underlying tree are shown in similar line styles: solid lines for the tree ((A,D),(B,C)), long dashes for ((A,C),(B,D)), and shorter dashes for ((A,B),(C,D)). Clearly if we knew that the underlying tree for the four taxa was that shown in Fig. 13.6, i.e., ((A,B),(C,D)), we could adjust each branch length independently and correctly account for multiple hits on each. But without knowing the tree, we would have to operate with our correction function on paths which may or may not intersect, effectively treating the path lengths as independent; in effect assuming that the paths do not intersect. This is clearly not going to be a consistent approach.

The Hadamard matrix can make all this much easier, for it uses the property that there are precisely the same number of splits of a set of n objects as there are even-ordered subsets of the set (an even-ordered subset is one with an even number of elements). It is these even-ordered subsets which correspond to path sets in a tree, and the splits correspond to branches.

Suppose we have the tree shown in Fig. 13.6, and that the taxa A–D are numbered 1–4 respectively. Let the branches of the tree be labelled according to the same number as the split which each induces on the taxa: thus the pendant branches are numbered 1, 2, 4 and 7 (for taxa A, B, C and D) and the internal branch is numbered 3. The other possible split numbers for a tree on 4 taxa are 5 (putting taxa 1 and 3 together) and 6 (putting taxa 2 and 3 together).

If we let the length of branch e be s_e, then the total path length between A and C in tree ((A,B),(C,D)) is $(s_1 + s_3 + s_4)$, the path length between A and D is $(s_1 + s_3 + s_7)$, etc. Then the conversion from distances to branches becomes:

$$\frac{-1}{4}\mathbf{H}_4 \begin{bmatrix} 0 \\ d_{1,4} \\ d_{2,4} \\ d_{1,2} \\ d_{3,4} \\ d_{1,3} \\ d_{2,3} \\ d_{1,2,3,4} \end{bmatrix} = \frac{-1}{4}\mathbf{H}_4 \begin{bmatrix} 0 \\ s_1+s_7 \\ s_2+s_7 \\ s_1+s_2 \\ s_4+s_7 \\ s_1+s_3+s_4 \\ s_2+s_3+s_7 \\ s_1+s_2+s_4+s_7 \end{bmatrix} = \begin{bmatrix} \sum\limits_{0<i<8} si \\ s_1 \\ s_2 \\ s_3 \\ s_4 \\ s_5 \\ s_6 \\ s_7 \end{bmatrix}$$

where $d_{\{1,2,3,4\}}$ is defined as the minimum total path length of the even-ordered subset $\{1,2,3,4\}$, so is $min\{d_{1,2} + d_{3,4}, \; d_{1,3} + d_{2,4}, \; d_{1,4} + d_{2,3}\}$. This definition is tree independent, but in the example above it becomes $d_{1,2} + d_{3,4} = s_1 + s_2 + s_4 + s_7$.

This is easily generalised to more taxa: the interested reader is directed to Hendy and Penny (1993).

Note that this multiplication is not tree-dependent: all support values for all splits are shown in the resulting array, not just those which can correspond to a specific tree. Armed with this array we can probe the information content of a character array (using the raw spectrum) or even a distance matrix (using the distance spectrum) for *any* tree. However we have not yet completely dealt with the problem of multiple hits, and we need the *Hadamard conjugation* for this purpose.

13.3.2 The Hadamard conjugation

The conversion described above between path sets and branches enables the Hadamard conjugation to work: we convert from observed path lengths (the observed spectrum) to 'observed' branch lengths, which we can then adjust for multiple hits, and then we convert back again. This conjugation technique, well known in mathematics, is of the form $q = f^{-1} g \ f(q')$, and in this case becomes

$$q = \mathbf{H}^{-1} \ln \mathbf{H}s;$$
$$s = \mathbf{H}^{-1} \exp\mathbf{H}q,$$

where s is the observed bipartition spectrum, and q is the spectrum of estimated branch lengths, having taken account of multiple hits with the natural log function ln (we drop the subscript from the H's for convenience and generality).

The use of the natural log function is consistent with the Jukes–Cantor 1-parameter model (1969). Other corrections exist for other models, including the Kimura 3ST model (Kimura, 1980), which uses the quadripartition form of the Hadamard (Hendy *et al.*, 1994).

13.4 The NNI neighbourhood

Another useful trait of spectral analysis is that related splits can be investigated, being for instance one perturbation from the current tree of interest. The *nearest neighbour interchange* or 'NNI' tree perturbation (Fig. 13.7) is a common one used in heuristic search strategies (Swofford, 1993) and in analyses where the precise detail of a tree is not as important as its general trends.

We can quickly get an impression of how robust an individual split is, and what might be prime contenders for possible alternative hypotheses, by finding the support values of splits which are just one NNI removed from our base tree. Cooper and Penny (1997) used this to good effect to show that within a small variation, their bird phylogeny was robust, and allowed them to conclude that most modern orders of birds were present before the Cretaceous–Tertiary boundary. Given a NNI neighbourhood such as is shown in Fig. 13.8(a) we may conclude that in general the tree is likely to be correct, or at least is consistent with a large part of the data, since the amount of support for similar trees is much smaller than for the tree itself. If however we were to obtain the NNI neighbourhood graph which is shown in Fig. 13.8(b) we would be much less sure that the base tree was reliable, since there is substantial support for similar yet alternative hypotheses.

13.5 Summary

We have presented here a brief outline of the uses of spectral analysis, and some of the simpler results therein. The process is computationally intense, but conceptually simple, and allows researchers to investigate phylogenetic information in an unbiased and transparent way, prior to choosing a 'best' tree, or even an optimality criterion by which to choose it.

Spectral analysis makes explicit the structure in the data, by showing precisely the support and conflict for each possible branch without recourse to a tree.

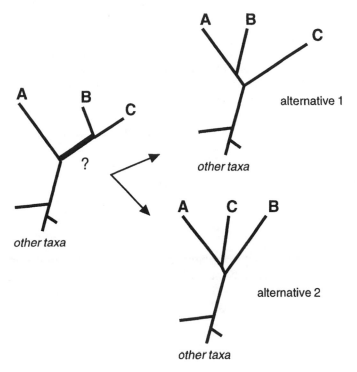

Figure 13.7 The operation of the Nearest Neighbour Interchange tree perturbation is equivalent to contracting an internal branch and re-expanding it in either of two possible new 'directions'; thus each internal branch has two NNI neighbours and a given unrooted binary tree on n taxa has $2(n-2)$ NNI neighbours.

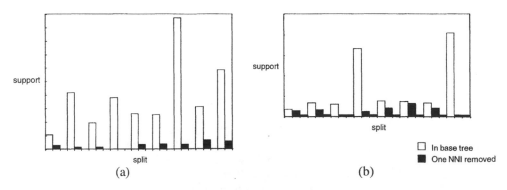

Figure 13.8 'Good' and 'bad' NNI neighbourhoods. In the figure on the left (a) we see much higher support for our reference tree than there is for any near contender, so can be reasonably confident that our tree is reliable. The right-hand figure (b) has no such promise since there are a number of support values in the data set for splits which are similar to, but different from, splits in our base tree.

It can be applied to both discrete and distance data, through the use of the Hadamard matrix H_n. This matrix is central to the *Hadamard conjugation*, which enables the individual inferred branch lengths to be adjusted to take into account multiple hits – where the stochastic evolutionary process gives rise to parallel mutations and reversals – in a consistent fashion, independent of any tree.

The ready evaluation of alternative trees is possible through the NNI neighbourhood, and for other tree perturbations in general, by simply summing the support for the splits in those alternative hypotheses.

Finally, the correlation between bootstrap values and the ratio between support and conflict values for a given split is known to be high (Lento *et al.*, 1995), which means we can get a quick guide to the robustness of different splits (internal branches) of a tree, sometimes a great deal more quickly than by performing an entire bootstrap analysis.

13.6 Spectrum 2.0

The analyses in this chapter were conducted using SPECTRUM 2.0β, a newly released beta-version Macintosh® program written by Charleston and Page (Charleston, 1998). It is free and available from the Centre for Taxonomy web site of Glasgow University, at http://taxonomy.zoology.gla.ac.uk/~mac/spectrum/spectrum.html.

REFERENCES

Charleston, M. A. (1994) *Factors affecting the performance of phylogenetic methods*. Ph.D. thesis, Massey University, Palmerston North, New Zealand.

Charleston, M. A. (1998) Spectrum: spectral analysis of phylogenetic data. *Bioinformatics*, **14**, 98–99.

Cooper, A. and Penny, B. (1997) Mass survival of birds across the Cretaceous–Tertiary boundary: Molecular evidence. *Science*, **275**, 1109–1113.

Efron, B., Halloran, E. and Holmes, S. (1996) Bootstrap confidence levels for phylogenetic trees. *Proceedings of the National Academy of Sciences USA*, **93**, 7085–7090.

Felsenstein, J. (1985) Confidence limits on phylogenies: an approach using the bootstrap. *Evolution*, **39**, 783–791.

Hendy, M. D. (1991) A combinatorial description of the closest tree algorithm for finding evolutionary trees. *Discrete Mathematics*, **96**, 51–58.

Hendy, M. D. and Charleston, M. A. (1993) Hadamard conjugation: a versatile tool for modelling sequence evolution. *New Zealand Journal of Botany*, **31**, 231–237.

Hendy, M. D. and Penny, D. (1993) Spectral analysis of phylogenetic data. *Journal of Classification*, **10**, 5–24.

Hendy, M. D., Penny, D. and Steel, M. A. (1994) A discrete Fourier analysis for evolutionary trees. *Proceedings of the National Academy of Sciences USA*, **91**, 3339–3343.

Jukes, T. H. and Cantor, C. R. (1969) Mammalian protein metabolism, in *Evolution of protein molecules* (ed. H. M. Munro) Academic Press, New York, pp. 21–132.

Kimura, M. (1980) A simple method for estimating the rates of base substitutions through comparative studies of nucleotide sequences. *Journal of Molecular Evolution*, **16**, 111–120.

Lento, G. M., Hickson, R. E., Chambers, G. K. and Penny, D. (1995) Use of spectral analysis to test hypotheses of the origin of pinnipeds. *Molecular Biology and Evolution*, **12**, 28–52.

Swofford, D. L. (1993) *PAUP: Phylogenetic Analysis Using Parsimony*, version 3.1.1. Computer program distributed by the Illinois Natural History Survey.

Waddell, P.J., Penny, D., Hendy, M. D. and Arnold, G. (1994) The sampling distributions and covariance matrix of phylogenetic spectra. *Molecular Biology and Evolution*, **11**, 630–642.

Chapter 14

Ribosomal DNA sequences and angiosperm systematics

M. A. Hershkovitz, E. A. Zimmer and W. J. Hahn

ABSTRACT

DNA sequences from ribosomal genes (rDNA) are the most widely exploited nuclear molecular markers in angiosperm systematics. In concert with various chloroplast DNA (cpDNA) sequences, ribosomal DNA sequences have revolutionised ideas on angiosperm evolution. The utility and popularity of ribosomal genes derives in part from the relative ease with which the sequences can be determined, which in turn derives from the large number of rDNA gene copies in the genome. More importantly, rDNA sequences collectively exhibit a broad range of phylogenetic signal. The relatively slowly-evolving 18S rDNA sequences in combination with the cpDNA genes *rbc*L and *atp*B have yielded unequivocal evidence for many angiosperm inter-familial relationships not predicted from decades of research using non-molecular characters. Sequences from the rapidly evolving rDNA 18S-26S internal transcribed spacers (ITS), currently the most widely exploited molecular marker in angiosperm systematics, have had similar impact on the understanding of interspecific and inter-generic relationships, while also yielding insights on speciation and biogeography. ITS sequences appear to exhibit an unusual property of containing phylogenetic traits specific to many levels of angiosperm divergence, from the most recent to the most ancient. Current research suggests that 26S sequences will become an invaluable source of phylogenetic information spanning the divergence gap between 18S and ITS, while the 18S-26S 5' external transcribed spacer (ETS) will augment greatly the systematic utility of ITS at the lowest taxonomic levels. Sequences from the 5S rDNA non-transcribed spacer (NTS) have proven useful for dissecting origins of hybrid species. Their phylogenetic utility notwithstanding, rDNA sequences exhibit patterns that suggest peculiar evolutionary processes, and these inferred processes must be considered in phylogenetic reconstruction. These patterns include differential levels of homogenisation of rDNA gene copies among both the different gene regions and different rDNA loci within a genome. Incomplete homogenisation of the most rapidly evolving sequences can lead to infragenomic polymorphism, which, if not detected, can yield incorrect phylogenetic interpretations. rDNA sequences also exhibit strong biases in patterns of substitution among bases and among sequence positions. Moreover, the patterns of bias sometimes appear to differ among members of a particular taxonomic sampling. These biases can mislead popular phylogenetic methods that assume no bias. Provided that rDNA evolutionary processes are considered, however, phylogenetic results from rDNA

In *Molecular systematics and plant evolution* (1999) (eds P. M. Hollingsworth, R. M. Bateman and R. J. Gornall), Taylor & Francis, London, pp. 268–326.

sequences can be as robust as those using any other DNA sequence. Besides facilitating phylogenetic study, the ever-expanding database of angiosperm rDNA sequences contribute to studies of molecular evolution in general and likely will contribute also to advances in the understanding of ribosomal function.

14.1 Introduction

Ribosomal characters are the most exploited source of molecular data for systematic studies of organismal relationships (Appels and Honeycutt, 1986; Troitsky and Bobrova, 1986). Early and (eventually) widespread exploitation of ribosomal characters in systematics can be attributed to four factors: critical function, ubiquity, variability and abundance. The discovery of the role of the ribosome in translation focused considerable and still persistent effort to resolve its structure and mechanisms. Biochemical genetic research work soon revealed that the ribosome and much of its underlying genetic organisation is conserved in all free-living cellular organisms. Nonetheless, specific regions of ribosomal genes were found to be conserved differentially, so that they collectively provided at least some diagnostic phylogenetic signal at most levels of the taxonomic hierarchy. Finally, ribosomal genes are expressed highly and, in eukaryotes, usually occur in high copy numbers. These features rendered ribosomal RNAs and eukaryotic ribosomal genes among the most readily isolated molecules from a lysed cell, and the most facile target for even the most primitive sequencing, cloning and enzymatic digestion techniques. In angiosperm systematics, the popularity of ribosomal characters has been rivalled only by the collective chloroplast genome (D. Soltis and Soltis, 1998), including one gene that was translocated evolutionarily to the nucleus (*rbc*S: the ribulose–bisphosphate carboxylase small subunit).

To estimate the proportion of research effort directed to nuclear ribosomal and other DNA-based approaches to plant systematics, we perused the 262 systematics and pteridological contributed paper and poster abstracts from the 1997 American Institute of Biological Sciences meeting in Montreal, Canada. Overall, 146 (56%) included analyses of DNA characters, and 81 (31%) of these included ribosomal (sometimes among other) DNA characters. Seventy (27%) included, specifically, rDNA internal transcribed sequence (ITS) data. These numbers are remarkable considering, that as recently as the 1985 meeting, the equivalent sessions included no DNA-based studies, and the first paper using ITS data was not presented until the 1991 meeting (Baldwin, 1991). Even 1995 publications included ITS among 'alternative' genes for plant phylogenetic analysis (Baldwin *et al.*, 1995; Soltis and Soltis, 1995).

14.2 Ribosomal gene exploitation in plant systematics

The exploitation of nuclear ribosomal data in plant systematics can be divided according to the functional genetic elements (Fig. 14.1), each of which contain phylogenetic signal considered especially applicable to particular levels of divergence (D. Soltis and Soltis, 1998). In eukaryotes, the rDNA cistron encodes the 18S, 5.8S and 26S rRNAs, separated by the two internal transcribed spacers (ITS1 and ITS2), and flanked by the 5′ and 3′ external transcribed spacers (5′-ETS and 3′-ETS). The 18S (*ca.* 1.8 kilobases, or kb) sequence forms the ribosomal small subunit (SSU), while the 5.8S (163–165 base pairs, or bp) and 26S (*ca.* 3.5 kb) form most of the large subunit (LSU). Angiosperm 18S and 26S sometimes have been referred to as, respectively, 17S and any of 25S, 27S or 28S, based on the names of non-angiospermous homologues that have slightly different sedimentation values. The ITS1 is homologous to the SSU-LSU spacer in non-eukaryotic and organellar rDNA, while ITS2 appears to have originated as a more recently evolved cleavage product of a hypervariable LSU sequence tract (Appels and Honeycutt, 1986; Troitsky and Bobrova, 1986). The cistron occurs in hundreds to thousands of copies, each separated by the intergenic spacer (IGS *sensu stricto*), at one to a few chromosomal loci (gene nomenclature from Singer and Berg (1991); the IGS sometimes is defined as including the ETS, and the non-transcribed IGS *sensu stricto* then is called NTS). The 5S rDNA forms the balance of the LSU. The 5S cistron also occurs in hundreds to thousands of copies that alternate with non-transcribed spacers (NTS) at 1–6 loci, usually on different chromosomes than the 18S-26S genes (Appels and Baum, 1992; Cronn *et al.*, 1996; Baum and Bailey, 1997; Li *et al.*, 1997). Some nomenclatural conveniences have been applied in distinguishing the cistrons and loci. For example, the 18S-26S gene family often is referred to simply as 'rDNA' or 'ribosomal DNA', while the 5S rDNA is ignored or termed '5S DNA' (e.g. Rieseberg and Brunsfeld, 1992). Alternatively, because the 18S-26S locus represents the classical nucleolar organising region (NOR), it has been called 'Nor rDNA' (Appels and Baum, 1992). However, functional 18S-26S loci that are not expressed in a particular genomic context or developmental stage will not form nucleoli. Moreover, transcriptionally active 5S loci are reported to be associated with nucleoli, and it is possible that 5S loci not so associated are inactive (Shi *et al.*, 1996).

By the early 1980s, all of the various coding and non-coding regions had been applied at least seminally to plant systematic studies. Over the past five or so years, however, most efforts involving rDNA have exploited sequence data from the 18S and ITS regions. Exploitation of 26S sequences has lagged behind 18S and ITS, but now is underway in earnest. The 5′-ETS sequence had been difficult to obtain for technical reasons, but studies within the past year indicate that its systematic

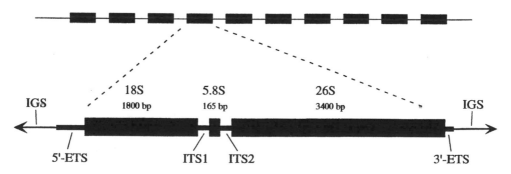

Figure 14.1 Generalised genetic map of angiosperm 18S-26S ribosomal DNA genes. Length of the 18S and 26S genes are variable and rounded up to the nearest 100 bp. The 5.8S gene is usually 163–165 bp. The structure of the 5S rDNA (not shown) is much simpler, consisting of alternating 5S coding regions and non-trasncribed spacers (NTS).

utility is considerable. 5S rDNA sequences are used seldomly in phylogenetic analysis because of high homoplasy related to functional constraints. Infragenomic poly-morphism limits the utility of variation in the 5S NTS. Patterns of organismal and molecular evolution evident from these rDNA sequence data will be emphasised in the present discussion. Other forms of rDNA data applied in plant systematics will not be reviewed, including restriction fragment length polymorphisms (RFLPs) of both the 18S-26S and 5S loci. RFLP approaches were popular relatively early in the molecular systematics era. Although largely displaced by sequence analyses, rDNA-RFLP approaches are not obsolete (e.g. Pillay, 1997; Warpeha *et al.*, 1998). The cytogenetic use of nuclear rDNA, specifically of the NOR as a chromosomal marker, pre-dates by decades modern understanding of ribosomal genetics. The popularity of cytogenetics has resurged with the advent of molecular technology that enables precise chromosomal localisation of the various rDNA loci, in turn facilitating a wide range of plant genetic and evolutionary analyses (e.g. Shi *et al.*, 1996; Kamstra *et al.*, 1997; Li *et al.*, 1997; Cerbah *et al.*, 1998; Fransz *et al.*, 1998; Leitch *et al.*, 1998; McGrath and Helgeson, 1998; Roose *et al.*, 1998). Finally, although widely exploited in systematic studies of other organisms, angiosperm organellar rDNA is useful only exceptionally (e.g. Duff and Nickrent, 1997; Goremykin *et al.*, 1997; Nickrent *et al.*, 1997).

14.2.1 The 18S rRNA gene

14.2.1.1 18S sequencing in angiosperms

The 18S region, which encodes the RNA of the small ribosomal subunit, along with its prokaryotic and organellar 16S homologue, is taxonomically the most broadly applied sequence in molecular phylogenetics. Among the historical reasons mentioned earlier, this is mostly because it is present in all organisms and, across the tree of life, exhibits similar phylogenetic signal at equivalent taxonomic ranks (Gerbi, 1985; Hillis and Dixon, 1991; Embley *et al.*, 1994; D. Soltis and Soltis, 1998).

The first major effort to apply 18S data to angiosperm phylogeny was under-taken in the late 1980s and based on direct RNA sequencing of portions of both the 18S and 26S regions (Hamby and Zimmer, 1988, 1992; Zimmer *et al.*, 1989). Subsequent studies (e.g. Nickrent and Franchina, 1990; Boulter and Gilroy, 1992; Bharathan and Zimmer, 1995) contributed to a growing plant small subunit data-base, with the trend toward complete 18S DNA sequences rather than partial 18S and 26S RNA sequences. In these early studies, general concordance was reported between phylogenies based on 18S and other molecular and morphological data (e.g. Doyle *et al.*, 1994), but several features of the 18S data, most notably the paucity of phylogenetically informative base positions, suggested the need for more extensive taxon and character sampling.

Nickrent and Soltis (1995) examined patterns of 18S variation in 62 species of flowering plants, compared this variation with that of the plastid gene *rbc*L for the same taxa, and thereby set the stage for further 18S surveys of the angiosperms. Soltis *et al.* (1997a, b, 1998) conducted a more comprehensive survey and maximum parsimony (MP) analyses of 223 angiosperm 18S sequences, generating results largely congruent with those observed in MP analyses of *rbc*L representing 499 angiosperm species (Chase *et al.*, 1993). In general, the amount of phylogenetic signal was lower for 18S. However, many base-calling errors were discovered in previously reported sequences, revealing that the amount of length variability in angiosperm 18S was less than previously thought and that most regions of the sequence could be aligned with little or no ambiguity.

14.2.1.2 Complementary value of 18S and rbcL

Although both sampling and phylogenetic signal were less compared to that for *rbc*L, the 18S analyses of Soltis *et al.* (1997a, b, 1998) represent a significant contribution to angiosperm phylogenetics. The 18S data not only corroborate phylogenetic results of *rbc*L analyses, but do so with greater independence than would two plastid genes (e.g. *rbc*L versus *atp*B) that are more likely to share misleading artifacts. Points of general agreement between the published angiosperm *rbc*L and 18S rDNA parsimony analyses (Soltis *et al.*, 1997a, b; 1998) include: (1) the first-branching families being those with uniaperturate pollen and ethereal oils (i.e. traditional woody magnoliid taxa), and (2) most of the taxa with triaperturate pollen, tannins and alkaloids forming a large terminal clade (the eudicots). Some uniaperturate magnoliid taxa (Winteraceae plus some palaeoherbs) occurred within the eudicots in the 18S tree, but Soltis *et al.* (1997) suggested that this placement was simply an artifact of inadequate taxon sampling. Also, Illiciales, which have triaperturate pollen, were nested among uniaperturate taxa, but morphological evidence indicates that Illiciales evolved triaperturate pollen independently of the eudicots (Doyle *et al.*, 1990).

The 18S data also provided critical corroboration for relationships in *rbc*L-based trees that were considered novel (hence, not previously corroborated) in the context of existing theories and classifications. Computational limitations and apparently limited support for some relationships (Rice *et al.*, 1997) had rendered these (and even some partially corroborated) *rbc*L results subject to debate. For example, the inclusion of *Drosera* and allies in a Caryophyllidae *s.l.* clade was a surprise result of the *rbc*L analysis; its corroboration by 18S (Soltis *et al.*, 1997b) and high bootstrap and jackknife support in combined-data analyses (Soltis *et al.*, 1997a, 1998) suggests that this result is not an artifact of the *rbc*L data, analysis, or evolution. The inclusion of Polygonaceae and Plumbaginaceae in Caryophyllidae in both *rbc*L and 18S trees was less unconventional, but non-molecular cladistic evidence for this relationship was at best weak (Bittrich, 1993), and the relatively high *rbc*L patristic distance between these families and Caryophyllales (Clement and Mabry, 1996) was disconcerting. Other relatively novel *rbc*L-based phylogenetic results corroborated by 18S include the composition and relationships of Saxifragales (including, e.g. Paeoniaceae and Hamamelidaceae, while excluding Hydrangeaceae) and Ericales (including, e.g. Theaceae and Polemoniaceae).

Beyond mere corroboration of phylogenies derived from *rbc*L and other data, a previously unappreciated value of 18S data lies in its role in analyses of combined data-sets. MP analyses combining angiosperm 18S with *rbc*L and/or *atp*B data apparently were much less computationally intensive than for the single genes, thereby allowing much faster analysis (Soltis *et al.*, 1997a, 1998). Moreover, the bootstrap and jackknife support for many nodes became convincingly strong, even though support may have been marginal or lacking in the bootstrap majority rule trees for the separate data-sets.

14.2.1.3 Differences between 18S and rbcL gene trees and patterns of sequence evolution

In the present section, we compare 18S and *rbc*L trees in the context of sequence evolutionary patterns and the assumptions of popular phylogenetics methods. Although angiosperm 18S and *rbc*L trees generally agree, the extent and significance of disagreements has not been explored adequately. Discussion of conflicts between 18S and *rbc*L has focused mainly on topological differences in the MP trees and possible explanations thereof. Some differences are to be expected and probably are not significant, considering that topological differences commonly occur even in analyses using a single molecule but different sampling and/or method. Obviously, in these cases, the 'conflict' cannot be in the data. Analysis of conflict between 18S and *rbc*L is thus a complex issue, as it must discriminate among several variables that could contribute to topological differences in gene trees. These variables include differences in molecular evolutionary processes that might yield gene-specific patterns of sequence variation. Molecular evolutionary processes implied by those patterns should affect the choice of phylogenetic method for separate and combined analyses of each gene, as well as evaluation of support for particular clades and of conflict/concordance of different loci (see section 3 below).

Examples of topological difference in phylogenetic resolution between published *rbc*L and 18S MP trees include the position of the Caryophyllidae *s.l.* among the Asteridae *s.l.* rather than a more commonly accepted rosid affinity. Re-analyses of the *rbc*L data-set have questioned this placement of the Caryophyllidae (Rice *et al.*, 1997), but in neither data-set is the resolution strongly supported. Other differences seen in the 18S analyses relative to the *rbc*L studies include paraphyly of the legumes, the position of *Sagittaria* and *Cuscuta*, and some of the relationships within palaeoherbs and woody Magnoliales. Support for these various placements is not particularly strong and, as pointed out by Soltis *et al.* (1997), many of the different topologies recovered might simply reflect differences in taxon sampling or errors in individual sequences. The early 18S analyses were particularly sparse in taxon sampling; even Soltis *et al.* (1997) included only half the number of taxa sampled for *rbc*L variation by Chase *et al.* (1993). Furthermore, many of the differences between the *rbc*L and 18S analyses involve the placement of groups associated with short branches within major radiations and/or long terminal branches, conditions often associated with uncertain or spurious resolutions (e.g. Rice *et al.*, 1997).

Quantitative comparisons of 18S trees with those from *rbc*L and other genes remain limited. Soltis *et al.* (1998) reported that 18S, *rbc*L, and *atb*B data-sets were significantly incongruent based on their analysis of partition homogeneity. They also indicated that, contrary to expectation, the two plastid genes were more incongruent with each other than either was with 18S. Despite these results, as mentioned above, the number of nodes in the bootstrap (and jackknife; Soltis *et al.*, 1997a) majority rule trees and the amount of support per node increased when 18S was combined with either or, especially, both of the other data-sets. Their analyses

revealed no cases of nodes strongly supported in one data-set and strongly contradicted in another. These results suggest that much of the topological conflict between
18S and chloroplast gene MP trees is mainly random error (i.e. arbitrary resolution at nodes where phylogenetic signal is lacking in one or another data-set). From
the compiled bootstrap data cited in Soltis *et al.* (1997a), we noticed that adding
18S data to either or both of the chloroplast genes decreased support for 19–26%
of the recovered clades, whereas adding either chloroplast gene to the other or the
other plus 18S decreased support for only 11–15% of the recovered clades.
Superficially, these observations suggest that 18S is the least congruent of the three
genes, but they might be an artifact of lower 18S variation. Specifically, Soltis *et
al.* (1997a) included the invariant and uninformative sites in their analysis (D. Soltis,
pers. comm.), and we have found that 'padding' data with uninformative sites can
decrease bootstrap support at substantial proportions of nodes. We do not address
here whether the decreases are significant (see Harshman, 1994); we only propose
that the relatively large number of uninformative sites in 18S might explain the
larger proportion of nodes with decreased bootstrap support.

As noted above, analyses of angiosperm 18S and other genes have been based
mainly on unweighted MP. This method assumes that all substitutions occur in
equal frequency, and that all positions in the sequence evolve independently
(Swofford *et al.*, 1996; Olmstead *et al.*, 1998). Soltis *et al.* (1997b; P. Soltis and
Soltis, 1998) described patterns of among-base and among-site substitution biases
in angiosperm 18S. Among-site variation patterns were examined with respect to
both position along the sequence (primary structure) and that corresponding
to stems versus loops in rRNA secondary structure. They found that MP character
weighting schemes that might counter these biases yielded less resolved, but not
contradictory, 18S trees. Still, these findings provide some indication that the true
18S gene tree is something other than an unweighted MP tree, hence that differences observed between 18S and other gene trees might be at least partially
systematic rather than completely random, i.e. the result of systematic violation of
the assumptions of unweighted MP.

As a means of improving understanding of 18S relative to *rbc*L evolution, we
have conducted preliminary analyses over a subset of angiosperms, specifically the
monocots. Published 18S and *rbc*L trees include many of the major monocot clades
(e.g. Alismatanae, Zingiberales, Poales) and family-level clades (e.g. Poaceae,
Arecaceae, Araceae), but differ in their arrangement of some of these clades. The
present analysis sought to clarify the source of these apparent differences.

Monocot taxon sampling has been greater for *rbc*L than 18S. Chase *et al.* (1995)
performed MP analysis on sequences from 172 genera, 109 of which were represented in the earlier studies of Duvall *et al.* (1993a, b). The *rbc*L MP analyses
indicated monophyly for the monocots and a basic tree structure of:

(*Acorus*,(alismatids/aroids,(dioscorioids,Liliales,Asparagales,commelinids))),

in which all of the named taxa are monophyletic. The commelinid assemblage
included Arecanae, Bromeliales, Poales, Commelinales, and Zingiberales (the ABCZ
clade of Davis *et al.*, 1998), plus some enigmatic members of traditional Asparagales.

All three analyses of monocot 18S presented to date conflict in topology
with the *rbc*L trees. The partial 18S MP analyses of Bharathan and Zimmer
(1995) included 26 genera of monocots dominated by Alismatanae, Aranae, and

Dioscoreales *s.l.*. In contrast, the 26 genera included in the angiosperm-wide 18S MP analyses of Soltis *et al.* (1997) were predominantly Zingiberales and Asparagales. Despite the differences in taxa sampled, there was considerable congruence between these two studies and shared conflicts with the *rbcL* analyses, most notably in the nesting of *Acorus* among the monocot outgroups, non-monophyly of Dioscoreales *s.l.*, and non-monophyly of the commelinid monocots. A further notable feature of the Soltis *et al.* (1997) tree was the position of the single orchid sampled (*Oncidium*) as sister to all other monocots. Hahn *et al.* (1996) analysed an 86-taxon 18S data-set, which reduced the conflict between the various 18S and *rbcL* phylogenies, notably in supporting monocot monophyly. Nonetheless, several conflicts remained, including non-monophyly of Dioscoreales *s.l.*, and the position of the orchids. These studies have not quantified rigorously the degree of conflict between 18S and *rbcL*, e.g. in terms of data support for conflicting clades. Nonetheless, all authors have noted the possible sources of conflict, namely sampling differences and, in the 18S analyses, erratic branch lengths and biases in patterns of base composition and substitution.

Partially to address the differential sampling issue and partially because of computational limitations, we analysed 18S and *rbcL* sequences from the same 40 monocot genera, usually the same species in each genus, representing most major taxa of monocots (Appendix). We also included *Asarum* as the single outgroup. All of our analytical procedures were performed using PAUP* 4.0 version d64 (Swofford, in press). Conflict in tree topologies was examined by cross-comparison of trees optimised using each data-set. We also performed statistical tests that evaluate whether such topological differences could arise by random error in sampling from either data-set. Conflict in the data (as opposed to the trees) was examined using the incongruence length difference (ILD) test (Farris *et al.*, 1994), as well as by bootstrapping. Systematic error was examined by performing analyses under two different models of sequence evolution. One model assumes that all sequence base positions (sites) have evolved independently and identically, with no substitution bias. These assumptions are inherent in the unweighted MP method. The other model assumes that the amount of evolution is correlated among the sites, that some sites consistently evolve faster than others, that some substitutions are more likely than others, and that substitution biases are influenced by base frequencies. Analysis using this model was carried out using maximum likelihood (ML), with the assumption that the values of the various parameters can be approximated adequately on the basis of substitution patterns observed across a 'reasonable' (in this case, an MP) tree topology (Yang, 1996; Yang *et al.*, 1995). Parameters were re-estimated using alternative trees, but these were found to be only trivially different and appeared to have no appreciable effect on the ML scores of the alternative trees. We also performed analyses on the combined data-sets. The bootstrap analysis of the combined data-set is useful for evaluating both random and systematic error in the separate data analyses. In particular, spurious groupings in the separate analyses result from error that causes homoplasy to be mistaken for homology. If the causes of error in the separate analyses are unrelated, then the impact of the error in the combined analysis should be reduced relative to the underlying phylogenetic signal, resulting in altered bootstrap values (cf. Hillis, 1996; Soltis *et al.*, 1998). Our analysis did not examine other types of systematic error, such as that resulting from compensatory mutation to maintain 18S rRNA secondary structure (P. Soltis and Soltis, 1998) or base compositional bias (see below).

Alignment for the *rbcL* data was unambiguous and included 1321 base positions, with 481 (36%) variable and 320 (24%) parsimony-informative sites. Alignment of 18S was substantially unambiguous, but required insertions/deletions (mostly single-base) at 38 positions, five of which involved only the outgroup, *Asarum*. Gap scores were treated as missing in our MP analysis. The four length-variable regions of 18S excluded from the broader angiosperm analysis (Soltis *et al.*, 1997b) appeared to be substantially alignable among the analysed monocots, hence we did not exclude them. The 18S data-set included 1721 base positions, with 466 (27%) variable and 215 (12%) parsimony-informative sites. Each data-set exhibited strong substitution biases and among-site rate heterogeneity. Likelihoods were significantly improved using six substitution parameters (a general time-reversible model; Swofford *et al.*, 1996) plus a discrete-gamma correction for among-site rate heterogeneity (Yang, 1996) compared to a model with fewer substitution parameters and/or no among-site rate heterogeneity (data not shown). The estimated reversible substitution rates used in the analysis are provided in the ML tree figure captions referred to below. For the *rbcL* data the rates were *ca.* 6:3:1 CT:AG:tv (where tv refers to each of the transversion rates), while the 18S rates exhibited a similar pattern but were more erratic and extreme (e.g. $AG \approx AT \approx 2CG$, the last two being transversions). The value of α, the gamma shape parameter, was *ca.* 0.21 and 0.18 for the *rbcL* and 18S data-sets, respectively.

Maximum parsimony searches for the *rbcL*, 18S, and combined data-sets were conducted using a heuristic search with simple taxon addition, a stepwise starting tree, and the tree bisection–reconnection (TBR) algorithm, but with no MAXTREES limit. We also searched using 100 or more replicates with random sequence addition. The MP bootstrap analysis was conducted using 1000 replicates with random sequence addition, 100 trees held at each addition step, and MAXTREES limited to 100 per replicate. ML searches were conducted using the TBR with starting trees derived three ways: neighbour-joining and both as-is and random sequence addition, using stepwise addition with one tree held at each addition step. The ML bootstrap used a random addition sequence with no branch swapping, using the substitution model estimated for the full data-set (Yang *et al.*, 1995). Despite the much reduced degree of tree optimisation, the combined data ML bootstrap required about 100 times as much computation time as the MP bootstrap, and the single data-set ML bootstraps required about 50 times as much. Results from several bootstrap searches initiated using different seeds were concatenated to yield the 1000 ML replicates.

The ILD test compares the sum of MP tree lengths of separate data-sets with that of data-sets of equal size formed by randomly partitioning the characters in the original data-sets (Farris *et al.*, 1994). If phylogenetic signal in the data-sets is incongruent, the tree length sum from the contrived data-sets sets will be greater on average than for the original data-sets. Our application (MP criterion, 1000 replicates, MAXTREES = 100) yielded a highly significant statistic for incongruence between *rbcL* and 18S ($p = 0.001$), consistent with results of Soltis *et al.* (1998) for angiosperms as a whole. This statistic alone does not indicate whether the incongruence reflects strong or trivial conflict, i.e. random data-sets might also show significant incongruence by this measure, even though neither set strongly supports any particular topology. The statistic does demonstrate, however, that the differences in the *rbc*L and 18S topologies cannot be attributed simply to poor resolvability of the latter (cf. Cunningham, 1997). Interestingly, partitioning the *rbc*L data into two comparably variable data-sets yielded a *p* value of 0.08, suggesting that the two *rbc*L partitions are congruent with each other only barely. Similar partitioning of the 18S data yielded a *p* value of 0.16.

Trees and bootstrap values from our analyses are shown in Figs 14.2 and 14.3 and are summarised in terms of major taxonomic groupings in Fig. 14.4 (cf. Appendix). For MP trees, Table 14.1 summarises and compares tree scores across the different methods and data-sets, as well as the results of tests of topological differences. For both the MP and ML criterion, these include the Kishino–Hasegawa test (Kishino and Hasegawa, 1989). For the MP criterion, these also include the non-parametric Templeton (Wilcoxan signed-rank; Templeton, 1983) and winning-sites (sign; Prager and Wilson, 1988) tests. These tests examine site-wise support for alternative topologies and, effectively, the degree to which the optimisation score difference between topologies reflects random error in sampling of the underlying data. In our application, we examined those suboptimal trees that happened to be optimal by a different data-set and/or optimisation criterion.

The default PAUP* heuristic search procedure found only 6/36 of the *rbc*L MP trees, 8/24 of the 18S trees, and 0/44 of the combined data trees. The random addition procedure yielded all the trees we found for each data-set. The homoplasy indices of the MP trees for the separate and combined data-sets are essentially identical (Fig. 14.2, caption). The strict consensus of the *rbc*L trees (Figs 14.2, 14.4) is similar to that seen in the more extensive studies of Duvall *et al.* (1993a, b). Although unresolved in the consensus, some of the trees include an ABCZ clade subtended by the lilioid grade (Liliales, Asparagales, Orchidaceae, Dioscoreales). The consensus of the 18S MP trees (Figs 14.2, 14.4) is resolved less than that for *rbc*L. Several of the higher taxonomic groups are non-monophyletic, most notably the sampled Commelinales, Dioscoreales, and, equivocally, Alismatanae. In the various resolutions, and in contrast to the results of Soltis *et al.* (1997b), *Acorus* is never sister to the remaining monocots; it is always sister to a clade comprising different combinations of clades emerging from the large unresolved node (Figs 14.2, 14.4), although never sister to the largest of these clades. One Orchidaceae member is always sister to the remaining monocots, with the other two sometimes diverging at the next node. The ABCZ clade is absent. The consensus of the combined data MP trees (Figs 14.2, 14.4) resembles but is resolved less than the *rbc*L consensus. The influence of 18S is evident in these trees, in which the ABCZ clade includes Dioscoreales. The computer search time for the combined data MP analysis was much less than for 18S alone, but greater than for *rbc*L alone. This contrasts with results of the angiosperm-wide analysis, in which run times for combined 18S/*rbc*L data were much less than for either data-set alone.

The *rbc*L and combined data ML searches each produced three trees, differing only in the position of *Heliconia* among the Zingiberales. The *rbc*L ML topology (Figs 14.3, 14.4) is similar to the MP tree, differing mainly in rearrangements of short branches, one of which results in non-monophyly of Liliales. In four-taxon simulations, ML plus gamma has been shown to perform poorly relative to MP in

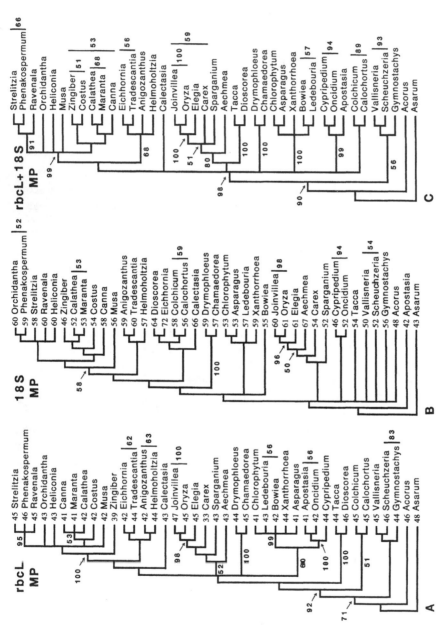

Figure 14.2 Maximum parsimony (MP, unweighted) consensus trees for *rbcL*, 18S rDNA, and combined data for selected monocots. Tree scores are compared in Table 14.1. Numbers along the branches are bootstrap values. Numbers to the right of the branch tips are %GC for the branch tips. A. Strict consensus of 36 trees for the *rbcL* data: consistency index excluding uninformative characters (CI) = 0.44, retention index (RI) = 0.47. B. Strict consensus of 24 trees for the 18S rDNA data: CI = 0.37, RI = 0.43. C. Strict consensus of 44 trees for the combined data: CI = 0.36, RI = 0.44.

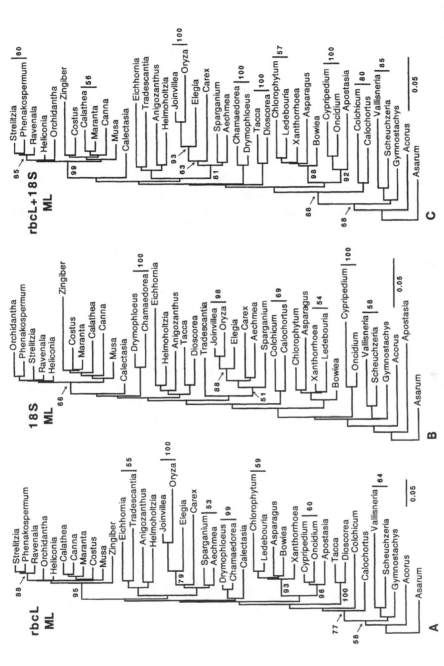

Figure 14.3 Maximum likelihood (ML) trees for *rbcL*, 18S rDNA, and combined data for selected monocots. Tree scores are compared in Table 14.1. A. One of three ML phylograms for the *rbcL* data. The substitution rates are: AC (1.07), AG (4.196), AT (0.702), CG (1.523), CT (6.423), GT (1.0); alpha (0.2641). B. ML phylogram for the 18S rDNA data. The substitution rates are: AC (0.881), AG(1.459), AT (1.532), CG (0.850), CT (6.361), GT (1.0); alpha (0.1799). C. One of three ML phylograms for the combined data. The substitution rates are: AC (1.004), AG (2.833), AT(1.039), CG (1.082), CT (6.082), GT (1.0); alpha (0.2124).

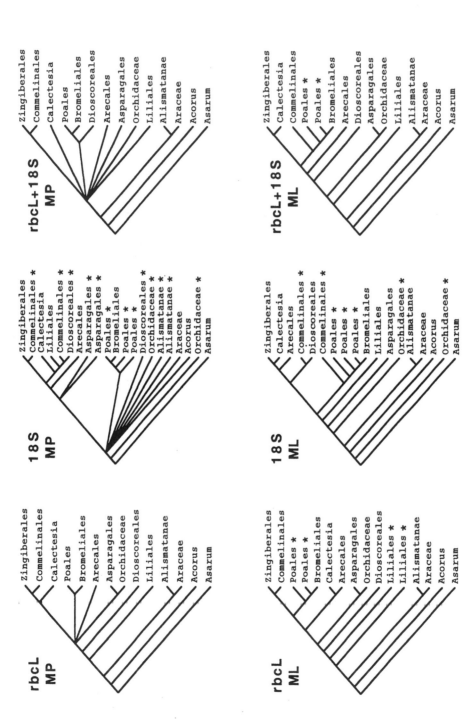

Figure 14.4 Summary of relationships in Figs 14.2 and 14.3. Relationships of higher taxa, classified according to Thorne (1992; Appendix) are indicated. Taxa that appear more than once in the trees (i.e. non-monophyletic taxa) are highlighted with asterisks.

Table 14.1 Comparison of MP and ML tree scores for all monocot *rbcL*, 18S, and combined data trees optimised using each data and optimisation criterion. Scores for MP criteria are numbers of steps (including autapomorphies); scores for ML criteria are negative log likelihoods (–lnL). For all heterologous data and/or optimisation comparisons, test statistics for the Kishino–Hasegawa (KH; MP and ML criteria), and Templeton and winning-sites (T and WS; MP only) tests are indicated. Test statistics for homologous comparisons (i.e., different optimal trees using the same data and optimisation) are not shown, but a few were slightly less than 1.0. Values of p ≤ 0.05 are underlined. Every optimal MP and ML tree was used as the test tree in separate comparisons. The range of test statistics shown represent the extreme values for all combinations of all test and compared trees for each criterion.

	rbcL-MP trees	*18S-MP* trees	*comb-MP* trees	*rbcl-ML* trees	*18S-ML* trees	*comb-ML* trees
rbcL-MP criterion	1498	1648 – 1672 KH p = < 0.0001 T p = < 0.0001 WS p = < 0.0001	1505 – 1513 KH p = 0.04 – 0.35 T p = 0.06 – 0.41 WS p = 0.06 – 0.58	1509 – 1510 KH p = 0.15 – 0.19 T p = 0.14 – 0.25 WS p = 0.23 – 0.52	1595 KH p = < 0.0001 T p = < 0.0001 WS p = < 0.0001	1511 KH p = 0.06 – 0.13 T p = 0.11 – 0.18 WS p = 0.09 – 0.18
18S-MP criterion	1246 – 1255 KH p = < 0.0002 T p = < 0.0001 WS p = < 0.0004	1192	1214 – 1222 KH p = 0.01 – 0.03 T p = 0.02 – 0.05 WS p = 0.03 – 0.12	1247 KH p = < 0.0001 T p = 0.0001 WS p = < 0.0003	1207 KH p = 0.15 – 0.20 T p = 0.16 – 0.18 WS p = 0.60 – 0.72	1226 KH p = 0.003 – 0.005 T p = 0.001 WS p = 0.01 – 0.03
comb-MP criterion	2743 – 2753 KH p = 0.01 – 0.16 T p = 0.02 – 0.21 WS p = 0.01 – 0.12	2840 – 2864 KH p = < 0.0001 T p = < 0.0001 WS p = < 0.0001	2727	2756 – 2757 KH p = 0.01 – 0.03 T p = 0.02 – 0.03 WS p = 0.02 – 0.03	2802 KH p = < 0.0001 T p = < 0.0001 WS p = < 0.0002	2737 KH p = 0.20 – 0.36 T p = 0.26 – 0.40 WS p = 0.22 – 0.61
rbcl-ML criterion	9340.29 – 9353.63 KH p = 0.25 – 0.49	9640.11 – 9679.95 KH p = < 0.0001	9358.41 – 9382.69 KH p = 0.02 – 0.18	9332.14	9537.27 KH p = < 0.0001	9347.28 KH p = 0.19
18S-ML criterion	8890.23 – 8906.82 KH p = < 0.0005	8818.34 – 8826.42 KH p = 0.30 – 0.39	8839.31 – 8850.50 KH p = 0.03 – 0.10	8892.26 KH p = < 0.0001	8799.73	8847.60 KH p = 0.01
comb-ML criterion	18491.58 – 18517.50 KH p = 0.01 – 0.05	18730.94 – 18781.80 KH p = < 0.0001	18462.90 – 18483.99 KH p = 0.13 – 0.48	18476.97 KH p = 0.11	18596.05 KH p = < 0.0001	18451.40

cases where long branches are sister in the true tree (Yang, 1998). The present results for Liliales may represent a practical example. The 18S ML search found one optimal tree. Ignoring the position of *Apostasia* and Dioscoreales, the basic structure is similar to the *rbcL* trees (the position of *Acorus*, monophyly of higher monocots, a lilioid grade, and an ABCZ clade). The basic structure of the combined data ML tree resembles the *rbcL* MP consensus more than the combined data MP consensus.

Comparison of the scores and tests of significant difference in topologies across the different data-sets and optimisation criteria (Table 14.1) bears out the impressions from visual inspection of the figures. These tests evaluate score differences in the optimisation of each character on the alternative trees (each optimal MP and ML tree for one data-set against, arbitrarily, each tree from the other data-sets and optimisation criteria). The tests do not evaluate conflict between the two data-sets *per se*, because there is likely to be a shared set of trees that these tests would not reject using any tree optimised using any combination of data-set and criterion. Moreover, the results should be interpreted in the context of the assumptions and limitations of each test (Templeton, 1983; Swofford *et al.*, 1996; Yang *et al.*, 1995). For example, the non-parametric Templeton and winning-sites tests count the numbers of optimal versus suboptimal character state reconstructions on alternative trees, and alternative trees may score similarly if homoplasy is high (Templeton, 1983). The parametric Kishino–Hasegawa test (Kishino and Hasegawa, 1989) squares each of the sitewise differences in optimisation score, hence it may be more intolerant of a tree with a few especially poor sitewise scores (ignoring many possible sources of error for these sites) than a tree with numerous but only slightly suboptimal sitewise scores. In any case, the tests consistently indicate that the *rbcL* topologies are highly improbable given the 18S trees/data and vice versa, regardless of optimisation criterion. The *rbcL* and 18S trees optimised according to one evolutionary model, however, are not significantly improbable given the trees/assumptions of the other. Although all 18S topologies are highly improbable given either *rbcL*-MP or *rbcL*-ML, for both criteria the 18S ML topology tree scores are markedly better than for the 18S MP topologies, consistent with the observation that the 18S ML topology more closely resembles *rbcL*-based topologies. Assuming that true relationships among the major monocot taxa are consistent more with the *rbcL* topologies, these results suggest that at least some of the difference between the 18S ML and 18S MP topologies is the result of systematic error in the latter. We caution that many of the relationships considered in our analysis cannot be considered known, but we note that the 18S topologies conflict with 4–5 nodes strongly supported in the combined data bootstrap analysis (Table 14.2). We included the test statistics for the combined data criteria and trees, but the interpretation of these results is problematic for two reasons. First, the combined data obviously are not independent of the respective separate data components. Second, the test statistics are borderline significant, which suggests that a sequential Bonferoni correction for multiple tests should be applied to examine whether the extreme values could have been derived by chance (Rice, 1989). However, we fully expect that the test statistics involving combined data or trees should be intermediate between those for the separate data and trees. We would be concerned with significance of statistics comparing MP and ML for combined data (there apparently is

Table 14.2 Bootstrap values for *rbcL*, 18S, and combined data MP and ML analyses. Groups recovered in any of the six bootstrap majority rule trees are shown (cf. Figs 14.2–4, Appendix). A hyphen denotes absence of the group in a bootstrap majority rule tree, an asterisk denotes groups also absent in the optimised MP consensus or ML tree, and parentheses denote one group present in a bootstrap but not in an optimised tree.

	rbcL MP / ML	*18S* MP / ML	*rbcL + 18s* MP / ML
Zingiberanae	100 / 95	58 / 66	99 / 99
Strelitzia/Phenakospermum/Ravenala	95 / 88	* / *	91 / 85
Strelitzia/Phenakospermum	– / –	* / *	66 / 60
Orchidantha/Phenakospermum	* / *	52 / –	* / *
Zingiber/Costus	* / *	* / *	51 / –
Calathea/Maranta/Canna	53 / –	* / *	– / –
Calathea/Maranta	– / *	53 / *	68 / 56
Zingiber/Costus/Calathea/Maranta/Canna	* / *	– / –	53 / –
2 Arecaceae	100 / 99	100 / 100	100 / 100
4 Commelinanae	– / –	* / *	68 / –
Anigozanthus/Helmoholtzia	53 / –	* / *	* / –
Eichornia/Tradescantia	62 / 55	* / *	56 / *
Aechmaea + Poales	52 / –	– / 51	80 / 81
Carex/Elegia/Joinvillea/Oryza	– / –	* / *	59 / 63
Aechmaea/Elegia/Joinvillea/Oryza	* / *	59 / *	* / *
Elegia/Joinvillea/Oryza	98 / 79	96 / 88	100 / 93
Joinvillea/Oryza	100 / 100	98 / 98	100 / 100
Sparganium/Aechmea	* / 53	* / *	* / –
2 Dioscoreales	100 / 100	* / –	100 / 100
2 Liliales	51 / –	59 / 69	89 / 80
Aparagales excl. Orchidaceae	99 / 93	– / –	100 / 98
Asparagales incl. Orchidaceae	60 / –	* / –	– / –
Bowiea/Ledebouria	56 / *	* / *	57 / *
Chlorophytum/Ledebouria	* / 59	* / *	* / 57
Xanthorrhoea/Ledebouria	* / *	– / 54	* / *
Chlorophytum/Ledebouria/Xanthorrhoea/Bowiea	* / *	* / *	– / (51)
3 Orchidaceae	100 / 96	* / *	99 / 92
Cypripedium/Oncidium	* / 60	94 / 100	94 / 100
Apostasia/Oncidium	56 / *	* / *	* / *
2 Alismatanae	83 / 64	54 / 58	93 / 85
Alismatanae + Araceae	– / –	* / –	56 / –
Monocots excl. Alismatanae, Araceae, *Acorus*	92 / 77	* / *	99 / 88
Monocots excl. *Acorus*	71 / 58	* / *	90 / 68

not any), but we interpret test statistics comparing combined data with single data results only in terms of trends. Under both the MP and ML trees/criteria for the combined data, 18S topologies fare worse than the *rbcL* topologies. Likewise, the combined data topologies fare worse for the 18S trees/data than for the *rbcL* trees/data.

The bootstrap analyses (Figs 14.2–4, Table 14.2) suggest that the bulk of the difference in *rbcL* and 18S topologies arises from the essentially randomised variation in each data-set. Specifically, relatively few nodes in either 18S tree are recovered in ≤ 50% bootstrap replicates. There are a few instances where bootstrap support of 50–60% using one data-set is countered by 50–60% support for a conflicting clade using the other. These conflicts probably are insignificant, especially considering that there are two instances of such conflict between the rbcL MP and ML bootstraps. The combined data bootstraps parallel the results of Soltis et al. (1998) in that the number of nodes supported and the degree of support is greater in the combined than in the separate data-sets. The rbcL MP and ML majority rule bootstrap trees share 12 nodes, the 18S MP and ML bootstraps trees seven, while the combined data bootstrap trees share 17. The increase in support can be substantial, e.g. for the Liliales and Bromeliales/Poales clades. The combined data bootstrap support for a Strelitzia/Phenakospermum clade is not so convincing, but still notable considering that the clade is absent in the separate data bootstraps and contradicted in all 18S topologies, with 52% support for a conflicting node in the 18S MP bootstrap. While nodes supported by rbcL dominate the combined data bootstrap, there are two cases (Liliales and Cypripedium/Oncidium) where 18S appears to provide the principal support. Support for monophyly of monocots minus Acorus is increased substantially in both combined data bootstraps, but is substantially less in the ML than in the MP bootstraps. This discrepancy might reflect the differences in our MP and ML bootstrap procedures (see above), but might also indicate that support for this relationship, even if correct, might be attributable partially to overestimation of support by MP (Sullivan et al., 1997).

Our results resemble those from other organisms in which trees optimised on small subunit rDNA data have deviated from more generally accepted and otherwise better supported topologies (e.g. Huelsenbeck and Bull, 1996; Sullivan et al., 1996). Based on our impression of the degree of topological discordance between 18S and rbcL MP trees for monocots compared to that for other angiosperms (Soltis et al., 1997b), the monocot discordance appears to be at least as much and possibly more. The fact that in monocots, addition of 18S to rbcL data actually slowed the MP search also indicates that 18S had some confounding effect. Our results also indicate that the topologies optimised with 18S versus rbcL truly are different, i.e. that the 18S trees are improbable given the rbcL data and vice versa. The data per se, however, probably are not truly contradictory.

The difference in optimised topologies probably is derived mainly from random error, as suggested by the bootstrap results. Random error also is suggested by branch lengths in Fig. 14.3. For both molecules, disproportionate amounts of divergence tend to occur on the outer (terminal) branches rather than the inner. This pattern amounts to the classic long-branch problem, in which the proportion of random (homoplastic) similarity between terminals becomes high relative to the proportion of homologous (synapomorphic) similarity (Swofford et al., 1996). The ratio of terminal to interior branch lengths appears on average to be greater for 18S than for rbcL. Divergences of 18S are generally about half that of rbcL, as indicated by the scale bar, but there are several cases where 18S divergences to the root of major groups amount to much less than half that for rbcL, e.g. Asparagales, Zingiberales, Dioscoreales. In other words, the 18S branches are shorter than would be expected if 18S and rbcL divergences were strictly in proportion. The ML trees also suggest that divergence is more erratic in 18S than rbcL, i.e. there are more cases where terminal branch lengths among closely related taxa are substantially different.

At least some of the topological differences appear to derive from systematic error in the MP analysis, suggested by both the strong violation of MP assumptions evident in the 18S data and the apparent improvement in 18S topology using an ML model that accounts for empirically observed data patterns. We did not identify the exact parameter(s) most responsible for the systematic error. Although the substitution and among-site rate heterogeneity biases evident in 18S are more extreme and erratic than those evident in *rbc*L (Fig. 14.3, caption), they are of similar pattern and magnitude. Yet, the ML model seems to have had much less effect on *rbc*L topology. Thus, an important component of the systematic error might be simply the MP assumption that sites evolve independently, which leads to underestimation of divergence and the classic long-branch problem (Swofford *et al.*, 1996). As we noted, our analyses did not test for other systematic errors that might contribute to topological conflict, including functional constraints on rRNA evolution or effects of base composition.

In conclusion, our results extend to angiosperms, earlier observations on the evolution of the ribosomal small subunit RNA and its performance in phylogenetic reconstruction under different methodological assumptions. Our results re-emphasise that phylogenies based on 18S should be interpreted with caution. In many instances, 18S may simply lack signal adequate to resolve problems of interest, but the sequence also appears to be especially sensitive to violation of methodological assumptions. Increasing taxon sampling might improve the 18S trees (but see Kim, 1998), but we expect that the general pattern of difference in performance of 18S and *rbc*L would persist. Our results also extend to monocots the observations of Soltis *et al.* (1998) that 18S can still provide signal that strongly complements data from other sequences in combined analyses. In monocots, however, the combined data that includes 18S became more sensitive to violation of methodological assumptions, as the combined 18S-*rbc*L MP trees are resolved less and apparently less accurately than MP trees based on *rbc*L alone. We have considered some, but not all, possible causes of the topological discrepancies. As for monocot phylogeny *per se*, our results suggest that even combined 18S-*rbc*L analyses may be inadequate for resolving some relationships. For example, monophyly of Commelinales and the ABCZ clade, as well as the composition (i.e. position of Dioscoreales) and divergence pattern of the lilioid grade, are not supported strongly by either data-set or their combination. Likewise, relationships within major groups, e.g. Zingiberales and of Bromeliaceae among Poales, are not resolved. Improved sampling might increase resolution, but these same relationships were poorly resolved in a *rbc*L data-set representing more than four times as many taxa (Chase *et al.*, 1995). Thus, future efforts at monocot phylogenetic reconstruction might be directed better towards analyses using additional genes.

14.2.2 The rDNA ITS region, including the 5.8S rRNA gene

The rDNA ITS region encodes the 5.8S rRNA and the flanking rRNA spacers that are cleaved or otherwise digested during the assembly of the ribosomal subunits. Selective constraints are thus lower on the spacers versus the coding regions, leading to higher intertaxon variability. Sequence and structural elements in the spacers are critical nonetheless for several steps in the production of functional ribosomes (Liang and Fournier, 1997; Peculis, 1997). As suggested above, rDNA-ITS probably has become the most widely exploited molecular marker in plant systematics, displacing *rbc*L for this distinction (cf. Davis *et al.*, 1998). In September, 1998, we searched GenBank (the sequence database maintained by the National Center for Biotechnology Information, or NCBI) and counted the number of angiosperm species for which there are ITS and/or *rbc*L sequence documents. We used the NCBI Entrez database browser to perform a Boolean string search to account for all of the various document descriptors (e.g. 'ITS' versus 'internal transcribed spacer', as well as 5.8S sequences with ITS not explicitly indicated). Through subsequent manipulation and screening of the output, we found that *ca.* 2900 angiosperm species were represented. A similar search for *rbc*L documents yielded *ca.* 2300 species. We also examined abstracts from the most recent major meetings, and we estimate that the number of species represented in unsubmitted or unreleased sequences is at least 25 times greater for ITS than for *rbc*L (data not shown; the numbers of abstracts mentioning ITS and *rbc*L are similar, but the ITS abstracts almost all refer to newly sampled taxa, whereas the *rbc*L abstracts almost all refer to updated analyses of pre-existing data).

Important issues that have been raised in considerations of ITS include (cf. Baldwin *et al.*, 1995; Buckler *et al.*, 1997): sequencing technology; quantity/quality of diagnostic signal; and paralogous and recombined polymorphisms. We will consider these points in turn:

14.2.2.1 Sequencing ITS

In some taxonomic groups, the ITS region reportedly has been amplified easily and sequenced cleanly using either manual or automated sequencing technology. In other groups, generation of ITS sequence has proven to be a Herculean exercise in patience. The GC content of angiosperm ITS is almost always higher than 50%, and in some cases exceeds 75% (Hershkovitz and Zimmer, 1996, 1997). Substantial portions of both ITS1 and ITS2 in rRNA transcripts are believed to fold into helices (double-stranded stems with terminal loops) or more complex structures (Conn and Draper, 1998) via intramolecular base pairing, and this folding potential might block polymerisation steps during amplification and sequencing. These problems can be confounded by infragenomic ITS polymorphism, which necessitates cloning of PCR products and sequencing of multiple clones. Even then, PCR and/or cloning steps can strongly favour one of the forms, sometimes an apparent pseudogene (as evidenced by anomalous divergences and extraordinary deletions and/or substitutions at otherwise highly conserved positions) or organismal contaminant, so that the constitutively functional sequences are not obtained (Hershkovitz and Lewis, 1996; Buckler *et al.*, 1997; M. Hershkovitz, unpublished).

For especially recalcitrant sequences, some of the sequencing problems have been ameliorated using dimethylsulfoxide (DMSO) in PCR reactions and advanced chemistry for automated cycle-sequencing (e.g. Buckler *et al.*, 1997; Hershkovitz and Zimmer, 1997). Still, we have found that positively and usually predictably incorrect base calls commonly occur in individual dye chromatographs, underscoring the necessity for sequencing both strands of PCR or cloning products (see Kellogg and Juliano (1997) regarding the evidence for and consequences of base-calling errors for the much less biochemically recalcitrant *rbc*L gene).

14.2.2.2 Quantity/quality of ITS phylogenetic signal

There can be no doubt as to the tremendous utility of ITS sequences and their contribution to angiosperm phylogenetics at lower taxonomic levels (Baldwin *et al.*, 1995). The accumulating data indicate that ITS is suited best for diagnosing relationships among closely related genera and infrageneric groups. In many cases, ITS divergence between closely related species is too little to resolve relationships with high statistical confidence or at all (Baldwin *et al.*, 1995; Baldwin and Markos, 1998). In particular, the low divergence is coupled with the small length of the spacers, usually 200–300 bases for ITS1 and 180–240 bases for ITS2. Divergence values between morphologically distinct species are sometimes < 1% (e.g. Hershkovitz and Zimmer, submitted ms.), which translates to fewer than five substitutions, some of which usually are unequivocal autapomorphies. Interspecific ITS divergences within a genus can be highly variable (0–30%), however, even among different taxa in a single data-set (Susanna *et al.*, 1995; Möller and Cronk, 1997; Pridgeon *et al.*, 1997; Hershkovitz and Zimmer, submitted ms.).

The upper bounds of divergence at which ITS holds practical phylogenetic utility are delimited by length rather than sequence variability. ITS base positions that vary among closely related species commonly are unalignable among more divergent taxa, because the most rapidly evolving regions also are those most prone to length variation. As a result, increasing divergence may not be accompanied by an equiproportional increase in the number of alignable informative sites. Although the insertion/deletion (indel) rate has never been quantified rigorously, we estimated from infrageneric comparisons among Portulacaceae (Hershkovitz and Zimmer, 1997, submitted ms.) that an indel accompanied every 8–10 substitutions (but with high variance). Given that ITS1 and ITS2 together typically comprise *ca.* 450 bases, this amounts to one indel per 2% divergence. Indel positions, however, vary in different lineages, so that indels can occur in a substantial proportion of positions in a multiple sequence alignment. For example, in a 45-sequence alignment representing the portulacaceous alliance (Hershkovitz and Zimmer, 1997), the aligned length of ITS1 plus ITS2 (excluding 5.8S) was 556. Of these 556 aligned sites, 376 (68%) included an indel in at least one taxon. Of the 205 informative sites (excluding four in 5.8S), only 62 (30%) were free of indels. Summarising nine data-sets, Baldwin *et al.* (1995) reported that the proportion of aligned sites with indels in *more* than one taxon ranged from < 1 to 40%, apparently approximately correlated with maximal sequence divergence in the data-set. Obviously, phylogenetic analysis of only indel-free sites (cf. Wheeler *et al.*, 1995; Swofford *et al.*, 1996) would constrain phylogenetic information severely. Such a data transformation probably is unnecessary, given that often only a minor proportion of taxa have an indel at a given site. In the portulacaceous data-set, alignable indels were used as phylogenetic characters, sequence regions ambiguously alignable across all taxa were aligned within taxon subsets, and substitution variation within these subsets was used to resolve and root subtrees (Hershkovitz and Zimmer, 1997).

Within the useful bounds of ITS divergence, phylogenetic patterns themselves may account for at least part of the failure of ITS to consistently resolve interspecific relations. In particular, the occurrence of short internal branches in phylograms is consistent with the prediction that rapid species radiations are common or even the norm in evolution (Futuyma, 1997). Such a pattern is evident among western American Portulacaceae (Fig. 14.5; Hershkovitz and Zimmer, submitted ms.), in which ITS diagnoses generic and subgeneric clades also diagnosed by morphology, but the species and clade branches tend to emerge from dense clusters of very short branches.

Although it is clear that ITS is a tool primarily for resolving low-level divergences (Baldwin *et al.*, 1995), the phylogenetic signal is not saturated at greater (familial and higher) level divergences (Hershkovitz and Lewis, 1996; Hershkovitz and Zimmer, 1996, 1997). In some cases, signal conservation is an artifact of classification. Some taxa traditionally classified in different families are diverged less than taxa classified in the same family. For example, Hershkovitz and Zimmer (1997) showed that certain Cactaceae were diverged less from certain Portulacaceae than most Portulacaceae genera are from each other (Fig. 14.6). In other cases, we have shown that lineages can retain diagnostic ITS2 sequence motifs through > 200 million years of divergence (Hershkovitz and Lewis, 1996; Hershkovitz and Zimmer, 1996, 1997). For the present paper, we re-examined our earlier angiosperm ITS2 analysis (Hershkovitz and Zimmer, 1996) in order to compare more recently released sequences representing a broader range of angiosperm taxa. We found that the five conserved ITS2 motifs reported previously occur in all of the newly examined taxa.

We also updated our angiosperm ITS2 guide tree analysis (Fig. 14.7; see Hershkovitz and Zimmer, 1996) to add 72 (for a total of 147) diverse angiosperm species. The guide tree is a neighbour-joining tree, based on pairwise optimal alignment scores, that determines sequence addition order for multiple alignment in CLUSTAL W (Thompson *et al.*, 1994). As in our earlier study, we again emphasise that the ITS2 guide tree is not intended as a phylogenetic estimate, but as a means of comparing ITS2 divergences across angiosperms and for inferring the existence of conserved sequence motifs. Moreover, the distance data are not recovered (although CLUSTAL program modification could enable this), so alternative trees cannot be evaluated. We examined a range of high gap insertion (10–20) combined with low gap extension (0.1–1.0) penalties so that shared sequence motifs would be aligned regardless of overall sequence length. In our earlier analysis, we used CLUSTAL W version 1.74. Penalty calculations have been modified in more recent versions, and we found that this affects guide tree calculation. To avoid the need for parameter re-optimisation, we used the older version for the present

Figure 14.5 Maximum likelihood (ML) tree of western American Portulacaceae ITS region sequences (Hershkovitz and Zimmer, submitted ms.). The taxonomy follows Hershkovitz (1993), except that the name *Parakeelya ptychosperma* is used for this member of an Australian genus previously referred to as *Rumicastrum* (Hershkovitz and Zimmer, 1997). The tree was derived using PAUP* 4.0 version d63 according to Hershkovitz and Zimmer (1997). A heuristic search using the tree bisection-reconnection algorithm was conducted using ML under a general time-reversible (GTR) substitution model with among-site rate heterogeneity. ML estimates of the GTR substitution parameters, gamma curve shape, and proportion of invariant sites were estimated over a maximum parsimony tree. Numbers along the branches are bootstrap percentages for 1000 replicates using MP. Numbers to the right of the tree are %GC among parsimony-informative sites. Permutation tail probability (Faith, 1991; MP, 100 replicates, 100 trees held during stepwise addition, MAXTREES = 100) tests were conducted over the entire data-set (p = 0.01) and for a subset including the underlined taxa (p = 0.36).

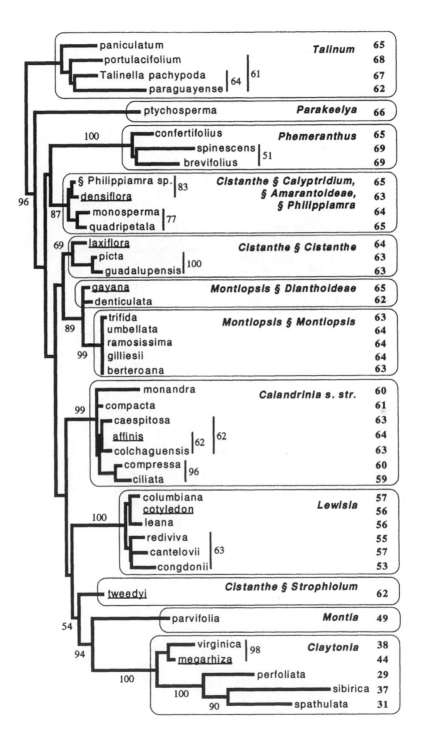

paniculatum	*Talinum*	65
portulacifolium		68
Talinella pachypoda	64	61	67
paraguayense		62
ptychosperma	*Parakeelya*	66
confertifolius	*Phemeranthus*	65
spinescens	51	69
brevifolius		69
§ Philippiamra sp.	83	*Cistanthe § Calyptridium,*	65
densiflora	*§ Amarantoideae,*	63
monosperma	77	*§ Philippiamra*	64
quadripetala		65
laxiflora	*Cistanthe § Cistanthe*	64
picta		63
guadalupensis	100	63
gayana	*Montiopsis § Dianthoideae*	65
denticulata		62
trifida	*Montiopsis § Montiopsis*	63
umbellata		64
ramosissima		64
gilliesii		64
berteroana		63
monandra	*Calandrinia s. str.*	60
compacta		61
caespitosa		63
affinis	62	62	64
colchaguensis		63
compressa	96	60
ciliata		59
columbiana	*Lewisia*	57
cotyledon		56
leana		56
rediviva		55
cantelovii	63	57
congdonii		53
tweedyi	*Cistanthe § Strophiolum*	62
parvifolia	*Montia*	49
virginica	98	*Claytonia*	38
megarhiza		44
perfoliata		29
sibirica	37
spathulata	31

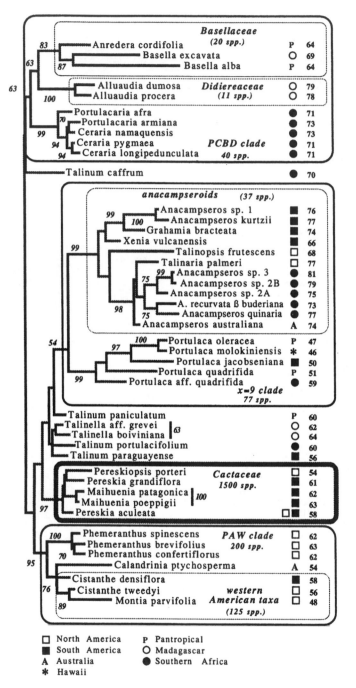

Figure 14.6 Maximum likelihood (ML) tree of portulacaceous ITS region sequences (reproduced from Hershkovitz and Zimmer, 1997). Numbers along the branches are bootstrap percentages for 1000 replicates using MP. Numbers to the right of the tree are %GC among parsimony-informative sites. Symbols to the right of the branches are native geographic distributions.

analysis. Consistent clustering of members of monophyletic groups suggests a conserved and probably unique ITS2 signal in those taxa. Inconsistent and/or incorrect clustering is difficult to interpret, given the simplicity of the method. It is possible that some portions of the sequence truly are saturated mutationally, but also that more sophisticated analytical methods (e.g. incorporating secondary structure information) would reveal additional features characteristic of particular higher-level taxa.

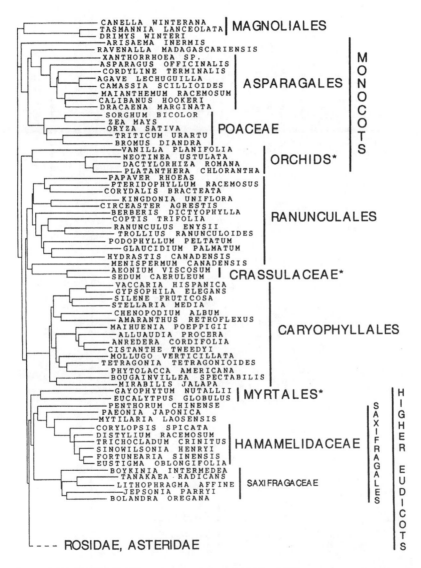

Figure 14.7 CLUSTAL W guide tree of angiosperm ITS2 sequences (continued overleaf). Higher-level classification follows Bremer et al. (1997). Some especially anomalously clustered taxa are denoted with an asterisk. Sequence accessions for the taxa analysed are available upon request from the first author.

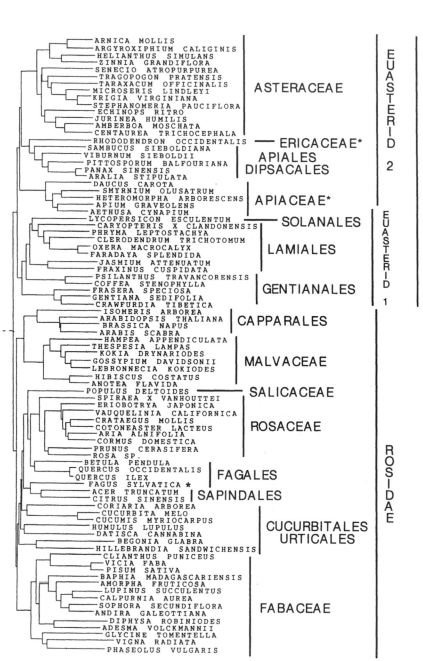

Figure 14.7 continued

The current ITS2 database includes a much broader sampling of non-Asteraceae asterids than before, and these sequences (including Ericaceae) consistently cluster (Fig. 14.7), suggesting the existence of an 'asterid ITS2' type. (Sequences for Polemoniaceae, Sarraceniaceae, and Actinidiaceae became available after we completed Fig. 14.7; in subsequent analyses, we found that these cluster with Ericaceae, consistent with 18S/*rbc*L/*atp*B evidence; cf. Soltis *et al.* 1997b, 1998). In addition, the 'euasterid 1' and 'euasterid 2' (Bremer *et al.*, 1997) ITS2's are clustered. The Apiaceae sequences are excluded anomalously from the Araliaceae/Pittosporaceae/Adoxaceae cluster (cf. Plunkett *et al.*, 1997). We suspect that base composition may have affected this result. Specifically, the five Apiaceae sequences average 27.4% T nucleotides, while the five sequences representing the other three families average 22.8%, and all 147 sequences average 21%.

Most rosid sequences cluster, but the composition of the cluster varies with sampling and parameter settings. In Fig. 14.7, the two Myrtales (Onagraceae, Myrtaceae) sequences are excluded from Rosidae. Using these same parameters but excluding the *Hillebrandia* (Begoniaceae) sequence, the Myrtales cluster with rosids, but *Begonia* and orchids form a cluster within eudicots but outside Rosidae. In the Fig. 14.7 tree, Begoniales (Begoniaceae, Datiscaceae, Coriaceae) are close neighbours, but are not clustered.

With some exceptions, sequences of Ranunculales (including Papaverales) and Saxifragales (including Penthoraceae, Hamamelidaceae, Paeoniaceae, Saxifragaceae) each cluster outside the rosid-asterid cluster, consistent with phylogenetic evidence. Menispermaceae (Ranunculales) and Crassulaceae (Saxifragales) sequences clustered apart from other members of these orders in Fig. 14.7 and in other guide trees generated using a range of alternative gap parameters.

Eudicots cluster apart from magnoliid and monocot sequences. Monocots cluster except for the orchids. Finally, as in our previous analysis, sequences of Caryophyllales cluster as tightly as or more tightly than many families (e.g. Fabaceae, Rosaceae, Asteraceae and, especially, Begoniaceae), as evidenced by the length of the subtending branch.

As noted above, ITS1 is unrelated evolutionarily to ITS2; it is distinct also structurally and functionally (Venema *et al.*, 1997), although evolutionary patterns in the two spacers (overall rates, base compositional biases, etc.) usually are parallel (Baldwin *et al.*, 1995). Higher-level conservation in ITS1 sequences appears to be considerably less than in ITS2. Only two broadly conserved motifs are distinct, one at the 5′ end (c1), and the other near the middle portion of the sequence (c2; Fig. 14.8). The c2 region corresponds to that recognised earlier by Liu and Schardl (1994). This sampling of angiosperms was extracted from an alignment of 131 ITS sequences representing most of the taxa included in Fig. 14.7. Only a few landmarks within the short conserved region appear to be truly alignable. The distal 'ragged' portions included in the figures are not alignable, but they appear to represent regions of conserved base composition, hence possible functional conservation. In most eudicots, the 3′ end of c1 is rich in A with few or no Gs. The 20 bases immediately 3′ to this (not shown) usually include 50–75% G. A-rich regions also occur in the 3′ and 5′ portions of c2, and the flanking regions (not shown) are C-rich. Across angiosperms, ITS1 appears to tolerate more length variability than ITS2. The maximum/minimum length ratio for ITS1 sequences we examined is

5'   cl

Taxon	Sequence		
Tragopogon	t.gaaa..tg .g----ag.a aga.ga..	--gtgaa.at	gtaattac.a atca..	49 (-95-)
Sambucus	t.gaaa..tg .a.---ag.a gaatga..	--g.gaa.t.	gttta.ata tt	46 (-71-)
Rhododendron	t.gaaa..tg .caa.aag.a gaaaa-.tt	--g.gaa.tt	gt.taata.a a.	47 (-103-)
Clerodendrum	t.gaaat.tg .aa.---ag.a aa-----..	--g.gaa.a	gtg.ttata .aaa	43 (-52-)
Lycopersicon	t.gaaa..tg .ca.---ag.a gaa.ga.c.	--g.gaa.t.	gttttaaa.a a.c.	47 (-63-)
Gentiana	t.gaat..tg .t.---aag.a ga--ga..	--gagaa.at	gtttat.g a.	45 (-72-)
Arabidopsis	t.gata..tg .t.c-aaa.a gaa.aa..	--g.gaa.ga	atgat.at.a a.t.t.	52 (-107-)
Lebronnecia	t.gaaa..tg .g---g.a gaa.ga..	--g.gaa.g.	gttgta.a.a aa.a.c.	50 (-117-)
Populus	t.gaaa..tg .rt.---ag.a gaa.ga..	--g.gaa.c.	atgg.atga .ca.	46 (-73-)
Prunus	tg.aaa..tg .t.---ag.a gaa.ga.c.	--gagaa.ta	gt.ttt-aaag	45 (-92-)
Quercus	t.gaaa..tg .a.---ag.a gaa.ga.c.	--g.gaattg	gtga.aa.c. ga.	47 (-61-)
Citrus	t.gaaa..tg .---ag.a gaa.ga..	--g.gaa.ca	gttgatat.c a..	47 (-92-)
Cucurbita	t.gatg.cta .aa.---at.ca -aa.gac.c.	--g.gaa.g.	gttttaaaa.a .aaa.	48 (-62-)
Lupinus	t.gaag..tc .a.---aag.a gtg.ga..	--gtgaattt	gtttta.ta .t.a.	49 (-80-)
Eucalyptus	t.gaat.-tg .c..---ag.a gaa.gac.a	--gagaa.g	gtaa.aaa.c t.aa.c.	48 (-86-)
Corylopsis	t.gaaa..tg .c.---ag.a gaa.ga..	--g.gaa.g.	gtaaaaaat .a.aa	48 (-120-)
Penthorum	t.gaaa..tg .ta.---ag.a gaa.ga.c.	--g.gtgaa.t	gtaaaa.aa t.	46 (-115-)
Tetragonia	t.gaaa.tg .c.t.---ag.a gaa.ga.c.	--gtgaa.a.	gtttat.ac t.c.	47 (-78-)
Papaver	-.gaaa..tg .c.---ag.a gaa.ga.c.	--gagaa.ca.	gtgaa.c.a a.	44 (-82-)
Drimys	.gaa.a.tg .g.---aaag.a g.g.gac.tg	gtgtgaa.t.	gt.aaatta .t.aaatt..	55 (-110-)
Caligana	t.gaga.c.c .-----gaat.- gga.gatt-	--gtgaa.c.	gt.gatg.t t.	41 (-102-)

(-v1-)

		c2				(-v2-)	3'
Tragopogon	44	a..aagttaa-aa-..gg-a.. gaa---tgtg ..aaggaaaa agaaaacaaa agaagg	198	(- 51-)	249
Sambucus	17	a..aaaa.aatgaa-..gg g. gaa---..tg.g ccaaggaatt ttta.tgaag ag	169	(- 58-)	227
Rhododendron	50	aa.aa.gaa-...gg.g. aaaa---g.g ..aaggataa ttgaa.aaag	196	(-101-)	297
Clerodendrum	97	..taaaaaa-t.gggg.g. ggaaa---tg.g. tcaaggaata tataaaagag	142	(- 68-)	210
Lycopersicon	10	a.-aa..ga---gg.g.g ggaaa---g-g .c.aaggaata .gta.aat.ga .ag	159	(- 95-)	254
Gentiana	19	.aaa.aa-aa..g.gggg.g. agaaa---gtg .caaggaaaa gtaaaaagg a	170	(- 66-)	236
Arabidopsis	59	atat.a.aaaa...gg.a.. gaaaa---gtg t.aaggaa-c atg.aa-taa.g	207	(- 66-)	273
Lebronnecia	69	.aaaa.gaa.aaa...gg.g. gaa---t.g.g .caaggaatc gaaa.gaaaa gggggg	226	(- 75-)	301
Populus	19	.aaa.gaa-...gg.g. aaga---a.g .caaggaaat tgagtactag gagg	167	(- 52-)	219
Prunus	37	.aaa.gaa-a..gg.g. gaa---ttg.g .caaggaa-c ttgaa.gaga gag	184	(- 65-)	249
Quercus	08	.aaa.gaa-a..gg.g. gga---atg -.aaggaaa- ttgaa.caag agag	154	(- 88-)	242
Citrus	39	aaa.aa.gaa.c..gg.g. gga---ctg.g .caaggaa- t.taa.gaga gag	189	(- 66-)	255
Cucurbita	10	ttaaacaaaa---g.g agg---t.g.g .caaggaa-c ttgaaatgaa	155	(- 95-)	240
Lupinus	29	aataa.gaa-...gg.g. gaa---g.g .caaggaaa- ttgaaat.gt ttag	179	(- 64-)	243
Eucalyptus	34	.caa.gaa-...gg.g. ggaa---tg.g .caaggaa-c ttgaa.caga gtg	180	(- 65-)	245
Corylopsis	68	.ataa.gaa-...gg-g. aaaa---g.g tcaaggaa-- t.taaaatga aagag	217	(- 44-)	261
Penthorum	61	.caa.aa.gaa-g.g.g ga.aa---g.g .caaggaa-- .tcaaaa.ga atgag	210	(- 48-)	258
Tetragonia	25	.caa.aa.gaa-...gg.g. ggaaa---g.g .caaggaa-- .atgaa.a.a tagta	175	(- 53-)	228
Papaver	26	aaaaa.gaa.caagg.g. ggtaa---g.g .caaggaaaa aa.aaatgga tg	175	(- 85-)	260
Drimys	65	aataa.gaa-...gg.g. aatgg--g.g .caaggaaat g.ag.agtag aa	213	(- 56-)	269
Caligana	43	aaa.aaagaa...gg.g. ggtgg--g.g .caaggaata gtgatgt.gg agag	182	(- 56-)	238

Figure 14.8 Alignment of conserved ITS1 motifs in selected angiosperms. Sequence accessions for the taxa compared are available upon request from the first author. The two conserved motifs (c1 and c2) alternate with the variable regions (v1 and v2). The 5' end of the alignment is the ITS1 5' end. The ragged regions at the c1–3' end and flanks of c2 are not aligned, but are included to demonstrate conserved base conservation patterns in these regions. Numbers immediately adjacent to the alignment are the base numbers of the adjacent base. Length of v1 and v2 and total length are indicated.

1.7; for ITS2 it is 1.4. This discrepancy is not apparent from typical low-divergence data-sets (Baldwin et al., 1995). The 21 ITS1 sequences in Fig. 14.8 exhibit only a 1.4 maximum/minimum length ratio, but the shortest angiosperm ITS1's (< 200 bp) are not represented. Among these 21 sequences, 2.3-fold length differences occur in both the first and second of the sequence-variable regions (v1 and v2). The length ratio v1/v2 ranges from 0.7 to 2.7 within a given species. The v1 appears to be generally longer, however, and more length-heterogeneous, because v2 is especially long in relatively few taxa. We attempted a guide tree analysis for ITS1 using a taxon complement similar to that in Fig. 14.7, but we were unable to generate a clustering pattern resembling plausible phylogenetic relationships. Either the deeper-level phylogenetic signal has been lost or it is not detected easily. Analysis of secondary structures probably will shed additional light on angiosperm ITS1 evolution. The considerably lower degree of length and sequence conservation of ITS1 versus ITS2 indicated in our angiosperm-wide comparisons is not reflected always in raw sequence divergence at lower taxonomic levels. ITS2 divergence has been reported to be similar or greater than ITS1 in some taxa (Baldwin et al., 1995).

Although 5.8S appears to have evolved somewhat faster than 18S (Hershkovitz and Lewis, 1996), the small size of this molecule (163–165 bases) limits its utility in angiosperm phylogenetics. The hundreds of 5.8S sequences available in GenBank mainly are a by-product of ITS studies, although the 5.8S commonly is not sequenced completely or not reported. A comparison of 5.8S sequences in nine diverse angiosperms revealed (excluding indels) 29 variable and 12 parsimony-informative sites (Hershkovitz and Lewis, 1996). Three of the nine sequences were derived using RNA, which is less accurate than DNA-based methodology, and hence might inflate our estimates of variability. Divergences among the sequences (estimated using ML and a GTR model corrected for among-site rate variation) ranged as high as 18% (between two of the RNA-based sequences), but usually was less than 10% (e.g. 4% between wheat and tomato). Divergence of 5.8S from the parasitic genus *Arceuthobium* from this angiosperm sample is 25–35%, reflecting the generally high genomic divergence of parasitic plants (Nickrent et al., 1994). Much of 5.8S variation (10/29 sites in the nine angiosperms) was concentrated in a 24-base helical region near the 3′ end of the molecule. While overall variability in 5.8S appears to be low, it might have utility in augmenting 18S and/or 26S data.

The total amount of 5.8S variation in typical ITS region data-sets apparently differs among taxa, but the proportion of variable sites in 5.8S appears to be about 5%. In the ITS data-set comprising portulacaceous taxa (including Cactaceae, Didiereaceae, and Basellaceae; Hershkovitz and Zimmer, 1997), in which total raw divergence ranged up to 25%, 16/299 base-variable sites occurred in 5.8S; 9/16 of which reflected unique states in one sequence. Only 4/188 sites were potentially parsimony-informative, yielding only two unambiguous synapomorphies involving only three of the 46 taxa. For an Asteraceae ITS data-set, Baldwin et al.'s (1995) illustration indicates 8/167 variable sites in 5.8S, none informative. From an illustration representing an Orchidaceae-Orchidoideae ITS data-set (Pridgeon et al., 1997), it appears that 29/552 variable sites occur in the 5.8S. This number of variable 5.8S sites is equal to that in our angiosperm sample above, but angiosperm-wide comparisons of 18S and ITS variation discussed above suggest that overall rDNA evolution in Orchidaceae has been relatively high. In all three data sets, 5.8S variation is especially high near the 3′ end, probably in the helix region (10/16 sites in the case of the portulacaceous data). The short region 3′ of the variable helix pairs transiently with the relatively conserved 5′ end of ITS2 and ultimately with the 26S rDNA (Hershkovitz and Zimmer, 1996; Liang and Fournier, 1997; Peculis, 1997). Possible coevolution of these paired sites has not been investigated.

While not especially informative for angiosperm phylogenetics, the 5.8S does provide an excellent indicator of the functionality of the amplified/cloned ITS copy, which often is not clear from the ITS sequence alone. Specifically, the 5.8S portion of the alignment can be screened for unusual characters (especially transversions and indels not present in the other sampled taxa and/or angiosperms in general) that might suggest that the rDNA copy is not expressed. This sort of comparison is useful considering that unexpressed and/or degenerate rDNA copies probably exist in all genomes (Buckler et al., 1997).

The range of phylogenetic signal present across the ITS region, combined with its small size, render it a powerful, rapidly-screened source of data for studies of angiosperm organismic and molecular diversification. We do not propose that ITS should be used as a marker for higher-level angiosperm phylogenetic analysis, although we suspect that the agreement between the ITS2 guide tree and other phylogenetic evidence would improve as additional poorly represented taxa are sampled (e.g. magnoliids, monocots besides grasses, palaeoherbs, Violales, Myrtales). Nevertheless, existing ITS data suggest that: (1) generally it should be possible to predict for any angiosperm ITS region sequence its phylogenetic relations at many hierarchical levels; (2) in view of the apparent degree of phylogenetic accuracy of angiosperm classifications at all taxonomic ranks published before the molecular era, the phylogenetic information contained in the ITS region appears to be comparable to that of all non-molecular characters combined; (3) the patterns of higher-level sequence conservation will provide insights into ITS function not apparent from lower-level variation patterns; (4) the phylogenetic detail with which ITS evolution can be traced from the angiosperm root to species-level terminals (analogous to, e.g. floral morphological evolution) should provide insights into evolution of ribosomal processing in angiosperms as well as a large statistical sample for studies of molecular evolutionary trends. In particular, ITS can be used to identify taxa with markedly different evolutionary patterns/rates compared to members of the broader group. For example, Fig. 14.6 implies that the rate of ITS evolution in orchids is much greater than in other monocots, Fig. 14.5 indicates that the base composition in *Portulaca* departs markedly from that in other portulacaceous taxa, and Fig. 14.4 indicates that both the rate and base composition have changed in *Claytonia*. Such evolutionary transitions would be less obvious using data from a more conservative marker.

14.2.2.3 ITS polymorphism

Application of ITS to analysis of relations among closely related species raises the concern that phylogenetic analysis might be misled by incomplete sampling or lineage sorting of polymorphism within genomes at individual to populational levels (Doyle, 1993; Buckler et al., 1997). ITS polymorphism has been reported to exist as non-recombined paralogues, as recombinant mosaic sequences, or a mixture of both within a genome (Suh *et al.*, 1993; Wendel *et al.*, 1995a, b; Buckler *et al.*, 1997; Campbell *et al.*, 1997). This polymorphism could result from either mutation or from hybridisation between genomes with different, yet functional, ITS sequences.

It has been supposed, albeit without proof, that the lax selection on spacer relative to coding sequence results in relatively high spacer variation not only among taxa, but also among spacers within individuals (Schlötterer, 1998). This does not appear to be the case for the 5S/NTS repeats, in which intertaxon variability of NTS is higher than 5S, while infragenomic variability in 5S and NTS is roughly equivalent (see below). Although polymorphism in ITS appears to be reported more commonly than for 18S-26S coding sequences, data relating to the latter are few, probably because the coding sequences are studied at divergence levels where variation is presumed to be much greater among than within taxa. Buckler *et al.* (1997) reported that among divergent ITS region paralogues in five diverse angiosperm species or species groups, the 5.8S sequences generally appeared to have relatively fewer substitutions than spacer sequences. Thus, they remarked that 'when two functional rDNA arrays diverge [within a genome], the ITS regions will evolve faster than the 5.8S.' In addition, homogenisation of rDNA loci via unequal crossing over and/or gene conversion is predicted to occur more quickly within than between rDNA loci (Buckler *et al.*, 1997; Karvonen and Savolainen, 1993; Schlötterer and Tautz, 1994). In this case, it seems that indeed the same selective forces that result in higher intertaxon spacer variability could also promote interlocus spacer polymorphism. Moreover, concerted evolution should be considered as a stochastic, rather than orthogenetic, process. In other words, any rDNA repeat might 'seed' one or more homogenisation events at a given time, and multiple seed repeats might be proliferating at the same time, according to a probabilistic function. Such a process might allow, if not favour, a dynamic equilibrium of spacer variability, even under strong selection for coding sequence uniformity. The critical role of the spacers in ribosomal processing (Venema *et al.*, 1997; Peculis, 1997) evokes a more speculative and less selectively neutral explanation relating high intertaxon to intra-individual spacer variation, in which polymorphism might serve to developmentally regulate ribosomal production and maturation. Selection for developmental characters then would drive selection for ITS variability both within and among organisms.

Most plant ITS literature indicate that homogeneous sequences were derived directly from PCR products amplified from genomic DNA. Homogeneity has been reported even in cases where polymorphism might be expected, as in recently formed allopolyploids derived from substantially diverged diploids (Ainouche and Bayer, 1997). When ITS polymorphism has been reported, it usually has been detected when PCR amplification from genomic DNA yields multiple products (multiple bands) or when multiple sequence signals are detected from apparently uniform PCR products (Baldwin *et al.*, 1995; Campbell *et al.*, 1997; Hershkovitz and Zimmer, 1997). In these cases, polymorphism can be dissected using simple cloning techniques. In general, it appears that ITS heterogeneity is not sought purposely if it is not detected using these simple protocols (i.e. if PCR products yield resolvable sequences). Theory and empirical evidence, however, suggests that ITS polymorphism is the rule but that it might be undetected using the simplest proto-cols (Buckler et al., 1997; Samuel et al., 1998). Because of amplification bias, even sequencing of multiple clones from genomic DNA amplification may fail to recover different morphs (Buckler et al., 1997). As an extreme and noteworthy example, sequencing of multiple ITS clones from multiple amplifications from several Mimulus taxa yielded only algal contaminant sequences (Hershkovitz and Lewis, 1996).

For practical purposes, the problem of ITS polymorphism is not so much whether it exists, but whether its pattern could mislead phylogenetic analysis. In general, this does not seem to be the case, at least based on the observation that ITS gene trees are not reported commonly to be discordant with independent evidence for organismal phylogeny. The reason for this is probably that, as noted above, ITS divergence among most closely related species tends to be inadequate for phyloge-netic resolution. Thus, at this level, it is unlikely that paralogues are differentiated sufficiently to strongly support an incorrect organismal tree node. At greater diver-gences, homogenisation is sufficient to fix intertaxon differences, such that intra-individual variants with a few substitutions or indels will not effect interpre-tation of organismal phylogeny (e.g. Sun et al., 1994; Buckler et al., 1997; Hershkovitz and Zimmer, 1997). If functional paralogues persist through multiple and substantial divergences (e.g. Suh et al., 1993; Waters and Schaal, 1996; Campbell et al., 1997), it seems likely that the variants represent major loci, such that standard amplification and sequencing protocols will reveal the polymorphism in at least some samples (but see Baldwin et al., 1995), and protocols can be adjusted (e.g. primer redesign) to search for complete paralogue sets. Buckler et al.'s (1997) finding that various amplification conditions differentially amplified divergent paralogues suggests that this procedure might provide a good screening system for polymorphism. Regardless of divergence level, Buckler et al.'s results suggest that some ITS morphs will be highly divergent and probably will represent pseudogenes. These can be discerned on the basis of unstable secondary structure (Buckler et al., 1997), or, perhaps more easily, on the basis of substitutions in otherwise highly conserved positions (Hershkovitz and Lewis, 1996; Hershkovitz and Zimmer, 1996; Buckler et al., 1997). The latter method probably is more reli-able also, because rDNA function might be disabled by even single mutations without substantially affecting secondary structure. Another approach, suggested for avoiding 18S pseudogenes (and which might also help for avoiding contami-nants) is sequencing of rRNA transcripts incorporated into cDNA libraries (Chaw et al., 1995). Besides the time and expense involved, however, the practicality of this method for ITS would be limited by the relatively low population of complete 18S-26S (i.e. uncleaved) rRNA transcripts and possibly also the inadequately under-stood regulation of expression of different rDNA loci (e.g. Donald et al., 1997; Cerbah et al., 1998). In any case, because of their extreme divergence, true pseudo-genes generally should not be a problem for analysis of the phylogenetic problem of interest. Non-recombined polymorphisms in hybrid taxa may not be problem-atic, provided that the paralogues are detected and they group with the parental species. Cases of lineage sorting, recombination from ancestral polymorphisms, and/or gene conversion or introgression to one or the other parental type remain problematic (Wendel et al., 1995a, b; Buckler et al., 1997; Comes and Abbott, 1999 – this volume). In such cases, hybrid origin may not be detectable from ITS data alone, and cladistic analysis can yield misleading results. ITS is not unique among nuclear sequences in this respect, however, and lineage sorting can also occur among non-recombining plastid sequences (Rieseberg and Morefield, 1995). Moreover, discordant phylogenies, from ITS or any source, cannot be considered misleading where hybridisation is involved. To the contrary, it should be considered no more or less misleading than data that fails to detect hybridisation.

14.2.3 The rRNA 5′ external transcribed spacer (5′-ETS)

During the late 1980s and earliest 1990s, one of the most exploited data sources in plant molecular systematics was RFLP's in the rDNA IGS-ETS region, with much of the variation detected in the IGS (Crawford, 1991). For both technical and theoretical reasons, sequence data from this region have not been exploited. In particular, the IGS evolves rapidly in both sequence and structure, and it is composed mainly of repeating motifs for which inference of homology among closely related species or even within a genome is difficult or impossible (King *et al.*, 1993; Nickrent and Patrick, 1998). Moreover, the IGS-ETS is commonly 3–6 kb long, which has challenged the limits for conventional amplification, cloning, and sequencing protocols, especially given the substantial degree of infragenomic polymorphism known for IGS.

The development of reliable protocols for amplification of 3–10 kb genomic sequences and sequencing of 700–800 bp has offered the opportunity to re-examine the utility of sequence variation in the IGS-ETS region, and preliminary studies show remarkable promise for the 3′ end of the 5′-ETS (immediately upstream of the 18S). Systematic analysis of partial 5′-ETS sequences in species of *Calycadenia* (Asteraceae: Baldwin and Markos, 1998) and *Medicago* (Fabaceae: Bena *et al.*, 1998) have yielded virtually identical patterns: the amplified region includes a sequence *ca.* 0.8–1.5 times the combined length of ITS1 and ITS2 that evolves 1.3–2.4 times as fast. Assuming these values are typical for angiosperms, this region would thus hold 1–3 times as much informative variation as the ITS region. Both studies found high concordance between ITS and the partial 5′-ETS gene trees and much improved resolution when the data were combined. Alignment of the partial 5′-ETS region was unambiguous. Although both short (one to a few bases) and long (up to 200 bp; Asteraceae only) indels occurred, the 5′-ETS lacked the poorly alignable regions that tend to characterise ITS. Baldwin and Markos (1998) developed 5′ amplification/sequencing primers for Asteraceae tribe Heliantheae (and possibly other Asteraceae) based on preliminary alignments of sequences extended from 3′ primers in a few phylogenetically diverse taxa. With this 5′ primer and a 3′ primer in the 18S sequence, a relatively small fragment (< 1 kb) could be thus amplified in all taxa. This approach can probably be extended to many if not all angiosperm groups of interest.

The 5′ end of the 5′-ETS (the transcription initiation site) was not identified and probably not included in either the Asteraceae or Fabaceae studies, for which reason Bena *et al.* (1998) coined the term ETSf[ragment] for the analysed region. This or similar terminology is useful to avoid eliciting and perpetuating a notion that the sequenced region is the entire 5′-ETS, contrary to terms and existing nomenclature. To complicate matters further, 'the 3′ region of the 5′-ETS' must be distinguished from the 3'-ETS, which occurs immediately 5' of the 26S. The systematically useful length of the 5'-ETS might vary more than the Asteraceae and Fabaceae data would suggest, because the 5'-ETS in some angiosperms is reportedly as long as 3 kb (*Nicotiana:* Borisjuk *et al.*, 1997). The longer solanaceous 5'-ETS regions appear to be 'padded' with repetitive DNA towards the 5' end, but the 3' end still features a 500–600 bp region that exhibits relatively high intergeneric similarity and includes regions of non-repetitive and subrepetitive sequence (Borisjuk *et al.*, 1997). A comparable pattern characterises ETS among seven legume species representing five genera (Nickrent and Patrick, 1998). Two subrepeats were shared by all species, but in all but one species, the subrepeats were termed 'cryptic' because their repetitive nature was obscured by superimposed mutations. All three of the examined *Vicia* species shared an additional region of repetitive DNA very close to the 3' end of the 5'-ETS.

For technical and analytical purposes, 5'-ETS data probably should be approached much as for ITS data, with additional effort to detect for polymorphism and non-independently evolving repeat motifs. As with other rDNA regions, probably it is critical that both strands be sequenced. Both the Asteraceae and Fabaceae ETS regions were sequenced directly and without strong evidence of severe infragenomic polymorphism (except possibly in one Asteraceae taxon (Baldwin and Markos, 1998). Nevertheless, given the high infragenomic polymorphism known for the immediately adjacent IGS and the apparently high interspecific functional tolerance of ETS sequence variability, the potential for polymorphism should be considered greater than for ITS. ETS polymorphism probably best is presumed for taxa known to be polymorphic for ITS, although no published data to this effect yet exist. If ITS polymorphism occurs (technically, even if it does not), it will be important to verify linkage between ETS and ITS paralogues. This could be accomplished by long-range amplification using a 5'-ETS primer with a 3' primer in the 5.8S or 26S region. This template could be sequenced directly or used to amplify separate, but linked, ETS and ITS fragments. Baldwin and Markos (1998) demonstrated multiple origins of repeating units in Asteraceae tribe Heliantheae, although they also showed that subsequent evolution in the repeats appeared to proceed independently. Nevertheless, the high potential for internal repeats (e.g. Borisjuk *et al.*, 1997) should elicit caution in sequence analysis. Dot-plot analyses (e.g. Bult *et al.*, 1995) probably should be performed routinely.

14.2.4 The 26S rRNA gene

The *ca.* 3.5 kb plant 26S rDNA consists of alternating conserved (totalling 2.5 kb) and variable (totalling 1 kb) regions (Bult *et al.*, 1995; Kuzoff *et al.*, 1998). The variable regions are known variously as 'expansion segments' or 'D' segments or 'divergent domains' (Bult *et al.*, 1995; De Rijk *et al.*, 1995). Because 26S is overall more variable and much longer than 18S, it has been considered a potentially rich source of phylogenetic information. Nonetheless, 26S exploitation has lagged far behind that for 18S, both for angiosperm (Kuzoff *et al.*, 1998) and 'tree of life'-level (De Rijk *et al.*, 1995) phylogenetics. By 1992, full-length 26S sequences had been published for seven angiosperm species, including three Brassicaceae (Bult *et al.*, 1995). The ensuing five years yielded only 15 additional full- or nearly full-length sequences, five of which belong to Saxifragaceae (Kuzoff *et al.*, 1998). Aside from minor (usually terminal) fragments known from clone libraries or flanks of ITS or IGS regions, partial 26S sequences are known from *ca.* 160 additional angiosperm species, including Ranunculales (Oxelman and Lidén, 1995a; Ro *et al.*, 1997), Caryophyllaceae (Oxelman and Lidén, 1995b), and the 60 assorted angiosperms sampled for partial 18S + 26S RNA sequences (Zimmer *et al.*, 1989; Hamby and Zimmer, 1992). Partial 26S sequences have been reported for 51 gymnospermous species, including a *ca.* 650 bp fragment from 39 species (Stefanovic *et al.*, 1998).

Popularity of 26S in angiosperm phylogenetics has been inhibited by a combination of technical and perceptual problems. Amplification of *ca.* 3.5 kb was technically difficult for many labs, at least until the development of longer range amplification chemistry. Even so, reliable amplification primers had not been available for the relatively variable 26S-3' end. Perhaps dampening efforts to overcome these obstacles was the fear that 26S data might lack useful phylogenetic signal, or, worse still, might contain biased variation that would mislead phylogenetic analysis. Preliminary analyses indicated that the conserved regions possibly were conserved more per length of sequence than 18S, and that the expansion regions were too variable to be useful at the same divergence level as the conserved regions (Bult *et al.*, 1995). Moreover, variation in the expansion regions potentially was biased by infragenomic sequence homogenisation within and among expansion segments, such that expansion segment DNA essentially would be repetitive. Subsequent point mutations can mask this simple pattern, resulting in 'cryptic sequence simplicity' (Bult *et al.*, 1995).

The most recent 26S research indicates that amplification difficulties are readily surmountable (Kuzoff *et al.*, 1998), and that the utility and versatility of 26S may equal or exceed that of any molecule currently applied to angiosperm phylogenetics. For example, a *ca.* 1066 bp alignment (including expansion segments D1–D4) from 35 Ranunculales (Ro *et al.*, 1997) revealed, exclusive of positions that were aligned ambiguously or in which gaps occurred, 203 variable and 132 parsimony-informative characters. This compares with 76 variable and 25 informative sites for 18S in 10 mostly ranunculalean taxa (Hoot *et al.* 1995; informative 18S sites increased to 74 in a data-set that included 15 Ranunculales but also three highly divergent outgroups: Hoot and Crane, 1995). ITS2 could not be aligned 'meaningfully' across ranunculalean families (Oxelman and Lidén, 1995a). Only 39 aligned 26S sites, or *ca.* 4% of the total aligned length, included alignment ambiguities or positions with gaps in some taxa. Among 81 Caryophyllaceae (including 76 Sileneae) species, an 822 bp 26S alignment (including expansion segments D1–D3) held 98 variable and 44 informative sites, compared with 306 variable and 134 informative for the *ca.* 450 bp of ITS1 and ITS2. Although the shorter ITS held about three times as much variability as the 26S segment, it included about seven times as many indels (34 vs. 5). Comparison of *rbc*L, 18S and nearly full-length 26S sequences from 21 angiosperm species (Kuzoff *et al.*, 1998) indicated that 26S evolutionary rate is about twice that of 18S and half to equal that of *rbc*L. Because 26S is longer than these other sequences, however, it provided two to three times as many parsimony informative characters.

These results illustrate that the 26S rDNA can contribute useful phylogenetic information over a broad range of divergences. The signal probably is most comparable to that of the chloroplast *ndh*F gene (Kuzoff *et al.*, 1998) and perhaps also the chloroplast *mat*K gene, especially if combined with the adjacent ITS region. The analyses of Ranunculales (Ro *et al.*, 1997) and angiosperms (Kuzoff *et al.*, 1998) also found that the trees produced using 26S data were largely congruent with plastome-derived trees. Thus, 26S rDNA may be destined to become one of the most widely exploited sequences for angiosperm phylogenetics.

14.2.5 The 5S rRNA gene

Comparison of the *ca.* 120 bp 5S rRNA at the tree-of-life level represents one of the seminal applications of nucleic acid sequences to plant phylogenetics (Kumazaki *et al.*, 1983; Hori and Osawa, 1986, 1987). It was also one of the first to be discredited on theoretical grounds. Specifically, the molecule was deemed too short, too conservative at sites constrained by secondary structure, and too variable at unconstrained sites, to yield accurate (much less accurate and well-supported) phylogenies (Steele *et al.*, 1991; Cronn *et al.* 1996). Otsuka *et al.* (1996) proposed a formula to correct for these problems, but their resultant tree is no more satisfactory than those derived previously. For example, their tree showed the chlorophytes (green algae plus land plants) as tetraphyletic, with the land plants arising in a clade of metazoans exclusive of green algae.

Current plant systematic research focuses on the phylogenetic utility of the 5S locus at interspecific levels (Baum and Johnson, 1996; Cronn *et al.*, 1996; Kellogg *et al.*, 1996; Baum and Bailey, 1997). Interspecific variation at 5S rDNA loci, especially in the non-transcribed spacer (NTS), first was explored using RFLPs (Scoles *et al.*, 1988; Crawford, 1991; Appels and Baum, 1992). These data were useful primarily for diagnosing genome donors in allopolypoid species. More recently, attention has returned to 5S locus sequence data, especially to the NTS. As in deeper level applications, the 5S gene itself appears to be virtually useless at interspecific levels, even though variation within species, populations, and even individuals can be substantial (Cronn *et al.*, 1996). Presumably, some sites are constrained highly functionally, while those that are free to vary do not exhibit fixed interspecific differences. Useful phylogenetic signal has been found, however, in the 100–700 bp NTS (Baum and Johnson, 1996; Cronn *et al.*, 1996; Baum and Bailey, 1997). Among wheat relatives (Poaceae tribe Triticeae), preliminary analyses indicated that 5S genes sorted into two forms, one long and one short (Appels and Baum, 1992; Baum and Johnson, 1996; Van Campenhout *et al.*, 1998). Baum and Bailey (1997) found five spacer types among 30 5S clones from the hexaploid *Kengyilia alatavica*. The shorter forms resembled other grass short spacers in length, but they clearly were related more closely to long spacers in particular taxa. Among the longer spacers recovered, only one appeared to be related to a spacer previously considered to be a 'long' grass form. In *Gossypium* (Cronn *et al.*, 1996), variants within individuals generally appeared to be closest relatives, but in some cases, paralogues appeared to have closest relatives in other individuals. Synthetic and natural AD genome polyploids possessed either or both A and D genome spacers. An introgressant D genome species from an ancient AD hybrid had a single and evidently hybrid spacer form.

The potential utility of 5S NTS sequence data in broader angiosperm interspecific systematics remains unclear. Homogenisation of NTS sequences within a genome apparently is slower than the rate of speciation, increasing the likelihood of confounding gene with species phylogeny. Phylogenetic resolution is possible by resolving gene trees for orthologous NTS copies, although cloning techniques would be necessary to extract the paralogous sequences from individuals. This approach would provide NTS with an important advantage over a single-copy or highly homogenised sequence: the species tree can be rooted at the point where the orthologues diverge. Short of thoroughly screening genomic libraries of each sampled organism, however, the orthology of infragenomic variants could be difficult to establish with certainty. For example, Baum and Bailey (1997) questioned whether the unique spacer forms in *Kengyilia* truly were absent in other taxa or merely overlooked. For that matter, their homologues could have been present in a common ancestor and differentially extinguished by lineage sorting. Thus, despite the availability of over 200 sequences from 35 species, Baum and Bailey remarked that, 'the Triticeae database is still too small and not yet suitable for making general inferences about the 5S gene.'

While the 5S NTS may not be as useful as ITS or plastome data for simple approximation of a species tree, its infra- and intergenomic patterns of variation can be informative in an evolutionary context. The potential for relatively long persistence of distinct heterologous NTS forms in descendants of hybrid taxa, especially in cases where parental 18S–26S genes have become recombined, homogenised or differentially extinguished (Adachi *et al.*, 1997), may render NTS a good choice for investigation of suspected hybridisation. NTS sequence data from hexaploid wheat lines appear to be consistent with other evidence in identifying the diploid progenitor genomes. Unfortunately, reduced 5S copy numbers in some natural allopolyploid *Gossypium* species suggest that in other cases the genome might actively root out 5S polymorphism (Cronn *et al.*, 1996). Thus, while the presence of multiple ancestral 5S forms can be positive evidence of hybridisation, the absence is not negative evidence: 5S copy number and expression inheritance evidently cannot be predicted from Mendelian segregation ratios (Zimmer *et al.*, 1988; Cronn *et al.*, 1996). In any case, comparison of 5S NTS and 18S-26S patterns in known hybrid taxa should help shed light on passive and active forces underlying concerted evolution (Cronn *et al.*, 1996; Baum and Bailey, 1997). Regardless of whether homogenisation occurred via gene conversion or unequal crossing-over, concerted evolution cannot be viewed as a simple stochastic process if there are clear biases in the homogenisation pattern and rate for different genes within the same genome and/or, for that matter, between the coding and non-coding region of the same 5S repeat region (Cronn *et al.*, 1996). Finally, whatever generalisations on 5S rDNA evolution might emerge from Triticeae and *Gossypium*, it would be premature to extend these to all angiosperms, because the complexity of evolution in these groups has tended to resist dissection from any singular approach (Cronn *et al.*, 1996; Kellogg *et al.*, 1996).

14.3 Advances in phylogenetic analysis relevant to rDNA systematics

General and case-specific peculiarities of rDNA sequence evolution affect phylogenetic interpretation and should be considered when applying analytical methods.

These peculiarities include the potential for infragenomic polymorphism (discussed above); the affect of length variability on homology assessment (i.e. on alignment); substitution patterns, including the effect of functional constraints (especially secondary structure) on sequence variation patterns; and the affect of divergence patterns on phylogenetic resolvability.

14.3.1 rDNA sequence alignment

Alignment is equivalent to homology assessment. It is part and parcel of phylogenetic analysis and is an issue separate from that of how the length-variable regions are treated in tree-building. The alignment of rDNA sequences necessarily precedes application of popular tree-building methods. Alignment methodology is particularly relevant to rDNA analysis because of the length variability characteristic of rDNA sequences, and because much of the phylogenetic signal is associated with length-variable regions. At least some reported rDNA length variability, however, probably reflects sequencing error (Hershkovitz and Lewis, 1996; Soltis *et al.*, 1997). Plant rDNA analyses have employed both manual and computational alignment methods. Those employing computational methods (e.g. CLUSTAL) have tended to optimise the computational result manually (i.e. 'by eye'). This approach has been criticised as overly subjective (Wheeler, 1994; Wheeler *et al.*, 1995), but satisfactory 'objective' methods remain elusive. When a purely computational alignment is subject to phylogenetic analysis, the choice of alignment method and/or alignment parameters has been found to affect phylogenetic inference even more than the choice of tree-building method (Morrison and Ellis, 1997). The purportedly objective alignment optimisation procedures in the MALIGN alignment package (Wheeler and Gladstein, 1994) nonetheless require subjective and/or arbitrary parameter settings, and the reliability of alignment optimised according to a MP tree and/or the majority-rule 'elision' alignment weighting procedure remain unproved. Thus, computational alignments should be treated as heuristic solutions that should be subject to re-evaluation in view of broader evidence.

Comparison of rRNA secondary structure has been advocated as a means of refining rDNA sequence alignment. As with alignment, the methods used to determine rRNA secondary structure are an issue separate from how structure is used in subsequent phylogenetic analysis. Two methods used to derive secondary structure are: analysis of substitution covariation, which can be interpreted as evidence of compensatory mutation and hence base-pairing in secondary structure (Gutell *et al.*, 1994); and analysis of minimum free-energy of folded rRNA (Zuker, 1989). In addition, sequences are sometimes folded to match a model proposed for the same sequence in a different organism. These methods generally give a good approximation of the correct structure because they incorporate biologically reasonably models: RNA molecules are expected to fold in accordance with physical laws, and mutations that might destabilise secondary structure should be restabilised by compensatory mutations. But, as with tree-building and alignment, these methods use simplified models (e.g. Conn and Draper, 1998); hence they should be considered as heuristics and the resulting solutions as estimates. Moreover, RNA secondary structure is apparently dynamic *in vivo* (Thirumalai, 1998; Wu and Tinoco, 1998) and presumably also evolutionary. The method of covariation analysis is basically phenetic, because the covariation is determined on the basis of the raw number of taxa possessing alternate potentially paired bases. The statistical significance of covariation has not been considered in a phylogenetic context, for example by determining the observed versus expected number of times in a phylogeny that potentially pairing bases have alternated. Structures derived using free-energy computation usually require manual adjustment to yield realistic models (Hershkovitz and Zimmer, 1996; Mai and Coleman, 1997). Several strikingly different structures may have similar free-energy, and, probably because rRNA has interactions *in trans*, the true model may not have the lowest free-energy *in cis* (see examples discussed in Hershkovitz and Lewis, 1996; Hershkovitz and Zimmer, 1996). Moreover, non-canonical base-pairing is not deduced readily using available algorithms. To encapsulate, secondary structure is estimated least reliably in precisely the taxonomically variable regions where this information would be most useful; for example in the 26S expansion regions (Chenuil *et al.*, 1997).

14.3.2 Implications of rDNA substitution patterns for phylogenetic analysis

Substitution patterns emerge from the observed base variation that occurs at any base position, both absolutely and relative to that at any and all other sites. These patterns suggest the existence of particular substitution processes that can bias the instantaneous frequency of particular substitution types (e.g. transitions), the relative frequency of substitution at different sites, and the change in substitution frequency over evolutionary time. The existence of such processes can be inferred, in some instances, from experimental evidence (e.g. by determining the *in vivo* effect of particular mutations), but the complexity of such processes renders experimental approaches impractical. Another approach is to examine independent evidence for structural and functional constraints on particular mutations (e.g. Kellogg and Juliano, 1997). Inadequate understanding of molecular function limits the utility of this approach to the most obvious examples, e.g. mutations that would result in a premature stop codon in a protein coding sequence (obviously absent in rDNA) or those involving a well-characterised protein binding or base-pairing site. A more practical approach has been to examine the fit of empirically observed sequence patterns to those that would be generated assuming particular substitution processes operating through evolutionary time. The inferred substitution processes can then be incorporated into the evolutionary model assumed by the phylogenetic method. We do not consider here philosophical disagreements concerning the role of models in phylogenetics (e.g. Siddall and Kluge, 1997), although we comment on practical aspects of different phylogenetic methods in section 14.3.3.

14.3.2.1 Among-site rate variation in rDNA sequence evolution

Among various types of possible substitution patterns that have been considered, simulation studies indicate that among-site evolutionary rate variation in a sequence most profoundly affects phylogenetic inference, probably more so than substitution biases (e.g. transition versus transversion rates: Jin and Nei, 1990; Yang *et al.*, 1994; Yang, 1996; Swofford *et al.*, 1996; Sullivan and Swofford, 1997; Hwang *et al.*, 1998). If unaccounted for, substantial among-site rate variation can cause all widely used methods (MP, ML, and distance) to become inconsistent; in other words, adding more data can cause the method to converge on an incorrect tree. Among-site variation also affects estimates of substitution bias (Yang, 1996). Among-site variation probably characterises evolution of most biosequences useful for angiosperm phylogenetics, (e.g. *rbc*L: reviewed in Kellogg and Juliano, 1997; 18S: Nickrent and Soltis, 1995; Soltis *et al.*, 1997; P. Soltis and Soltis, 1998; ITS: Baldwin *et al.*, 1995; Hershkovitz and Zimmer, 1996; 26S: Bult *et al.*, 1995; Kuzoff *et al.*, 1998; 5S: Steele *et al.*, 1991). Unlike protein-coding sequences, however, rDNA does not have codon positions that would suggest obvious rate classes among sites. In the case of rDNA, evidence that paired positions in secondary structure helices coevolve via compensatory mutation has led to the idea that helix positions should be weighted differently from non-helix positions, effectively delineating two discrete rate classes (Wheeler and Honeycutt, 1988; Tillier and Collins, 1998). In angiosperm 18S, the most conserved regions tend to occur in helix regions (which would suggest the opposite variation pattern as predicted by the compensatory mutation model), but some loop positions also are conserved (P. Soltis and Soltis, 1998). Interestingly, the conservation in helices appears to be substantially asymmetrical, i.e. the most highly conserved helix regions are base-paired with regions apparently not so conserved (P. Soltis and Soltis, 1998: 196, Fig. 7.5). The most variable 18S domains generally correspond to terminal helix/loop regions in the secondary structure and include both helix and loop positions. Soltis *et al.* (1997b; P. Soltis and Soltis, 1998) also examined the importance of compensatory mutation in 18S evolution. While 73% of the observed stem substitutions were compensatory, only 19% involved double mutations, i.e. paired bases mutating to alternative paired bases. Base-pairing was destroyed by 27% of stem substitutions. Quantitative patterns in rDNA among-site rate variation may be compounded by qualitative. For example, comparisons of angiosperm ITS sequences (see Fig. 14.2B and Hershkovitz and Zimmer, 1996) indicate that base compositional biases can be localised to different parts of the sequence, implying that qualitative substitution probabilities vary in a site-specific manner.

14.3.2.2 Base substitution bias in rDNA

Ribosomal DNA usually violates a common methodological assumption (i.e. that of unweighted MP) that all bases occur in equal frequencies and evolve to other bases at comparable rates. Presumably because of constraints for maintaining stems in secondary structure, rDNA tends to have > 50% overall GC content. This bias usually but not always is maintained at the more variable positions, as we illustrated for 18S (Fig. 14.2) and ITS (Figs 14.5, 14.6; see also Hershkovitz and Zimmer, 1996). The CT transition rate appears to be (sometimes 2–3 times) higher than the AG rate, while the AT rate can be substantially higher than for the other transversions (see above; Hershkovitz and Zimmer, 1997, submitted ms.; Pridgeon *et al.*, 1997). The exaggerated CT rate may reflect the interchangeability of C and U in pairing with G, and it seems that both AT and CT rates might be associated with methylation/demethylation trends in different lineages (Buckler *et al.*, 1997).

14.3.2.3 Non-stationarity in rDNA evolution

An apparent feature of rDNA evolution that probably impacts phylogenetic inference is non-stationarity, which can be defined as changes in substitution processes
over time and across lineages. These processes might operate at the genomic, organismic, and/or environmental level. Ribosomal DNA is probably especially prone to
non-stationarity because of the myriad of rRNA molecular interactions *in trans*,
e.g. ribosomal proteins, nucleolar RNAs and proteins, and protein translation
factors. In other words, lineage-specific evolution in any of these *trans* factors might
correlate with lineage-specific rDNA evolution. Several types of non-stationarity can
occur independently or in concert, including change in overall evolutionary rate, in
substitution pattern (which yields differences in base composition: Swofford *et al.*,
1996), and in pattern of among-site rate-heterogeneity. Examples of non-stationarity of rDNA evolutionary rate and substitution pattern are evident in Figs 14.3,
14.5, and 14.6. Whereas changes in overall evolutionary rate also are evident for
rbcL (cf. Wilson, *et al.*, 1990; Gaut, *et al.*, 1992), the base composition (as evident
from GC content) appears to be much more stable than for 18S (Fig. 14.2). Changes
in patterns of among-site rate variation do not appear to be proven but probably
exist, given that different regions of each rDNA molecule may be associated with
different functions, hence might evolve with some degree of independence. As an
example, in an admittedly small sample of taxa, Hershkovitz and Lewis (1996)
found that of the 56 variable sites in a 5.8S rDNA alignment of angiosperms and
chlorophycean green algae, only eight varied within both groups. The remainder
varied in only one group or between the two groups. Another possible example is
the occasional departure from the rule that ITS1 is more variable than ITS2,
discussed above. The pattern of among-site rate variation in 18S appears to be
more stationary, i.e. it is similar in different angiosperm lineages (P. Soltis and
Soltis, 1998).

14.3.2.4 Accommodating complex substitution patterns in phylogenetic analysis

Corrections for among-site rate heterogeneity and substitution bias are achieved in
different ways for different methods. For MP, correction is accomplished via
weighting (Swofford *et al.*, 1996; Olmstead *et al.*, 1998). For ML and distance
methods, correction is accomplished via likelihood approximation (Swofford *et al.*,
1996).

Among-site rate variation has been recognised in systematic comparisons of
angiosperm sequences, but methodological adjustments have been rare in practice.
The predominant optimisation criterion in angiosperm molecular phylogenetics has
been MP, usually without weighting (Olmstead *et al.*, 1998). Among-site rate variation could be accommodated in MP by differentially weighting specific sites, for
example according to evidence for molecular functional constraints. Such evidence
usually is insufficient to justify *a priori* weighting (Kellogg and Jualiano, 1997).
Weighting according to codon position is a form of MP correction for among-site
rate variation, but this method still ignores the variation that may occur within
each position. Moreover, and aside from the questions of how weights should be
determined, empirical evidence indicates that the assumption of especially high

homoplasy of the third codon position is not justified always (Kellogg and Juliano, 1997; Olmstead *et al.*, 1998). In the case of rDNA, weights sometimes have been applied according to knowledge of secondary structure, with helix positions weighted less to accommodate compensatory mutation, i.e. downweighting sites constrained to coevolve with paired sites in the rRNA secondary structure (Wheeler and Honeycutt, 1988; Tillier and Collins, 1998). As with codon position weighting, this procedure does not adjust for rate variation within each position class (helix and non-helix) and empirical evidence indicates that neither a high rate of compensatory mutation nor a fixed ratio of helix to non-helix rates can be assumed *a priori* (Hwang *et al.*, 1998). For example, as discussed above, P. Soltis and Soltis (1998) determined that global downweighting of angiosperm 18S helix positions in MP analysis was not warranted.

A more general approach to MP correction for among-site rate variation involves multiple rounds of MP analysis with successive reweighting of sites according to sitewise homoplasy indices. The method assumes that the more variable sites also will exhibit more homoplasy than the less variable sites. Success of the method clearly is dependent upon the accuracy of the initial MP tree, however, as this determines the initial pattern of homoplasy. MP underestimation of homoplasy tends to be more severe for the most rapidly evolving sites, however (Yang, 1996), and the most rapidly evolving sites are likely to contribute to long-branch attraction, which would result in a concomitant overestimation of the amount of homoplasy of more slowly evolving sites. Effectively, the successive approximation approach to among-site rate variation correction can be self-defeating (Swofford *et al.*, 1996; Cunningham, 1997; Sullivan *et al.*, 1997; Olmstead *et al.*, 1998), especially under conditions empirically observed for rDNA evolution (extreme among-site rate variation and tendency to exhibit high branch-length heterogeneity). An alternative, but related, approach is to assign a weight of zero (i.e. remove) to sites exceeding some homoplasy threshold. Both intrinsic (i.e. the tree for those data and taxa) and extrinsic (i.e. trees for other taxa using a homologous sequence) criteria can be used to identify the highly homoplastic sites. This method has an advantage over successive weighting in that the set of most homoplastic sites will tend to be the same over a range of nearly optimal trees, and, in any case, erroneous estimation can affect only the sites near the threshold, rather than all of the sites. At the same time, the method will tend to reduce resolution near the tree terminals (Olmstead *et al.*, 1998).

'The standard approach to characterising among-site rate variation is to use a statistical distribution' (Yang, 1996; cf. Kellogg and Juliano, 1997). The statistical approach is applied in ML and distance analysis. The rate variation parameter is alpha (α), the shape of the gamma distribution of the proportion of sites that evolve at a given rate (Swofford *et al.*, 1996; Yang, 1996). Another statistical approach is to estimate a proportion of sites that are invariant (ρ or p_i) with the remainder assumed to vary either at the same rate or according to a gamma distribution. Alpha and the invariant sites proportion can be estimated using ML and a 'reasonable' tree. An MP approach can also be used to estimate alpha, but the estimate is considered less reliable than an ML-based estimate (Yang, 1996). The statistical significance of the rate variation estimates (whether α or p_i or a combined α/p_i estimate explain the data better than the assumption of no rate variation) can be determined using a likelihood ratio test. All of these procedures can be carried out using PAUP*. The advantage of the statistical approach over MP successive weighting is that it is much less sensitive to initial tree topology. It is, however, sensitive to the estimate of substitution bias (Yang, 1996).

Correction methods for substitution bias parallel those for among-site rate variation, i.e. correction is achieved by weighting in MP (e.g. Cunningham, 1997; Olmstead *et al.*, 1998) and statistically in ML and distance (Swofford *et al.*, 1996). The substitution bias corrections behave more similarly in MP and ML, because the bias parameters are constants factored into all substitutions and are not site-specific. In other words, the bias for a particular substitution at one site does not depend directly upon the bias at a different site. ML correction differs from MP weighting, however, in that branch length is a factor, such that uncommon substitutions are more likely as the amount of total substitution increases. Phylogenetic analysis appears to be less sensitive to error in parameters for substitution bias than for among-site rate variation (e.g. Yang, 1998).

The effect of various forms of non-stationarity on different methods using different data-sets is not explored well. MP is sensitive to change in overall evolutionary rates because of long-branch effects, whereas ML is considered insensitive (Swofford *et al.*, 1996). All methods except minimum evolution using log-det distance may be sensitive to changes in base substitution pattern (as detected by pronounced differences in base composition), but log-det is sensitive to among-site rate heterogeneity (Swofford *et al.*, 1996). While log-det seems to be a good method for examining effects of differing base compositional biases, the method appeared to perform poorly in two particular rDNA analyses (Hershkovitz and Lewis, 1996; Hershkovitz and Zimmer, 1997). For these data-sets, all methods except log-det (MP, ML, and time-reversible distance) yielded topologies reasonably concordant with each other and with non-molecular evidence. At present, there does not appear to be a program that corrects for possible non-stationarity in patterns of among-site rate variation. Both statistical and successive weighting approaches might be sensitive to this form of non-stationarity. In this case, ignoring among-site rate variation might even be advantageous. Finally, we note that precision and accuracy in estimation of stationarity might be difficult to achieve. This requires a demonstration that the substitution pattern in one portion of a tree is significantly different than in another. Sullivan *et al.* (in press) generated and resolved 40-sequence data-sets using specified α and p_i values and then re-estimated these parameters in all possible portions of the resulting trees. They found that estimates of these known parameter values became increasingly inaccurate as the number of taxa in the subtree was decreased below 20. They concluded that for parameter estimation in real data-sets, the effect of sampling error might mimic the effect of true non-stationarity among taxon partitions. This result is disconcerting considering, for example that suspected non-stationarity cases highlighted in the present paper involve fewer than 10 sequences (Figs 14.2, 14.5, 14.6).

14.3.3 Resolvability of rDNA data-sets

The various rDNA sequences usually are exploited on the basis of the relative amount of divergence among the taxa. The pattern of divergence is important also.

The divergence pattern of a data-set is a function of the distribution of molecular divergence in relation to the tree branches. If divergence is concentrated on terminal branches, resolvability of internal branches will be poor, even if divergences among terminals appears to be in the phylogenetically useful range. A pattern of long and short branches might reflect the pattern of organismal divergence or the tendency of the molecule to evolve multimodally (rather like punctuated equilibria). Either way, a tree with branchings evenly dispersed is easier to resolve than one with branches clumped in various ways (Kim, 1998). If the organismal divergence pattern reflects rapid radiations, however, a sequence might appear to have poor resolving power, yet still be the best sequence for that particular problem.

Understanding of angiosperm rDNA divergence patterns is rudimentary at best. For example, as discussed above, it is clear that divergence amounts of angiosperm 18S are less than for *rbc*L. Fig. 14.5 illustrates that the divergence patterns of these two molecules can differ markedly (e.g. in Orchidaceae). Thus, the *rbc*L data provides some indication that the resolvability of 18S reflects the pattern of molecular rather than organismal divergence. Less is known about the quality of ITS and 26S relative to other molecules. Moreover, not even divergence amounts for angiosperm rDNA sequences are characterised adequately. As discussed above, angiosperm sequence analyses rarely consider substitution pattern, which affects estimates of divergence amounts. Uncorrected divergences are known mainly from branch lengths in MP trees, and these can be underestimated grossly (Huelsenbeck, 1998). Finally, estimates of 'true' divergence patterns, which should be based on a propensity of evidence (including morphology and palaeontology), are not available generally, either because data are lacking or simply not yet synthesised.

Methods that have been applied to evaluate both the absolute and relative resolving power of rDNA, as well as the ability of a data set to discriminate among different topologies, include resampling (bootstrap and jackknife) analyses, data randomisation and permutation tests, homoplasy estimation, as well as the tests of topological differences we applied in the monocot analysis above (Huelsenbeck and Hillis, 1996). Detailed discussion of these and other evaluation methods is beyond the scope of this chapter, except to the degree that they have been and can be applied in angiosperm rDNA analyses. Each of these methods can yield misleading results.

The 'bootstrap' commonly applied in phylogenetic analysis is a non-parametric resampling of the data and re-analysis of replicates using the same method. The method corrects for potential bias in the original data sample relative to a hypothetical universe of data. Interpretation of bootstrap values remains a complex issue (Swofford *et al.*, 1996). Regardless of the subjective or objective significance attached to these values, we emphasise Swofford *et al.*'s (1996) point that the values themselves are only as valid as the inference method. Using one method or model, the bootstrap may indicate strong support for a node only weakly supported using a different method or model. The effect is dramatically illustrated by Hwang *et al.* (1998: 475, Table 3), in which five models/methods yielded 79–99% support for a node lacking in the bootstrap majority rule consensus of a sixth. In this case, the five analyses indicating strong support applied the more commonly used models/methods, whereas the sixth (ML plus gamma correction), better fits most data-sets, yet has rarely been applied (see also Sullivan and Swofford, 1997). Thus, regardless of the quantitative relationship between bootstrap support and confidence, it seems advisable to apply the bootstrap using different models and methods. It is not clear how many strongly supported nodes in the body of published angiosperm unweighted MP bootstrap trees would collapse if ML plus gamma were applied. In the case of our monocot analysis above, it is reassuring that the two models yielded very similar bootstrap values.

The parametric bootstrap (reviewed in Huelsenbeck *et al.*, 1996) has several applications in assessing not only resolvability and resolution *per se*, but also methodological bias in assessing resolvability and resolution (Sullivan and Swofford, 1997; Hwang *et al.*, 1998; Huelsenbeck, 1998). The most fundamental difference between parametric and non-parametric bootstraps is that the parametric does not re-sample the original data: it re-samples the universe for hypothetical data that conform to the model assumed in analysing the original data. Thus, one can evaluate a score for an optimal tree in relation to the expected distribution of tree scores given the tree and the assumed model of substitution. The method is computationally intensive, and does not appear to have been exploited thus far in angiosperm analyses.

Phylogenetic signal content in data-sets often have been examined using skewness (data randomisation: Hillis and Huelsenbeck, 1992) and PTP (permutation tail probability: Faith and Trueman, 1996) tests. In practice, 18S-26S ribosomal sequence data-sets typically will exhibit significant phylogenetic signal using these methods. This should not be interpreted to mean that signal is uniform over the entire data-set, nor that a tree derived from the data is correct (see also Lamboy, 1996; Carpenter *et al.*, 1998). As indicated in Fig. 14.5, signal for the western American Portulacaceae ITS data-set is significantly structured according to the PTP test, yet a data-set comprising one sequence from each of eight non-nested clades showed no significant structure. In this case, the five-fold increase in taxon sampling does not improve resolvability among these clades. In other words, ITS is useful for resolving some but certainly not all relationships among these sequences. Note that this should not be viewed as a failure of rDNA; rather, it invites explanation of why the signal is inadequate. Morphological data also are resolvable poorly in this case (Hershkovitz, 1993), suggesting that the ITS data simply are reflecting a rapid radiation of these clades. In other words, the ITS data are recovering the 'true' organismal divergence pattern and, unlike morphology, doing so using characters largely homologous across all of the sampled taxa. We found a similar loss of skewness when sampling up to 12 taxa scattered along the monocot 18S tree (data not shown). In this case, the *rbc*L data suggests that organismal divergence is not the primary cause, because the same sampling of *rbc*L sequences always was skewed highly.

Resolvability raises two practical issues: the amount of support needed before a particular tree node can be believed (or at least considered credible) and the best way to improve resolution. In phylogenetic practice, the criterion of topology optimisation (regardless of method) appears to have been accepted routinely as sufficient to impart credibility, if not believability, to the topology. Believability generally seems to increase, but not always decrease, in proportion to the degree of statistical support and independent corroboration. For example, we cited above earlier results of 18S analyses of monocot relationships. Statistical support for these relationships had not been investigated previously, and our present results indicate that the support is not significant for most of them. We showed that methodological adjustments can bring 18S evidence more in accord with *rbc*L evidence, but this exercise improved only the apparent accuracy, not the degree of support for relationships. If, at the other extreme, credibility of molecular evidence for phylogenetic relationships was limited to those nodes strongly supported by bootstrap values, much of past discussion on 18S evidence for monocot relationships, and some of what we present here, as well as a non-trivial proportion of the results of all molecular phylogenetic studies, might be dismissed as folly.

For improving phylogenetic resolution, our results contribute to the evidence that increasing taxon and character sampling offers much more than increasing analytical intensity. As we noted, increasing analytical intensity (i.e. applying complex substitution models using maximum likelihood) only increased the presumed accuracy of the 18S and combined data trees, thereby possibly demonstrating systematic error in our MP analysis. Again, we did not identify the exact source of error, nor examine whether computationally easier methods (distance or weighted parsimony) would have yielded a comparable correction. In fact, we could not even demonstrate that the ML and MP topologies were significantly different. For *rbc*L, it is not clear whether the MP or ML tree is more accurate. Bootstrap support under MP and ML did not appear to be substantially different for any data-set. In contrast, improvement in resolution was substantial when 18S was combined with *rbc*L data, regardless of method. The practical significance of these results is obvious when one considers the computational intensity required for ML. Our ML analysis, even with methodological corners cut relative to our MP analysis, would not have been feasible but for our access to 10 dedicated high-speed computers.

If the goal of a phylogenetics investigation is to compare the reliability of trees generated using different methods, an analytically intensive approach is necessary. If the goal of the program is to maximise the number of clades that can be considered resolved in terms of support, it might be more efficient to defer ML searches, concentrate effort on data gathering, analyse data using the faster and reasonably robust (Yang, 1998) MP methods, and use ML primarily as a tool for examining the potential MP pitfalls. This suggestion follows from other recently published results. Huelsenbeck (1998; Huelsenbeck, 1997; see also Hwang *et al.*, 1998) re-examined Whiting *et al.*'s (1997) data for insect rDNA: 'Searching the space of trees for the maximum likelihood tree proved difficult, especially for the complex models of DNA substitution . . .'. Huelsenbeck's three-parameter model is simpler than the six-parameter model we used in our monocot analysis, but because the the number of sequences analysed, tree space was explored less thoroughly. Huelsenbeck demonstrated that the Strepsiptera and Diptera branches were sufficiently long so that MP would attach them no matter what their relative position in the true tree (notwithstanding the prediction that the ML model might tend to detach them even if they truly were attached). The ML trees for 18S and combined 18S + 28S also attached these branches, but Huelsenbeck stressed that neither monophyly nor non-monophyly was supported statistically under the ML criterion. Statistical support for monophyly under the MP criterion, however, was not claimed by Whiting *et al.* (1997), was not tested by Huelsenbeck, and seems unlikely given the small MP tree length differences (< 1%). In this case, if the resolvability criteria were equivalent, Whiting et al.'s (1997) and Huelsenbeck's (1998) conclusions would have been the same, at least for the 18S and combined 18S + 28S data. Hwang et al. (1998) also examined this phylogenetic question. As alluded to above, several methods yielded strong bootstrap support for the Strepsiptera + Diptera clade but the support vanished using ML plus gamma. Again, however, they were not able to reject monophyly of Strepsiptera + Diptera. Another example is Sullivan and Swofford's (1997) re-analysis of a mammalian multigene mtDNA data set. The original analysis (D'Erchia et al., 1996) had yielded allegedly strong evidence (boot-strap support high for MP and distance, although modest for ML) that rodents are polyphyletic, but re-analysis showed that the support vanished when the empiri-cally estimated degree of among-site rate variation was incorporated into the analytical model. This represents a textbook case of inconsistency: the large amount of analysed data (covering 10 kb of sequence), intuitively believed to enhance reli-ability of the conclusions, is precisely what amplified erroneous support for an essentially random resolution. As in the insect example, the optimal ML tree in Sullivan and Swofford's analysis agreed with the original trees but lacked statis-tical support. Taxon sampling in this case was especially poor, again suggesting that the obstacle to resolution was lack of taxa rather than lack of analysis. In all of these cases, ML analysis was valuable heuristically, but the degree of positive phylogenetic resolution could not be considered improved. Huelsenbeck noted, however, that branch length estimates are much more accurate using the ML model. If accurate branch lengths are desired, they can be estimated relatively easily and quickly using ML over any topology, and Huelsenbeck's results suggest that the original MP topology would have sufficed. Thus, likelihood ratio tests can be used to examine nested evolutionary models that include additional parameters. ML can be used to detect extreme branch lengths. If branch attraction is suspected, alter-native MP constraints can be devised. The Kishino–Hasagawa test can be used to compare alternative MP trees (constraint trees, incrementally suboptimal trees, and trees derived using substitution or successive weighting) under a reasonable ML model. One can then consider whether a full ML tree search will yield a topology significantly better supported than alternative trees. In our monocot analyses and in Sullivan and Swofford's (1997) and Huelsenbeck's (1998) analyses, it did not. One should also keep in mind that even if one model appears to fit the data signif-icantly better than another, it is probably not the 'true' model (e.g. stochastic ML models applied in practice usually do not incorporate empirical evidence for secondary structure or other empirically evident constraints), that there is a range of models that are not significantly worse that the optimised model (because of variance), and that the estimation of parameters is sensitive to parameters not esti-mated (Yang, 1996). If one adopts a probabilistic approach, one should follow through with a probabilistic interpretation of the results. The criterion of strict parsimony that the optimised tree is better than any suboptimal tree is inapplic-able to ML. An optimised ML tree cannot be considered statistically better than trees ML cannot reject. If the goal of phylogenetic analysis is to confidently reject wrong trees, the economics of more analysis versus more data is worth consid-ering. We close with two caveats. First, as sequencing technology continues to improve, increasing amounts of data might be expected to aggravate the practical problem of statistical inconsistency in phylogenetic methods. At the same time, it might not, for example if long branches in one data-set are not long proportionally in another. Second, computation technology and ML search algorithms are likely to improve also. For example, the quartet puzzling (QP) algorithm for ML tree searching (Strimmer and von Haeseler, 1996) is much faster than the global branch swapping algorithms in other programs. While QP might be less stringent then global branch swapping, a QP ML tree optimised using a particular model should serve as a reasonable approximation for comparing trees derived using different models and methods (e.g., Hwang *et al.*, 1998). Thus, for larger data-sets, QP appears to offer a reasonably efficient means for exploring ML trees.

14.4 Future prospects for rDNA in angiosperm systematics

Ribosomal DNA data are demonstrably useful in angiosperm phylogenetics. They have provided, in addition to powerful phylogenetic evidence, an easily obtained character set for advancing the understanding of molecular evolution at all phylogenetic levels. The contributions of rDNA to angiosperm phylogenetics undoubtedly will increase as additional data accumulate, especially for 26S and ITS.

Advances in molecular evolutionary theory and phylogenetic methodology carry strong implications for past, present, and future applications of rDNA data to angiosperm phylogenetics. Emphasis in angiosperm rDNA systematics has been on generating data, running a program that finds 'the tree' (Swofford *et al.*, 1996), and discussion of 'what goes where' taxonomically. The 'utility' of rDNA data was evaluated mainly in terms of the tree resolution and concurrence with non-rDNA trees. Current phylogenetic theory and methods demonstrate that phylogenetic results can be swayed by sometimes subtle differences in analytical method, and that statistically well-supported resolution (and, consequently, conflict between different data-sets) is generally less than apparent in optimal trees. Thus, current applications of rDNA data, as well as considerations of previous work, should include increased emphasis on the assumptions of the method, whether the data fit the assumptions, and how violations of assumptions might affect the results. This consideration requires only understanding of the different methods. Application of the more computationally intensive methods may not be required if it appears that the problem can be resolved more easily with additional data. For the future, angiosperm rDNA phylogenetics would benefit from a strategic shift in approach. While phylogenetic reconstruction should remain a major motivation, we should accept that fully and unequivocally resolved phylogenies generally are not to be expected from single or even multiple data-sets. The emphasis should shift from resolving trees *per se* to focusing on and critically analysing results of particular interest. Ribosomal DNA should not be viewed merely as phylogenetic data, but as a suite of characters essential to life. Ribosomal DNA variation patterns should be viewed in the context of natural history, i.e. as information about biotic diversity and evolutionary history. The extraction of this information and its biological significance will compensate for less than complete and strongly supported cladistic resolution. As an example, relatively few nodes were resolved confidently by our portulacaceous ITS data-set (Hershkovitz and Zimmer, 1997), yet this resolution was sufficient to yield novel observations, hypotheses, and conclusions on portulacaceous phylogeny, taxonomy, biogeography, palaeoecology, and molecular evolution. As a more extreme example, as discussed above, even the completely unresolvable *Gossypium* 5S data (Cronn *et al.*, 1996) document genetic and evolutionary phenomena that beg explanation. Ribosomal DNA data undoubtedly have made significant and indelible contributions to angiosperm systematics. Viewed through new and more critical perspectives, they promise to contribute a great deal more.

ACKNOWLEDGEMENTS

We thank Dave Swofford for permission to use test versions of PAUP* 4.0, and Richard Bateman, Michael Möller, Richard Olmstead, Douglas Soltis, and Jonathan Wendel for review and helpful suggestions.

REFERENCES

Adachi, J., Watanabe, K., Fukui, K., Ohmido, N. and Kosuge, K. (1997) Chromosomal location and reorganization of the 45S and 5S rDNA in the *Brachyscome lineariloba* complex (Asteraceae). *Journal of Plant Research*, **110** (1099), 371–377.

Ainouche, M. L. and Bayer, R. J. (1997) On the origins of the tetraploid *Bromus* species (section *Bromus*, Poaceae): insights from internal transcribed spacer sequences of nuclear ribosomal DNA. *Genome*, **40**, 730–743.

Appels, R. and Baum, B. (1992) Evolution of the NOR and 5S DNA loci in the Triticeae, in *Molecular systematics of plants*, (eds P. S. Soltis, D. E. Soltis and J. J. Doyle), Chapman and Hall, New York, pp. 92–116.

Appels, R. and Honeycutt, R.L. (1986) rDNA: evolution over a billion years, in *DNA systematics, vol. 2, plants*, (ed. S. K. Dutta), CRC Press, Boca Raton, Florida, pp. 81–135.

Baldwin, B. G. (1991) Nuclear ribosomal DNA phylogenetics of the tarweeds (Madiinae: Asteraceae). *American Journal of Botany*, **78** (suppl.), 165–166.

Baldwin, B. G. and Markos, S. (1998) Phylogenetic utility of the external transcribed spacer (ETS) of 18S-26S rDNA: congruence of ETS and ITS trees of *Calycadenia* (Compositae). *Molecular Phylogenetics and Evolution*, **10**, 449–463.

Baldwin, B. G., Sanderson, M. J., Porter, J. M., Wojciechowski, M. F., Campbell, C. S. and Donoghue, M. J. (1995) The ITS region of nuclear ribosomal DNA: a valuable source of evidence on angiosperm phylogeny. *Annals of the Missouri Botanical Garden*, **82**, 247–277.

Baum, B. R. and Bailey, L.G. (1997) The molecular diversity of the 5S rRNA gene in *Kengyilia alatavica* (Drobov) J. L. Yang, Yen & Baum (Poaceae: Triticeae): potential genomic assignment of different rDNA units. *Genome*, **40**, 215–228.

Baum, B. R. and Johnson, D. A. (1996) The 5S rRNA gene units in ancestral two-rowed barley (*Hordeum spontaneum* C. Koch) and bulbous barley (*H. bulbosum* L.): sequence analysis and phylogenetic relationships with the 5S rDNA units of cultivated barley (*H. vulgare* L.). *Genome*, **39**, 140–149.

Bena, G., Jubier, M. -F., Olivieri, I. and Lejeune, B. (1998) Ribosomal external and internal transcribed spacers: combined use in phylogenetic analysis of *Medicago* (Leguminosae). *Journal of Molecular Evolution*, **46**, 299–306.

Bharathan, G. and Zimmer, E. A. (1995) Early branching events in monocotyledons: partial 18S ribosomal DNA sequence analysis, in *Monocotyledons: systematics and evolution*, (eds P. J. Rudall, P. J. Cribb, D. F. Cutler and C. J. Humphries), Royal Botanic Gardens, Kew, pp. 81–107.

Bittrich, V. (1993) Introduction to Centrospermae, in *The families and genera of vascular plants, vol. 2*, (eds K. Kubitzki, V. Bittrich and J. Rohwer), Springer, Berlin, pp. 13–19.

Borisjuk, N. V., Davidjuk, Y. M., Kostishin, S. S., Miroshnichenco, G. P., Velasco, R., Hemleben, V. and Volkov, R. A. (1997). Structural analysis of rDNA in the genus *Nicotiana*. *Plant Molecular Biology*, **35**, 655–660.

Boulter. D. and Gilroy, J. S. (1992) Partial sequences of 18S ribosomal RNA of two genera from each of six flowering plant families. *Phytochemistry*, **31**, 1243–1246.

Bremer, K., Bremer, B. and Thulin, M. (1997) *Introduction to the phylogeny and systematics of flowering plants*, 3rd ed. Department of Systematic Botany, Uppsala University, Uppsala.

Buckler, E. S. 4th, Ippolito, A. and Holtsford, T. P. (1997) The evolution of ribosomal DNA: divergent paralogues and phylogenetic implications. *Genetics*, **145**, 821–832.

Bult, C. J., Sweere, J. A. and Zimmer, E. A. (1995) Cryptic sequence simplicity, nucleotide composition bias and molecular coevolution in the large subunit of ribosomal DNA in plants: implications for phylogenetic analyses. *Annals of the Missouri Botanical Garden*, **82**, 235–246.

Campbell, C. S., Wojciechowski, M. F., Baldwin, B. G., Alice, L. A. and Donoghue, M. J. (1997) Persistent nuclear ribosomal DNA sequence polymorphism in the *Amelanchier* agamic complex (Rosaceae). *Molecular Biology and Evolution*, **14**, 81–90.

Carpenter, J. M., Goloboff, P. A. and Farris, J. S. (1998) PTP is meaningless, T-PTP is contradictory: a reply to Trueman. *Cladistics*, **14**, 105–116.

Cerbah, M., Couland, J. and Siljak-Yakovlev, S. (1998) rDNA organization and evolutionary relationships in the genus *Hypochaeris* (Asteraceae). *Journal of Heredity*, **89**, 213–318.

Chase, M. W., Soltis, D. E., Olmstead, R. G., Morgan, D., Les, D. H., Mishler, B. D. *et al.* (1993) Phylogenetics of seed plants: an analysis of nucleotide sequences from the plastid gene rbcL. *Annals of the Missouri Botanical Garden*, **80**, 528–580.

Chase M. W., Duvall, M. R., Hills, H. G., Conran, J. G., Cox, A. V., Eguiarte, L. E. *et al.* (1995) Molecular phylogenetics of Lilianae, in *Monocotyledons: systematics and evolution*, (ed P. J. Rudall, P. J. Cribb, D. F. Cutler and C. J. Humphries), Royal Botanic Gardens, Kew, pp. 109–137.

Chaw, S. M., Sung, H. M., Long, H., Zharkikh, A. and Li, W.-H. (1995) The phylogenetic positions of the conifer genera *Amentotaxus*, *Phyllocladus*, and *Nageia* inferred from 18S rRNA sequences. *Journal of Molecular Evolution*, **41**, 224–230.

Chenuil, A., Solignac, M. and Bernard, M. (1997) Evolution of the large-subunit ribosomal RNA binding site for protein L23/25. *Molecular Biology and Evolution*, **14**, 578–588.

Clement, J. S. and Mabry, T. J. (1996) Pigment evolution in the Caryophyllales: a systematic overview. *Botanica Acta*, **109**, 360–367.

Comes, H. P. and Abbott, R. J. (1999) Reticulate evolution in the Mediterranean species complex of *Senecio* sect. *Senecio*: Uniting phylogenetic and population-level approaches, in *Molecular systematics and plant evolution*, (eds P. M. Hollingsworth, R. M. Bateman and R. J. Gornall), Taylor & Francis, London, pp. 171–198.

Conn, G. L. and Draper, D. E. (1998) RNA structure. *Current Opinions in Structural Biology*, **8**, 278–285.

Crawford, D. J. (1991) *Plant molecular systematics*. John Wiley and Sons, New York.

Cronn, R. C., Zhao, X., Paterson, A. H. and Wendel, J. F. (1996) Polymorphism and concerted evolution in a tandemly repeated gene family: 5S ribosomal DNA in diploid and allopolyploid cottons. *Journal of Molecular Evolution*, **42**, 685–705.

Cunningham, C. W. (1997) Is incongruence between data partitions a reliable predictor of phylogenetic accuracy? Empirically testing an iterative procedure for choosing among phylogenetic models. *Systematic Biology*, **46**, 464–478.

Davis, J. I., Simmons, M. P., Stevenson, D. W. and Wendel, J. W. (1998) Data decisiveness, data quality, and incongruence in phylogenetic analysis: an example from the monocotyledons using mitochondrial atpA sequences. *Systematic Biology*, **47**, 282–310.

D'Erchia. A. M., Gissi, C., Pesole, G., Saccone, C. and Arnason, U. (1996) The guinea-pig is not a rodent. *Nature*, **381** (6583), 597–600.

De Rijk, P., Van de Peer, Y., Van den Broeck. I. and De Wachter, R. (1995) Evolution according to large ribosomal subunit RNA. *Journal of Molecular Evolution*, **41**, 366–375.

Donald, T. M., Houben, A., Leach, C. R. and Timmis, J. N. (1997) Ribosomal RNA genes specific to the B chromosomes in *Brachycome dichromosomatica* are not transcribed in leaf tissue. *Genome*, **40**, 674–681.

Doyle, J. J. (1993) DNA, phylogeny, and the flowering of plant systematics. *BioScience*, **43**, 380–389.

Doyle, J. A., Jutton, C. L. and Ward, J. V. (1990) Early Cretaceous tetrads, zonasulculate pollen, and Winteraceae. II. Cladistic analysis and implications. *American Journal of Botany*, **77**, 1558–1568.

Doyle, J. A., Donoghue, M. J. and Zimmer, E. A. (1994) Integration of morphological and ribosomal RNA data on the origin of angiosperms. *Annals of the Missouri Botanical Garden*, **81**, 419–450.

Duff, R. J. and Nickrent, D. L. (1997) Characterization of mitochondrial small-subunit ribosomal RNAs from holoparasitic plants. *Journal of Molecular Evolution*, **45**, 631–639.

Duvall, M. R., Clegg, M. T., Chase, M. W., Clark, W. D., Kress, W. J., Zimmer, E. A. *et al.* (1993a) Phylogenetic hypotheses for the monocotyledons constructed from *rbcL* sequence data. *Annals of the Missouri Botanical Garden*, **80**, 607–619.

Duvall, M. R., Learn, G. H. Jr., Eguiarte, L. E. and Clegg, M. T. (1993b) Phylogenetic analysis of *rbcL* sequences identifies *Acorus calamus* as the primal extant monocotyledon. *Proceedings of National Academy of Sciences USA*, **90**, 4641–4644.

Embley, M. T., Hirt, R. P. and Williams, D. M. (1994) Biodiversity at the molecular level: the domains, kingdoms, and phyla of life. *Philosophical Transactions of the Royal Society*, **B345** (1311), 21–31.

Faith, D. P. and Trueman, J. W. H. (1996) When the topology-dependent permutation test (T-PTP) for monophyly returns significant support for monophyly, should that be equated with (a) rejecting a null hypothesis of non-monophyly, (b) rejecting a null hypothesis of 'no structure,' (c) failing to falsify a hypothesis of monophyly, or (d) none of the above? *Systematic Biology*, **45**, 580–586.

Farris, J. S., Källersjo, M., Kluge, A. G. and Bult, C. (1994) Testing the significance of congruence. *Cladistics*, **10**, 315–319.

Fransz, P., Armstrong, S., Alonso-Blanco, C., Fischer, T. C., Torres-Ruiz, R. A. and Jones, G. (1998) Cytogenetics for the model system *Arabidopsis thaliana*. *Plant Journal*, **16**, 867–876.

Futuyma, D. J. 1997. *Evolutionary biology*, 3rd ed. Sinauer, Sunderland Massachusetts.

Gaut, B. S., Muse, S. V., Clark, W. D. and Clegg, M. T. (1992) Relative rates of nucleotide substitution at the *rbcL* locus in monocotyledonous plants. *Journal of Molecular Evolution*, **35**, 292–303.

Gerbi, S. A. (1985) Evolution of ribosomal RNA, in *Molecular evolutionary genetics*, (ed. R. J. MacIntyre), Plenum Press, New York, pp. 419–518.

Goremykin, V., Bobrova, V., Pahnke, J., Troitsky, A., Antonov, A. and Martin, W. (1997) Noncoding sequences from the slowly evolving chloroplast inverted repeat in addition to *rbcL* data do not support gnetalean affinities of angiosperms. *Molecular Biology and Evolution*, **13**, 383–396.

Gutell, R. R., Larsen, N. and Woese, C. R. (1994) Lessons from evolving rRNA: 16S and 23S rRNA structures from a comparative perspective. *Microbiological Review*, **58**, 10–26.

Hahn, W J., Kress, W.J. and Zimmer, E.A. (1996) 18S nrDNA sequence phylogenetics of the monocots. *American Journal of Botany*, **83** (suppl.), 211–212.

Hamby, R. K. and Zimmer, E. A. (1988) Ribosomal RNA sequences for inferring phylogeny within the grass family (Poaceae). *Plant Systematics and Evolution*, **160**, 29–37.

Hamby, R. K. and Zimmer, E. A. (1992) Ribosomal RNA as a phylogenetic tool in plant systematics, in *Molecular systematics of plants*, (eds P. S. Soltis, D. E. Soltis and J. J. Doyle), Chapman and Hall, New York, pp. 50–101.

Harshman, J. (1994) The effect of irrelevant characters on bootstrap values. *Systematic Biology*, **43**, 419–424.

Hershkovitz, M. A. (1993) Revised circumscription and sectional taxonomy of *Calandrinia* Kunth and *Montiopsis* Kuntze (Portulacaceae) with notes on phylogeny of the portulacaceous alliance. *Annals of the Missouri Botanical Garden*, **80**, 333–365.

Hershkovitz, M. A. and Lewis, L. A. (1996). Deep-level diagnostic value of the rDNA-ITS region. *Molecular Biology and Evolution*, **13**, 1276–1295.

Hershkovitz, M. A. and Zimmer, E. A. (1996) Conservation patterns in angiosperm ITS2 sequences. *Nucleic Acids Research*, **24**, 2857–2867.

Hershkovitz, M. A. and Zimmer, E. A. (1997) On the evolutionary origins of the cacti. *Taxon*, **46**, 217–232.

Hillis, D. M. (1996) Inferring complex phylogenies. *Nature*, **383** (6596), 130.

Hillis, D. M. and Dixon, M. T. (1991) Ribosomal DNA: Molecular evolution and phylogenetic inference. *Quarterly Review of Biology*, **66**, 411–453.

Hillis, D. M. and Huelsenbeck, J. P. (1992) Signal, noise, and reliability in molecular phylogenetic analyses. *Journal of Heredity*, **83**, 189–195.

Hoot, S. B. and Crane, P. R. (1995) Inter-familial relationships in the Ranunculidae based on molecular systematics, in *Systematics and evolution of the Ranunculiflorae*, (eds U. Jensen and J. W. Kadereit), *Plant Systematics and Evolution*, Suppl. **9**, 119–131.

Hoot, S. B., Culham, A. and Crane, P. R. (1995) The utility of atpB gene sequences in resolving phylogenetic relationships: Comparison with rbcL and 18S ribosomal DNA sequences in the Lardizabalaceae. *Annals of the Missouri Botanical Garden*, **82**, 194–207.

Hori, H. and Osawa, S. (1986) Evolutionary change in 5S rRNA secondary structure and a phylogenic tree of 352 5S rRNA species. *Biosystems*, **19**, 163–172.

Hori, H. and Osawa, S. (1987) Origin and evolution of organisms as deduced from 5S ribosomal RNA sequences. *Molecular Biology and Evolution*, **4**, 445–472.

Huelsenbeck, J. P. (1997) Is the Felsenstein zone a fly trap? *Systematic Biology*, **46**, 69–74.

Huelsenbeck, J. P. (1998) Systematic bias in phylogenetic analysis: is the Strepsiptera problem solved? *Systematic Biology*, **47**, 519–137.

Huelsenbeck, J. P. and Bull, J. J. (1996) A likelihood ratio test to detect conflicting phylogenetic signal. *Systematic Biology*, **45**, 92–98.

Hwang, U. W., Kim, W., Tautz, D and Friedrich, M. (1998) Molecular phylogenetics in the Felsenstein zone: approaching the Strepsiptera problem using 5.8S and 28S rDNA sequences. *Molecular Phylogenetics and Evolution*, **9**, 470–480.

Jin, L. and Nei, M. (1990) Limitations of the evolutionary parsimony method of phylogenetic analysis. *Molecular Biology and Evolution*, **7**, 82–102.

Kamstra, S. A., Kuipers, A. G., De Jeu, M. J., Ramanna, M. S. and Jacobsen, E. (1997) Physical localisation of repetitive DNA sequences in *Alstroemeria*: karyotyping of two species with species-specific and ribosomal DNA. *Genome*, **40**, 652–658.

Karvonen, P. and Savolainen, O. (1993) Variation and inheritance of ribosomal DNA in *Pinus sylvestris* L. (Scots pine). *Heredity*, **71**, 614–622.

Kellogg, E. A. and Appels, R. A. (1995) Intraspecific and interspecific variation in 5S RNA genes are decoupled in diploid wheat relatives. *Genetics*, **140**, 325–343.

Kellogg, E. A. and Juliano, N. D. (1997) The structure and function of RuBisCo and their implications for systematic studies. *American Journal of Botany*, **84**, 413–428.

Kellogg, E. A., Appels, R. and Mason-Gamer, R. J. (1996) When genes tell different stories: the diploid genera of Triticeae (Gramineae). *Systematic Botany*, **21**, 321–348.

Kim, J. (1998) Large-scale phylogenies and measuring the performance of phylogenetic estimators. *Systematic Biology*, **47**, 43–60.

King, K., Torres, R. A., Zentgraf, U. and Hemleben, V. (1993) Molecular evolution of the intergenic spacer in the nuclear ribosomal RNA genes of Cucurbitaceae. *Journal of Molecular Evolution*, **36**, 144–152.

Kishino, H. and Hasegawa, M. (1989) Evaluation of the maximum likelihood estimate of the evolutionary tree topologies from DNA sequence data, and the branching order in Hominoidea. *Journal of Molecular Evolution*, **29**, 170–179.

Kress, W. J., Hahn, W. J., Evans, T. M. and Zimmer, E. A. (1995) Unraveling the evolutionary radiation of the families of the Zingiberanae using morphological, molecular and fossil characters. *American Journal of Botany*, **82** (suppl.), 142.

Kumazaki, T., Hori, H. and Osawa, S. (1983) Phylogeny of protozoa deduced from 5S rRNA sequences. *Journal of Molecular Evolution*, **19**, 411–419.

Kuzoff, R. K., Sweere, J. A., Soltis, D. E., Soltis, P. S. and Zimmer, E. A. (1998) The phylogenetic potential of entire 26S rDNA sequences in plants. *Molecular Biology and Evolution*, **15**, 251–263.

Lamboy, W. F. (1996) Morphological characters, polytomies, and homoplasy indices: response to Wiens and Hillis. *Systematic Botany*, **21**, 243–253.

Leitch, A. R., Lim, K. Y., Leitch, I. J., O'Neill, M., Chye, M. and Low, F. (1998) Molecular cytogenetic studies in rubber, *Hevea brasiliensis* Muell. Arg. *Genome*, **41**, 464–467.

Li, X., Guo, R., Pedersen, C., Hayman, D. and Langridge, P. (1997) Physical localization of rRNA genes by two-colour fluorescent in-situ hybridization and sequence analysis of the 5S rRNA gene in *Phalaris coerulescens*. *Hereditas*, **126**, 289–294.

Liang, W. Q. and Fournier, M. J. (1997) Synthesis of functional eukaryotic ribosomal RNAs *in trans*: development of a novel in vivo rDNA system for dissecting ribosome biogenesis. *Proceedings of the National Academy of Sciences USA*, **94**, 2864–2868.

Liu, J. S. and Schardl, C. L. (1994) A conserved sequence in internal transcribed spacer 1 of plant nuclear rRNA genes. *Plant Molecular Biology*, **26**, 775–778.

Mai, J. C. and Coleman, A. W. (1997) The internal transcribed spacer 2 exhibits a common secondary structure in green algae and flowering plants. *Journal of Molecular Evolution*, **44**, 258–271.

Maley, L. E. and Marshall, C. R. (1998) The coming of age of molecular systematics. *Science*, **279** (5350), 505–506.

McGrath, J. M. and Helgeson, J. P. (1998) Differential behavior of *Solanum brevidens* ribosomal DNA loci in a somatic hybrid and its progeny with potato. *Genome*, **41**, 435–439.

Möller, M. and Cronk, Q. C. B. (1997) Origin and relationships of *Saintpaulia* (Gesneriaceae) based on ribosomal DNA internal transcribed spacer sequences. *American Journal of Botany*, **84**, 956–965.

Morrison, D. A. and Ellis, J. T. (1997). Effects of nucleotide sequence alignment on phylogeny estimation: a case study of 18S rDNAs of apicomplexa. *Molecular Biology and Evolution*, **14**, 428–41.

Nickrent, D. L. and Franchina, C. R. (1990) Phylogenetic relationships of the Santalales and relatives. *Journal of Molecular Evolution*, **31**, 294–301.

Nickrent, D. L. and Patrick, J. A. (1998) The nuclear ribosomal DNA intergenic spacers of wild and cultivated soybean have low variation and cryptic subrepeats. *Genome*, **41**, 183–192.

Nickrent, D. L. and Soltis, D. E. (1995) A comparison of angiosperm phylogenies from nuclear 18S rDNA and rbcL sequences. *Annals of the Missouri Botanical Garden*, **82**, 208–234.

Nickrent, D. L., Schuette, K. P. and Starr, E. M. (1994) A molecular phylogeny of *Arceuthobium* (Viscaceae) based on nuclear ribosomal DNA internal transcribed spacer sequences. *American Journal of Botany*, **81**, 1149–1160.

Nickrent, D. L., Duff, R. J. and Konings, D. A. (1997) Structural analyses of plastid-derived 16S rRNAs in holoparasitic angiosperms. *Plant Molecular Biology*, **34**, 731–743.

Olmstead, R. G., Reeves, P. A. and Yen, A. C. (1998) Patterns of sequence evolution and implications for parsimony analysis of chloroplast DNA, in *Molecular systematics of plants II: DNA sequencing*, (eds D. E. Soltis, P. S. Soltis and J. J. Doyle), Kluwer Academic Publishers, Boston, pp. 164–164.

Otsuka, J., Nakano, T. and Terai, G. (1996) A theoretical study on the nucleotide changes under a definite functional constraint of forming stable base-pairs in the stem regions of ribosomal RNAs: its application to the phylogeny of eukaryotes. *Journal of Theoretical Biology*, **184**, 171–186.

Oxelman, B. and Lidén, M. (1995a) The position of *Circaeaster* – evidence from nuclear ribosomal DNA, in Systematics and Evolution of the Ranunculiflorae (eds U. Jensen and J. W. Kadereit), *Plant Systematics and Evolution*, Suppl. **9**, 189–193.

Oxelman, B. and Lidén, M. (1995b) Generic boundaries in the tribe Sileneae (Caryophyllaceae) as inferred by nuclear rDNA sequences. *Taxon*, **44**, 525–542.

Peculis, B. (1997) The sequence of the 5' end of the U8 small nucleolar RNA is critical for 5.8S and 28S rRNA maturation. *Molecular and Cellular Biology*, **17**, 3701–3713.

Pillay, M. (1997) Variation of nuclear ribosomal RNA genes in *Eragrostis tef* (Zucc.) Trotter. *Genome*, **40**, 815–821.

Plunkett, G. M., Soltis, D. E. and Soltis, P. S. (1997) Clarification of the relationship between Apiaceae and Araliaceae based on *mat*K and *rbc*L sequence data. *American Journal of Botany*, **84**, 565–580.

Prager, E. M. and Wilson, A. C. (1988) Ancient origin of lactalbumin from lysozyme: analysis of DNA and amino acid sequences. *Journal of Molecular Evolution*, **27**, 326–335.

Pridgeon, A. M., Bateman, R. M., Cox, A. V., Hapeman, J. R. and Chase, M. W. (1997) Phylogenetics of subtribe Orchidinae (Orchidoideae, Orchidaceae) based on nuclear ITS sequences. 1. Intergeneric relationships and polyphyly of *Orchis sensu lato. Lindleyana*, **12**, 89–109.

Rice, W. R. (1989) Analyzing tables of statistical tests. *Evolution*, **43**, 223–225.

Rice, K. A., Donoghue, M. J. and Olmstead, R. G. (1997) Analyzing large data-sets: *rbc*L 500 revisited. *Systematic Biology*, **46**, 554–563.

Rieseberg, L. H. and Brunsfeld, S. J. (1992) Molecular evidence and plant introgression, in *Molecular systematics of plants*, (eds P. S. Soltis, D. E. Soltis and J. J. Doyle), Chapman and Hall, New York, pp. 151–176.

Rieseberg, L. H. and Morefield, J. D. (1995) Character expression, phylogenetic reconstruction and the detection of reticulate evolution, in *Experimental and molecular approaches to plant biosystematics*, (eds P. C. Hoch and A. G. Stephenson), Missouri Botanical Garden, St. Louis, pp. 333–350.

Ro, K.-E., Keener, C. S. and McPheron, B. A. (1997) Molecular phylogenetic study of the Ranunculaceae: Utility of the nuclear 26S ribosomal DNA in inferring intrafamilial relationships. *Molecular Phylogenetics and Evolution*, **8**, 117–127.

Roose, M. L., Schwarzacher, T. and Heslop-Harrison, J. S. (1998) The chromosomes of *Citrus* and *Poncirus* species and hybrids: identification and physical mapping of rDNA chromosomes and physical mapping of rDNA loci using *in situ* hybridization and fluorochrome banding. *Journal of Heredity*, **89**, 83–86.

Samuel, R., Bachmair, A. and Ehrendorfer, F. (1998) ITS sequences from nuclear rDNA suggest unexpected phylogenetic relationships between Euro-Mediterranean, East Asiatic and North American taxa of *Quercus* (Fagaceae). *Plant Systematics and Evolution*, **211**, 129–140.

Schlötterer, C. (1998) Ribosomal DNA probes and primers, in *Molecular tools for screening biodiversity*, (eds A. Karp, P. G. Isaac and D. S. Ingram), Chapman and Hall, London, pp. 267–276.

Schlötterer, C. and Tautz, D. (1994) Chromosomal homogeneity of *Drosophila* ribosomal DNA arrays suggests intrachromosomal exchanges drive concerted evolution. *Current Biology*, **4**, 777–783.

Scoles, G. J., Gill, B. S., Xin, Z. -Y., Clarke, B. C., McIntyre, C. L., Chapman, C. *et al.* (1988) Frequent duplication and deletion events in the 5S RNA genes and the associated spacer regions of the Triticeae. *Plant Systematics and Evolution*, **160**, 105–122

Shi, L., Zhu, T., Morgante, M., Rafalski, J. A. and Keim, P. (1996) Soybean chromosome painting: a strategy for somatic cytogenetics. *Journal of Heredity*, **87**, 308–313.

Siddall, M. E. and A. G. Kluge (1997) Probabilism and phylogenetic inference. *Cladistics*, **13**, 313–336.

Singer, M. and Berg, P. (1991) *Genes and genomes*. University Science Books, Mill Valley, California.

Soltis, D. E. and Soltis, P. S. (1998) Choosing an approach and an appropriate gene for phylogenetic analysis, in *Molecular systematics of plants II: DNA sequencing*, (eds D.E. Soltis, P. S. Soltis and J. J. Doyle), Kluwer Academic Publishers, Boston, pp. 1–42.

Soltis, D. E., Hibsch-Jetter, C., Soltis, P. S., Chase, M. W. and Farris, J. S. (1997a) Molecular phylogenetic relationships among angiosperms: an overview based or *rbc*L and 18S rDNA sequences, in *Evolution and diversification of land plants* (eds K. Iwatsuki and P. H. Raven), Springer, Berlin, pp. 157–178.

Soltis, D. E., Soltis, P. S., Nickrent, D. L., Johnson, L. A., Hahn, W. J., Hoot, S. B. *et al.* (1997b) Angiosperm phylogeny inferred from 18S ribosomal DNA sequences. *Annals of the Missouri Botanical Garden*, **84**, 1–49.

Soltis, D. E., Soltis, P. S., Mort, M. E., Chase, M. W., Savolainen, V., Hoot, S. B. *et al.* (1998) Inferring complex phylogenies using parsimony: an empirical approach using three large DNA data-sets for angiosperms. *Systematic Biology*, **47**, 32–42.

Soltis, P. S. and Soltis, D. E. (1995) Introduction. *Annals of the Missouri Botanical Garden*, **82**, 147.

Soltis, P. S. and Soltis, D. E. (1998) Molecular evolution of 18S rDNA in angiosperms: implications for character weighting in phylogenetic analysis, in *Molecular systematics of plants II: DNA sequencing*, (eds D. E. Soltis, P. S. Soltis and J. J. Doyle), Kluwer Academic Publishers, Boston, pp. 188–210.

Steele, K. P., Holsinger, K. E., Jansen, R. K. and Taylor, D. W. (1991) Assessing the reliability of 5S rRNA sequence data for phylogenetic analysis. *Molecular Biology and Evolution*, **8**, 240–248.

Stefanovic, S., Jager, M. and Masselot, M. (1998) Phylogenetic relationships of conifers inferred from partial 28S rRNA gene sequences. *American Journal of Botany*, **85**, 688–703.

Strimmer, K and von Haeseler, A. (1996) Quartet puzzling – a quartet maximum likelihood method for reconstructing tree topologies. *Molecular Biology and Evolution*, **13**, 964–969.

Suh, Y., Thien, L. B., Reeve, H. and Zimmer, E. A. (1993) Molecular evolution and phylogenetic implications of internal transcribed spacer sequences of ribosomal DNA in Winteraceae. *American Journal of Botany*, **80**, 1042–1055.

Sullivan, J. and Swofford, D. L. (1997) Are guinea pigs rodents? The importance of adequate models in molecular phylogenies. *Journal of Mammalian Evolution*, **4**, 77–86.

Sullivan, J., Holsinger, K. E. and Simon, C. (1996) The effect of topology on estimates of among-site rate variation. *Journal of Molecular Evolution*, **42**, 308–312.

Sullivan, J., Market, J. A. and Kilpatrick, C. W. (1997) Phylogeography and molecular systematics of the *Peromyscus aztecus* species group (Rodentia: Muridae) inferred using parsimony and likelihood. *Systematic Biology*, **46**, 426–440.

Sullivan, J., Swofford, D. L. and Navin, G. P. The effect of taxon sampling on estimating rate-heterogeneity parameters of maximum-likelihood models. *Molecular Biology*, in press.

Sun, Y., Skinner, D. Z., Liang, G. H. and Hulbert, S. H. (1994) Phylogenetic analysis of sorghum and related taxa using internal transcribed spacers of nuclear ribosomal DNA. *Theoretical and Applied Genetics*, **89**, 26–32.

Susanna, A., Nuria Garcia, J. and Soltis, P. S. (1995) Phylogenetic relationships in the tribe Cardueae (Asteraceae) based on ITS sequences. *American Journal of Botany*, **82**, 1056–1068.

Swofford, D. L., Olsen, G. J., Waddell, P. J. and Hillis, D. M. (1996) Phylogenetic inference, in *Molecular systematics*, 2nd edn., (eds D. M Hillis, C. Moritz and B. K. Mable), Sinauer Assoc., Sunderland, Massachusetts, pp. 407–514.

Templeton, A. R. (1983) Phylogenetic inference from restriction endonuclease cleavage site maps with particular reference to the evolution of humans and apes. *Evolution*, **367**, 221–244.

Thirumalai, D. (1998) Native secondary structure formation in RNA may be a slave to tertiary folding. *Proceedings of the National Academy of Sciences USA*, **95**, 11506–11508.

Thompson, J. D., Higgins, D. G. and Gibson, T. J. (1994) CLUSTAL W: improving the sensitivity of progressive multiple sequence alignment through sequence weighting, position specific gap penalties and weight matrix choice. *Nucleic Acids Research*, **22**, 4673–4680.

Thorne, R. (1992) Classification and geography of the flowering plants. *Botanical Review*, *(Lancaster)*, **58**, 225–348.

Tillier, E. R. and Collins, R. A. (1998) High apparent rate of simultaneous compensatory base-pair substitutions in ribosomal RNA. *Genetics*, **148**, 1993–2002.

Troitsky, A.V. and Bobrova, V. K. (1986) 23S rRNA-derived small ribosomal RNAs: their structure and evolution with references to plant phylogeny, in *DNA systematics, vol. 2, plants*, (ed. S. K. Dutta), CRC Press, Boca Raton, Florida, pp. 137–170.

Van Campenhout, S., Aert, R. and Volckaert, G. (1998) Orthologous DNA sequence variation among 5S ribosomal RNA gene spacer sequences on homoeologous chromosomes 1B, 1D, and 1R of wheat and rye. *Genome*, **41**, 244–255.

Venema, J., Bousquet-Antonelli, C., Gelugne, J. P., Caizergues-Ferrer, M. and Tollervey, D. (1997) Rok1p is a putative RNA helicase required for rRNA processing. *Molecular and Cellular Biology*, **17**, 3398–3407.

Warpeha, K. M., Gilliland. T. J. and Capesius I. (1998) An evaluation of rDNA variation in *Lolium* species. *Genome*, **41**, 307–311

Waters, E. R. and Schaal, B. A. (1996) Biased gene conversion does not occur in the *Brassica* triangle. *Genome*, **39**, 150–154.

Wendel, J. F., Schnabel, A. and Seelanan, T. (1995a) An unusual ribosomal DNA sequence from *Gossypium gossypioides* reveals ancient, cryptic, intergenomic introgression. *Molecular Phylogenetics and Evolution*, **4**, 298–313.

Wendel, J. F., Schnabel, A. and Seelanan, T. (1995b) Bidirectional interlocus concerted evolution following allopolyploid speciation in cotton (*Gossypium*). *Proceedings of the National Academy of Sciences USA*, **92**, 280–284.

Wheeler, W. C. (1994). Sources of ambiguity in nucleic acid sequence alignment, in *Molecular zoology: Advances, strategies and protocols*, (eds J. D. Ferraris and S. R. Palumbi), John Wiley and Sons, New York, pp. 323–352.

Wheeler, W. C. and Gladstein, D. (1994). MALIGN: a multiple sequence alignment program. *Journal of Heredity*, **85**, 417.

Wheeler, W. C. and Honeycutt, R. L. (1988) Paired sequence difference in ribosomal RNAs: evolution and phylogenetic implications. *Molecular Biology and Evolution*, **5**, 90–96.

Wheeler, W. C., Gatesy, J. and DeSalle, R. (1995) Elision: a method for accommodating multiple molecular sequence alignments with alignment-ambiguous sites. *Molecular Phylogenetics and Evolution*, **4**, 1–9.

Whiting, M. F., Carpenter, J. C., Wheeler, Q. D. and Wheeler, W. C. (1997) The Strepsiptera problem: phylogeny of the holometabolous insect orders inferred from 18S and 28S ribosomal sequences and morphology. *Systematic Biology*, **46**, 1–68.

Wilson, M. A., Gaut, B. and Clegg, M. T. (1990) Chloroplast DNA evolves slowly in the palm family (Arecaceae). *Molecular Biology and Evolution*, **7**, 303–314.

Wu, M. and Tinoco, I., Jr. (1998) RNA folding causes secondary structure rearrangement. *Proceedings of the National Academy of Sciences USA*, **95**, 11555–11560.

Yang, Z. (1996) Among-site rate variation and its impact on phylogenetic analyses. *Trends in Ecology and Evolution*, **11**, 367–372.

Yang, Z. (1998) On the best evolutionary rate for phylogenetic analysis. *Systematic Biology*, **47**, 125–133.

Yang, Z., Goldman, N. and Friday, A. (1994) Comparison of models for nucleotide substitution used in maximum-likelihood phylogenetic estimation. *Molecular Biology and Evolution*, **11**, 316–324.

Yang, Z., Goldman, N. and Friday, A. (1995) Maximum likelihood trees from DNA sequences: a peculiar statistical problem. *Systematic Biology*, **44**, 384–399.

Zimmer, E. A., Jupe, E. R. and Walbot, V. (1988) Ribosomal gene structure, variation and inheritance in maize and its ancestors. *Genetics*, **120**, 1125–1136.

Zimmer, E. A., Hamby, R. K., Arnold, M. L., Leblanc, D. A. and Theriot, E. C. (1989) Ribosomal RNA phylogenies and flowering plant evolution, in *The hierarchy of life*, (eds B. Fernhilm, K. Bremer and J. Jörnvall), Elsevier Science Publishers, Amsterdam, pp. 205–214.

Zuker, M. (1989) On finding all suboptimal foldings of an RNA molecule. *Science*, **244** (4990), 48–52.

APPENDIX   Classification of taxa analysed in Figs 14.2 and 14.3 with GenBank sequence accession numbers or specimen voucher if sequence is not accessioned. Taxa are arranged according to Thorne (1992). Non-accessioned *rbc*L sequences were provided by M. W. Chase from the data-set used in Chase *et al.* (1995).

TAXON	18S / *rbc*L
DICOTYLEDONS	
Aristolochiaceae	
Asarum canadense	L24043 / L14290
MONOCOTYLEDONS	
ARANAE	
Acoraceae	
Acorus calamus	L24078 / M901625
Araceae	
Gymnostachys anceps	AF069204 / M91629
ALISMATANAE	
Hydrocharitaceae	
Vallisneria americana	AF069201 / U03726
Scheuchzeriaceae	
Scheuchzeria palustris	AF069202 / U03728

LILIANAE – DIOSCOREALES
  Dioscoreaceae
    Dioscorea bulbifera   AF069203 / D28327
  Taccaceae
    Tacca chantrieri   U42063 / *Chase 175* (NCU)
LILIANAE – LILIALES
  Calochortaceae
    Calochortus albus   AF069204 / –
    Calochortus minimus   – / Z77263
  Colchicaceae
    Colchicum autumnale   U42072 / –
    Colchicum speciosum   – / L12673
LILIANAE – ASPARAGALES
  Anthericaceae
    Chlorophytum nepalense   U42065 / –
    Chlorophytum comosum   – / L05031
  Asparagaceae
    Asparagus falcatus   AF069205 / –
    Asparagus officinalis   – / L05028
  Hyacinthaceae
    Bowiea volubilis   U42067 / Z69237
    Ledebouria socialis   AF069206 / L05038
  Orchidaceae
    Apostasia stylidioides   AF069207 / Z73705
    Cypripedium calceolus   AF069208 / –
    Cypripedium irapeanum   – / Z73706
    Oncidium excavatum   U42791 / AF074201
  Xanthorrhoeaceae
    Xanthorrhoea sp.   U42064 / –
    Xanthorrhoea hostilis   – / Z73710
COMMELINANAE – INCERTAE SEDIS
  Calectasiaceae
    Calectasia intermedia   AF069209 / *Chase 456* (K)
COMMELINANAE – ARECALES
  Arecaceae
    Chamaedorea siefrizii   AF069210 / –
    Chamaedorea costaricana   – / *Gaut s.n.*
    Drymophloeus subdisticha   AF069211 / M81812
COMMELINANAE – BROMELIALES
  Bromeliaceae
    Aechmea sp.   AF069212 / –
    Aechmea chantinii   – / L19978
COMMELINANAE – COMMELINALES
  Commelinaceae
    Tradescantia ohiensis   AF069213 / –
    Tradescantia zebrina   – / L05042

Haemodoraceae
    Anigozanthus flavidus   AF069214 / *Chase 159* (NCU)
Phylidraceae
    Helmoholtzia sp.   U42074 / *Chase s.n.*
Pontederiaceae
    Eichhornia crassipes   AF069215 / U41574
COMMELINANAE – POALES
    Cyperaceae
        Carex sp.   AF069216 / –
        Carex hostiana   – / L12672
    Joinvilleaceae
        Joinvillea ascendens   AF069217 / –
        Joinvillea plicata   – / L01471
    Poaceae
        Oryza sativa   AF069218 / D00207
    Restionaceae
        Elegia sp.   AF069219 / L12675
    Sparganiaceae
        Sparganium eurycarpum   AF069220 / –
        Sparganium americanum   – / M91633
COMMELINANAE – ZINGIBERALES
    Cannaceae
        Canna indica   AF069221 / L05445
    Costaceae
        Costus barbatus   AF069222 / U42080
    Heliconiaceae
        Heliconia indica   U42082 / –
        Heliconia paka   – / L05452
    Lowiaceae
        Orchidantha fimbriata   AF069223 / L05456
    Marantaceae
        Calathea loesneri   AF069224 / L05444
        Maranta bicolor   AF069225 / U42079
    Musaceae
        Musa acuminata   AF069226 / U42079
    Strelitziaceae
        Phenakospermum guyannense   AF069227 / L05458
        Ravenala madagascariensis   AF069228 / L05459
        Strelitzia nicolai   AF069229 / L05461
    Zingiberaceae
        Zingiber gramineum   U42081 / L05465

Chapter 15

Proteins encoded in sequenced chloroplast genomes: an overview of gene content, phylogenetic information and endosymbiotic gene transfer to the nucleus

B. Stoebe, S. Hansmann, V. Goremykin,
K. V. Kowallik and W. Martin

ABSTRACT

From 12 completely sequenced plastid genomes, the presence or absence of 254 protein-coding genes was tabulated. Only 11 of these genes (all encoding ribosomal proteins) were found to be common to all twelve genomes. A phylogeny of major plastid lineages was inferred with a data set of 11,521 amino acids per species for the 46 proteins common to nine genomes of algae and metaphytes. Rooting to the homologous proteins from a cyanobacterial genome revealed the cyanelle of *Cyanophora paradoxa* as basal among plastid lineages sampled. Plotting of the presence or absence of genes in plastid genomes across the tree inferred revealed that parallel gene losses from chloroplast DNA in independent plastid lineages outnumber phylogenetically unique gene losses by a ratio of about five to one. Functional nuclear homologues of 47 chloroplast genes were identified that were successfully transferred to the nucleus during evolution, providing an intriguing glimpse of endosymbiotic genomic processes in plants.

15.1 Introduction

Plants possess a notoriously incomplete fossil record, yet they also possess chloroplast DNA, a molecule well-suited to studying plant evolution. In theory (Nei, 1996) and in practice (Graur *et al.*, 1996), the larger the molecular data matrix that is analyzed for a group of organisms, the more reliably can their history be inferred. Zoologists have effectively used the information contained in completely sequenced mitochondrial genomes to reconstruct phases of vertebrate evolution that are poorly documented in the fossil record (Janke *et al.*, 1997; Penny and Hasegawa, 1997). The information contained in completely sequenced chloroplast genomes has, however, remained largely untapped for the purpose of reconstructing plant evolution. A recent contribution on that topic revealed that the completely sequenced chloroplast genomes of five land plants (liverwort, tobacco, rice, pine and maize) possess 58 genes in common that are also present in the cpDNA of the red alga *Porphyra purpurea*. The information contained in that data matrix was ample to recover the widely accepted branching order for those species (Goremykin *et al.*, 1997).

In *Molecular systematics and plant evolution* (1999) (eds P. M. Hollingsworth, R. M. Bateman and R. J. Gornall), Taylor & Francis, London, pp. 327–352.

Several chloroplast genomes have been sequenced that were not considered in the previous contribution. Among these are the sequences from the diatom *Odontella sinensis* (Kowallik *et al.*, 1995), the glaucocystophyte *Cyanophora paradoxa* (Stirewalt *et al.*, 1995), the euglenophyte *Euglena gracilis* (Hallick *et al.*, 1993), the nonphotosynthetic, parasitic angiosperm *Epifagus virginiana* (Wolfe *et al.*, 1992), the curious plastid (McFadden *et al.*, 1996; Köhler *et al.*, 1997) of the human apicocomplexan parasite *Plasmodium falciparum* (Williamson *et al.*, 1994), and the chlorophytic alga *Chlorella vulgaris* (Wakasugi *et al.*, 1997).

Utilizing the information in completely sequenced chloroplast genomes has several simple practical limitations. First, gene content across chloroplast genomes differs dramatically, many genes having been lost altogether in some lineages, others having been successfully transferred from cpDNA to the nucleus (Martin *et al.*, 1998). Thus, the common set of genes for a given taxon sample must be identified before they can be retrieved, aligned and used for phylogenetic inference. Second, the nomenclature of genes across various chloroplast genomes is not uniform, hindering easy retrieval from various databases. Third, some chloroplast genes encode highly conservative proteins, whereas others possess large regions of sequence variability and/or length heterogeneity (indels), and the influence of these regions upon phylogenetic inference is not always predictable in a straightforward manner, thereby casting a shadow of uncertainty across the information contained within cpDNA-encoded proteins.

The purpose of this paper is to underscore the utility of proteins encoded in chloroplast genomes for phylogenetic inference. We tabulate the gene content of previously sequenced chloroplast genomes, suggest a unified nomenclature for homologous genes previously designated differently in various genomes, and expand the existing list (Reardon and Price, 1995) of hypothetical chloroplast open reading frames (*ycfs*) that occur in more than one chloroplast genome, or occur in one chloroplast genome and in the sequenced genome of the cyanobacterium *Synechocystis* PCC6803 (Kaneko *et al.*, 1996). We also present the results of phylogenetic inference using a dataset of 46 proteins, encompassing 11,521 amino acid positions per taxon common to the chloroplast genomes of five land plants, as well as *Odontella*, *Porphyra*, *Euglena* and *Cyanophora*, rooted to their homologues from *Synechocystis*. To address the amount of phylogenetic information contained within these genomes and the effect of highly variable or poorly aligned amino acid positions upon inference, we introduce a novel sorting and sampling procedure that has promising properties for estimating these parameters. We also plot the fate of genes that have been lost from chloroplast DNA across the tree inferred from the 46-protein phylogeny, revealing that parallel gene losses in independent lineages greatly outnumber phylogenetically unique gene losses. Finally, we examine the process of genome migration from the chloroplast to the nucleus by reporting 47 functional homologues of chloroplast genes that have undergone successful endosymbiotic gene transfer during plant evolution.

15.2 Determining common sets of genes across chloroplast genomes

Before one can use cpDNA-encoded proteins for phylogenetics, one has to know which ones are available for that purpose, and one would like to know whether one has identified all that are common to a given taxon sample. This is more tedious than it might appear at first sight. Complete sequences of *Zea mays* (X86563: Maier *et al.*, 1995), *Oryza sativa* (X15901: Hiratsuka *et al.*, 1989), *Nicotiana tabacum* (S54304: Shinozaki *et al.*, 1986), *Pinus thunbergii* (D17510: Wakasugi *et al.*, 1994), *Marchantia polymorpha* (X04465: Ohyama *et al.*, 1986), *Euglena gracilis* (Z11874), *Porphyra purpurea* (U38804: Reith and Munholland, 1995), *Odontella sinensis* (Z67753), *Chlorella vulgaris* (AB001684), *Epifagus virginiana* (M81884), *Plasmodium falciparum* (X95275, X95276) and *Cyanophora paradoxa* (U30821) chloroplast genomes were retrieved from GenBank. Starting with the *Porphyra* sequence, BLAST searches were performed with all designated encoded proteins and annotated open reading frames. Homologues from the other eight cpDNA genomes and from *Synechocystis* PCC6803 (Kaneko *et al.*, 1996) were retrieved. Case-by-case evaluation of the results for all genes and reading frames present in the 12 chloroplast genomes mentioned above that occur in either a) at least two cpDNAs or b) one completely sequenced cpDNA and in *Synechocystis* DNA are given in Table 15.1. ORFs present in only one chloroplast genome with no detectable similarity to other sequences were excluded from the table. The putative or known functions of the encoded products are mostly taken from the designations in Kaneko *et al.* (1996), though in a few cases we have deviated slightly from those definitions.

For most cpDNA-encoded proteins, the BLAST searches were quite unambiguous. Among the best hits were cpDNA-encoded orthologues, followed in order by cyanobacterial orthologues, then in some cases transferred plant nuclear genes of plastid origin (see section 15.7), and then various bacterial homologues. In such cases, the results simply indicated that the chloroplast genes had been directly inherited from cyanobacteria. This is not surprising, although there have been reports that gene transfer to chloroplasts occurs in evolution (Palmer, 1995; Delwiche and Palmer, 1996), as is known for mitochondrial genomes, where uptake of wayward DNA is very commonplace in the case of angiosperms (Unseld *et al.*, 1997). The BLAST queries of GenBank with all 254 chloroplast proteins surveyed revealed that for 235, the best hit among non-chloroplast sequences was a cyanobacterial homologue (usually from *Synechocystis*), indicating that > 92% of these chloroplast genes were inherited by chloroplasts via direct filiation from cyanobacteria.

There were 21 exceptions comprising genes present in more that one cpDNA that did not directly reveal a cyanobacterial ancestry. Fifteen of these, *ycf1*, *ycf2*, *ycf15* and *ycf67–78*, remain of currently indiscernable origins for lack of reference sequences and have no significant similarity to any known sequence outside of chloroplasts. Another exception, *matK* of land plant chloroplasts, has no orthologues in eubacteria and shares greatest similarity with an intron-encoded protein found in the 18S rDNA gene of *Marchantia* mitochondrial DNA (accession number M68929). *MatK* is similar but paralogous to the reverse transcriptase-like, intron-encoded protein of *Scenedesmus* cpDNA (Kück, 1989), that in turn appears to have an orthologue among cyanobacterial group II introns (Fontaine *et al.*, 1997). An orthologue of *matK* may eventually be detected in cyanobacterial genomes. Other exceptions are *minE*, which is only found in *Chlorella* among plastid genomes surveyed here and has 22% identity to an *E. coli* homologue, and *petL*, which is probably too short (31 aa) to detect its homologues among cyanobacteria using BLAST searches.

Table 15.1 Protein-coding genes in sequenced chloroplast genomes.

Gene	Zea	Ori	Nic	Pin	Mar	Eug	Chl	Odo	Por	Cpa	Pla	Epi	Syn	Transferred Nuclear Homologue	Probable or known function, Comments
accA	–	–	–	–	–	–	–	–	+	–	–	–	+	U40979	Acetyl-CoA carboxylase α SU
accB	–	–	–	–	–	–	–	–	+	–	–	–	+	U40666	Acetyl-CoA carboxylase biotin carrier SU
accD	–	–	+	+	–	+	–	+	–	–	+	+	+	–	Acetyl-CoA carboxylase carboxytransferase β SU (ycf11, zfpA)
acpP	–	–	–	–	–	–	–	+	+	+	–	–	+	P07854	Acyl carrier protein
apcA	–	–	–	–	–	–	–	–	+	+	–	–	+	–	Allophycocyanin α SU
apcB	–	–	–	–	–	–	–	–	+	+	–	–	+	–	Allophycocyanin β SU
apcD	–	–	–	–	–	–	–	–	+	+	–	–	+	–	Allophycocyanin γ SU
apcE	–	–	–	–	–	–	–	–	+	+	–	–	+	–	Phycobilisome core linker prot.
apcF	–	–	–	–	–	–	–	–	+	+	–	–	+	–	Allophycocyanin 18 kDa β SU
argB	–	–	–	–	–	–	–	–	+	–	–	–	+	–	Acetylglutamate kinase
atpA	+	+	+	+	+	+	+	+	+	+	–	–	+	–	ATP synthase CF1 α SU
atpB	+	+	+	+	+	+	+	+	+	+	–	–	+	–	ATP synthase CF1 β SU
atpD	–	–	–	–	–	–	–	+	+	+	–	–	+	–	ATP synthase CF1 δ SU
atpE	+	+	+	+	+	+	+	+	+	+	–	–	+	–	ATP synthase CF1 ε SU
atpF	+	+	+	+	+	+	+	+	+	+	–	–	+	–	ATP synthase CF0 SU I
atpG	–	–	–	–	–	–	–	+	+	+	–	–	+	P31953	ATP synthase CF0 SU II
atpH	+	+	+	+	+	+	+	+	+	+	–	–	+	–	ATP synthase CF0 SU III
atpI	+	+	+	+	+	+	+	+	+	–	–	–	+	–	ATP synthase CF0 SU IV
basI	–	–	–	–	–	–	–	+	+	–	–	–	+	AB000405	Thiol-specific antioxidant prot. (ycf42)
bioY	–	–	–	–	–	–	–	–	–	+	–	–	+	–	Cpa ORF188, Syn slr1365 biotin synthase
carA	–	–	–	–	–	–	–	–	–	+	–	–	+	U73175	Carbamoyl phosphate synthetase small SU
cbbX	–	–	–	–	–	–	–	–	+	+	–	–	–	–	Orthologues in red type Calvin cycle operons of proteobacteria (cfxQ, cbxX)[1]
ccsA	+	+	+	+	+	–	+	+	+	+	–	–	+	–	Heme attachment to plastid cyt c (ycf5)
cemA	+	+	+	+	+	–	+	–	+	–	–	–	+	–	Envelope membrane prot. (ycf10, hbp)
chlB	–	–	+	+	+	–	+	–	+	+	–	–	+	–	Protochlorophyllide reductase ChlB SU
chlI	–	–	–	–	–	+	+	+	+	+	–	–	+	P16127	Protochlorophyllide reductase magnesium chelatase SU
chlL	–	–	–	+	+	–	+	–	+	+	–	–	+	–	Protochlorophyllide reductase ATP-bind. SU (frxC)
chlN	–	–	–	+	+	–	+	–	+	+	–	–	+	–	Protochlorophyllide reductase ChlN SU
clpC	–	–	–	–	–	–	–	+	+	–	+	–	+	P31542	Clp protease ATP-bind. SU
clpP	+	+	+	+	+	–	+	–	–	+	–	+	+	–	Clp protease proteolytic SU[2]
cpcA	–	–	–	–	–	–	–	–	+	+	–	–	+	–	Phycocyanin α SU
cpcB	–	–	–	–	–	–	–	–	+	+	–	–	+	–	Phycocyanin β SU
cpcG	–	–	–	–	–	–	–	–	+	+	–	–	+	–	Phycobilisome rod-core linker polypeptide
cpeA	–	–	–	–	–	–	–	–	+	–	–	–	+	–	Phycoerythrin α SU
cpeB	–	–	–	–	–	–	–	–	+	–	–	–	+	–	Phycoerythrin β SU
crtE	–	–	–	–	–	–	–	–	–	+	–	–	+	P34802	Geranylgeranyl pyrophosphate synthetase hom.
cysA	–	–	–	–	+	–	+	–	–	–	–	–	+	–	Probable transport prot. (mbpX)
cysT	–	–	–	–	+	–	+	–	–	–	–	–	+	–	Probable transport prot. (mbpY)
dfr	–	–	–	–	–	–	–	–	–	+	–	–	+	–	Syn sll0698 drug sensory prot. A (ycf26)
dnaB	–	–	–	–	–	–	–	–	+	+	–	–	+	–	DNA-replication helicase SU
dnaK	–	–	–	–	–	–	–	+	+	+	–	–	+	Q02028	Hsp70-type chaperone
dsbD	–	–	–	–	–	–	–	–	+	–	–	–	+	–	Por ORF240, Syn thiol:disulfide interchange prot. DsbD sll0621
fabH	–	–	–	–	–	–	–	–	+	–	–	–	+	Q07510	β-ketoacyl-acyl carrier prot. synthase III
fdx	–	–	–	–	–	–	–	–	+	–	–	–	+	–	Por ORF75a, Syn ssr3184, 2[4Fe-4S] ferredoxin
ftrB	–	–	–	–	–	–	–	–	+	–	–	–	+	P41349	Ferredoxin-thioredoxin reductase β SU

Table 15.1 Continued

	Zea	Ory	Nic	Pin	Mar	Eug	Chl	Odo	Por	Cpa	Pla	Epi	Syn	Transferred Nuclear Homologue	Probable or known function, Comments
ftsH	–	–	–	–	–	–	–	+	+	–	–	–	+	X99808	Cell division prot., ATPase, protease (*ycf25*)
ftsW	–	–	–	–	–	–	–	–	–	+	–	–	+	–	Putative cell/organelle division prot.
glnB	–	–	–	–	–	–	–	–	+	–	–	–	+	–	Nitrogen regulatory prot. PII
gltB	–	–	–	–	–	–	–	–	+	–	–	–	+	P23225	Glutamate synthase (GOGAT)
groEL	–	–	–	–	–	–	–	+	+	–	–	–	+	–	Chaperonin, 60 kDa
groES	–	–	–	–	–	–	–	–	+	–	–	–	+	Q02073	Chaperonin, 10 kDA
hemA	–	–	–	–	–	–	–	–	+	–	–	–	+	P42804	5-Aminolevulinic acid synthase
hisH	–	–	–	–	–	–	–	–	+	–	–	+	+	–	Histidinol-phosphate aminotransferase
ilvB	–	–	–	–	–	–	–	–	+	–	–	–	+	–	Acetohydroxyacid synthase large SU
ilvH	–	–	–	–	–	–	–	–	+	–	–	–	+	–	Acetohydroxyacid synthase small SU
infA	+	+	+	+	–	+	–	–	–	–	+	–	+	–	Translational initiation factor 1
infB	–	–	–	–	–	–	–	–	+	–	–	–	+	–	Translational initiation factor 2
infC	–	–	–	–	–	–	–	–	+	–	–	–	+	–	Translational initiation factor 3
matK	+	+	+	+	–	–	–	–	–	–	+	–	–	–	Probable intron maturase (*ycf14*)
minD	–	–	–	–	–	–	+	–	–	–	–	–	+	–	Septum-site determining prot.
minE	–	–	–	–	–	–	+	–	–	–	–	–	–	–	22% identity to *E. coli minE*, no hom. in *Syn* or cpDNA, tentative designation
mntA	–	–	–	–	–	–	–	–	–	+	–	–	+	–	*Cpa* ORF244, manganese transport system ATP-bind. prot.
mntB	–	–	–	–	–	–	–	–	–	+	–	–	+	–	*Cpa* ORF299, manganese transport system membrane prot.
moeB	–	–	–	–	–	–	–	–	+	–	–	–	+	–	*Por* ORF382, molybdopterin biosynthesis prot. *Syn* sll1536
nadA	–	–	–	–	–	–	–	–	–	+	–	–	+	–	Quinolinate synthase
nblA	–	–	–	–	–	–	–	–	+	–	–	–	+	–	Phycobilisome degradation prot. (*ycf18*)
ndhA	+	+	+	–	+	–	–	–	–	–	–	–	+	–	NADH-plastoquinone oxidoreductase SU 1
ndhB	+	+	+	–	+	–	–	–	–	–	–	–	+	–	NADH-plastoquinone oxidoreductase SU 2
ndhC	+	+	+	–	+	–	–	–	–	–	–	–	+	–	NADH-plastoquinone oxidoreductase SU 3
ndhD	+	+	+	–	+	–	–	–	–	–	–	–	+	–	NADH-plastoquinone oxidoreductase SU 4
ndhE	+	+	+	–	+	–	–	–	–	–	–	–	+	–	NADH-plastoquinone oxidoreductase SU 4L
ndhF	+	+	+	–	+	–	–	–	–	–	–	–	+	–	NADH-plastoquinone oxidoreductase SU 5
ndhG	+	+	+	–	+	–	–	–	–	–	–	–	+	–	NADH-plastoquinone oxidoreductase SU 6
ndhH	+	+	+	–	+	–	–	–	–	–	–	–	+	–	NADH-plastoquinone oxidoreductase 49 kDa SU
ndhI	+	+	+	–	+	–	–	–	–	–	–	–	+	–	NADH-plastoquinone oxidoreductase SU I
ndhJ	+	+	+	–	+	–	–	–	–	–	–	–	+	–	NADH-plastoquinone oxidoreductase SU J
ndhK	+	+	+	–	+	–	–	–	–	–	–	–	+	–	NADH-ubiquinone oxidoreductase SU K (*psbG*)
ntcA	–	–	–	–	–	–	–	–	+	–	–	–	+	–	Global nitrogen transcriptional regulator (*ycf28*)
odpA	–	–	–	–	–	–	–	–	+	–	–	–	+	U80185	Pyruvate dehydrogenase E1 component, A SU
odpB	–	–	–	–	–	–	–	–	+	–	–	–	+	U80186	Pyruvate dehydrogenase E1 component, B SU
pbsA	–	–	–	–	–	–	–	–	+	–	–	–	+	–	Heme oxygenase
petA	+	+	+	+	–	+	+	+	+	–	–	–	+	–	Apocytochrome *f*
petB	+	+	+	+	+	+	+	+	+	–	–	–	+	–	Cytochrome b_6/f apoprotein
petD	+	+	+	+	+	–	+	+	+	–	–	–	+	–	Cytochrome b_6/f complex SU IV
petF	–	–	–	–	–	–	+	+	+	–	–	–	+	P00221	Ferredoxin
petG	+	+	+	+	+	+	+	+	+	–	–	–	+	–	Cytochrome b_6/f complex SU V
petJ	–	–	–	–	–	–	–	–	+	–	–	–	+	P08197	Cytochrome c_{553}
petL	+	+	+	+	–	+	+	+	+	–	–	–	(+)	–	*Syn* sll1845, cyt. b_6/f complex SU VI (*ycf7*)[3]
petM	–	–	–	–	–	–	+	+	+	–	–	–	+	–	Cytochrome b_6/f complex SU VII (*ycf31*)
pgmA	–	–	–	–	–	–	–	–	+	–	–	–	+	–	Phosphoglycerate mutase[4]
preA	–	–	–	–	–	–	–	+	+	–	–	–	+	–	Prenyl transferase, related to *crtE*
psaA	+	+	+	+	+	+	+	+	+	–	–	–	+	–	PSI P700 apoprotein A1

Table 15.1 Continued

	Zea	Ory	Nic	Pin	Mar	Eug	Chl	Odo	Por	Cya	Pla	Epi	Syn	Transferred Nuclear Homologue	Probable or known function, Comments
psaB	+	+	+	+	+	+	+	+	+	+	–	–	+	–	PSI P700 apoprotein A2
psaC	+	+	+	+	+	+	+	+	+	+	–	–	+	–	PSI iron-sulphur center FA/FB containing SU VII
psaD	–	–	–	–	–	–	–	+	+	–	–	–	+	P12353	PSI ferredoxin-bind. prot. SU II
psaE	–	–	–	–	–	–	–	+	+	+	–	–	+	–	PSI SU IV, 18–20 kDa
psaF	–	–	–	–	–	–	–	+	+	+	–	–	+	–	PSI SU III, plastocyanin-bind.
psaI	+	+	+	+	+	–	+	+	+	+	–	–	+	–	PSI SU VIII
psaJ	+	+	+	+	+	+	+	+	+	+	–	–	+	–	PSI SU IX
psaK	–	–	–	–	–	–	–	+	–	–	–	–	+	P36886	PSI SU X
psaL	–	–	–	–	–	–	–	+	+	–	–	–	+	P23993	PSI SU XI, reaction center
psaM	–	–	–	+	+	+	+	+	+	–	–	–	+	–	PSI M-polypeptide
psbA	+	+	+	+	+	+	+	+	+	+	–	–	+	–	PSII D1 reaction center prot.
psbB	+	+	+	+	+	+	+	+	+	+	–	–	+	–	PSII CP47 chlorophyll apoprotein
psbC	+	+	+	+	+	+	+	+	+	+	–	–	+	–	PSII CP43 chlorophyll apoprotein
psbD	+	+	+	+	+	+	+	+	+	+	–	–	+	–	PSII D2 reaction center prot.
psbE	+	+	+	+	+	+	+	+	+	+	–	–	+	–	PSII cytochrome b_{559} α SU
psbF	+	+	+	+	+	+	+	+	+	+	–	–	+	–	PSII cytochrome b_{559} β SU
psbH	+	+	+	+	+	+	+	+	+	+	–	–	+	–	PSII 10 kDa phosphoprotein
psbI	+	+	+	+	+	+	+	+	+	+	–	–	+	–	PSII I polypeptide
psbJ	+	+	+	+	+	+	+	+	+	+	–	–	+	–	PSII J protein
psbK	+	+	+	+	+	+	+	+	+	+	–	–	+	–	PSII K protein
psbL	+	+	+	+	+	+	+	+	+	+	–	–	+	–	PSII L protein
psbM	+	+	+	+	+	–	+	–	–	+	–	–	+	–	PSII M protein
psbN	+	+	+	+	+	+	+	+	+	+	–	–	+	–	PSII N protein
psbT	+	+	+	+	+	+	+	+	+	+	–	–	+	–	PSII T protein (ycf8)
psbV	–	–	–	–	–	–	–	+	+	+	–	–	+	–	PSII cytochrome c550 (petK)
psbW															Gene nomenclature problem: see ycf79 and footnote[5]
psbX	–	–	–	–	–	–	–	+	+	+	–	–	+	–	PSII X 4.1 kDa protein
rbcLg	+	+	+	+	+	+	+	–	–	+	–	–	+	–	Rubisco large SU, green type
rbcLr	–	–	–	–	–	–	–	+	+	–	–	–	(+)	–	Rubisco large SU, red type
rbcR	–	–	–	–	–	–	–	+	+	+	–	–	+	–	Syn sll0998 LysR family of transcriptional regulators (ycf30)[6]
rbcSg	–	–	–	–	–	–	–	–	+	–	–	–	+	many	Rubisco small SU, green type
rbcSr	–	–	–	–	–	–	–	+	+	–	–	–	(+)	–	Rubisco small SU, red type
rdpO	–	–	–	–	–	+	–	–	–	–	–	–	(+)	–	Eug ORF506 S34497, Scenedesmus obliquus RT rdpO S05341, Calothrix PCC7601 S40013
rne	–	–	–	–	–	–	–	+	+	+	–	–	+	–	RNA component of RNAseE, RNA processing
rpl1	–	–	–	–	–	–	–	+	+	+	–	–	+	–	Ribosomal protein L1
rpl2	+	+	+	+	+	+	+	+	+	+	+	+	+	–	Ribosomal protein L2
rpl3	–	–	–	–	–	–	–	+	+	+	–	–	+	–	Ribosomal protein L3
rpl4	–	–	–	–	–	–	–	+	+	–	+	–	+	–	Ribosomal protein L4
rpl5	–	–	–	–	–	+	+	+	+	–	–	–	+	g2191128	Ribosomal protein L5
rpl6	–	–	–	–	–	–	–	+	+	+	+	–	+	–	Ribosomal protein L6
rpl9	–	–	–	–	–	–	–	+	–	–	–	–	+	P11894	Ribosomal protein L9
rpl11	–	–	–	–	–	–	–	+	+	+	–	–	+	P31164	Ribosomal protein L11
rpl12	–	–	–	–	+	+	+	+	+	–	–	–	+	P02398	Ribosomal protein L12[7]
rpl13	–	–	–	–	–	–	–	+	+	–	–	–	+	P12629	Ribosomal protein L13
rpl14	+	+	+	+	+	+	+	+	+	+	+	–	+	–	Ribosomal protein L14
rpl16	+	+	+	+	+	+	+	+	+	+	+	+	+	–	Ribosomal protein L16
rpl18	–	–	–	–	–	–	–	+	+	+	–	–	+	–	Ribosomal protein L18
rpl19	–	–	–	–	–	+	+	+	+	–	–	–	+	–	Ribosomal protein L19

Table 15.1 Continued

	Z e a	O r y	N i c	P i n	M a r	E u g	C h l	O d o	P o r	C l a	P l a	E p i	S y n	Transferred Nuclear Homologue	Probable or known function, Comments
rpl20	+	+	+	+	+	+	+	+	+	–	+	+	+	–	Ribosomal protein L20
rpl21	–	–	–	–	+	–	–	+	+	+	–	–	+	P24613	Ribosomal protein L21
rpl22	+	+	+	+	+	+	–	+	+	+	–	–	+	–	Ribosomal protein L22
rpl23	+	+	+	+	+	+	+	+	+	–	+	–	+	–	Ribosomal protein L23
rpl24	–	–	–	–	–	–	–	–	+	+	–	–	+	P11893	Ribosomal protein L24
rpl27	–	–	–	–	–	–	–	+	+	–	–	–	+	P30155	Ribosomal protein L27
rpl28	–	–	–	–	–	–	–	–	+	+	–	–	+	P30956	Ribosomal protein L28
rpl29	–	–	–	–	–	–	–	+	+	–	–	–	+	–	Ribosomal protein L29
rpl31	–	–	–	–	–	–	–	+	+	–	–	–	+	–	Ribosomal protein L31
rpl32	+	+	+	+	+	+	+	+	+	–	–	–	+	–	Ribosomal protein L32
rpl33	+	+	+	+	+	–	–	+	+	+	–	+	+	–	Ribosomal protein L33
rpl34	–	–	–	–	–	–	–	+	+	+	–	–	+	–	Ribosomal protein L34
rpl35	–	–	–	–	–	–	–	+	+	+	–	–	+	P23326	Ribosomal protein L35
rpl36	+	+	+	+	+	+	+	+	+	+	+	+	+	–	Ribosomal protein L36
rpoA	+	+	+	+	–	+	+	+	+	–	–	–	+	–	RNA polymerase α SU
rpoB	+	+	+	+	+	+	+	+	+	+	+	–	+	–	RNA polymerase β SU
rpoC1	+	+	+	+	+	+	+	+	+	+	–	–	+	–	RNA polymerase β′ SU
rpoC2	+	+	+	+	+	+	+	+	+	+	–	–	+	–	RNA polymerase β″ SU[8]
rps1	–	–	–	–	–	–	–	–	+	–	–	–	+	P29344	Ribosomal protein S1
rps2	+	+	+	+	+	+	+	+	+	+	+	+	+	–	Ribosomal protein S2
rps3	+	+	+	+	+	+	+	+	+	+	+	+	+	–	Ribosomal protein S3
rps4	+	+	+	+	+	+	+	+	+	+	+	+	+	–	Ribosomal protein S4
rps5	–	–	–	–	–	–	+	+	+	+	–	–	+	gl70700	Ribosomal protein S5
rps6	–	–	–	–	–	–	–	+	+	+	–	–	+	–	Ribosomal protein S6
rps7	+	+	+	+	+	+	+	+	+	+	+	+	+	–	Ribosomal protein S7
rps8	+	+	+	+	+	+	+	+	+	+	+	+	+	–	Ribosomal protein S8
rps9	–	–	–	–	+	+	+	+	+	+	–	–	+	–	Ribosomal protein S9
rps10	–	–	–	–	–	–	–	+	+	+	–	–	+	–	Ribosomal protein S10
rps11	+	+	+	+	+	+	+	+	+	+	+	+	+	–	Ribosomal protein S11
rps12	+	+	+	+	+	+	+	+	+	+	+	+	+	–	Ribosomal protein S12
rps13	–	–	–	–	–	–	–	+	+	+	–	–	+	P42732	Ribosomal protein S13
rps14	+	+	+	+	+	+	+	+	+	+	–	+	+	–	Ribosomal protein S14
rps15	+	+	+	+	+	–	–	–	–	–	–	–	+	–	Ribosomal protein S15
rps16	+	+	+	–	–	–	–	+	+	+	–	–	+	–	Ribosomal protein S16
rps17	–	–	–	–	–	–	–	+	+	+	+	–	+	Z11151	Ribosomal protein S17
rps18	+	+	+	+	+	+	+	+	+	+	–	+	+	–	Ribosomal protein S18
rps19	+	+	+	+	+	+	+	+	+	+	+	+	+	–	Ribosomal protein S19
rps20	–	–	–	–	–	–	+	+	+	–	–	–	+	–	Ribosomal protein S20
secA	–	–	–	–	–	–	–	+	+	–	–	–	+	Z49124	Preprotein-translocase SU A
secY	–	–	–	–	–	–	–	+	+	+	–	–	+	Z54351	Preprotein-translocase SU Y
syfB	–	–	–	–	–	–	–	+	+	–	–	–	+	–	Phenylalanine tRNA synthetase
syh	–	–	–	–	–	–	–	+	+	–	–	–	+	–	Histidine tRNA synthetase
thiG	–	–	–	–	–	–	–	+	+	–	–	–	+	–	ThiG protein, thiamine biosynthesis
trpA	–	–	–	–	–	–	–	–	+	+	–	–	+	P42390	Tryptophane synthase α SU
trpG	–	–	–	–	–	–	–	–	+	+	–	–	+	L22585	Anthranilate synthase component II, glutamine amidotransferase
trxA	–	–	–	–	–	–	–	–	+	+	–	–	+	P48384	Thioredoxin
tsf	–	–	–	–	–	–	–	–	+	+	–	–	+	–	Translational elongation factor Ts
tufA	–	–	–	–	–	+	+	+	+	+	+	–	+	P41342	Translational elongation factor Tu
upp	–	–	–	–	–	–	–	–	+	+	–	–	+	Y11210	Por ORF198, uracil phosphoribosyltransferase sll1035

Table 15.1 Continued

	Z e a	O r y	N i c	P i n	M a r	E u g	C h l	O d o	P o r	C p a	P l a	E p i	S y n	Transferred Nuclear Homologue	Probable or known function, Comments
ycf1	−	−	+	+	+	−	−	−	−	−	−	+	−	−	* see also *ycf77*[9]
ycf2	−	−	+	+	+	−	−	−	−	−	−	+	−	−	* sim. FtsH-like proteins, see also *ycf78*[10]
ycf3	+	+	+	+	+	−	+	+	+	+	−	−	+	−	*Syn* slr0823
ycf4	+	+	+	+	+	+	+	+	+	+	−	−	+	−	*Syn* sll0226
ycf6	+	+	+	+	+	−	−	+	+	+	−	−	+	−	*Syn* sml0004
ycf9	+	+	+	+	+	+	+	+	+	+	−	−	+	−	*Syn* sll1281
ycf12	−	−	−	+	+	+	+	+	+	+	−	−	+	−	*Syn* sll0047
ycf13	−	−	−	−	−	+	−	−	−	−	−	−	(+)	−	*Anabaena* U13767 intron-encoded ORF (Bauer et al., 1997), in *Astasia* P14761
ycf15	+	−	+	−	−	−	−	−	−	−	−	−	−	−	* *Zea* ORF99
ycf16	−	−	−	−	−	−	−	+	+	+	−	−	+	−	*Syn* slr0075 abc-transporter
ycf17	−	−	−	−	−	−	−	−	+	+	−	−	+	P14896	*Syn* ssl1633, slr0839, ssr2595, ssl2542, ssr1789 ELIP superfamily (Grimm et al., 1989)
ycf19	−	−	−	−	−	−	−	+	−	−	−	−	+	−	*Syn* ssr2142
ycf20	−	−	−	−	−	−	−	+	−	−	−	−	+	−	*Syn* sll1509
ycf21	−	−	−	−	−	−	−	+	+	−	−	−	+	−	*Syn* sll1797
ycf22	−	−	−	−	−	−	−	+	−	−	−	−	+	−	*Syn* sll0751, sll1002
ycf23	−	−	−	−	−	−	−	+	+	−	−	−	+	−	*Syn* slr2032
ycf24	−	−	−	−	−	−	−	+	+	+	+	−	+	−	*Syn* slr0074 abc-transporter
ycf27	−	−	−	−	−	−	−	+	+	−	−	−	+	−	*Syn* slr0115, slr0947 regulatory component of sensory transduction system
ycf29	−	−	−	−	−	−	−	+	+	−	−	−	+	−	*Syn* slr1783 regulatory component of sensory transduction system
ycf32	−	−	−	−	−	−	−	+	+	+	−	−	+	−	*Syn* sml0007
ycf33	−	−	−	−	−	−	−	+	+	+	−	−	+	−	*Syn* ssl1417
ycf34	−	−	−	−	−	−	−	+	+	−	−	−	+	−	*Syn* ssr1425
ycf35	−	−	−	−	−	−	−	+	+	+	−	−	+	−	*Syn* sll0661
ycf36	−	−	−	−	−	−	−	+	+	−	−	−	+	−	*Syn* sll0584
ycf37	−	−	−	−	−	−	−	+	+	−	−	−	+	−	*Syn* slr0171
ycf38	−	−	−	−	−	−	−	+	+	−	−	−	+	−	*Syn* sll0760
ycf39	−	−	−	−	−	−	−	+	+	+	−	−	+	−	*Syn* sll1218, slr0399
ycf40	−	−	−	−	−	−	−	+	+	−	−	−	+	−	*Syn* ssr0102
ycf41	−	−	−	−	−	−	−	+	+	−	−	−	+	−	*Syn* slr1034
ycf43	−	−	−	−	−	−	−	+	+	−	−	−	+	−	*Syn* sll0194
ycf44	−	−	−	−	−	−	−	+	+	−	−	−	+	−	*Syn* slr2087
ycf45	−	−	−	−	−	−	−	+	+	−	−	−	+	−	*Syn* slr0692
ycf46	−	−	−	−	−	−	−	+	+	−	−	−	+	−	*Syn* slr0480, slr0374 sim. SU 4 of 26S proteasome[11]
ycf47	−	−	−	−	−	−	−	+	+	−	−	−	+	−	*Syn* ssr3307
ycf48	−	−	−	−	−	−	−	−	−	+	−	−	+	−	*Cpa* ORF333, *Syn* slr2034
ycf49	−	−	−	−	−	−	−	−	−	+	−	−	+	−	*Cpa* ORF102, *Syn* sll0608
ycf50	−	−	−	−	−	−	−	−	−	+	−	−	+	−	*Cpa* ORF108, *Syn* slr2073
ycf51	−	−	−	−	−	−	−	−	−	+	−	−	+	−	*Cpa* ORF163, *Syn* sll1702
ycf52	−	−	−	−	−	−	−	−	+	−	−	−	+	−	*Por* ORF174, *Syn* sll0286
ycf53	−	−	−	−	−	−	−	−	+	−	−	−	+	−	*Por* ORF238, *Syn* sll0558
ycf54	−	−	−	−	−	−	−	−	+	−	−	−	+	−	*Por* ORF108, *Syn* slr1780
ycf55	−	−	−	−	−	−	−	−	+	−	−	−	+	−	*Por* ORF320, *Syn* sll1879
ycf56	−	−	−	−	−	−	−	−	+	−	−	−	+	−	*Por* ORF263a, *Syn* slr0050
ycf57	−	−	−	−	−	−	−	−	+	−	−	−	+	−	*Por* ORF114, *Syn* slr1417 sim. prot. in *nifU* region
ycf58	−	−	−	−	−	−	−	−	+	−	−	−	+	−	*Por* ORF149, *Synechococcus* M95288, *Syn* slr2049

Table 15.1 Continued

	Z e a	O r y	N i c	P i n	M a r	E u g	C h l	O d o	P o r	C p a	P l a	E p i	S y n	Transferred Nuclear Homologue	Probable or known function, Comments
ycf59	–	–	–	–	–	–	–	–	+	–	–	–	+	U75599	Por ORF349, Syn sll1214 leucine zipper protein
ycf60	–	–	–	–	–	–	–	–	+	–	–	–	+	–	Por ORF203, Syn sll1737
ycf61	–	–	–	–	–	–	–	–	+	–	–	–	+	–	Por ORF75b, Syn ssl2982
ycf62	–	–	–	–	–	–	–	–	+	–	–	–	+	–	Por ORF327, Syn slr1278
ycf63	–	–	–	–	–	–	–	–	+	–	–	–	+	–	Por ORF263b, Syn slr1045
ycf64	–	–	–	–	–	–	–	–	+	–	–	–	+	–	Por ORF107, Syn slr1846[12]
ycf65	–	–	–	–	–	–	–	–	+	–	–	–	+	–	Por ORF99, Synechococcus U62737, Syn slr0923
ycf66	–	–	–	–	+	–	–	+	–	–	–	–	+	–	Odo ORF99, Syn slr0503, Mar ORF135 A05060
ycf67	–	–	–	–	–	+	–	–	–	–	–	–	–	–	* Eug ORF161 P48337, Ast ORF170 P34778
ycf68	+	+	–	+	–	–	–	–	–	–	–	–	–	–	* ORF in trnI intron Pin ORF75a P52807, Zea ORF134 P03938 Ory ORF133b P12173
ycf69	+	+	+	–	–	–	–	–	–	–	–	–	–	–	* Nic ORF131 A05210, Ory ORF72 JQ0277, Zea ORF58 S58628
ycf70	+	+	–	–	–	–	–	–	–	–	–	–	–	–	* Ory ORF91 JQ0209, Zea ORF69 S58539
ycf71	+	+	–	–	–	–	–	–	–	–	–	–	–	–	* Ory ORF82 JQ0268, Zea ORF75 S58641
ycf72	+	+	–	–	–	–	–	–	–	–	–	–	–	–	* Ory ORF137 JQ0269, Zea ORF137 S58640
ycf73	+	+	–	–	–	–	–	–	–	–	–	–	–	–	* Ory ORF249 JQ0274, Zea ORF173 S58632
ycf74	+	+	–	–	–	–	–	–	–	–	–	–	–	–	* ORF in trnA intron Ory ORF109 JQ0280, Zea ORF49 S58608
ycf75	+	+	–	–	–	–	–	–	–	–	–	–	–	–	* Ory ORF63a JQ0283, Zea ORF63 S58623
ycf76	+	+	–	–	–	–	–	–	–	–	–	–	–	–	* Ory ORF85b JQ0278, Zea ORF85 S58627
ycf77	–	–	–	–	–	–	+	–	–	–	–	–	–	–	Chl ORF1720, in positions 950–1150 residual similarity to ycf46 and Syn fth3 P73437[9]
ycf78	–	–	–	–	–	–	+	–	–	–	–	–	–	–	* Chl ORF819, hom. in Chlamydomonas reinhardtii cpDNA X92726[10]
ycf79	–	–	–	–	–	–	+	+	+	–	–	–	+	–	Syn sll1398, annotated as 13 kD protein in PSII[5]

Notes

Plus and minus signs indicate presence and absence, respectively of the gene in the given chloroplast genome. A plus sign in parentheses (+) indicates that the gene is not found in BLAST searches in the Synechocystis genome, but was scored for specific reasons as being cyanobacterial in origin and present in the ancestral plastid genome. Synonyms are indicated in parentheses, except in the case of ORF designations. Accession numbers refer to PIR, GenBank or SwissProt data bases. In cases where several nuclear homologues were found, only one is indicated in the corresponding column. Reading frames designated as pseudogenes in Epifagus cpDNA (Wolfe et al., 1992) were scored as absent here. The last revision of ycf nomenclature ended with ycf47 (Reardon and Price, 1995). The gene names suggested here have been approved by the commission on plant gene nomenclature (Stoebe et al., 1998).

Abbreviations:

Zea, Zea mays; Ory, Oryza sativa; Nic, Nicotiana tabacum; Pin, Pinus thunbergii; Mar, Marchantia polymorpha; Eug, Euglena gracilis; Odo, Odontella sinensis; Por, Porphyra purpurea; Chl, Chlorella vulgaris; Pla, Plasmodium falciparum; Cpa, Cyanophora paradoxa; Syn, Synechocystis PCC6803; Epi, Epifagus virginiana; sim., similar to; hom., homologue; SU, subunit; bind., binding; prot., protein; RT, reverse transcriptase.

* No similarity to prokaryotic genes was detected in BLAST searches.

[1] cbbX is located in proteobacterial cbb operons (rbcLr-rbcSr-cbbX), not to be confused with rbcX in cyanobacterial cbb operons (rbcLg-rbcX-rbcSg; Li and Tabita, 1997).

[2] Cyanophora cpDNA encodes two distinct clpP genes that are 36% identical, whereby Synechocystis also possesses two distinct clpP genes that are however 74% identical.

[3] petL, a 31 amino acid long component of the cytochrome b_6/f complex, is too short to detect its Synechocystis homologue (gi1653694) among the highest BLAST scores, but it was counted as being of cyanobacterial origin.

Table 15.1 Continued

[4] *pgmA* has related but paralogous nuclear homologues for cytosolic phosphoglycerate mutase in many higher plants (*e.g. Nic* P35494).

[5] There is a severe problem with psbW nomenclature. The gene designated in previous studies of chloroplast genomes (including Martin et al., 1998) as *psbW* shares no similarity with the nuclear-encoded gene for the psbW protein functionally characterized from spinach chloroplasts (Lorkovic et al., 1995). The spinach gene designation has priority. The chloroplast-encoded protein previously designated as psbW is similar to *Synechocystis* sll1398 (GenBank 1651690), annotated as a 13 kDa PSII protein. Since a function for the chloroplast gene previously described as *psbW* has not been reported, we designate it here as *ycf79*.

[6] Closely related to the Calvin cycle transcriptional regulator *cbbR* (Qian and Tabita, 1996).

[7] Chloroplast *rpl12* has in some cases also been designated as *rpl7*.

[8] *rpoC2* is sometimes also designated as *rpoD* in cpDNA, but the gene designated *rpoD* in *Syn* is a sigma factor, the designation *rpoC2* for the chloroplast gene is less ambiguous and hence preferable.

[9] We found no significant similarity between cpDNA genes previously designated as *ycf1* (Reardon and Price, 1995) and the gene tentatively designated as *ycf1* in the *Chlorella* genome (Wakasugi et al., 1997), that is hence designated as *ycf77* here.

[10] We found no significant similarity between cpDNA genes previously designated as *ycf2* (Reardon and Price, 1995) and the gene tentatively designated as *ycf2* in the *Chlorella* genome (Wakasugi et al., 1997), that is hence designated as *ycf78* here.

[11] *ycf46* has a related but paralogous nuclear homologue in *Ory* D17789 (Suzuka et al., 1994).

[12] *ycf64* has similarity to glutaredoxin-like proteins, nuclear homologues are found in yeast and *Plasmodium*, a distant homologue is nuclear in *Arabidopsis* (e327479 in Z97339).

Another exception is *cbb*X of *Odontella* and *Porphyra*, that has no known cyanobacterial homologue, but is clearly orthologous to *cbb*X found in some proteobacterial Calvin cycle (*cbb*) operons (Gibson and Tabita, 1996). *Cbb*X is not a Calvin cycle enzyme (Gibson and Tabita, 1997; Martin and Schnarrenberger, 1997), and all known proteobacterial *cbb*X genes are encoded as a -*rbc*L-*rbc*S-*cbb*X- cluster (Gibson and Tabita, 1997). This cluster is preserved in *Porphyra* (Reith and Munholland, 1995), but not in *Odontella* (Kowallik *et al.*, 1995) chloroplast DNA. The final two exceptions are the red type Rubisco subunits of *Odontella* and *Porphyra*. Notably, the adjacent *rbc*Lr and *rbc*Sr genes of rhodophytic and derivative plastids ('red' type Rubisco: Lr_8Sr_8) are paralogous to cyanobacterial, glaucocystophyte and chlorophyte *rbc*Lg and *rbc*Sg genes ('green type' Rubisco: Lg_8Sg_8) (Martin and Schnarrenberger, 1997). The widespread occurrence of fully duplicate, differentially regulated and distinct proteobacterial *cbb* operons on plasmids (Bowien *et al.*, 1993; Gibson and Tabita, 1996), and the documented functional requirement of *cbb*X for Calvin cycle function with Lr_8Sr_8 Rubisco (Gibson and Tabita, 1997) suggest functional association between the gene products of *rbc*Lr/*rbc*Sr and *cbb*X. This suggests to us that differential loss of paralogous *rbc*LS clusters in the ancestor of plastids (one containing -*rbc*Lr-*rbc*Sr-*cbb*X-, one containing -*rbc*Lg-*rbc*Sg-) is behind eukaryotic class I (L_8S_8) Rubisco disparity (Martin *et al.*, 1992; Martin *et al.*, 1993; Morse *et al.*, 1995; Martin and Schnarrenberger, 1997), rather than horizontal gene transfer of *rbc*Lr/*rbc*Sr genes from proteobacteria to plastids. That the vast majority of protein coding genes in plastids were demonstrably inherited from cyanobacteria is of little surprise, and critical inspection of putative cases of gene acquisition may seem belaboured, but these considerations suggest that foreign genes have stumbled into these chloroplast genomes rarely at best, and possibly not at all. But it is too early to generalize, since analyses of chloroplast genomes in other algal lineages may eventually provide a different picture of uptake of foreign genes by cpDNA.

15.3 A rooted phylogeny for nine plastid genomes

Table 15.1 summarizes the distribution of 254 genes across 13 genomes and provides access to synonyms (also accession numbers where relevant). Starting with the total number of 254, we will filter them to see how many remain for meaningful analysis. Six genes in the table are present in only one of the plastid genomes surveyed, but have putative homologues in some other cpDNA or prokaryotic genome: *minE*, *rdpO*, *ycf13*, *ycf67*, *ycf77*, and *ycf78*, leaving 248. Eighteen of these have no identifiable homologues in the *Synechocystis* genome, but occur in at least two of the 12 plastid genomes listed, leaving 230 that are present in *Synechocystis* and at least one plastid genome surveyed. Of these, only 11 are present in all 13 genomes, among which the very low gene content of the *Plasmodium* and *Epifagus* genomes are obviously the limiting factors. All of these encode ribosomal proteins: *rpl2*, *rpl16*, *rpl36*, *rps2*, *rps3*, *rps4*, *rps7*, *rps8*, *rps11*, *rps12* and *rps19*. The remaining 219 protein-coding genes are distributed in some manner across the genomes, and are missing in at least one. Seven are missing in only one genome, 26 are missing in two genomes, 11 are missing in three, two are missing in four, seven in five, two in six, eight in seven, 18 in eight, 31 in nine, 40 in ten, and 67 are missing in 11 of the 12 plastid genomes surveyed. A glance at the table reveals that the phylogenetic pattern of gene loss is not simple, a matter to which we will return later.

Obviously, there is a trade-off between the number of genomes that one could subject to phylogenetic inference and the number of proteins that would be common to each for providing a large data set to study. Here we will consider only one example involving the five land plants, together with *Odontella*, *Porphyra*, *Euglena* and *Cyanophora*. There are 46 genes in the table that have homologues in these nine genomes and *Synechocystis*. These proteins were retrieved, aligned in individual files with PILEUP of the GCG package (Genetics Computer Group, 1994), then written into PHYLIP (Felsenstein, 1993) format with CLUSTALW (Thompson *et al.*, 1994). The individual alignments were merged into one large, concatenated alignment that contained 11,521 amino acid positions per genome for each of 10 operational taxonomic units (OTUs). This alignment was analyzed with the neighbour-joining (NEIGHBOR) (Saitou and Nei, 1987) and parsimony (PROTPARS) routines of the PHYLIP package. For NJ, both the Dayhoff distance (Schwartz and Dayhoff, 1978) and the Kimura distance (Kimura, 1983) options of PHYLIP programs were used. All three methods produced the same tree, and all seven branches in the tree receive 100 bootstrap proportion (BP) support with all three methods (Fig. 15.1). In protein maximum likelihood analysis (Adachi and Hasegawa, 1996b), branches 2–7 receive full support and branch 1 receives 78% support (M. Hasegawa, pers. comm. 1997). Branch numbers 1–7 are indicated (see text).

The protein sequences were obtained by direct translation from DNA, distances estimated are therefore the sum of substitutions plus differential RNA editing events per site. Editing is much more rare in chloroplasts than in mitochondria, and among chloroplasts it is more common in higher plants than in algae (Yoshinaga *et al.*, 1996). In maize, it is estimated that about 25 sites in the complete chloroplast genome are edited (Maier *et al.*, 1995), in black pine and tobacco the total number is somewhat higher (Wakasugi *et al.*, 1996). Although the degree of transcript

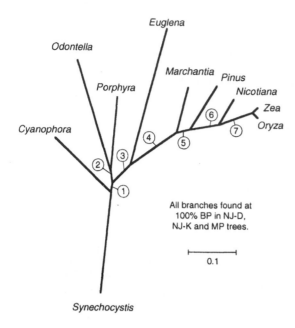

Figure 15.1 Rooted nine species neighbour joining (NJ) tree of Dayhoff distances for 11521 amino acid positions from the set of all 46 proteins common to these chloroplast genomes. All seven branches receive 100% bootstrap support in maximum parsimony (PROTPARS of PHYLIP) and NJ analysis using either Kimura or Dayhoff distances. The alignment can be obtained via anonymous ftp from 134.169.70.80/usr/local/ftp/pub/incoming/ 11521.infile.

editing is not known for all OTUs considered here, it will only affect comparison of differentially edited codons at the small fraction of edited sites, and was therefore neglected. There is one – but only one – case of deep paralogy in the data: *rbc*L. Its topological effect in this taxon sample is not expected to be severe, since paralogues are shared only by *Odontella* and *Porphyra*; *rbc*L constitutes only 4.2% of the positions and was therefore not excluded from the 10 OTU alignment. Thus, these simple methods of inference yield what appears to be a fully resolved, properly rooted, nine-genome phylogeny.

15.4 Individual proteins in comparison to the concatenate proteins

Most previous studies of plastid evolution have been conducted with one or a few genes (Martin *et al.*, 1992; Palenik and Haselkorn, 1992; Urbach *et al.*, 1992; Chase *et al.*, 1993; Bhattacharya and Medlin, 1995; Helmchen *et al.*, 1995; Henze *et al.*, 1995; Cavalier-Smith *et al.*, 1996; Van de Peer *et al.*, 1996: for a recent review see Melkonian, 1996). Thus, for comparison, one might ask how well the 46 proteins fare in recovering the topology of Fig. 15.1 when analyzed individually for the same ten taxa using a simple (Dayhoff distance) but robust substitution model? In consensus trees of 100 replicates constructed with the NJ method using the Dayhoff matrix, only two proteins recovered the topology of Fig. 15.1 in the single NJ-tree: *rps11* and *psa*B (Table 15.2). In general, the more conservatively evolving proteins (upper portion of Table 15.2) tended to recover the deeper branches 1–3 in Fig. 15.1 somewhat more often than the less conservatively evolving proteins did (lower portion of Table 15.2). This is not surprising, since one might expect more conservative proteins to have preserved more evidence for ancient evolutionary events than have more variable proteins. One would also expect the converse for more recent evolutionary events, notwithstanding that terms such as 'conservative' and 'variable' are relative, and that we make no effort to quantify them here.

Unexpectedly, sister genes that derived from ancient duplication (*psa*A/*psa*B, *psb*A/*psb*D, *psb*E/*psb*F, *atp*A/*atp*B) did not reveal identical topologies, even though their products are functionally related and even though within each pair similar length and degree of divergence is observed. This underscores the limitations of single-gene phylogenies. That only two proteins recovered the topology of Fig. 15.1 was quite surprising, since the complete data set of 11,521 positions unambiguously supports that topology. This suggested to us that the fraction of sites in the concatenated data set that support the topology in Fig. 15.1 might be sufficiently dispersed across genes that their phylogenetic signal is obscured in individual alignments, perhaps by stochastically similar sites. The general picture obtained from Table 15.2 also suggested that different types of sequence data support different portions of the topology in Fig. 15.1, whereas the total data support only that topology. The distictness of the phylogenetic behaviour of the subunits of the plastid-encoded RNA polymerase (*rpo*-genes; see Table 15.2) was previously noted (Martin *et al.*, 1998).

15.5 The information content of the data

Given these observations, we wished to estimate how much of the available data is necessary to support various regions of the topology in Fig. 15.1 (particularly the more difficult-to-recover deep branches) at a given BP interval. To estimate such parameters from real rRNA sequence data, Lecointre *et al.* (1994) jackknifed subsamples of non-autapomorphic polymorphic sites and bootstrapped each subsample. Their results from sampling in upper portions of the tree permitted estimation of the amount of data needed in theory to resolve deeper branches (Lecointre *et al.*, 1994). Our approach to this protein data was similar, but our problem was different: glaucocystophytes, euglenophytes, rhodophytes, chromophytes and chlorophytes probably represent the full depth, but not the entire breadth, of plastid evolution. Instead of jackknifing from a phylogenetically restricted number of polymorphic positions to extrapolate BP support for deeper branches (Lecointre *et al.*, 1994), we started from a well-supported tree and took samples of increasing size from the polymorphic positions available, interpolating what fraction of the available data in practice provides 95% BP support for branches.

To facilitate sampling, we used a sorting approach (Fig. 15.2a). Positions of the concatenate alignment were rearranged according to the number of states (different amino acids) that occur at a given position, whereby gaps are scored as different from every other amino acid or gap at the position (Goremykin *et al.*,

Table 15.2 Summary of results of individual proteins with NJ–D.

Protein	No. sites	Fig. I Topol.?	BP for Branches in Fig. I 1	2	3	4	5	6	7	Tree Length*
psbA	343	no	1	53	12	100	59	97	30	0.41
psaC	82	no	0	6	0	0	0	2	98	0.44
psbD	353	no	5	99	53	100	38	77	93	0.51
atpH	81	no	30	26	56	80	56	44	67	0.55
petB	215	no	16	54	8	98	1	97	31	0.58
psaA	754	no	6	97	93	100	100	100	100	0.73
psaB	738	yes	84	100	100	100	96	100	100	0.74
atpB	491	no	4	46	81	100	58	99	100	0.79
rps12	123	no	45	0	7	100	76	74	100	0.79
psbC	460	no	71	100	81	100	17	91	95	0.80
psbE	72	no	79	100	23	99	55	84	91	0.80
psbF	40	no	56	61	86	83	20	26	11	0.82
psbB	507	no	18	100	92	100	97	100	100	0.88
psbL	39	no	45	43	24	72	71	91	26	0.88
atpA	494	no	0	0	99	100	55	100	100	0.98
rbcL	482	no	45	100	90	99	90	70	90	1.19
psbI	36	no	21	40	60	27	10	73	6	1.26
rpl14	124	no	39	45	31	99	87	91	100	1.28
petG	38	no	0	0	37	78	62	97	32	1.30
psbT	31	no	27	35	0	99	71	52	26	1.40
rpl36	38	no	16	26	48	37	21	58	89	1.40
psbH	58	no	1	97	97	59	71	46	54	1.48
psbJ	42	no	11	13	6	99	38	15	14	1.50
rpl16	134	no	74	44	39	94	81	74	100	1.52
rps19	94	no	0	0	9	100	38	28	100	1.83
psbK	46	no	21	48	95	9	3	45	98	1.85
psaJ	38	no	11	59	35	88	0	97	95	1.93
rpl2	277	no	8	53	42	100	100	100	100	1.93
psbN	43	no	0	44	6	94	56	94	73	2.00
rps11	146	yes	33	73	95	99	77	64	100	2.09
rps18	60	no	86	14	92	91	96	90	100	2.09
rps7	156	no	11	33	19	100	93	96	100	2.25
rps4	211	no	7	0	32	100	47	94	100	2.38
rps14	103	no	16	1	5	100	93	91	100	2.41
rps2	225	no	71	4	54	100	89	99	100	2.52
rps8	144	no	0	2	70	99	99	84	100	2.55
rpoB	1098	no	11	0	83	100	100	100	100	2.77
rps3	250	no	95	5	15	100	88	99	100	2.81
rpoC1	689	no	65	1	99	100	98	100	100	2.93
rpl20	115	no	54	40	50	98	67	99	100	2.99
rpl22	111	no	21	5	19	96	36	94	100	3.03
atpE	137	no	6	0	77	99	88	100	100	3.31
ycf4	175	no	11	54	18	100	96	100	100	3.34
ycf9	58	no	45	25	55	99	6	93	100	3.43
rpoC2	1388	no	43	0	35	100	100	100	100	4.18
atpF	182	no	5	89	94	98	100	99	100	5.18

* Tree length is total branch length (substitutions per site) in the NJ–D tree. It gives a rough measure of the variability of the protein.

1997). Briefly, the number of amino acid states occurring at each position of a PHYLIP infile (interleaved format) is counted. Each position is assigned a priority value according to its variability, with invariant positions receiving the highest priority and the most polymorphic positions receiving the lowest. Positions are sorted into a new file according to their priority using a standard queue. For inference algorithms that assume that sites evolve independently, sorted and unsorted alignments produce identical results. Positions with no phylogenetic information (identical sites), potentially misleading positions (extremely variable sites) and potentially useful positions are nested in the sorted alignment. Since gaps are scored as different from every other amino acid or gap at the position, positions with many gaps are sorted into the more variable fractions of sites. Window lengths (L_w) of 500, 1000, 1500, 2000 and 2500 amino acids were moved in steps of 500 positions across the sorted alignment, the window range (R_w) indicating the position number at mid-window (Fig. 15.2b). For each window, 100 bootstrap samples for neighbour-joining with the Dayhoff distance (NJ–D) were taken. Samples from windows in which R_w fell within invariant sites were disregarded.

First, we examined the average BPs of each branch in Fig. 15.1 for increasing L_w (Fig. 15.2c). For determining these average BPs, we excluded results from windows $R_w < 5746$ or $R_w > 10,254$, i.e. the midpoints of which lay (a) within the 6-state or greater, or (b) within the autapomorphic class. Inspection of the data in such sites (Fig. 15.2a) reveals that this is a justifiable exclusion of data for determining averages, particularly for short windows. Branches 4–7 received on average $\geqslant 98\%$ support even for windows of length 500. Since 5000 sites are uniformly sampled for $L_w = 500$ and $5746 < R_w < 10,254$, this suggests that the excess of data in chloroplast proteins over that needed to support branches 4–7 with the simple but robust Dayhoff substitution model and a robust tree inference method (Nei, 1996) is ten-fold or greater. Branch 3, which is not detected in most rRNA trees and receives no statistical support from rRNA data, even with sophisticated substitution models (Lockhart et al., 1994; Van de Peer et al., 1996), receives an average of $\geqslant 98\%$ support for $L_w = 1000$ (Fig. 15.2c), in which 5500 sites are sampled (though not uniformly), indicating that a roughly five-fold excess of data in chloroplast proteins exists over that needed for support of the branch. Average BP for branch 2 reaches $> 95\%$ for $L_w = 1500$ and $\geqslant 98\%$ for $L_w = 2500$, suggesting an excess of available data for support of about three-fold. Thus, branches 2–7 are overwhelmingly supported by the data, leaving only three of the 2,027,025 possible rooted nine species trees for these plastids to examine, i.e. the position of Cyanophora.

Branch 1 is more difficult. Average BP for $5746 < R_w < 10254$ increases only to 69.7% for $L_w = 2500$ (Fig. 15.2c) and reaches 95.2% only at $L_w = 5000$. It is evident in Fig. 15.2d, where individual BP for each R_w is shown, that informative sites contain very little support for branch 1. Fig. 15.2e reveals why: informative two-state sites – but only they – tend to support a topology conflicting with branch 1, namely one that contains the grouping (syn(odo,por)) (Fig. 15.2e; for OTU abbreviations see Table 15.1). But of the total of 1171 informative two-state sites, only 25 support (syn(odo,por)) and only 26 support branch 1. By comparison, 48 support branch 2 (excluding the 78 informative 2:8 sites in rbcL paralogues alone that support branch 2; see below) and 89 support branch 3. But 628 informative sites

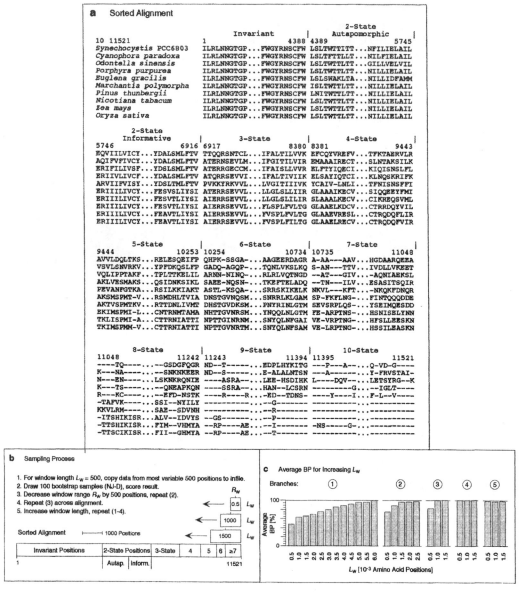

Figure 15.2 Amount and nature of phylogenetic data in chloroplast encoded proteins. **(a)** Segment of the alignment sorted by positional variability. The 11,521 sites break down into 4388 invariant (sites 1–4388 of the alignment), 2528 2-state (1357 autapomorphic, 1171 informative), 1464 3-state, 1063 4-state, 810 5-state, 481 6-state, and 787 highly variable sites with greater than 7-states each (sites 10735–11521 of the sorted alignment). **(b)** Sampling process from sorted data. Borders between positional classes in the sorted alignment, window lengths, window ranges and the scale bar are indicated. NJ-D: neighbour joining of Dayhoff distances. **(c)** Average BPs for branches 1–5 across values of $5746 < R_w < 10254$ for increasing window length (vertical bars). Continued overleaf.

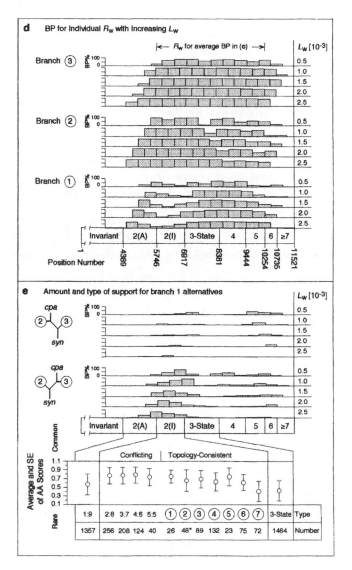

Figure 15.2 continued **(d)** BP of branches 1, 2, and 3 of Fig. 15.1 for individual R$_w$ (vertical bars, scale at left) and different window lengths (right). The range for averaging BPs in (c) is indicated (top), as are position numbers and positional class boundaries (bottom). **(e)** Results as in (d) for topologies indicated that compete with branch 1. Lower portion: Normalized Dayhoff frequencies (Gribskov and Burgess, 1986) of amino acid substitution types (average ± SE) occurring in various 2- and 3-state sites. Circled integers 1–7 indicate values for the fraction of informative sites supporting the respective branch (topology-consistent). Values for informative sites of the 2:8, 3:7, 4:6, and 5:5 type that do not support branches 1 through 7 (conflicting) are indicated. The number of topology-consistent and conflicting sites for each type is given (bottom). An asterisk indicates that 78 informative sites are contained in the deep *rbcL* paralogues alone that support branch 2 which were excluded from averaging of substitution frequency types.

support branches that conflict with unambiguously supported branches in Fig. 15.1, among them 79 sites that conflict the widely accepted topology of land plants sampled, including 11 that support branching of *Cyanophora* among land plants.

The interpretation of these findings is straightforward: at least 628 sites appear to be informative in the sense of having preserved evidence for evolutionary process, but in fact they are not. They are merely toggling between two amino acids over time. For practical purposes, there are no two-state informative sites left in the data to support any specific position of *Cyanophora*. This conclusion is substantiated by the relative frequencies of amino acid subsitution types that occur at informative sites (Fig. 15.2e, bottom). They consist to a very large extent of substitutions involving structurally and functionally very similar amino acids (e.g. K/R, I/V, L/I, M/L, and L/V) *except* in branch 7, the most recent branch in the tree, where rare substitution types have apparently been well preserved among informative sites. By contrast, both three-state and autapomorphic sites in the data contain a lower proportion of such frequent subsitution types (Fig. 15.2e), and are evidently evolving, on the whole, more slowly than two-state sites. This is a direct manifestation of the difficulty of estimating the variation of substitution rate across sites (Rzhetsky and Nei, 1994; Yang *et al.*, 1994; Gu and Li, 1996).

These findings indicate that studies involving more distantly related plastid relatives (cyanobacteria) with data sets of this size will mandate more sophisticated substitution models, possibly utilizing substitution matrices estimated from the alignment (Adachi and Hasegawa, 1996a), in order to resolve deeper branches. Nonetheless, even for branch 1, the deepest and shortest branch of the topology, an excess of information appears to exist in the complete set of homologous chloroplast proteins over that required to achieve 95% BP support using simple but robust models (and including highly variable sites in analysis). The excess is, however, very slight – much less than two-fold. One further finding from Fig. 15.2 is noteworthy: the topology does not derive support from the most variable fraction of sites, indicating that the tree in Fig. 15.1 is neither an artefact of questionable alignment, nor of stochastic similarity at highly variable sites. From the foregoing analyses we conclude that chloroplast genomes contain enough – but only just enough – information to infer the rooted phylogeny of these nine photosynthetic organelles.

As previously argued (Martin *et al.*, 1998), the complete chloroplast genome data do not discriminate between the possibilities of a single or multiple origin of plastids, resolution of that issue will require analysis of further cyanobacterial rather than plastid genomes. For the purposes of this paper, we assume that plastids (Martin *et al.*, 1998) and mitochondria (Martin and Müller, 1998) had a single origin.

15.6 Predominance of parallel gene losses in independent lineages

It is known that genes are lost from cpDNA over time, but pattern and process of loss have not been viewed at the comparative genome level. Presence or absence of a gene (or many genes) in cpDNA should, in principle, be a reasonably good tool for inferring phylogeny. To investigate this, we scored gene presence or absence as Dollo data (that is, character state data where reversals are not allowed) and used it to infer phylogeny with parsimony (Swofford, 1993). The single shortest, properly rooted, loss-only tree by that model was quite inconsistent with the topology in Fig. 15.1. It clearly misplaced two OTUs within the land plant portion of the tree and required 306 steps: ((((((((zea(ory,nic))mar)(pin,eug))odo)cya)por)syn). The topology in Fig. 15.1 is 98 steps (32%) longer, requiring 404 losses; it is among the shortest 13,768 trees (top 0.8%) for this ten species case. This was surprising, because for this data conditions would seem quite good for parsimony to work since (a) the ancestral state is known (gene present in cyanobacteria), (b) we can directly observe an unambiguously scoreable character pattern, and (c) *a priori*, loss events might be expected to be rare.

We therefore mapped every gene occurrence onto the topology of Fig. 15.1 for the 235 genes that existed in the plastid genome ancestral to those surveyed (Table 15.1). The result was extremely surprising (Fig. 15.3). Gene loss events that are unique to individual lineages account for the fate of only 68 genes across the tree. The majority of genes (122) have undergone parallel loss in independent lineages. Forty-five have been lost twice independently, 62 genes have undergone three parallel losses, and 15 genes have been lost four times in independent lineages (note that all of these changes are inferred from a survey of only nine genomes). If we make the unlikely but event-minimizing assumption that loss occurred in the form of the gene blocks indicated, the numbers of unique (7) and parallel (44) events suggest a vast excess of parallel losses. But judging from the many pseudogenes in the *Epifagus* plastid genome (Wolfe *et al.*, 1992) and the many *ndh* pseudogenes in *Pinus* cpDNA (Wakasugi *et al.*, 1994), it would appear that genes are lost individually, rather than as blocks. If we count each loss of an individual gene as one event, the ratio of parallel to unique events becomes 4.9 (68 unique vs. 336 parallel) – still an overwhelming excess of parallelisms. When gene content in the more highly reduced genomes of *Epifagus* and *Plasmodium* in Table 15.1 are considered, it is clear that many more additional parallel losses must have occurred in chloroplast genome evolution beyond those summarized in Fig. 15.3. Large numbers of parallel losses have also been noted to have occurred in mitochondrial genome evolution (Palmer, 1997) where the overall reduction in genome size and gene content has proceeded to a much further extent (Martin and Müller, 1998; Martin and Herrmann, 1998) than in plastids. Should such massive parallelism warrant further caution (Stewart, 1993) concerning the strength of the parsimony principle? The gene loss data from chloroplast genomes tell a reasonably straightforward story: the model of parsimony does not approximate the evolutionary process underlying the contemporary phylogenetic distribution of chloroplast genes.

15.7 Transfer of functional genes from the plastid to the nucleus

In the process of BLAST searching, we noted several significant hits with previously characterized nuclear genes of higher plants. We initially extended these searches to EST databases, but that was quickly abandoned due to the presence of numerous, clearly chloroplast-derived sequences in those data, rendering the search results uninterpretable. We set simple criteria for indicating the presence of a functional

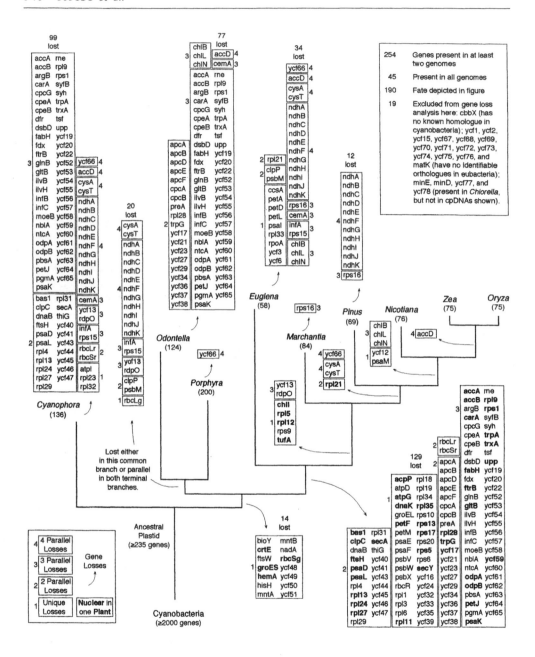

Figure 15.3 Phylogenetic distribution of gene loss from chloroplast genomes. Keys designating frequency of parallel gene losses are given at the lower left. Numbers below species names indicate the number of protein coding genes and *ycf*s that exist in the respective chloroplast genome (see also text and Table 15.1). *rne* encodes an RNA, but was included in the analysis. Numbers above gene columns represent the number of genes lost and accountable in the figure for the given genome. Genes for which nuclear homologues were identified are given in bold. BLAST results apply to GenBank on November 15, 1997.

nuclear gene: (a) expression (more-or-less full size cDNA sequence characterized), (b) presence of a functional or putative transit peptide, and (c) greater similarity between cpDNA and nuclear homologues than between cyanobacterial and nuclear homologues. By these criteria, we identified 47 plant nuclear genes among those examined that descend from cyanobacterial genomes (these are indicated in bold in Fig. 15.3; accession numbers given in Table 15.1). These genes were transferred from cpDNA to the nucleus during the course of evolution, where they came under the regulatory hierarchy of the eukaryotic transcription machinery, and acquired a transit peptide for re-import into the organelle of their genetic origin (Martin and Schnarrenberger, 1997; Martin et al., 1998). Many of the nuclear homologues identified belong to the classes of genes that were lost independently in different lineages. In such cases, either (a) independent gene transfer and integration occurred in distinct lineages, or (b) a single transfer occurred in the common ancestor with subsequent co-existence of functional nuclear and cpDNA copies such that gene loss, but not necessarily gene transfer, occurred independently (Baldauf et al., 1990). In some cases, however, the chloroplast protein may have been lost altogether and functionally replaced by a re-routed homologue of mitochondrial origin. This has been demonstrated for several enzymes of the Calvin cycle (Martin et al., 1996; Martin and Schnarrenberger, 1997). Fig. 15.3 is a convenient road-map to the study of endosymbiotic gene transfer from chloroplasts.

15.8 The phylogeny of plastids reflected in complete genomes

What can the topology in Fig. 15.1 tell us about the evolution of photosynthetic organelles? Branch 3 substantiates the chlorophytic provenance of *Euglena*'s plastids suggested by Gibbs (1978). Common ancestry of the euglenoid and kinetoplastid (trypanosomes and relatives) nucleocytoplasmic lineages is supported by 18S rRNA and protein data (Levasseur et al., 1994; Navazio et al., 1998), and by many cytological traits (Walne and Kivic, 1989; Cavalier-Smith, 1993). But the elucidation of *Euglena*'s plastid ancestry has been extremely difficult to resolve with rRNA data. Most analyses group *Euglena*'s plastids with rhodophytes, whereas a few detect branching with chlorophytes, and then with only minimal statistical support (Lockhart et al., 1994; Helmchen et al., 1995; Van de Peer et al., 1996). Similarly weak results are also observed with protein genes (Henze et al., 1995; Delwiche et al., 1995). In the case of rRNA, this is largely due to base compositional bias (Lockhart et al., 1994), but as shown in Table 15.2, it is also attributable to the limits of information contained in any individual gene. The present whole-genome data resoundingly support the chlorophytic ancestry of *Euglena*'s three-membrane-bound plastids.

The plastid protein data also support a rhodophytic affinity of *Odontella*'s four-membrane-bound plastids. Since molecular and cytological evidence indicate a common ancestry of *Odontella*'s nuclear lineage with oomycetes (Cavalier-Smith et al., 1996), this can be taken as strong support for a secondary symbiotic origin of *Odontella*'s plastids, congruent with previous interpretations on the basis of overall plastid gene content and structure (Kowallik et al., 1995; Kowallik, 1997; for recent overviews of secondary symbiosis see McFadden et al., 1997 and Gilson et al., 1997).

The phylogenetic position of the chlorophyll *a* and phycobilin-containing cyanelles of *Cyanophora paradoxa* has, as in the case of *Euglena*, also been virtually impossible to resolve with single gene phylogenies. Yet this organelle, characteristic of the glaucocystophytes, is critical to understanding plastid evolution (Löffelhardt and Bohnert, 1994; Helmchen *et al.*, 1995; Löffelhardt *et al.*, 1997) because it has retained not only phycobilisomes, but also a eubacterial peptidoglycan wall. Previous molecular studies have refuted early views that cyanelles might represent recently or independently captured cyanobacteria, but the long suspected basal status of cyanelles among contemporary plastids is simply not recovered with data from single genes (Helmchen *et al.*, 1995; Cavalier-Smith *et al.*, 1996; Van de Peer *et al.*, 1996). The suggestion that instability of cyanelle position in previous molecular studies might reflect a simultaneous radiation of rhodophyte, chlorophyte and glaucocystophyte plastids (Helmchen *et al.*, 1995) is not supported by the present data. Rather, our overall findings and specific analyses of branch 1 indicate (a) that cyanelles are indeed basal to both the rhodophyte and chlorophyte plastid lineages, (b) that strong support for this position is contained within the data, but (c) that the amount of data within the whole genome alignment necessary to achieve high bootstrap support for this position is very close to the amount available. This finding underscores the difficulty of inferring the phylogeny of plastids, even with very large amounts of molecular data.

15.9 Conclusions

Using simple models of amino acid substitution and phylogenetic inference, proteins in chloroplast genomes may contain just enough information to resolve the rooted nine species tree of a glaucocystophyte, a chromophyte, a euglenophyte, a rhodophyte and five chlorophyte plastids. They contain a vast excess of information for inference of higher plant evolution and can therefore provide plant molecular systematics with a steadfast backbone. Given the difficulty of obtaining support for the position of *Cyanophora* with these data, we wonder whether comparable matrices will be sufficient to identify possible sisters of plastids among contemporary cyanobacteria, and whether chloroplast DNA may be a model for bacterial evolutionary genomics. With more sophisticated inference methods and substitution models (Yang *et al.*, 1994; Lockhart *et al.*, 1994; Adachi and Hasegawa, 1996a; Lake, 1997), and with a defined tree to recover, it should eventually be possible to investigate substitution parameters of individual proteins and thereby increase their phylogenetic utility.

ACKNOWLEDGEMENTS

We thank Masami Hasegawa for suggesting the star topology to estimate positional variability and for communicating PROTML results with this data and the Rechenzentrum der Universität Braunschweig for generous use of their computer facilities. WM and KK acknowledge the Deutsche Forschungsgemeinschaft (DFG) for financial support. BS is the recipient of a post-doctoral stipend from the DFG, VG was the recipient of a Ph.D. stipend from the DAAD.

REFERENCES

Adachi, J. and Hasegawa, M. (1996a) Model of amino acid substitution in proteins encoded by mitochondrial DNA. *Journal of Molecular Evolution*, **42**, 459–468.

Adachi, J. and Hasegawa, M. (1996b) MOLPHY Version 2.3: *Programs for molecular phylogenetics based on maximum likelihood*. Computer science monographs, No. 28. Institute of Statistical Mathematics, Tokyo.

Baldauf, S., Manhart, J. and Palmer, J. (1990) Different fates of the chloroplast *tufA* gene following its transfer to the nucleus in green algae. *Proceedings of the National Academy of Sciences USA*, **87**, 5317–5321.

Bauer, C. C., Ramaswamy, K. S., Endley, S., Scappino, L. A., Golden, J. W. and Haselkorn, R. (1997) Suppression of heterocyst differentiation in *Anabaena* PCC 7120 by a cosmid carrying wild-type genes encoding enzymes for fatty acid synthesis. *FEMS Microbiology Letters*, **151**, 23–30.

Bhattacharya, D. and Medlin, L. (1995) The phylogeny of plastids: a review based on comparisons of small-subunit ribosomal RNA coding regions. *Journal of Phycolology*, **31**, 489–498.

Bowien, B., Bednarski, R., Kusian, B., Windhövel, U., Freter, A., Schäferjohann, J. *et al.* (1993) Genetic regulation of CO_2 assimilation in chemoautotrophs, in *Microbial growth on C1 compounds*, (eds J. C. Murrel and D. P. Kelley), Intercept Scientific, Andover UK, pp. 481–491.

Cavalier-Smith, T. (1993) Kingdom Protozoa and its 18 phyla. *Microbiological Reviews*, **57**, 953–994.

Cavalier-Smith, T., Couch, J. A., Thorstenstein, K. E., Gilson, P., Deane, J. A., Hill, D. R. A. *et al.* (1996) Cryptomonad nuclear and nucleomorph 18S rRNA phylogeny. *European Journal of Phycology*, **31**, 315–328.

Chase, M. W., Soltis, D. E., Olmstead, R. G., Morgan, D., Les, D. H., Mishler, B.D. *et al.* (1993) DNA sequence phylogenetics of seed plants: an analysis of nucleotide sequences from the plastid gene *rbcL*. *Annals of the Missouri Botanical Garden*, **80**, 528–580.

Delwiche, C. F. and Palmer, J. D. (1996) Rampant horizontal transfer and duplication of Rubisco genes in eubacteria and plastids. *Molecular Biology and Evolution*, **13**, 873–882.

Delwiche, C. F., Kuhsel, M. and Palmer, J. D. (1995) Phylogenetic analysis of *tufA* sequences indicates a cyanobacterial origin of all plastids. *Molecular Phylogenetics and Evolution*, **4**, 110–128.

Felsenstein, J. (1993) PHYLIP (Phylogeny inference package) manual, version 3.5c. Distributed by the author – University of Washington, Seattle: Department of Genetics.

Fontaine, J. -M., Goux, D., Kloareg, B. and Loiseaux-de Goër, S. (1997) The reverse-trancriptase-like proteins encoded by group II introns in the mitochondrial genome of the brown alga *Pylaiella littoralis* belong to two different lineages which apparently coevolved with the group II ribozyme lineages. *Journal of Molecular Evolution*, **44**, 33–42.

Genetics Computer Group (1994) *Program manual for Version 8*, 575 Science Drive, Madison, Wisconsin, 53711, USA.

Gibbs, S. H. (1978) The chloroplast of *Euglena* may have evolved from symbiotic green algae. *Canadian Journal of Botany*, **56**, 2883–2889.

Gibson, J. L. and Tabita, F. R. (1996) The molecular regulation of the reductive pentose phosphate pathway in proteobacteria and cyanobacteria. *Archives of Microbiology*, **166**, 141–150.

Gibson, J. L. and Tabita, F. R. (1997) Analysis of the *cbbXYZ* operon in *Rhodobacter sphaeroides*. *Journal of Bacteriology*, **179**, 663–669.

Gilson, P., Maier, U. -G. and McFadden, G. I. (1997) Size isn't everything: lessons in genetic miniaturization from nucleomorphs. *Current Opinion in Genetics and Development*, **7**, 800–806.

Goremykin, V., Hansmann, S. and Martin, W. (1997) Evolutionary analysis of 58 proteins encoded in six completely sequenced chloroplast genomes: revised molecular estimates of two seed plant divergence times. *Plant Systematics and Evolution*, **206**, 337–351.

Graur, D., Duret, L. and Gouy, M. (1996) Phylogenetic position of the order Lagomorpha (rabbits, hares and allies). *Nature*, **379**, 333–335.

Gribskov, M. and Burgess, R. R. (1986) Sigma factors from *E.coli*, *B. subtilis*, phage SP01 and phage T_4 are homologous proteins. *Nucleic Acids Research*, **14**, 6745–6763.

Grimm, B., Kruse, E. and Kloppstech, K. (1989) Transiently expressed early light-inducible thylakoid proteins share transmembrane domains with light-harvesting chlorophyll binding proteins. *Plant Molecular Biology*, **13**, 583–593.

Gu, X. and Li, W. -H. (1996) A general additive distance with time-reversibility and rate variation among nucleotide sites. *Proceedings of the National Academy of Sciences USA*, **93**, 4671–4676.

Hallick, R. B., Hong, L., Drager, R. G., Favreau, M., Monfort, A., Orsat, B. *et al.* (1993) Complete sequence of *Euglena gracilis* chloroplast DNA. *Nucleic Acids Research*, **21**, 3537–3544.

Helmchen, T.A., Bhattacharya, D. and Melkonian, M. (1995) Analyses of ribosomal RNA sequences from glaucocystophyte cyanelles provide new insights into the evolutionary relationships of plastids. *Journal of Molecular Evolution*, **41**, 203–210.

Henze, K., Badr, A., Cerff, R., Wettern, M. and Martin, W. (1995) A nuclear gene of eubacterial origin in *Euglena* reflects cryptic endosymbioses during protist evolution. *Proceedings of the National Academy of Sciences USA*, **92**, 9122–9126.

Hiratsuka, J., Shimada, H., Whittier, R., Ishibashi, T., Sakamoto, M., Mori, M. *et al.* (1989) The complete sequence of the rice (*Oryza sativa*) chloroplast genome: intermolecular recombination between distinct tRNA genes accounts for a major plastid DNA inversion during the evolution of the cereals. *Molecular and General Genetics*, **217**, 185–194.

Janke, A., Xu, X. and Arnason, U. (1997) The complete mitochondrial genome of the wallaroo (*Macropus robustus*) and the phylogenetic relationship among Monotremata, Marsupialia, and Eutheria. *Proceedings of the National Academy of Sciences USA*, **94**, 1276–1281.

Kaneko, T., Sato, S., Kotani, H., Tanaka, A., Asamizu, E., Nakamura, Y. *et al.* (1996) Sequence analysis of the genome of the unicellular cyanobacterium *Synechocystis* sp. strain PCC6803. II. Sequence determination of the entire genome and assignment of potential protein-coding regions. *DNA Research*, **3**, 109–136.

Kimura, M. (1983) *The neutral theory of molecular evolution*. Cambridge University Press, Cambridge.

Köhler, S., Delwiche, C.F., Denny, P.W., Tilney, L.G., Webster, P., Wilson, R.J.M. *et al.* (1997) A plastid of probable green algal origin in apicomplexan parasites. *Science*, **275**, 1485–1489.

Kowallik, K.V. (1997) Origin and evolution of chloroplasts: current status and future perspectives in *Eukaryotism and symbiosis intertaxonic combination versus symbiotic adaptation*, (eds H. E. A. Schenk, R. G. Herrmann, K. W. Jeon, N. E. Müller, W. Schwemmler), Springer Verlag, Berlin, pp. 3–23.

Kowallik, K. V., Stoebe, B., Schaffran, I., Kroth-Pancic, P. and Freier, U. (1995) The chloroplast genome of a chlorophyll *a* + *c*-containing alga, *Odontella sinensis*. *Plant Molecular Biology Reporter*, **13**, 336–342.

Kück, U. (1989) The intron of a plastid gene from a green alga contains an open reading frame for a reverse transcriptase-like enzyme. *Molecular and General Genetics*, **218**, 257–265.

Lake, J. A. (1997) Phylogenetic inference: how much evolutionary history is knowable? *Molecular Biology and Evolution*, **14**, 213–219.

Lecointre, G., Phillipe, H., Le, H. L. V. and Le Guyader, H. (1994) How may nucleotides are required to resolve a phylogenetic problem? The use of a new statistical method applicable to available sequences. *Molecular Phylogenetics and Evolution*, **3**, 292–309.

Levasseur, P. J., Meng, Q. and Bouck, B. (1994) Tubulin genes in the algal protist *Euglena gracilis*. *Journal of Eukaryotic Microbiology*, **41**, 468–477.

Li, L. -A. and Tabita, F. R. (1997) Maximum activity of recombinant ribulose 1,5-bisphosphate carboxylase/oxygenase of *Anabaena* sp. Strain CA requires the product of the *rbc*X gene. *Journal of Bacteriology*, **197**, 3793–3796.

Lockhart, P. J., Steel, M. A., Hendy, M. D. and Penny, D. (1994) Recovering evolutionary trees under a more realistic model of sequence evolution. *Molecular Biology and Evolution*, **11**, 605–612.

Löffelhardt, W. and Bohnert, H. J. (1994) Molecular biology of cyanelles, in *The molecular biology of Cyanobacteria*, (ed D. A. Bryant), Kluwer, Doordrecht, pp. 56–89.

Löffelhardt, W., Bohnert, H. J. and Bryant, D. (1997) The cyanelles of *Cyanophora paradoxa*. *Critical Reviews in Plant Science*, **16**, 393–413.

Lorkovic, Z. J., Schröder, W. P., Pakrasi, H. B., Irrgang, K. D., Herrmann, R. G. and Oelmüller, R. (1995) Molecular characterization of PsbW, a nuclear-encoded component of the photosystem II reaction center complex in spinach. *Proceedings of the National Academy of Sciences USA*, **92**, 8930–8934.

Maier, R.M., Neckermann, K., Igloi, G. L. and Kössel, H. (1995) Complete sequence of maize chloroplast genome: gene content, hotspots of divergence and fine tuning of genetic information by transcript editing. *Journal of Molecular Biology*, **251**, 614–628.

Martin, W. and Herrmann, R. G. (1998) Gene transfer from organelles to the nucleus: how much, what happens and why? *Plant Physiology*, **118**, 9–17.

Martin, W. and Müller, M. (1998) The hydrogen hypothesis for the first eukaryote. *Nature*, **392**, 37–41.

Martin, W. and Schnarrenberger, C. (1997) The evolution of the Calvin cycle from prokaryotic to eukaryotic chromosomes: a case study of functional redundancy in ancient pathways through endosymbiosis. *Current Genetics*, **32**, 1–18.

Martin, W., Somerville, C. C. and Loiseaux-de Goër, S. (1992) Molecular phylogenies of plastid origins and algal evolution. *Journal of Molecular Evolution*, **35**, 385–403.

Martin, W., Jouannic, S. and Loiseaux-de Goër, S. (1993) Molecular phylogeny of the *atp*B and *atp*E genes of the brown alga *Pylaiella littoralis*. *European Journal of Phycology*, **28**, 111–113.

Martin, W., Mustafa, A. -Z., Henze, K. and Schnarrenberger, C. (1996) Higher plant chloroplast and cytosolic fructose-1,6-bisphophosphatase isoenzymes: origins *via* duplication rather than prokaryote-eukaryote divergence. *Plant Molecular Biology*, **32**, 485–491.

Martin, W., Stoebe, B., Goremykin, V., Hansmann, S., Hasegawa, M. and Kowallik, K. V. (1998) Gene transfer to the nucleus and the evolution of chloroplasts. *Nature*, **393**, 162–165.

McFadden, G. I., Gilson, P. R., Douglas, S. E., Cavalier-Smith, T., Hofmann, C. J. and Maier, U. -G. (1997) Bonsai genomics: sequencing the smallest eukaryotic genomes. *Trends in Genetics*, **13**, 46–49.

McFadden, G. I., Reith, M. E., Munholland, J. and Lang-Unnasch, N. (1996) Plastid in human parasites. *Nature*, **381**, 482.

Melkonian, M. (1996) Systematics and evolution of the algae: endocytobiosis and the evolution of the major algal lineages. *Progress in Botany*, **57**, 281–311.

Morse, D., Salois, P., Markovic, P. and Hastings, J. W. (1995) A nuclear-encoded form II RuBisCO in dinoflagellates. *Science*, **268**, 1622–1624.

Navazio, L., Nardi, C., Baldan, B., Dainese, P., Fitchette, A. C., Martin, W. *et al.* (1998) Functional conservation of a calcium homeostat component of the endoplasmic reticulum: purification, characterization and cloning of calreticulin from *Euglena gracilis*. *Journal of Eukaryotic Microbiology*, **45**, 307–313.

Nei, M. (1996) Phylogenetic analysis in molecular evolutionary genetics. *Annual Review of Genetics*, **30**, 371–403.

Ohyama, K., Fukuzawa, H., Kohchi, T., Shirai, H., Sano, T., Sano, S. *et al.* (1986) Chloroplast gene organisation deduced from complete sequence of liverwort *Marchantia polymorpha* chloroplast DNA. *Nature*, **322**, 572–574.

Palenik, B. and Haselkorn, R. (1992) Multiple evolutionary origins of prochlorophytes, the chlorophyll *b*-containing prokaryotes. *Nature*, **355**, 265–267.

Palmer, J. D. (1995) Rubisco rules fall; gene transfer triumphs. *BioEssays*, **17**, 1005–1008.

Palmer, J. D. (1997) The mitochondrion that time forgot. *Nature*, **387**, 454–455.

Penny, D. and Hasegawa, M. (1997) The platypus put in its place. *Nature*, **387**, 549–550.

Qian, Y. and Tabita, F. R. (1996) A global signal transduction system regulates aerobic and anaerobic CO_2 fixation in *Rhodobacter sphaeriodes*. *Journal of Bacteriology*, **178**, 12–18.

Reardon, E. M. and Price, C. A. (1995) Plastid genomes of three non-green algae are sequenced. *Plant Molecular Biology Reporter*, **13**, 320–326.

Reith, M. and Munholland, J. (1995) Complete nucleotide sequence of the *Porphyra purpurea* chloroplast genome. *Plant Molecular Biology Reporter*, **13**, 333–335.

Rzhetsky, A. and Nei, M. (1994) Unbiased estimates of the number of nucleotide substitutions when substitution rate varies among different sites. *Journal of Molecular Evolution*, **38**, 295–299.

Saitou, N. and Nei, M. (1987) The neighbor-joining method: a new method for reconstructing phylogenetic trees. *Molecular Biology and Evolution,* **4,** 406–425.

Schwartz, R. M. and Dayhoff, M. (1978) Matrices for detecting distant relationships, in *Atlas of protein sequence and structure,* National Biomedical Research Foundation, Washington, D.C. pp. 353–358.

Shinozaki, K., Ohme, M., Tanaka, M., Wakasugi, T., Hayashida, N., Matsubayashi, T. *et al.* (1986) The complete nucleotide sequence of tobacco chloroplast genome: its gene organization and expression. *EMBO Journal,* **5,** 2043–2049.

Stewart, C.B. (1993) The powers and pitfalls of parsimony. *Nature,* **361,** 603–607.

Stirewalt, V. L., Michalowski, C. B., Löffelhardt, W., Bohnert, H. J. and Bryant, D. B. (1995) Nucleotide sequence of the cyanelle genome from *Cyanophora paradoxa. Plant Molecular Biology Reporter,* **13,** 327–332

Stoebe, B., Martin, W. and Kowallik, K. (1998) Distribution and nomenclature of protein-coding genes in 12 sequenced chloroplast genomes. *Plant Molecular Biology Reporter,* **16,** 243–255.

Suzuka, I., Koga-Ban, Y., Sasaki, T., Minobe, Y. and Hashimoto, J. (1994) Identification of cDNA clones for rice homologs of the human immunodeficiency virus-1 Tat binding protein and subunit 4 of human 26S protease (proteasome). *Plant Science,* **103,** 33–40.

Swofford, D. L. (1993) *PAUP 3.1: Phylogenetic Analysis Using Parsimony.* Smithsonian Institution Press, Washington DC.

Thompson, J. D., Higgins, D. G. and Gibson, T. J. (1994) CLUSTAL W: improving the sensitivity of progressive multiple sequence alignment through sequence weighting, position-specific gap penalties and weight matrix choice. *Nucleic Acids Research,* **22,** 4673–4680.

Unseld, M., Marienfeld, J. R., Brandt, P. and Brennicke, A. (1997) The mitochondrial genome of *Arabidopsis thaliana* contains 57 genes in 366924 nucleotides. *Nature Genetics,* **15,** 57–61.

Urbach, E., Robertson, D. L. and Chisolm, S. W. (1992) Multiple evolutionary origins of prochlorophytes within the cyanobacterial radiation. *Nature,* **355,** 267–270.

Van de Peer, Y., Rensing, S., Maier, U. -G. and De Wachter, R. (1996) Substitution rate calibration of small subunit RNA identifies chlorarachniophyte endosymbionts as remnants of green algae. *Proceedings of the National Academy of Sciences USA,* **93,** 7744–7748.

Wakasugi, T., Hirose, T., Horihata, M., Tsudzuki, T., Kössel, H. and Sugiura, M. (1996) Creation of a novel protein-coding region at the RNA level in black pine chloroplasts: the pattern of RNA editing in the gymnosperm chloroplast is different from that in angiosperms. *Proceedings of the National Academy of Sciences USA,* **93,** 8766–8770.

Wakasugi, T., Nagai, T., Kapoor, M., Sugita, M., Ito, M., Ito, S. *et al.* (1997) Complete nucleotide sequence of the chloroplast genome from the green alga *Chlorella vulgaris:* the existence of genes possibly involved in chloroplast division. *Proceedings of the National Academy of Sciences USA,* **94,** 5967–5972.

Wakasugi, T., Tsudzuki, J., Ito, S., Nakashima, K., Tsudzuki, T. and Sugiura, M. (1994) Loss of all *ndh* genes as determined by sequencing the entire chloroplast genome of black pine *Pinus thunbergii. Proceedings of the National Academy of Sciences USA,* **91,** 9794–9798.

Walne, P. L. and Kivic, P. A. (1989) Phylum Euglenida, in *Handbook of Protoctista Vol. 1,* (eds L. Margulis, J. O. Corliss, M. Melkonian and D. J. Chapman), Jones and Bartlett, Boston, pp. 270–287.

Williamson, D. H., Gardner, M. J., Preise, P., Moore, D. J., Rangachari, K. and Wilson, R.J. (1994) The evolutionary origin of the 35 kb circular DNA of *Plasmodium falciparum:* new evidence supports a possible rhodophyte ancestry. *Molecular and General Genetics,* **243,** 249–252.

Wolfe, K. H., Morden, C. W. and Palmer, J. D. (1992) Function and evolution of a minimal plastid genome from a nonphotosynthetic parasitic plant. *Proceedings of the National Academy of Sciences USA,* **89,** 10648–10652.

Yang, Z., Goldman, N. and Friday, A. (1994) Comparison of models for nucleotide substitution used in maximum-likelihood estimation. *Molecular Biology and Evolution,* **11,** 316–324.

Yoshinaga, K., Iinuma, H., Masuzawa, T. and Uedal, K. (1996) Extensive RNA editing of U to C in addition to C to U substitution in the *rbcL* transcripts of hornwort chloroplasts and the origin of RNA editing in green plants. *Nucleic Acids Research,* **24,** 1008–1014.

Chapter 16

Phylogenetics and diversification in *Pelargonium*

F. T. Bakker, A. Culham and M. Gibby

ABSTRACT

Correlations between phylogeny, biogeography and growth form were studied in the small-chromosome species group of *Pelargonium* using a phylogenetic hypothesis based on sequence data from the chloroplast DNA encoded *trnL* (UAA) 5′ exon – *trnF* (GAA) exon region. *Pelargonium* is a typical element of the South African fynbos and succulent karoo biomes, all in mediterranean climate regions. Within the small-chromosome species group of *Pelargonium* two main clades A and B were identified. Contrasting evolutionary patterns between clades A and B can be summarised as follows: clade A, representing roughly 60% of the genus, contains nine growth forms and is largely confined to the South African Cape winter rainfall region. Levels of karyological differentiation are low, with only polyploids having 'escaped' to regions with summer rainfall. Clade B, representing roughly 15% of the genus, has five different basic chromosome numbers, contains only herbaceous annuals and (sub)shrubs with or without succulent stems, and is widely distributed with oceanic and African/Australian disjunctions. Evolution in clade A has probably been driven by a combination of adaptive response to strong selective forces (Pliocene aridification in the winter rainfall region and varied soil types) and reproductive barriers maintained by pollinator dependence and poor dispersal potential. In clade B dispersal capacity is high and dependence on pollinators is generally low, indicating that ecological and geographical isolation mechanisms may have been more important in its evolution.

Evolution of growth form in clade A yielded predominantly geophytes, stem succulents and woody (sub)shrubs. Succulent growth forms make up the most speciose clade which evolved possibly in adaptive response to aridification in the south-western Cape winter rainfall region. Clade B contains herbaceous annuals and stem succulents that have evolved via at least two independent lineages.

16.1 Introduction

Pelargonium (Geraniaceae) is an example of a highly speciose and morphologically diverse genus that radiated in the South African Cape, especially in the fynbos and succulent karoo biomes within the winter rainfall region. The Mediterranean climate prevailing there is considered to be of recent origin, possibly late Pliocene (Deacon

In *Molecular systematics and plant evolution* (1999) (eds P. M. Hollingsworth, R. M. Bateman and R. J. Gornall), Taylor & Francis, London, pp. 353–374.

et al., 1992). *Pelargonium* is divided into two groups based on chromosome size which is supported by data from *rbc*L sequences (Price, unpublished). Within the small-chromosome species group of *Pelargonium*, which comprises c. 80% of all species, nine different growth forms can be distinguished. Some of the more xerophytic forms are confined largely to the Cape winter rainfall region whereas other forms are more widely distributed.

The purpose of this chapter is to explore correlations between phylogeny, biogeography and growth form using the small-chromosome species group of *Pelargonium* as an example. Our phylogenetic hypothesis is based on sequence data from the chloroplast DNA encoded *trnL* (UAA) 5′ exon – *trnF* (GAA) exon region (referred to hereafter as '*trnL–F* region'). Between clades within this species group, contrasting patterns of evolution were found that have generated hypotheses relating to speciation and biogeographic processes.

16.2 Mediterranean-type vegetation and the Cape Floristic Region

Mediterranean-type vegetation is known from the Mediterranean Basin (maquis), California (chaparral), central Chile (matorral), southern and south-western Australia (kwongan) and from the South African Cape region (fynbos and succulent karoo). These so-called mediterranean-type climatic regions with wet winters and dry, hot summers have produced ecophysiologically similar vegetation types that exhibit high levels of both endemism and ecological convergence (Cowling *et al.*, 1994). Regions of mediterranean-type vegetation are further characterised by high plant-species diversity, mostly low soil nutrient levels and frequent occurrence of fire.

The Cape Floristic Region (CFR) (Goldblatt, 1978) with fynbos as its most dominant biome, is recognised as one of the six botanical Kingdoms of the world (Good, 1974; Takhtajan, 1986; WWF and IUCN, 1994). It has the highest plant diversity of them all, with extraordinarily high levels of endemism including 68% of its 8600 species, 19.5% of its 955 genera and six of its ten largest families (Bond and Goldblatt, 1984). High species-level diversity, rather than generic or family level diversity, was reported by Linder *et al.*, (1992) and interpreted as being caused substantially by a massive burst of speciation associated with range extensions of surviving vegetation after the drastic climatic changes at the Tertiary–Pleistocene boundary. Cowling and Holmes (1992) gave examples of evolutionary radiations in this region in the Mesembryanthema clade of the *Aizoaceae* (sensu Hartmann, 1991) comprising 660 spp., *Erica* (526 spp.), *Proteaceae* (320 spp.) and *Restionaceae* (310 spp.). The mechanisms that lead to these 'explosive' speciation events in such a limited geographic area are not fully understood and pose interesting questions regarding the ecology, evolution and systematics of the groups involved (e.g. Cowling, 1992). What is needed here are sound phylogenetic frameworks describing species relationships within proliferated monophyletic groups in order to provide a basis for comparative evolutionary studies. Phylogenetic studies would also help in providing clues about the relationships between centres of endemism within the CFR and the history of the Cape flora (Linder *et al.*, 1992). Published phylogenetic studies using cladistic methods on fynbos plants have so far been limited to *Restionaceae*, which represents an old, probably Gondwanan lineage (Linder, 1987), *Compositae* (Anderberg and Bremer, 1991), *Orchidaceae* (Kurzweil *et al.*, 1991) and *Iridaceae* (Goldblatt and Manning, 1996; Goldblatt and Le Thomas, 1997).

An important factor in understanding patterns of diversity in the Cape Floristic Region is knowledge of the Cape climate and its history. At present, the amount of rainfall varies considerably from north to south and from west to east across southern Africa. This system is thought to be driven by differences in sea surface temperatures between the South Atlantic and Indian Oceans (Stokes *et al.*, 1997), resulting in a winter and a summer rainfall region in the Cape. Climatic variation, combined with the complex geology and range of soil types present, provides many ecological gradients (Goldblatt, 1997). The dominant fynbos and succulent karoo biomes within the Cape winter rainfall vegetation are habitats with relatively high selection pressures, i.e. nutrient-poor soils, recent aridification and the occurrence of fire cycles. During the late Miocene, the Cape region is hypothesised to have been covered with (sub)tropical forest (Coetzee, 1983) that retreated after the Pliocene increase in aridity, caused probably by changing patterns in ocean currents (Siesser, 1980; Deacon *et al.*, 1992). The establishment of a mediterranean climate in the Cape is believed to have been associated with the fixation of a South Atlantic high pressure cell that subsequently blocked summer precipitation in the fynbos region, probably in late Miocene (Deacon *et al.*, 1992). The hyper-diverse fynbos vegetation may then have become widespread throughout the Cape region. This is supported by pollen data showing strong development of fynbos elements by the Pliocene (Coetzee, 1983) and indirectly by the first direct evidence of fire coinciding with this period (Hendey, 1983). However, these data do not provide evidence to show whether the evolution of the component species happened before, or concurrently with the inferred spread of fynbos vegetation. Later, during the Pleistocene, glaciations are believed to have had a different impact on western and eastern Cape regions, resulting, in the eastern Cape, in the establishment of a summer rainfall climate and replacement of fynbos by grassland, whereas in the south-western Cape fynbos could have persisted uninterrupted in a mediterranean-type climate much like today (Cowling and Holmes, 1992). This is supported by pollen profiles from the Cedarberg Mountains that indicate no vegetational change over the last 14,500 years (Meadows and Sugden, 1991).

16.3 *Pelargonium*

Pelargonium is one of the groups thought to have proliferated 'explosively' in the South African Cape region and has been listed as the seventh largest genus in the region (Bond and Goldblatt, 1984). Given the palaeoclimatic changes outlined above, it is possible either that *Pelargonium* proliferated within the winter rainfall region, perhaps in response to the Pliocene aridification, or that it reflects more ancient radiation events across southern Africa, prior to the development of fynbos, and subsequent adaptation to this change in situ.

Pelargonium contains about 270 species in a range of growth forms, centred in the fynbos and succulent karoo biomes of the South African Cape Province and also occurring in the grassland biome within the Eastern Cape to the Northern

Transvaal region. The genus extends its distribution into tropical eastern and north-eastern Africa (15 spp.), the Arabian peninsula (two spp.) and Asia Minor (two spp.). A few species occur in Australia (seven spp.), Madagascar (two spp.) and on the South Atlantic islands of St Helena and Tristan da Cunha (one on each). Within southern Africa the highest diversity of *Pelargonium* species coincides largely with the winter rainfall area (Fig. 16.1) where in one instance as many as 70 species per square degree have been recorded (van der Walt and Vorster, 1983). Local endemism of *Pelargonium* species is high with individual distribution areas sometimes as small as a single hill (van der Walt and Vorster, 1988). Monophyly of the genus is supported by the presence in all species of a nectar spur adnate to the pedicel, unique within the Geraniaceae, and by *rbcL* sequence data (Price and Palmer, 1993; Price, unpublished).

In the genus *Pelargonium*, chromosome size has been shown to be highly significant taxonomically. Eighty per cent of the species, which belong to 11 of the 17 currently recognised sections, have small chromosomes, mostly less than 1.5µm long, whereas the remainder have larger chromosomes, 1.5–3.0 µm long (Albers

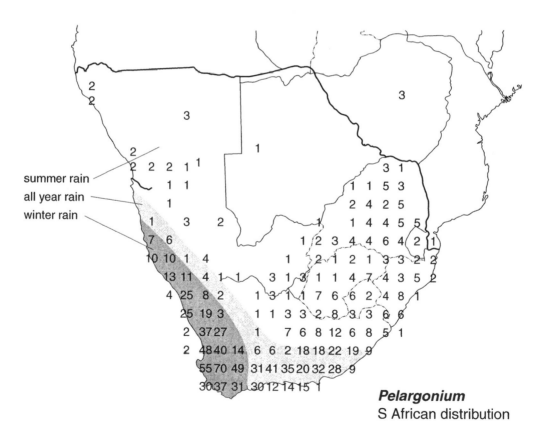

Pelargonium
S African distribution

Figure 16.1 Distribution and concentration of *Pelargonium* in South Africa (redrawn from van der Walt and Vorster (1983)). Numbers indicate numbers of species of *Pelargonium* per square degree.

and van der Walt, 1984; Gibby and Westfold, 1986; Gibby, 1990). Most individual sections can be defined by a combination of chromosome size and basic chromosome number. Changes in the basic chromosome number appear to have taken place more than once within the genus (Gibby *et al.*, 1996), but the change in chromosome size appears to have occurred only once, a hypothesis supported by molecular studies (Price, unpubl. *rbcL* sequence data).

16.3.1 Growth forms in Pelargonium

In addition to its biogeography, *Pelargonium* also provides a good opportunity to study the evolution of growth forms since the genus contains a wide range, from woody shrubs up to 3m tall to short-lived, herbaceous annuals that spend most of the year as seed. There is also a range of xerophytic forms including stem succulents and geophytes that re-sprout from large subterranean tubers. These xerophytic forms, with some 130 species grouped in six sections, represent about half the genus.

Growth form has largely been the basis for taxonomic classification of *Pelargonium*, the last major revision being by Knuth (1912) who recognised 15 sections. Three further sections have been added during the past two decades and others re-defined, based on evidence from cytology, phytochemistry, karyology and morphology (e.g. Gibby and Westfold, 1986; van der Walt, 1990; Albers *et al.*, 1992; Dreyer *et al.*, 1992; van der Walt *et al.*, 1995; van der Walt *et al.*, 1997). The species with large chromosomes include woody, herbaceous or subsucculent-stemmed shrubs (sectt. *Jenkinsonia, Ciconium, Subsucculentia*), subshrubs (sect. *Myrrhidium*) and two annual herbs (*P. redactum, P. senecioides*). A few species have tubers.

Among the *Pelargonium* species with small chromosomes, growth form is more diverse and nine different forms can be distinguished that are based on extensive field and culture observations and a range of published morphological studies (see references below). These forms relate closely to current sectional classification. They are represented schematically in Fig. 16.2, and are described below. Owing to a paucity of studies of stem anatomy in *Pelargonium* (Jones and Price, 1996), no anatomical characters could be included to define these forms. Names of the growth forms listed refer to the currently recognised sections of which they are typical:

a) Pelargonium form, 24 species (Van der Walt, 1985): woody evergreen shrubs, usually profusely branched and reaching heights of up to two metres. Seven species of section *Glaucophyllum* are included here; these are similar to species of sect. *Pelargonium*, but only the basal part of the stem becomes woody.

b) Ligularia form, 9 species (Albers *et al.*, 1992): deciduous subshrubs with fleshy stems and stem succulents, covered with persistent stipules and extended, woody subterranean parts.

c) Campylia form, 9 species (van der Walt and Van Zyl, 1988; van der Walt *et al.*, 1990; van der Walt *et al.*, 1991): small deciduous rosette subshrubs usually with extensive, unswollen underground parts.

d) Cortusina form, 7 species (Dreyer *et al.*, 1992): stem-succulent shrubs with entire leaves; *P. cotyledonis* from section *Isopetalum*, *P. drummondii* from sect. *Peristera* and *P. album* from sect. *Reniformia* are also included in this form.

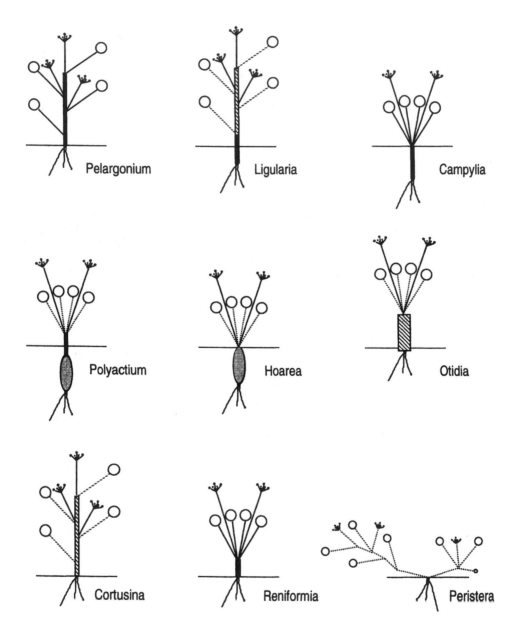

Figure 16.2 Life forms among sections of *Pelargonium* with small chromosomes. Solid lines indicate perennial branches and petioles; dotted lines indicate deciduous branches and petioles. Stems, indicated by rectangles, are: woody (black), herbaceous (grey), or succulent (hatched). Tubers are indicated by shaded ovals. Horizontal lines indicate soil surface.

e) Otidia form, 11 species (Vorster, 1990): stem succulents with highly swollen stems and small or succulent leaves.

f) Hoarea form, c. 80 species (Marais, 1994): geophytes with a series of small subterranean tubers, thickly covered in paper-like sheaths, no stems present, leaves and inflorescences frequently do not occur simultaneously.

g) Polyactium form, 13 species (Maggs *et al.*, 1995): geophytes with massive tubers lacking enveloping sheaths, stems present, leaves and inflorescences are carried simultaneously.

h) Reniformia form, 7 species (Dreyer *et al.*, 1992): tuft-like subshrubs with herbaceous stems, leaves entire.

i) Peristera form, c. 20 species (Hellbrügge, 1998; Bakker *et al.*, 1998): dwarf herbaceous annuals with small flowers and fruits.

The schematic representation of the range in growth forms of *Pelargonium* given above inevitably suffers from oversimplification but we feel that, nevertheless, the main character combinations have been captured and that the forms depicted represent discrete entities.

16.4 cpDNA *trnL-F* sequences in plant phylogeny reconstruction

Chloroplast DNA markers have been used extensively over the past decade for phylogeny reconstruction. They share with markers encoded by other organellar genomes such as mitochondrial DNA the advantage of uniparental inheritance, which ensures that the confounding effect of recombination on phylogenetic signal in ancestor-descendant relationships is alleviated (Clegg and Zurawski, 1992; Doyle, 1992). Furthermore, the high copy number of chloroplast genomes generally present in plant cells makes PCR amplification of chloroplast DNA regions straightforward.

The cpDNA *trnL* (UAA) 5′ exon – *trnF* (GAA) region has been used extensively as a source of phylogenetic markers (Kita *et al.*, 1995; Gielly and Taberlet, 1996; Böhle *et al.*, 1996; Bakker *et al.*, 1998, 1999). The region includes the 5′ *trnL* exon, the *trnL* intron, the 3′ *trnL* exon, the intergenic spacer and the *trnF* exon regions. Taberlet *et al.*, (1991) described a set of primer sites based on the highly conserved tRNA encoding exons with which *trnL–F* regions can be amplified from a wide range of green plants. The intergenic spacer region usually contains high levels of length variation, caused either by deletion or addition of sequence motifs by slipped strand mispairing (Levinson and Gutman, 1987), a mechanism involving *intra* helical rather than *inter* helical mispairing followed by replication. Indels, after recoding as either binary or multistate characters (Gielly and Taberlet, 1996; Oxelman *et al.*, 1997), have been included in cladistic analyses and found to be a potentially useful source of phylogenetic markers (Johnson and Soltis, 1995).

16.4.1 Data and analysis

Our data set, described in Bakker *et al.*, (1999), comprises DNA sequences of the *trnL–F* regions from 73 species of *Pelargonium*, representing the morphological range within the 11 sections with small chromosomes; five species with large chromosomes, representing two sections, were selected as the outgroup. The *trnL–F* alignment (921 positions) contained 155 (18%) variable and 140 (15%) phyloge-netically informative positions, as well as 22 phylogenetically informative indels that varied in length from two to 56 nucleotides and were flanked by at least six unambiguously aligned positions on either side. For two of the species, *P. iocastum* and *P. buysii*, only 130 consecutive homologous positions (containing eight of the 30 phylogenetically informative substitutions) could be assigned in the *trnL–F* align-ment, the rest of these sequences being absent (Bakker, unpubl.). An heuristic parsimony search of the *trnL–F* alignment (1000 cycles of random stepwise addi-tion without branch swapping) with informative indels included, yielded 66 equally most parsimonious trees (MPTs) of 332 steps with CI = 0.774 and RC = 0.712, the strict consensus of which is shown in Fig. 16.3.

Only three of the 22 informative indels included in the analysis showed parallel changes in different main clades (see Fig. 16.3), the rest being synapomorphic for single main clades. The RC of the 22 informative indels on the 66 MPTs was 0.699 and exclusion of these characters from the analysis (search conditions as above) increased the number of MPTs found to 289 which had CI = 0.781 and RC = 0.719. The strict consensus tree topology of these 289 trees was incongruent with the 'indels-included' strict consensus topology with respect to the outgroup, and was less well resolved with respect to the ingroup.

Clade support was estimated by 10000 replications of parsimony jackknifing (Farris *et al.*, 1996). The jackknife tree with informative indels included is almost identical to the strict consensus tree which implies that overall tree support is high and ambiguity in the data set is low. Excluding informative indels from the jack-knife analysis (not shown) did not change the overall jackknife tree topology but decreased resolution in some terminal clades as well as in the outgroup.

16.4.2 Patterns of substitution and insertion/deletion in trnL-F

The ratio of transitions (changes between pyrimidines or between purines) to trans-versions (changes between pyrimidines and purines), was estimated using the 'treescores/likelihood' option in PAUP*4 (Swofford, pers. comm.) on the 66 MPTs for the ingroup only. Substitution patterns were estimated using the general time-reversible model (Yang, 1994) which produces a symmetrical rate matrix in which both multiple substitution correction, among-site rate heterogeneity and base frequencies are taken into account. The following likelihood settings were used: rate heterogeneity assumed to follow a gamma distribution with estimated shape parameter; empirical nucleotide frequencies; proportion invariable sites estimated; discrete gamma approximation; number of rate categories = 4, average rate for each category represented by median. Following this approach the *Pelargonium trnL–F* data set was measured to have a transition/transversion ratio of 0.97–0.98. This value is rather low and is usually considered to be indicative of saturation in the transitional fraction of substitutions and consequently of reduced phylogenetic signal (Holmquist, 1983; DeSalle *et al.*, 1987; Hillis *et al.*, 1993; Wakely, 1996). However, when transitions and transversions are plotted against the proportion of sites differing (p-difference), no apparent saturation is observed and transitions are still accumulating linearly at sequence divergences > 10% (Fig. 16.4). Notably, all

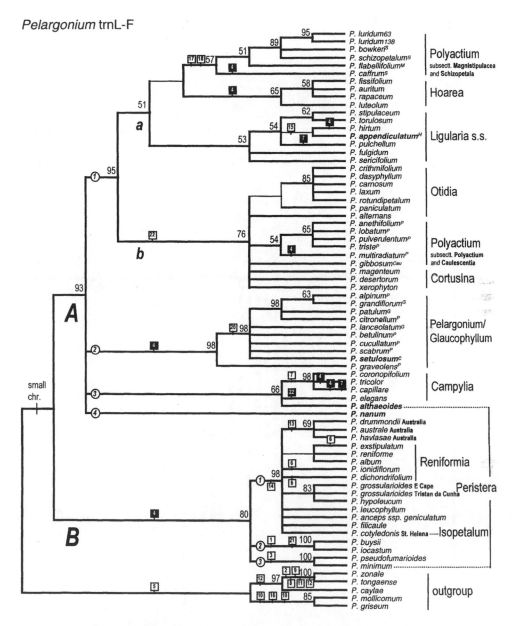

Pelargonium trnL-F

Figure 16.3 Pelargonium trnL-F: Strict consensus tree of 66 equally most parsimonious trees calculated by heuristic search with informative indels included (see text). Thin lines indicate internodes that collapse in the parsimony jackknife tree. Jackknife values (10000 replicates) are above the nodes. Bold script letters and numbers in circles are for clade reference. Numbers in squares refer to the occurrence of informative indels on the strict consensus with filled squares indicating parallel occurrences or reversals. 'Small chr.' refers to the presence of small chromosomes as a synapomorphy for the ingroup. Taxon names in boldface indicate deviation from the currently recognised sectional assignments which are indicated with vertical lines on the right: C = sect. *Campylia*, Cau = sect. *Polyactium* subsect. *Caulescentia*, H = sect. *Hoarea*, M = sect. *Polyactium* subsect. *Magnistipulacea*, P = sect. *Polyactium* subsect. *Polyactium*, S = sect. *Polyactium* subsect. *Schizopetala*. Geographic origin of non-African taxa is indicated in bold.

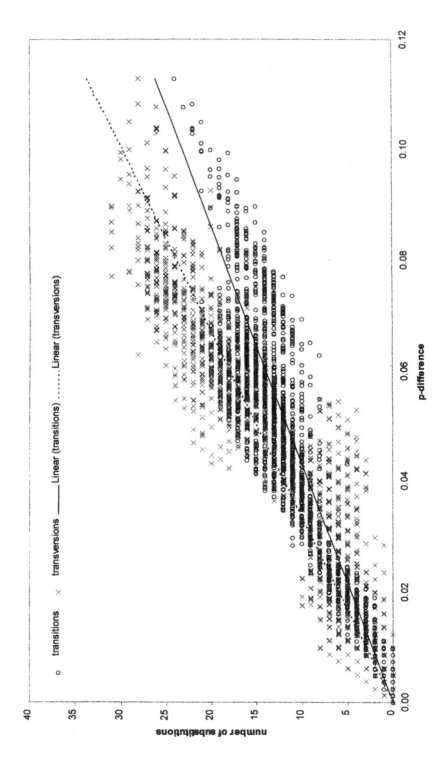

Figure 16.4 Pelargonium trnL–F: Numbers of transitions (circles) and numbers of transversions (crosses) plotted against p-difference (proportions of positions differing). The linear regressions were constrained to have an x and y intercept of zero. Transitions: $R^2 = 0.8568$, $y = 232.1x$; transversions: $R^2 = 0.8746$, $y = 298.3x$.

transition/transversion ratios > 4 are from comparisons between four species within the *Campylia* clade A3 which indicates that substitution patterns can differ substantially between closely related clades.

Substitution patterns in the separate intron and spacer alignments were then evaluated on most parsimonious trees calculated for the total alignments. The transition/transversion ratio in the intron alignment was estimated to be 1.15 whereas it was 0.85 in the spacer alignment (the rate heterogeneity shape parameter α was estimated to be 1.50 for the intron and 0.87 for the spacer alignment). Relative substitution rates in the intron and spacer regions were similar (Fig. 16.5) with A↔T and G↔C transversions occurring significantly less than expected. The rarity of A↔T transversions is somewhat unexpected and needs further explanation, given the relatively high A + T content (64%) for both regions.

A possible factor in explaining the different transition/transversion ratios in the intron and spacer is a different requirement for maintenance of secondary structure encoded in these regions. This has been suggested to have caused a T↔C transition bias in other RNA encoding genes, enabling substitution of complementary base pairs through a (less stable) G-T intermediate (Rousset *et al.*, 1991; Berbee and Taylor, 1992; Dixon and Hillis, 1993). The *trnL–F* region would provide a suitable test case since it contains within the intron region a highly conserved secondary structure whereas no such structure has been reported from the intergenic region. However, our data set contained no substitutions in regions encoding double stranded secondary structural elements, therefore the difference in transition/transversion ratio between the intron and spacer regions probably reflects other factors relating to underlying substitution dynamics.

As compared with sequences from for example animal mtDNA (Brown *et al.*, 1982) or 12S rRNA genes (Sullivan *et al.*, 1996), the *Pelargonium trnL–F* data exhibit rather different patterns of substitution. It shows that a low transition/transversion ratio and a strong phylogenetic signal as well as the presence of many well-supported groups in the strict consensus tree need not be mutually exclusive. Further studies are needed, however, especially of whether the transversional bias occurs at the level of mutation or of fixation.

There appears to be little homoplasy in the pattern in which indels evolved across the *Pelargonium trnL–F* phylogeny. As stated above, the ensemble RC of the 22 informative indels on the 66 MPTs was 0.699 which is comparable with RC values for nucleotide substitutions in this region. Some indels, however, are highly homoplasious and have occurred in parallel several times. In total, the 22 informative indels were found to increase total tree support by 18% (i.e. from 52.0 to 61.5 effective uncontradicted characters) as measured by parsimony jackknife analysis (Bakker *et al.*, 1999).

16.5 Phylogeny, karyology, growth form and biogeography in *Pelargonium*

We consider the best representation of our data to be the one that is based on maximal evidence, in this case the strict consensus tree depicted in Fig. 16.3 which includes all indel characters. The *trnL–F* phylogeny for species of *Pelargonium* with small chromosomes has two main clades (A and B). Clade A, comprises four

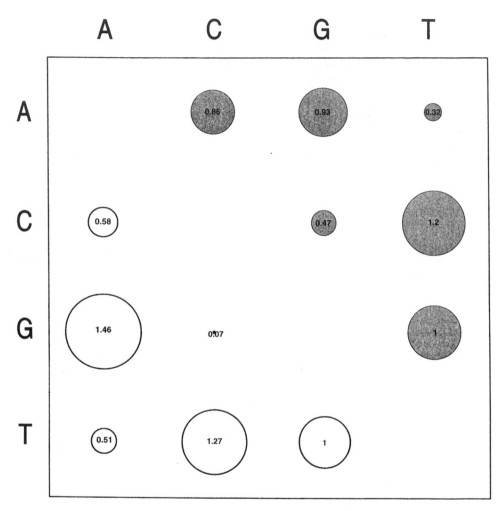

Figure 16.5 Pelargonium trnL–F: relative rates of substitution classes estimated on 66 most parsimonious trees using the general time-reversible model with rate heterogeneity (rates of changes between T and G are set to I). The diameter of the circles is proportional to the rate; values for the *trnL–F* intron are in open circles (lower left), those for the spacer region are in filled circles (upper right).

subclades of which clade A1 represents c. 50% of all *Pelargonium* species and five of its sections and clade A4 has only *P. nanum*, a dwarf herbaceous annual species from the Cape winter rainfall region. Clearly, most of the range of morphological variation in *Pelargonium* is accommodated by species represented by clade A.

Clade B, in contrast, comprises predominantly 'weedy', dwarf, herbaceous annual species, previously assigned to section *Peristera*. It also includes section *Reniformia* as well as the Australian pelargoniums and both species known from the South Atlantic islands of St Helena and Tristan da Cunha. In a previous analysis (Bakker *et al.*, 1998), a nrDNA ITS data set was produced complementing the *trnL–F* data set and both data sets were analysed simultaneously. The resulting semi-strict consensus topology is congruent with the clade B topology in the *trnL–F* phylogeny (Fig. 16.3), but is more resolved. It groups the section *Reniformia* species and the St Helena endemic *P. cotyledonis* in one clade, but support for this arrangement is low, indicating lack of resolution at this level.

Based on our 'best tree', the following evolutionary patterns of growth form, karyology and distribution can be inferred (Figs. 16.3 and 16.6).

Clade A1 (x = 11), representing half the species of *Pelargonium*, can be considered a true xerophytic clade. The two major groups of geophytes, sections *Polyactium* and *Hoarea*, are within clade A1. Section *Polyactium* is polyphyletic, with two subsections, *Magnistipulacea* and *Schizopetala*, forming a monophyletic group within clade A1a, subsection *Polyactium* forming a monophyletic group within clade A1b and the position of *Caulescentia* being unresolved. Section *Hoarea* includes more than 70 species (Marais, 1994) and within this section there is more variation in overall length of karyotype (range 1.3–2.0 μm, Gibby *et al.*, 1996) than in other sections. Species of section *Ligularia* s.s. (Albers *et al.*, 1992) in clade A1a are mostly stem succulents (e.g. *P. fulgidum*) or subshrubs with fleshy stems (e.g. *P. sericifolium*). With the exception of subsection *Polyactium*, all the species in clade A1b are stem succulents. Most species in clade A1a have x = 11, but within section *Hoarea*, whilst x = 11 is common, x = 10 and x = 9 occur at lower frequencies. It has been suggested that reduction in basic chromosome number in this section has occurred through Robertsonian translocation, $11 \rightarrow 10 \rightarrow 9$ (Gibby *et al.*, 1996).

Within clade A1a, the majority of species of section *Hoarea* and all of section *Ligularia* s.s. are confined to the winter rainfall region where they occur predominantly in succulent karoo; the major exception in clade A1a is section *Polyactium*, subsection *Magnistipulacea* and *Schizopetala*, where all the species are from the grassland biome in the summer rainfall areas. *P. luridum* (subsection *Magnistipulacea*) is distributed from South Africa north to Tanzania. Half of these species are polyploids (Gibby, unpublished). Clade A1b is confined largely to the winter rainfall region, with the exception of one species of section *Otidia*, a diploid, that occurs in the summer rainfall region of the eastern Cape. *Pelargonium pulverulentum* is the only species of subsection *Polyactium* that occurs in the summer rainfall region, and has diploid, tetraploid and hexaploid forms (Maggs *et al.*, in press). Section *Cortusina* is entirely from the winter rainfall region of the Cape and Namibia where it occurs in the succulent karoo biome.

Clade A2 (x = 11) comprises the evergreen (sub)shrubs of sections *Pelargonium* and *Glaucophyllum*, and includes 32 species. The suggestion by Albers and van

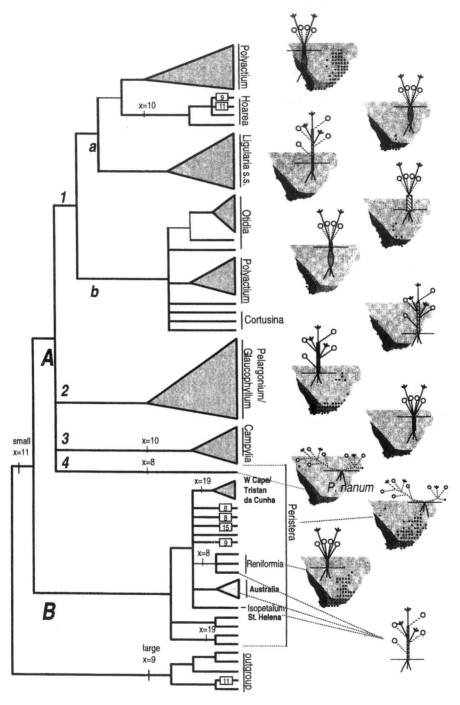

Figure 16.6 The *Pelargonium trnL-F* phylogeny showing sectional assignment, basic chromosome numbers (either on the branches or boxed), life forms and South African distributions of main clades. Clade numbering follows Figure 16.3. Life forms follow Figure 16.2. The shaded areas on the distribution maps refer to the winter rainfall region (dark), all year rainfall region (medium) and summer rainfall region (light). Non-African distributions (in clade B) are indicated with bold lettering.

der Walt (1984) that this growth form is the most primitive of the genus can neither be confirmed nor rejected by our data. Diploid taxa of clade A2 are confined exclusively to the south-western Cape region, mainly in fynbos. The derived polyploid species are distributed further east, in areas that receive rain in both winter and summer, or in summer rainfall regions (Albers and van der Walt, 1984).

Clade A3 (x = 10) comprises some species of section *Campylia*, a group of herbaceous subshrubs and *P. althaeoides*, a herbaceous annual species traditionally grouped in section *Peristera* of clade B. One species of section *Campylia*, *P. setulosum*, groups with section *Pelargonium* in clade A2. Species from clade A3 are all distributed in the winter rainfall area in the south-western Cape area, many of them associated with the Cape Agulhas sand-flats and all of them occurring in fynbos.

Clade A4 (x = 8) comprises a single species, the dwarf herbaceous *P. nanum*, that is unusual in its bi-coloured flowers and it has a distinctive leaf form, size and pubescence (van der Walt and Vorster, 1988). *P. nanum* is probably limited to the winter rainfall area, occurring in both the succulent karoo and fynbos biomes (van der Walt and Vorster, 1988).

Clade B (x = 8, 9, 11, 15 and 19) is in clear contrast with Clade A in that, in addition to having a range of basic chromosome numbers, it consists predominantly of weedy herbaceous annual species. The few non-weedy species include herbaceous subshrubs from section *Reniformia*, and some taxa with the *Cortusina* growth form, *P. album* of sect. *Reniformia*, *P. drummondii*, an Australian species of section *Peristera*, and *P. cotyledonis*, the St Helenan endemic. In a simultaneous analysis of a combined rDNA ITS/*trnL–F* data set for clade B species (Bakker *et al.*, 1998) the section *Reniformia* species and *P. cotyledonis* group together which would suggest a single origin of stem succulence within clade B. However, as stated in section 16.5, support for this topology is low. The weedy species from clade B are distributed throughout the Cape region. The section *Reniformia* species occur predominantly in summer rainfall areas.

The majority of species in clade A have x = 11. Most are diploids, and all have small chromosomes. Species of clade A rarely hybridise in the wild; each has its own well-defined niche and this ecological separation may be reinforced by pollinators (Struck, 1997), time of flowering, etc. But artificial hybridisation is very successful and many hybrids show some fertility – hence the explosion of interest shown by the early *Pelargonium* breeders, as illustrated in Sweet (1820–30). Hybridisations are successful not only within sections, e.g. *P. hirtum* × *P. fulgidum* (section *Ligularia* s.s.), but also between sections, *P. fulgidum* × *P. cucullatum* (section *Pelargonium*) (and see Table 3 in Albers *et al.*, 1992). Hybrids can also form between species with differing basic chromosome number, x = 11 and x = 10 (Sweet, 1824–26: 3:237; Gibby, 1994). Despite the great range in growth forms and numbers of taxa within clade A, there has been relatively little cytological differentiation in this group that may indicate relatively recent speciation on a large scale. In comparison, *Pelargonium* species with large chromosomes, the sister group to clade A and clade B species, cannot be crossed so successfully and the hybrids have low fertility, even when involving intra-specific taxa where chromosome number is the same (Gibby, 1994).

Within clade B a contrasting pattern is found with higher levels of cytological differentiation and five different basic chromosome numbers. Although there is inter-fertility within section *Reniformia* (Albers *et al.*, 1992), none is known for other species in the clade. Clade B species with x = 11 from the South African Cape Province are restricted to the south-western Cape, whereas species with deviating chromosome numbers have wider distribution areas throughout summer and winter rainfall regions (Hellbrügge and Albers, pers. comm.). The Australian clade B species are all diploids with x = 11 and with large distribution areas.

The contrasting cytological patterns seen in clades A and B could have resulted from different evolutionary mechanisms. Radiation in clade A was probably mainly based on differences in breeding-systems and pollinator dependence (Struck, 1997) whereas for the annuals and dispersers in clade B, ecological or geographical isolation mechanisms may have been more important.

16.6 Evolutionary trends in the small-chromosome group of *Pelargonium*

There is a strong correlation between membership of clade A and distribution in the Cape winter rainfall region (Fig. 16.6) and it can therefore be considered a true 'winter rainfall clade'. The exception to this pattern is *Polyactium* subsections *Magnistipulacea* and *Schizopetala* (five species in total) which are distributed in the summer rainfall region. The clade as a whole may have proliferated in situ after the establishment of the Cape winter rainfall climate, 2–3 My ago. Alternatively, clade A could have evolved before that time, possibly became widespread and subsequently went extinct from the non-winter rainfall regions once the winter rainfall climate became established. However, this scenario would necessitate the following extra assumptions when compared with the in situ scenario, making it a far less parsimonious explanation of the observed pattern: (1) past widespread biogeographic distributions; (2) extinction of the majority of species in clades A1–4 from the summer rainfall region; (3) the selective extinction of geophytic growth forms of *Polyactium* subs. *Polyactium*, and not subsections *Magnistipulacea* and *Schizopetala*, from the summer rainfall region, and (4) shifts in pollinator-dependence (but see Manning and Goldblatt (1997) for an example of a recent shift in pollinator in species of *Iridaceae*). The scenario involving in situ proliferation of clade A requires one migration to the summer rainfall regions and no mass extinctions need be hypothesised.

Within clade A the most speciose subclades A1a and A1b radiated after the main clades A1–4 were formed. This implies that the explosive radiation of *Pelargonium* species in the South African Cape winter rainfall region can be interpreted as a secondary radiation, the primary radiation being among clades A1–4. In order to trace back the origin of *Pelargonium* our phylogenetic exploration of the genus is currently being extended, accommodating species from tropical East Africa and Asia Minor and using other Geraniacean representatives as outgroups. Within clades A1a and A1b a vast array of morphology and growth form was achieved (i.e. two types of geophytes and stem succulents) whereas clades A2–A4 contain one growth form each. A key innovation in the proliferation of clade A1 has probably been the generation of succulence which could have triggered speciation and range expansion into many vacant niches after the late Pliocene western Cape aridifications. Likewise, the development of tubers which occurred only in clade A1 enabled survival in the arid conditions prevailing in the succulent karoo, and, through a resprouting-strategy, of the frequent occurrence of fire in fynbos.

The *trnL–F* data used to reconstruct *Pelargonium* phylogeny do not support a resolved phylogenetic hypothesis, i.e. branching order among clades A1–4 remains undecided. Therefore we cannot make inferences about the ancestral growth form for clade A. However, within the genus the succulent forms (clade A1a and A1b) are most derived in terms of topological position, i.e. developed after the split from the fynbos clades A2–4.

A clear correlation exists within clade A between polyploidy and geography: with the exception of three species of clade A1a and one from A1b, all clade A members that occur outside the winter rainfall area are polyploid. Since chromosome size is similar among clade A species, it is reasonable to assume that polyploids are derived from their diploid relatives in the winter rainfall region. This biogeographic pattern is therefore consistent with the younger age of the summer rainfall climate in the Cape, a result of the different impact that Pleistocene glaciations had on the south-western and eastern Cape regions (Cowling and Holmes, 1992). A comparable northern hemisphere situation exists, for example, in European *Asplenium* (Pteridophyta) where polyploid taxa occur far outside the southern European glacial refuges in which predominantly diploids are found (Vogel, 1995; Vogel *et al.*, 1999).

The acquisition of 'weediness' was fixed only in clade B as, apparently, was dispersal ability. In marked contrast with clade A, species within clade B show a broad geographic distribution. Species of section *Reniformia* occur in both winter and summer rainfall regions in South Africa, species of section *Peristera* are represented throughout South Africa, in Australia and Tristan da Cunha, and section *Isopetalum* (*P. cotyledonis*) is endemic to St Helena. The Australian pelargoniums, which generally exhibit a high resemblance to their Cape relatives, comprise 7 species distributed in mainland Australia (5, one shared with Tasmania), Tasmania (2), and New Zealand (1) where they occur mainly in coastal dune systems and on granite rock outcrops (Carolin, 1961). Most Australian pelargoniums are weedy, short lived annuals or perennials with small flowers, some of them reproducing autogamously (Meve, 1995).

African/Australian disjunctions in some woody plant groups have been explained by Gondwanan vicariance (Johnson and Briggs, 1974), mainly based on a fossil record and (sub)tribe level divisions. The same disjunctions in herbaceous plants are poorly understood, mainly because of the lack of a fossil record. In *Pelargonium*, the species level disjunction between Africa and Australia suggests that recent long-range dispersal rather than old vicariance is probably a more likely explanation for its origin. Indeed, levels of rDNA ITS sequence divergence between the African and Australian pelargoniums were found to be < 4% and interpreted to be consistent with a recent dispersal scenario rather than old Gondwanan vicariance (Bakker *et al.*, 1998). Following the same line of reasoning the split between *Pelargonium* species of the South Atlantic islands of St Helena and Tristan da Cunha and their African continental relatives was found to be of even more recent origin.

P. grossularioides from Tristan da Cunha is a weedy annual species which also occurs in the South African Cape, whereas *P. cotyledonis* on St Helena is an endemic stem succulent with white, virtually actinomorphic flowers which are unique to the genus. The case of *P. drummondii*, a stem succulent occurring in Australia, is a less extreme example of rapid change in growth form. This indicates an underlying potential for rapid change in growth form within clade B, in which the plesiomorphic condition was probably a short-lived annual form.

16.7 General summary and conclusions

In *Pelargonium*, the use of *trnL–F* sequence data has proven valuable in elucidating patterns of morphological and karyological variation as well as generating biogeographical hypotheses. Within the small-chromosome species group of *Pelargonium* two main clades A and B were identified. Contrasting evolutionary patterns between clades A and B can be summarised as follows: clade A is highly speciose (one of its subclades represents in itself > 50% of the genus), has a low level of karyological differentiation, contains nine growth forms and is largely confined to the South African Cape winter rainfall region, with only polyploids having 'escaped' to regions with summer rainfall. In contrast, clade B, representing roughly 15% of the genus, has five different basic chromosome numbers, contains only herbaceous annuals (the Peristera form) and subshrubs with or without succulent stems (the Reniformia and Cortusina forms), and is widely distributed with oceanic and African/Australian disjunctions. In clade A evolution has probably been driven by a combination of adaptive response to strong selective forces (aridification in the winter rainfall region) and, on the other hand, reproductive barriers maintained by pollinator dependence and poor dispersal potential. In clade B dispersal capacity is high and dependence on pollinators is generally low. Therefore ecological and geographical isolation mechanisms may have been more important in its evolution.

Patterns of growth form evolution in the small-chromosome species of *Pelargonium* can be summarised as follows: clade A contains predominantly geophytes, stem succulents and woody (sub)shrubs with succulent forms having evolved only in clade A1, which is the most speciose clade of the genus. Succulence in clade A1 can be regarded as a key innovation, possibly in an adaptive response to Pliocene aridification in the south-western Cape winter rainfall region. Herbaceous annuals have only evolved in clade B, *P. nanum* in clade A being the only exception. Stem succulents have evolved in clade B via at least two independent lineages, one containing *P. album* (section *Reniformia*) and *P. cotyledonis*, the other containing *P. drummondii* in Australia.

As stated above, the *trnL-F* sequence data have proven invaluable in studying broad contrasting patterns of evolution in *Pelargonium*. Upon further resolution of a phylogenetic hypothesis for this genus, it will become possible to address questions relating to phylogenetic constraints on growth form development, co-evolution between floral characteristics and pollinators, significance of polyploidy for adaptation to different climates and the acquisition of dispersability in greater detail.

ACKNOWLEDGEMENTS

We wish to thank two anonymous reviewers for their constructive comments on the manuscript. This work was supported by NERC grant GST\02\1169 to AC and MG.

REFERENCES

Albers, F. and van der Walt, J. J. A. (1984) Untersuchungen zur Karyologie und Mikrosporogenese von *Pelargonium* sect. *Pelargonium* (Geraniaceae). *Plant Systematics and Evolution*, **147**, 177–188.

Albers, F., Gibby, M. and Austmann, M. (1992) A reappraisal of *Pelargonium* sect. *Ligularia* (Geraniaceae). *Plant Systematics and Evolution*, **179**, 257–276.

Anderberg, A. A. and Bremer, K. (1991) Parsimony analysis and cladistic reclassification of the *Relhania* generic group (Asteraceae-Gnaphalieae). *Annals of the Missouri Botanical Garden*, **78**, 1061–1072.

Bakker, F. T., Hellbrügge, D., Culham, A. and Gibby, M. (1998) Phylogenetic relationships within *Pelargonium* sect. *Peristera* (Geraniaceae) inferred from nrDNA and cpDNA sequence comparisons. *Plant Systematics and Evolution*, **211**, 273–287.

Bakker, F. T., Culham, A., Daugherty, L. and Gibby, M. (1999) A *trnL–F* based phylogeny for species of *Pelargonium* (Geraniaceae) with small chromosomes. *Plant Systematics and Evolution*, **216**, 309–324.

Berbee, M. L. and Taylor, J. W. (1992) Convergence in ascospore discharge mechanism among Pyrenomycete fungi based on 18S ribosomal RNA gene sequence. *Molecular Phylogenetics and Evolution*, **1**, 59–71.

Böhle, U., Hilger, H. H. and Martin, W. F. (1996) Island colonization and evolution of the insular woody habit in *Echium* L. (Boraginaceae). *Proceedings of the National Academy of Sciences of the USA*, **93**, 11740–11745.

Bond, P. and Goldblatt, P. (1984) *Plants of the Cape flora*. National Botanic Gardens, Cape Town.

Brown, W. M. Prager, E. M., Wang, A. and Wilson, A. C. (1982) Mitochondrial DNA sequences of primates: the tempo and mode of evolution. *Journal of Molecular Evolution*, **18**, 225–239.

Carolin, R. C. (1961) The genus *Pelargonium* L'Hér. ex Ait. in Australia. *Proceedings of the Linnaean Society of New South Wales*, **86(3)**, 280–294.

Clegg, M. T. and Zurawski, G. (1992) Chloroplast DNA and the study of plant phylogeny, in *Molecular systematics of plants*, (eds P. S. Soltis, D. E. Soltis and J. J. Doyle), Chapman and Hall, London, pp. 1–13.

Coetzee, J. A. (1983) Intimations on the Tertiary vegetation of southern Africa. *Bothalia*, **14**, 345–354.

Cowling, R. M. (1992) *The ecology of fynbos: nutrients, fire and diversity*. Oxford University Press, Cape Town.

Cowling R. M. and Holmes, P. M. (1992) Flora and vegetation, in *The ecology of fynbos*, (ed. R. M. Cowling), Oxford University Press, Cape Town, pp. 23–61.

Cowling, R. M., Witkowski, E. T. F., Milewski, A. V. and Newbey, K. R. (1994) Taxonomic, edaphic and biological aspects of narrow plant endemism on matched sites in mediterranean South Africa and Australia. *Journal of Biogeography*, **21**, 651–664.

Deacon, H. J., Jury, M. R. and Ellis, F. (1992) Selective regime and time, in *The ecology of fynbos*, (ed. R. M. Cowling), Oxford University Press, Cape Town, pp. 6–22.

DeSalle, R., Freedman, T., Prager E. M. and Wilson, A. C. (1987) Tempo and mode of sequence evolution in mitochondrial DNA of Hawaiian *Drosophila*. *Journal of Molecular Evolution*, **26**, 157–164.

Dixon, M. T. and Hillis, D. M. (1993) Ribosomal RNA secondary structure: compensatory mutations and implications for phylogenetic analysis. *Molecular Biology and Evolution*, **10**, 256–267.

Doyle, J. J. (1992) Gene trees and species trees: molecular systematics as one-character taxonomy. *Systematic Botany*, **17**, 144–163.

Dreyer, L. L., Albers; F., van der Walt, J. J. A. and Marschewski, D. E. (1992) Subdivision of *Pelargonium* sect. *Cortusina* (Geraniaceae). *Plant Systematics and Evolution*, **183**, 83–97.

Farris, J. S., Albert, V. A., Källersjö, M., Lipscomb, D. and Kluge, A. G. (1996) Parsimony jackknifing outperforms neighbor-joining. *Cladistics*, **12**, 99–124.

Gibby, M. (1990) Cytological and morphological variation within *Pelargonium alchemilloides* s. l. (Geraniaceae). *Mitteilungen aus dem Institut für Allgemeine Botanik Hamburg*, **23b**, 707–722.

Gibby, M. (1994) Problems in the delimitation of species in *Pelargonium* (Geraniaceae), in *Proceedings of the XIIIth Plenary Meeting AETFAT, Malawi*, (eds J. H. Seyani and A. C. Chikuni), **1**, 421–427.

Gibby, M. and Westfold, J. (1986) A cytological study of *Pelargonium* section *Eumorpha* (Geraniaceae). *Plant Systematics and Evolution*, **153**, 205–222.

Gibby, M., Hinnah, S., Marais, E. M. and Albers, F. (1996) Cytological variation and evolution within *Pelargonium* section *Hoarea* (Geraniaceae). *Plant Systematics and Evolution*, **203**, 111–142.

Gielly, L. and Taberlet, P. (1996) A phylogeny of the European gentians inferred from chloroplast *trnL* (UAA) intron sequences. *Botanical Journal of the Linnean Society*, **120**, 57–75.

Goldblatt, P. (1978) An analysis of the flora of southern Africa: its characteristics, relationships and origins. *Annals of the Missouri Botanical Garden*, **65**, 369–436.

Goldblatt, P. (1997) Floristic diversity in the Cape Flora of South Africa. *Biodiversity and Conservation*, **6**, 359–377.

Goldblatt, P. and Le Thomas, A. (1997) Palynology, phylogenetic reconstruction, and classification of the Afro-Madagascan genus *Aristea* (Iridaceae). *Annals of the Missouri Botanical Garden*, **84**, 263–284.

Goldblatt, P. and Manning, J. C. (1996) Phylogeny and speciation in *Lapeirousia* subgenus *Lapeirousia* (Iridaceae, Ixioideae). *Annals of the Missouri Botanical Garden*, **83**, 346–361.

Good, R. (1974) *The geography of the flowering plants*. Longmans, Green & Co., London.

Hartmann, H. E. K. (1991) *Mesembryanthema. Contributions from the Bolus Herbarium*, **13**, 75–157.

Hellbrügge, D. (1998) Ph.D. thesis, University of Münster, Germany.

Hendey, Q. B. (1983) Cenozoic geology and paleogeography of the fynbos region, in *Fynbos palaeoecology: a preliminary synthesis* (eds H. J. Deacon, Q. B. Hendey and J. J. N. Lamprechts), South African National Programmes Report 75, CSIR, Pretoria.

Hillis D. M., Allard M. W. and Miyamoto, M. M. (1993) Analysis of DNA sequence data: phylogenetic inference. In *Molecular evolution: producing the biochemical data*, (eds E. A. Zimmer, T. J. White, R. L. Cann and A. C. Wilson), *Methods in Enzymology*, **224**, 456–490.

Holmquist, R. (1983) Transitions and transversions in evolutionary descent: an approach to understanding. *Journal of Molecular Evolution*, **19**, 134–144.

Johnson, L. A. S. and Briggs, B. G. (1974) Three old southern families – Myrtaceae, Proteaceae and Restionaceae, in *Ecological biogeography of Australia*, (ed. A. Keast), Utrecht, pp. 427–469.

Johnson, L. A. and Soltis, D. E. (1995) Phylogenetic inference in Saxifragaceae sensu stricto and *Gilia* (Polemoniaceae) using matK sequences. *Annals of the Missouri Botanical Garden*, **82**, 149–175.

Jones, C. S. and Price, R. A. (1996) Diversity and evolution of seedling Baupläne in *Pelargonium* (Geraniaceae). *Aliso*, **14**, 281–295.

Kita, Y., Ueda, K. and Kadota, Y. (1995) Molecular phylogeny and evolution of the Asian *Aconitum* subgenus *Aconitum* (Ranunculaceae). *Journal of Plant Research*, **198**, 429–442.

Knuth, R. (1912) Geraniaceae, in *Das Pflanzenreich* IV. 129, (ed. H. G. A. Engler), Berlin, pp. 1–640.

Kurzweil, H., Linder, P. and Chesselet, P. (1991) The phylogeny and evolution of the *Pterygodium–Corycium* complex (Coryciinae, Orchidaceae). *Plant Systematics and Evolution*, **175**, 161–223.

Levinson, G. and Gutman, G. A. (1987) Slipped-strand mispairing: a major mechanism for DNA sequence evolution. *Molecular Biology and Evolution*, **4**, 203–221.

Linder, P. (1987) The evolutionary history of the Poales/Restionales – a hypothesis. *Kew Bulletin*, **42**, 297–318.

Linder, P., Meadows, M. E. and Cowling, R. M. (1992) History of the Cape flora, in *The ecology of fynbos*, (ed. R. M. Cowling), Oxford University Press, Cape Town, pp. 113–134.

Maggs, G. L., Vorster, P. J. and van der Walt, J. J. A. (1995) Taxonomy of the genus *Pelargonium* (*Geraniaceae*): the section *Polyactium*. 1. Circumscription and intrasectional classification. *South African Journal of Botany*, **61**, 53–59.

Maggs, G. L, Vorster, P., van der Walt, J. J. A. and Gibby, M. Taxonomy of the genus *Pelargonium* (Geraniaceae): the section *Polyactium*. 3. The subsection *Polyactium*. *South African Journal of Botany*, in press.

Manning, J. C. and Goldblatt, P. (1997) The *Moegistorhynchus longirostris* (Diptera: Nemestrinidae) pollination guild: long-tubed flowers and a specialized long-proboscid fly pollination system in southern Africa. *Plant Systematics and Evolution*, **206**, 51–69.

Marais, E. M. (1994) *Taxonomic studies in* Pelargonium, *section* Hoarea *(Geraniaceae)*. Ph. D. thesis, University of Stellenbosch, 297pp.

Meadows, M. E. and Sugden, J. M. (1991) A vegetation history of the last 14000 years on the Cedarberg, south-western Cape Province. *South African Journal of Science*, **87**, 34–43.

Meve, U. (1995) Autogamie bei *Pelargonium*-Wildarten. *Der Palmengarten*, **59**, 100–108.

Oxelman, B., Lidèn, M. and Berglund, D. (1997) Chloroplast rps16 intron phylogeny of the tribe *Sileneae* (Caryophyllaceae). *Plant Systematics and Evolution*, **206**, 393–410.

Price, R. A. and Palmer, J. D. (1993) Phylogenetic relationships of the Geraniaceae and Geraniales from rbcL sequence comparisons. *Annals of the Missouri Botanical Garden*, **80**, 661–671.

Rousset, T., Pelandakis, M and Solignac, M. (1991) Evolution of compensatory substitutions through G. U intermediate state in *Drosophila* rRNA. *Proceedings of the National Academy of Sciences of the USA*, **88**, 10032–10036.

Siesser, W. G. (1980) Late Miocene origin of the Benguela upwelling system off northern Namibia. *Science*, **208**, 283–285.

Stokes, S., Thomas, D. S. G. and Washington, R. (1997) Multiple episodes of aridity in southern Africa since the last interglacial period. *Nature*, **388**, 154–158.

Struck, M. (1997) Floral divergence and convergence in the genus *Pelargonium* (Geraniaceae) in southern Africa: ecological and evolutionary considerations. *Plant Systematics and Evolution*, **208**, 71–97.

Sullivan, J., Holsinger, K. E. and Simon, C. (1996) The effect of topology on estimates of among-site rate variation. *Journal of Molecular Evolution*, **42**, 308–312

Sweet, R. 1820–1822. *Geraniaceae*, vol. 1. Ridgeway, London.

Sweet, R. 1822–1824. *Geraniaceae*, vol. 2. Ridgeway, London.

Sweet, R. 1824–1826. *Geraniaceae*, vol. 3. Ridgeway, London.

Sweet, R. 1826–1828. *Geraniaceae*, vol. 4. Ridgeway, London.

Sweet, R. 1828–1830. *Geraniaceae*, vol. 5. Ridgeway, London.

Taberlet, P., Gielly, L., Pautou, G. and Bouvet, J. (1991) Universal primers for amplification of three non-coding regions of chloroplast DNA. *Plant Molecular Biology*, **17**, 1105–1109.

Takhtajan, A. (1986) *Floristic regions of the world*. University of California Press, Berkeley.

Van der Walt, J. J. A. (1985) A taxonomic revision of the type section of *Pelargonium* L'Her. (Geraniaceae). *Bothalia*, **15**, 345–385.

Van der Walt, J. J. A. (1990) Taxonomic revision of *Pelargonium*: contributions of the Stellenbosch research team, in *Proceedings of the International Geraniaceae Symposium*, (ed. P. J. Vorster), University of Stellenbosch, RSA.

Van der Walt, J. J. A. and Van Zyl, L. (1988) A taxonomic revision of *Pelargonium* section *Campylia* (Geraniaceae). *South African Journal of Botany*, **54**, 145–171.

Van der Walt, J. J. A. and Vorster, P. J. (1983) Phytogeography of *Pelargonium*. *Bothalia*, **14**, 517–523.

Van der Walt, J. J. A. and Vorster, P. J. (1988) *Pelargoniums of Southern Africa*, vol. 3. National Botanic Gardens, Kirstenbosch.

Van der Walt, J. J. A., McDonald, D. J. and van Wyk, N. (1990) A new species of *Pelargonium* with notes on its ecology and pollination biology. *South African Journal of Botany*, **56**, 467–470.

Van der Walt, J. J. A., Venter, H. J. T., Verhoven, R. and Dreyer, L. L. (1991). The transfer of *Erodium incarnatum* into the genus *Pelargonium*. *South African Journal of Botany*, **56**, 560–564.

Van der Walt, J. J. A., Albers, F., Gibby., M., Marschewski, D. E. and Price, R. A. (1995) A biosystematic study of *Pelargonium* section *Ligularia*: 1. A new section *Subsucculentia*. *South African Journal of Botany*, **61**, 331–338.

Van der Walt, J. J. A., Albers, F., Gibby., M., Marschewski, D. E. and Hellbrügge, D. (1997) A biosystematic study of *Pelargonium* section *Ligularia:* 3. Reappraisal of section *Jenkinsonia*. *South African Journal of Botany*, **63**, 4–21.

Vogel, J. C. (1995) *Multiple origins of polyploids in European* Asplenium *(Pteridophyta)*. Ph. D. thesis, University of Cambridge, 499 pp.

Vogel, J. C., Rumsey, F. J., Schneller, J. J., Barrett, J. A. and Gibby, M. (1999) Where are the glacial refugia in Europe? Evidence from ferns. *Biological Journal of the Linnean Society*, **66**, 23–37.

Vorster, P. J. (1990) Taxonomy of the genus *Pelargonium*: review of the section *Otidia*, in *Proceedings of the International Geraniaceae Symposium*, (ed. P. J. Vorster), University of Stellenbosch, RSA.

Wakely, J. (1996) The excess of transitions among nucleotide substitutions: new methods of estimating transition bias underscore its significance. *Trends in Ecology and Evolution*, **11**, 158–163.

WWF and IUCN (1994) *Centres of plant diversity. A guide and strategy for their conservation.* Volume 1. IUCN Publications Unit, Cambridge, U. K.

Yang, Z. B. (1994) Estimating the pattern of nucleotide substitution. *Journal of Molecular Evolution*, **39**, 105–111.

Chapter 17

Integrating molecular phylogenies and developmental genetics: a Gesneriaceae case study

M. Möller, M. Clokie, P. Cubas and
Q. C. B. Cronk

ABSTRACT

We have analysed and characterised the phylogenetic potential of a nuclear developmental gene, *cycloidea* (originally isolated from *Antirrhinum*), involved in the development of floral zygomorphy. We have compared the evolution of part of this putative single copy gene in Old World Gesneriaceae with two contrasting DNA sequence regions, using two sets of data (a 'genus' data set and a 'species' data set); the chloroplast *trnL*(UAA) intron and the spacer between the *trnL* (UAA) 3′ exon and *trnF* (GAA) were relatively conserved and suitable for phylogenetic reconstruction at genus level. The multicopy internal transcribed spacers (ITS1 and ITS2) of nuclear ribosomal DNA in contrast appear to be evolving about five times faster and are suitable for resolution at the species level. The putative homologue of *cycloidea* (*Gcyc*) has an intermediate substitution rate about three times faster than the chloroplast intron/spacer region. However, the level of pairwise sequence divergence of *Gcyc* is higher than that of ITS at very low levels of divergence. This difference in apparent rate of molecular evolution between ITS and *Gcyc* at different levels of the taxonomic hierarchy we attribute to the process of molecular drive in the multicopy ITS. At lower levels of divergence (e.g. between closely related species) fixation of genetic changes in the multicopy ribosomal DNA acts as a restraint on evolutionary rate, whereas third codon position changes in coding single copy nuclear (scnDNA) genes are unconstrained. However, at high levels of divergence (e.g. between genera), scnDNA evolution is more functionally constrained than that of ITS and *Gcyc* therefore varies less. The small restraining effect of concerted evolution is not noticeable at these levels of sequence divergence. All three regions appear to evolve in a clock-like manner and are found to be suitable for phylogenetic reconstruction by parsimony, resulting in the same or similar topologies. We have examined the *Gcyc* sequences of three species that have reverted to actinomorphy from a zygomorphic condition. The gene appears to be intact and therefore, by implication, functional in these species. Furthermore, in one of these clades there has been a reversion back to zygomorphy which also implies that the gene is intact. We therefore suggest that in naturally occurring actinomorphic Gesneriaceae *Gcyc* continues to have a functional role, but zygomorphy is reduced by modifying genes. There is no convincing evidence that *Gcyc* evolves faster in actinomorphic lineages.

In *Molecular systematics and plant evolution* (1999) (eds P. M. Hollingsworth, R. M. Bateman and R. J. Gornall), Taylor & Francis, London, pp. 375–402.

17.1 Introduction

Greater access to single copy nuclear genes is a continuing goal for molecular systematists (Doyle and Doyle, 1999 – this volume). Their advantage over commonly used multicopy genes lies in their apparent higher rate of evolution and their potential to resolve relationships at lower taxonomic levels than the ribosomal internal transcribed spacer (ITS), the fastest evolving sequence commonly used today.

Since the invention of PCR and automated sequencing techniques there has been an almost exponential upsurge in the number of publications in molecular systematics. The great majority of investigations have been carried out using chloroplast or multicopy ribosomal DNA sequence data (Hershkovitz et al., 1999, this volume). Depending on the level of taxonomic distance, different types of DNA sequences are used. Chloroplast genes are generally more conserved than nuclear genes, due to their mode of inheritance (usually uniparental but with exceptions) and lower mutation rate (Hagemann and Schröder, 1989; Reboud and Zeyl, 1994; Tilney-Bassett, 1994; Mogensen, 1996; Ennos et al., 1999 – this volume). Coding genes, due to their functional constraints, are more conserved than intron or spacer sequences. Multicopy genes appear more conserved than single copy genes due to gene conversion events and the forces of concerted evolution (Dover, 1986; Hillis et al., 1991; Elder and Turner, 1995).

Studies of broad scale relationships amongst seed plants are often based on the coding chloroplast gene rbcL (Chase et al., 1993). This gene allows good resolution at the family level. At generic level, other chloroplast sequences are often chosen, such as the intron and spacer between transfer RNA the trnL (UAA) 5′ exon, trnL (UAA) 3′ exon and trnF (GAA) (Taberlet et al., 1991). For analyses at species level, multicopy nuclear ITS sequences are frequently used (Baldwin, 1992; Möller and Cronk, 1997a; Hershkovitz et al., 1999 – this volume). However, these sequences have their limitations in resolving relationships amongst very closely related taxa, as is shown in the case of Saintpaulia species, where seven species had identical ITS sequences. Thus, the ITS sequence analysis failed to resolve the relationships amongst members of the Saintpaulia ionantha-complex (Möller and Cronk, 1997b).

Access to single copy nuclear genes may provide further resolution here, as concerted evolution is not required for establishing mutations. This may result in higher sequence divergence compared to multicopy genes. Recently the gene involved in the expression of zygomorphy of Antirrhinum flowers, cycloidea (cyc), has been isolated (Luo et al., 1996). Its homologue in Gesneriaceae appears to be single copy. Sequence data for the Antirrhinum cyc gene and sequence data of a Saintpaulia clone allowed the design of Gesneriaceae-specific primer pairs amplifying a single product comprising ~70% of the open reading frame (Fig 17.1). However, it should be noted that as there are two related genes in Antirrhinum: cyc and dichotoma (dich) (Luo et al., 1996) and several cyc-like ESTs in Arabidopsis (Luo et al., 1996; Doebley et al., 1997), as well as a similar gene in maize, teosinte branched 1 (tb1) (Doebley et al., 1997), it is likely that cyc is part of a family of genes.

We chose cyc partly because in Gesneriaceae flower zygomorphy (and number of fertile stamens) is variable. Flower zygomorphy has evolved and been lost several times independently in angiosperms, e.g. in the Solanaceae and Boraginaceae (Coen and Nugent, 1994). In Gesneriaceae zygomorphy is ancestral, and has been lost

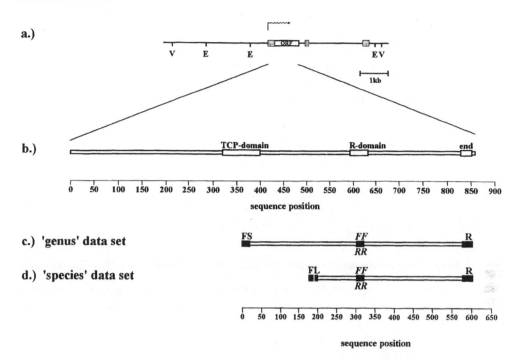

Figure 17.1   a) Structure and map of the *cyc* locus (after Luo *et al.*, 1996). Exons and predicted open reading frame (ORF) are indicated in rectangles; the arrow indicates the direction of transcription; restriction enzyme sites: E – *Eco*RI, V – *Eco*RV. b) The *cyc* ORF and the location of conserved regions (open boxes). Fragment of *cyc* amplified, and the PCR primer positions (closed boxes) for c) the 'genus' and d) for the 'species' data set.

several times independently, for instance in the Asiatic *Conandron*, and in the European *Ramonda* (Möller and Cronk, unpublished). This provides an ideal system for investigating differential gene evolution, and the relationship between sequence divergence of *cyc* homologues in Gesneriaceae and changes in flower morphology.

In this paper we compare sequence data from three DNA sources: from chloroplast DNA (the intron/spacer *trnL+F*), from nuclear DNA (the multicopy nuclear rDNA ITS) and also from the Gesneriaceae homologues of the single copy developmental gene *cyc*. It is essential that there is strict gene homology of the DNA fragments to be compared. We have ensured this by using the following checks: (1) the use of a specific primer pair; (2) sequence similarity including that of shared sequence motifs (the *TCP* domain, *R*-domain and *end*-box [Cubas *et al.*, 1999]) within the open reading frame of *Gcyc*; and (3) checking for congruence between phylogenies derived from independent (nuclear and chloroplast) genes with that derived from *Gcyc*. Any major departure from the topology confirmed by other nuclear and chloroplast genes may indicate homology problems. Two data sets, at the genus level and at the species level, were chosen, firstly to illustrate differences in sequence characteristics; secondly to test the suitability of *Gcyc* sequences for phylogenetic reconstructions; and lastly to investigate the possibility that gene sequence evolution is linked to floral morphology.

17.2 Materials and methods

17.2.1 Plant material

Plant material was from living plants cultivated at the Royal Botanic Garden Edinburgh (RBGE). Identifications were kindly confirmed by B. L. Burtt. For all taxa analysed, voucher herbarium specimens were prepared, flowers were preserved in Kopenhagen mixture in a spirit collection (both deposited in herb. E), and photographs of flowering specimens were taken, and deposited in the RBGE library. For this study a single individual was used to represent each species, which makes the 'species' and 'genus' data sets more directly comparable. Further studies on gene evolution at the species level would ideally choose to use more than one individual per species, as intra-specific variation in DNA sequences is now well known. However, this does not affect the main purpose of the present paper, which focuses on gene evolution rather than species evolution.

17.2.1.1 The 'genus' data set

The ingroup taxa were all from the Didymocarpeae, the largest tribe within Gesneriaceae subfamily Cyrtandroideae (Table 17.1). Representative taxa for zygomorphic flower morphology were the European *Haberlea rhodopensis* and *Jankaea heldreichii*, the Asiatic *Primulina tabacum*, and the African *Saintpaulia velutina*. Additionally, species representing the major growth forms of *Streptocarpus* were selected: the caulescent *Streptocarpus holstii*, the rosulate *Streptocarpus primulifolius* and the unifoliate *Streptocarpus dunnii*. Taxa representing actinomorphic flower types were the Asiatic *Conandron ramondioides*, and two European taxa, *Ramonda myconi* and *Ramonda nathaliae*. Recent classifications of the Gesneriaceae recognise two major subfamilies, the neotropical Gesnerioideae and the chiefly Old

Table 17.1 Accessions of eleven Gesneriaceae taxa examined for sequence variation: 'genus' data set.

No. Taxon	Origin: distribution	RBGE Accession no.[a]
1 *Sinningia schiffneri* Fritsch	South America	1978 1514
2 *Ramonda myconi* (L.) Rchb	Spain: Pyrenees	1971 1477
3 *Ramonda nathaliae* Pancic & Petrovic, white form	S. Jugoslavia, N. Albania, N.C. Greece	1978 4020
4 *Jankaea heldreichii* Boiss.	Greece: Mt. Olympus	1979 1939
5 *Haberlea rhodopensis* Friv. Symond's form	N.E. Greece, C. and S.C. Bulgaria	1975 4106
6 *Primulina tabacum* Hance	China: Gnangdong, Lian River	1995 1540
7 *Conandron ramondioides* Siebold & Zucc.	Japan:	1969 1267
8 *Saintpaulia velutina* B. L. Burtt	Tanzania: W Usambara Mts., Balangai	1987 2179
9 *Streptocarpus holstii* Engl.	Tanzania: E. Usambara Mts.	1959 2272
10 *Streptocarpus dunnii* Hook. f.	South Africa: Swaziland; N. Mbabane	1994 1745
11 *Streptocarpus primulifolius* Gand.	South Africa: E. Cape, Igoda River	1991 2192

[a] These numbers were also used as voucher numbers.

World Cyrtandroideae (Burtt and Wiehler, 1996). The phylogenetic relationships within and between the subfamilies, based on molecular and morphological cladistic analyses are complicated (Smith, 1996; Smith *et al.*, 1997). Therefore the neotropical *Sinningia schiffneri* was chosen as an outgroup for the genus comparison.

17.2.1.2 The 'species' data set

Nineteen species of *Saintpaulia*, representing all areas of geographical distribution, were chosen for this study (Table 17.2). The outgroup taxon for the species comparison was *Streptocarpus holstii*. This was based on recent results from molecular data where the closest known relatives of *Saintpaulia* are in fact caulescent African species of *Streptocarpus* (Möller and Cronk, 1997a; Smith *et al.*, 1997).

Table 17.2 Accessions of *Streptocarpus* and *Saintpaulia* examined for sequence variation: 'species' data set.

No.	Taxon	Origin: distribution	RBGE Accession no.[a]
1	*Saintpaulia brevipilosa* B. L. Burtt	Tanzania: Nguru Mts., Lulaga, Mt. Kanga	1970 0909
2	*Saintpaulia difficilis* B. L. Burtt	Tanzania: E Usambara Mts., Sigi River, Monga	1987 2176
3	*Saintpaulia diplotricha* B. L. Burtt	Tanzania: NE Usambara Mts., Maweni, Tanga	1987 2172B
4	*Saintpaulia grandifolia* B. L. Burtt	Tanzania: W Usambara Mts., Lutindi	1985 0678
5	*Saintpaulia grotei* Engl.	Tanzania: E Usambara Mts., Amani, Mt Mlinga	1987 2171
6	*Saintpaulia goetzeana* Engl.	Tanzania: Uluguru Mts., Lukwangule Plateau	1997 1201
7	*Saintpaulia intermedia* B. L. Burtt	Tanzania: E Usambara Mts., Kigongoi	1997 0101
8	*Saintpaulia* cf. *ionantha* H. Wendl.	Tanzania: Tanga, Sigi Caves	1971 0860
9	*Saintpaulia magungensis* E. Roberts	Tanzania: E Usambara Mts., Magunga, Mt. Mlinga	1992 3187
10	*Saintpaulia magungensis* var. *minima* B. L. Burtt	Tanzania: E Usambara Mts., Mavoera estate, Amani	1959 4352
11	*Saintpaulia magungensis* var. *occidentalis* B. L. Burtt	Tanzania: W Usambara Mts.	1985 0680
12	*Saintpaulia nitida* B. L. Burtt	Tanzania: Nguru Mts., Mkobwe, Turiani,	1992 3186
13	*Saintpaulia orbicularis* var. *purpurea* B. L. Burtt	Tanzania: W Usambara Mts., Ambangulu	1958 3586
14	*Saintpaulia pendula* var. *kizarae* B. L. Burtt	Tanzania: NE Usambara Mts., Mt. Mtai, Kizara	1997 0103
15	*Saintpaulia rupicola* B. L. Burtt	Kenya: Kaloleni	1997 0094
16	*Saintpaulia shumensis* B. L. Burtt	Tanzania: W Usambara Mts., Shume	1996 2088
17	*Saintpaulia* Sigi Falls	Tanzania: Tanga, Sigi River	1992 3183
18	*Saintpaulia teitensis* B. L. Burtt	Kenya: Teita Hills, Mbololo Hill	C 3771
19	*Saintpaulia tongwensis* B. L. Burtt	Tanzania: E Usambara Mts., Tongwe Mts.	1985 0668
20	*Saintpaulia velutina* B. L. Burtt	Tanzania: W Usambara Mts., Balangai	1987 2179
21	*Streptocarpus holstii* Engl.	Tanzania: E Usambara Mts.	1959 2272

[a] These numbers were also used as voucher numbers.

17.2.2 DNA extraction

Fresh leaf material was used for total DNA extraction using a modified CTAB procedure of Doyle and Doyle (1987), with no further purification.

17.2.3 PCR amplification and conditions

Chloroplast gene intron and spacer: chloroplast DNA (cpDNA) was PCR amplified using primers c and f of Taberlet *et al.* (1991), amplifying the *trnL*(UAA) intron and the intergenic spacer between the *trnL* (UAA) 3' exon and *trnF* (GAA). Ribosomal DNA (rDNA) spacer: the complete ITS region, including the 5.8S rDNA gene and both flanking internal transcribed spacers, was PCR amplified, using modified primers (Möller and Cronk, 1997a), based on published data by White *et al.* (1990). Single copy developmental gene (scnDNA): part of the open reading frame (ORF) of *Gcyc* was amplified using forward primer *Gcyc*FS (ATG CTA GGT TTC GAC AAG CC) and the reverse primer *Gcyc*R (ATG AAT TTG TGC TGA TCC AAA ATG) (Fig. 17.1), designed and modified after conserved motifs in *Antirrhinum cyc* sequences (Luo *et al.*, 1996) and a cloned *Saintpaulia* sequence. In *Saintpaulia* and *Streptocarpus* taxa a gene closely related to *Gcyc* was found. A selective forward primer *Gcyc*FL (CAC CCG GAT TCG AGA AAA TC) was designed that in combination with the reverse primer *Gcyc*R exclusively amplified *Gcyc* (Fig 17.1). The PCR reaction mixture and PCR cycle parameters, amplicon quantification and purification are described elsewhere (Möller and Cronk, 1997a).

17.2.4 Sequencing protocol

Cycle sequencing and analysis protocols followed Möller and Cronk (1997a). For each taxon, forward and reverse sequencing reactions were performed for sequence confirmation. Sequencing primers were identical to those used for PCR. Additionally, two shorter reactions were set up using internal primers anchored in highly conserved regions, sequencing the cpDNA intron and intergenic spacer with primers d and e (Taberlet *et al.*, 1991), or ITS 1 and ITS 2 with primers 2G and 3P (Möller and Cronk, 1997a), or *Gcyc* with primers *Gcyc*RR (CTT GAT GCA CAT TTT CTC CTT) and *Gcyc*FF (AAG GAG AAA ATG TGC ATC AAG) from within the PCR amplified products.

17.2.5 Sequence analysis

Sequence boundaries of both rDNA internal transcribed spacers of all taxa were determined as described previously (Möller and Cronk, 1997a). The full length of the PCR amplified fragments of the cpDNA, including the transfer RNA gene, or the whole *Gcyc* gene fragment amplified (~70% of the ORF), including the *TCP* and *R*-domains and the *end*-box, were included in the respective matrices. All matrices were aligned using the CLUSTAL option in the multiple alignment program Sequence Navigator™, version 1.0.1 software package (Perkin Elmer, Applied Biosystems Division, Foster City, CA, USA), followed by manual optimisation. The G + C content was determined by inspection, and transition/transversion ratios calculated using MacClade, version 3.01 (Maddison and Maddison, 1992).

Synonymous and non-synonymous substitutions were determined using the program Molecular Evolutionary Genetics Analysis, version 1.01 (MEGA; Kumar *et al.*, 1993). Sequence divergence among taxa was calculated using the DISTANCE MATRIX option in PAUP, version 3.1.1 (Swofford, 1993), based on unambiguously alignable regions, with adjustment for gaps in pairwise comparisons. Graphics and statistical analyses were produced using the program Statgraphic, version 1.03. Sequences used in this study are available from the authors on request.

17.2.6 Phylogenetic analysis

Phylogenetic trees were generated from unordered character states using PAUP, version 3.1.1 (Swofford, 1993). The genus comparison was analysed using the exhaustive search mode. In view of the large number of taxa included in the species comparison, the following heuristic search strategy was employed to find the most parsimonious trees: 500 replicates of RANDOM addition sequence with no swapping. This was followed by TBR swapping on the resulting trees (Möller and Cronk, 1997a). The options MULPARS, STEEPEST DESCENT, COLLAPSE, and ACCTRAN optimisation were selected.

Bootstrap analyses (Felsenstein, 1985) were performed using PAUP, set to HEURISTIC search option and SIMPLE addition sequence. Bootstrap values were calculated using 200 to 1000 replicates with MAXTREE set to 1000. Decay indices (DI) (Bremer, 1988; Donoghue *et al.*, 1992) for individual clades were obtained by comparing the strict consensus of all equal-length trees up to a maximum of 42 steps longer than the shortest tree, using SIMPLE addition sequence and TBR in PAUP. Descriptive statistics reflecting the amount of phylogenetic signal in the parsimony analyses were given by the consistency index (CI) (Kluge and Farris, 1969), retention index (RI) (Farris, 1989), and the resulting rescaled consistency index (RC) (Swofford, 1993).

All three data sets were used for the 'genus'-level comparison (cpDNA was omitted from the 'species'-level study as the cpDNA showed hardly any variation at this level).

For cpDNA and rDNA sequences, only combined spacer/intron sequence data were subjected to phylogenetic analyses. For one analysis of genus comparison cpDNA, rDNA and scnDNA sequence data matrices were combined. For simplicity, differential weighting schemes were not carried out, and in all analyses character state changes were weighted equally, and gaps were treated as missing data (Soltis and Kuzoff, 1995; Susanna *et al.*, 1995; Downie and Katz-Downie, 1996). Ambiguous regions that allowed alternative alignment interpretations were excluded from phylogenetic analyses (Wojciechowski *et al.*, 1993; Downie and Katz-Downie, 1996).

17.3 Results

17.3.1 The 'genus' data set

17.3.1.1 Sequence comparison between trnL+F, ITS and cyc

The distribution of character changes, base substitutions and indel events, for the cpDNA was higher in the spacer region than in the intron (Fig. 17.2a). The ITS data matrix showed a similar pattern, although with some more conserved regions within both spacers (Fig. 17.2b), where conservation of secondary structure is observed amongst angiosperms (Liu and Schardl, 1994). These are presumed to be recognition sites important during post-transcriptional processes. In the part of the *Gcyc* gene amplified, the distribution of variable sites was relatively evenly spread, except for approximately 20 to 40bp at the beginning and end of the fragment, representing the loop of the conserved *TCP*-domain found in related genes (Cubas *et al.*, 1999), and a conserved *end*-box, respectively. An extended *R*-domain (Cubas *et al.*, 1999) can be found in the Gesneriaceae, stretching from position 289 to 369, including considerably fewer changes (Fig 17.2c, Fig 17.3). Seemingly less variable regions beyond the *R*-domain up to position 453 and from 613 to 642 were the result of larger insertions in few taxa or the outgroup, respectively (Fig 17.3).

The alignment matrix of the *trnL+F* intron/spacer sequences required the insertion of 25 gaps of 1–65 bp. A 65 bp deletion in the cpDNA spacer was found only in *Streptocarpus primulifolius*. Previous analyses investigating the effects of large deletions on tree topologies found little effect of complete removal of the deletion sites from the matrix (Möller and Cronk, 1997a). As inclusion or exclusion did not affect tree topology, the cpDNA positions were left in the matrix to prevent loss of potential information amongst the other taxa at those positions. The matrix contained 88.2% constant sites, 7.6% autapomorphic sites, but only 4.2% informative sites (Table 17.3).

Due to the taxonomic distance of the taxa included in the 'genus' data set, alignment of the combined ITS 1 and ITS 2 matrix was difficult. Due to alternative alignment interpretation 115 sites had to be excluded. Optimised alignment required the insertion of 51 gaps of 1–10 bp length. Of the remaining 423 sites, 47.0% were constant, 27.9% were unique to individual taxa and 25.1% were informative phylogenetically (Table 17.3).

The amplified *Gcyc* fragment of the diverse genera varied from 530 to 656 bp, and the aligned matrix was 740 bp long, with 39 indels of 3–66 bp length, of which most were informative (34). Of all sites, 64.5% were constant, 18.2% autapomorphies, and 17.3% informative, which was intermediate between the cpDNA and rDNA sequences. Translation to amino acids resulted in a 246 codon matrix, with an increased percentage of variable sites, of which 23.6% were uninformative and 23.2% informative (Table 17.3).

17.3.1.2 Phylogenetic reconstruction using three genes

Parsimony analysis of aligned cpDNA sequences resulted in one most parsimonious tree (Fig. 17.4a). When all sites were included the tree had a length of 133 steps, with a high CI of 0.947, an RI of 0.908 and an RC of 0.860. The average number of nucleotide substitutions per site was low with 0.134, with only two out of 990 sites changing three times, indicating a very low saturation of base mutations. It is therefore unlikely that phylogenetic signal has become obscured by multiple substitutions. Forty-five base substitutions separated the outgroup taxon *Sinningia schiffneri* from the ingroup (Fig. 17.4a). The ingroup taxa form four clades, one consisting of the European taxa *Ramonda nathaliae* and *Jankaea heldreichii* as sister

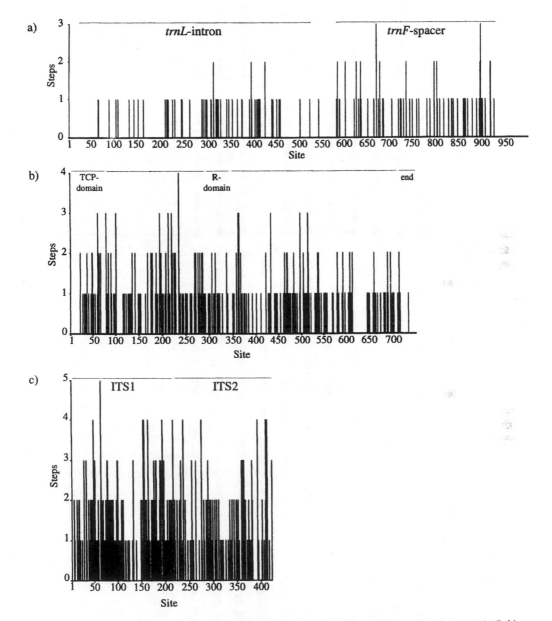

Figure 17.2 The number of steps per base position for the 'genus' data set of a) the *trnL+F*, b) the *Gcyc*, and c) the ITS sequence matrix, illustrating the variation across the gene sequences.

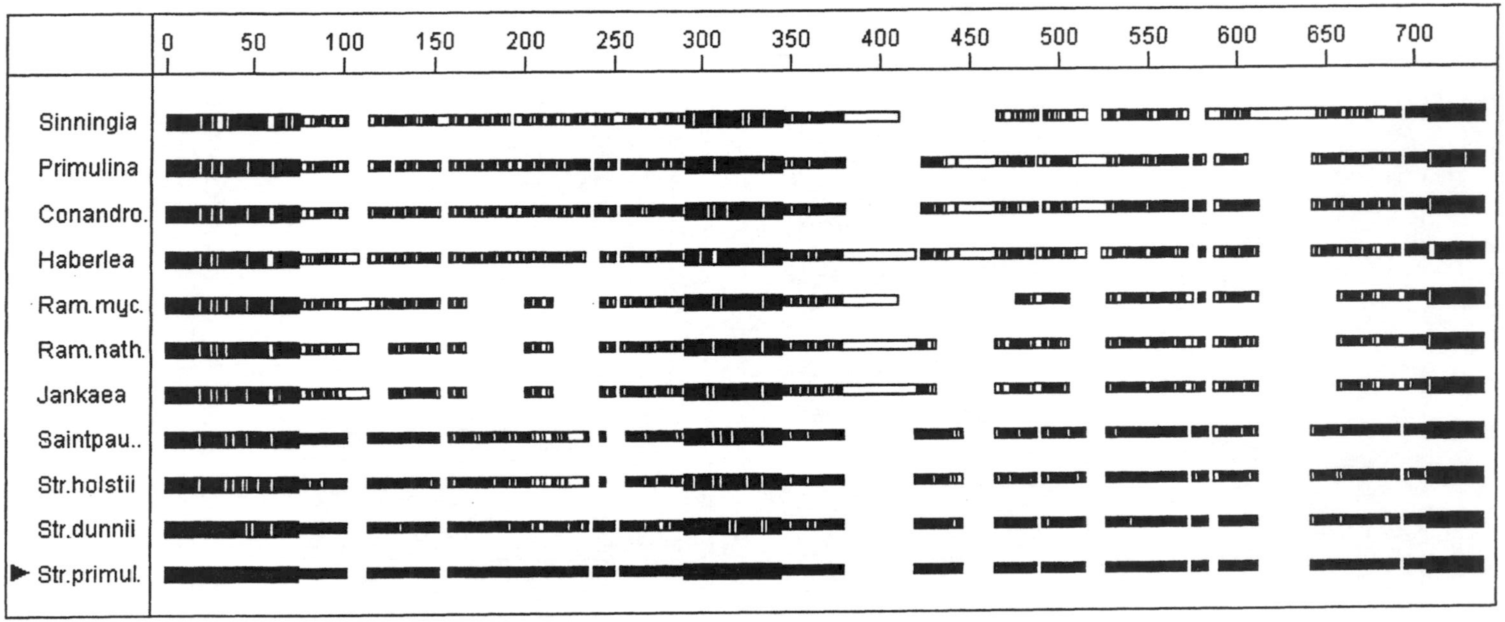

Figure 17.3 Conservation of *Gcyc* DNA sequences between *Streptocarpus primulifolius* and other taxa included in the 'genus' data set. Black = conserved, white = not conserved. Gaps: deletion events. Drawn with MACAW, vers. 2.0.5 (Schuler, 1995).

Table 17.3 Sequence characteristic of *trnL+F*, ITS 1 and ITS 2, and *Gcyc* regions of eleven taxa of Gesneriaceae: 'genus' data set.

Parameter	*trnL+F*	*ITS 1 and ITS 2*	*Gcyc nucleotides*	*Gcyc amino acids*
Length range, bp	870–970	380–390	530–656	176–218
Length mean, bp	946.0	384.4	597.6	198.5
Aligned length, bp	990	538	740	246
Number of sites excluded, bp	–	115	–	–
G+C content mean, %	36.5	56.99	41.0	–
Sequence divergence (in–outgroup)	45–68	109–126	93–136	43–52
Sequence divergence, (in–outgroup) %	4.8–7.0	30.1–34.5	18.4–24.0	23.2–31.0
Sequence divergence (ingroup)	2–45	16–87	27–98	10–46
Sequence divergence, (ingroup) %	0.2–4.8	4.2–23.6	4.9–17.0	5.8–24.2
Number of indels, (informative)	25 (14)	51 (17)	39 (34)	39 (34)
Size of indels, bp	1–65	1–10	3–66	1–22
Number of constant sites (%)	873 (88.2)	199 (47.0)	477 (64.5)	131 (53.2)
Number of variable sites (%)	117 (11.8)	224 (53.0)	263 (35.5)	115 (46.8)
Number of autapomorphic sites (%)	75 (7.6)	118 (27.9)	135 (18.2)	58 (23.6)
Number of informative sites (%)	42 (4.2)	106 (25.1)	128 (17.3)	57 (23.2)
Transitions (min–max)	67–70	218–237	200–213	–
Transversions (min–max)	63–66	143–161	174–187	–
Transitions/transversions	1.07	1.56	1.17	–
Average number of steps per character	0.134	0.898	0.523	0.801

taxa (bootstrap value [BS] = 62%, DI = +1) and *Ramonda myconi* (BS = 90%, DI = +2). *Haberlea rhodopensis* is situated between this clade (BS = 90%, DI = +2) and the rest of the taxa. The Asiatic taxa *Primulina tabacum* and *Conandron ramondioides* form a distinct clade (BS = 100%, DI = +9) and are sister to the African taxa. Within the African group (BS = 100%, DI = +7), the unifoliate *Streptocarpus dunnii* and the rosulate *Streptocarpus primulifolius* (BS = 90%, DI = +2) form a sister clade to the caulescent *Streptocarpus holstii* and *Saintpaulia velutina* (BS = 100%, DI = +7).

One most parsimonious tree was retained after parsimony analysis of unambiguously aligned sequences of both ITS spacers (Fig. 17.4b). The tree length was 380 steps and the CI was 0.779, lower than for cpDNA sequences, indicating more homoplastic changes. The higher homoplasy was also reflected in the higher number of steps per site; 11 sites have changed four times, and one site five times over the tree, with an average of 0.898, indicating a potential saturation and reversals across the matrix. The tree topology, however, was identical to the cpDNA tree (Fig. 17.4a), with similar BS and DI values, except for branches separating *Haberlea rhodopensis* from the other European taxa and the rest of the taxa, and the branch grouping the Asiatic taxa, which collapsed after bootstrap analysis.

Parsimony analysis of the *Gcyc* data matrix resulted in one most parsimonious tree of 387 steps. Estimates of homoplasy were intermediate between cpDNA and rDNA matrices (Fig. 17.4c). Five sites changed four times, and the average number of steps per site was 0.518. The topology of the *Gcyc* tree was similar to the previous trees, except of the position of *Haberlea rhodopensis* which was sister to the other European taxa (BS = 88%, DI = +4) (Fig. 17.4c). Branch support, BS and DI, for residual clades was similar to cpDNA data.

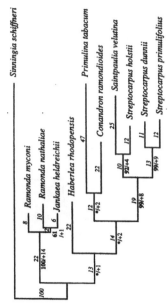

a) *trnL+F* tree: (133 steps; CI=0.947; RI=0.908)

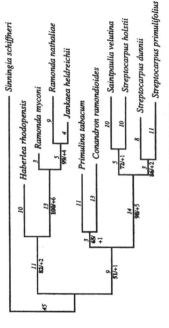

b) ITS tree: (380 steps; CI=0.779; RI=0.649)

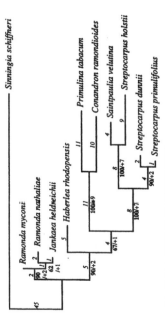

c) *Gcyc* tree (nucleotides): (387 steps; CI=0.845; RI=0.774)

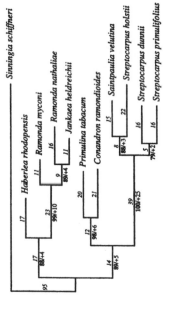

d) *Gcyc* tree (amino acids): (197 steps; CI=0.878; RI=0.782)

Figure 17.4 Most parsimonious trees for the 'genus' data set based on a) *trnL+F*, b) ITS, c) *Gcyc* nucleotides, and d) *Gcyc* amino acid sequences. Numbers above branches (italic) indicate branchlengths. Numbers below branches (bold) are bootstrap values of 1000 replicates, and numbers in normal print are decay indices.

Phylogenetic analysis of the translated *Gcyc* amino acids by parsimony gave one tree of 197 steps, with similar descriptive values compared to the nucleotide tree (Fig. 17.4d), except for slightly higher consistency values and a higher average number of steps per site, similar to ITS data, indicating a higher saturation to be present compared to the nucleotide data matrix. However, the topology was identical to the nucleotide tree topology. Branch support (as BS) was similar, except for the branches grouping the Asiatic taxa and the Asiatic and African taxa from the rest.

Combining all three data sets gave a matrix of 2275 sites. Parsimony analysis resulted in one most parsimonious tree of 904 steps (Fig. 17.5). The CI was 0.829 and the RC 0.607 (RI = 0.733). The topology was identical to the cpDNA and rDNA trees. Branch support was high (BS = 97–100%, DI = +7–42), except for the branch between the *Ramonda/Jankaea* clade and *Haberlea rhodopensis* and the rest of the taxa (BS = 54%, DI = +1).

17.3.1.3 Comparative pairwise sequence divergence

A translation of *Gcyc* nucleotides into *Gcyc* amino acids increased both the number of informative sites and the divergence in pairwise comparisons, ranging from 5.8% (*Ramonda myconi – Jankaea heldreichii*) to 31.0% (*Sinningia schiffneri – Jankaea heldreichii*) (Table 17.3). This was due to the relation between nucleotide and amino acid divergence in pairwise comparisons (r = 0.946; P > 0.001) that indicated disproportionally high values for amino acid changes (Fig. 17.6). This is unexpected, as it is believed that most of the variation would be silent, with synonymous sites not altering the amino acid sequence. However, a closer look at the relationship between

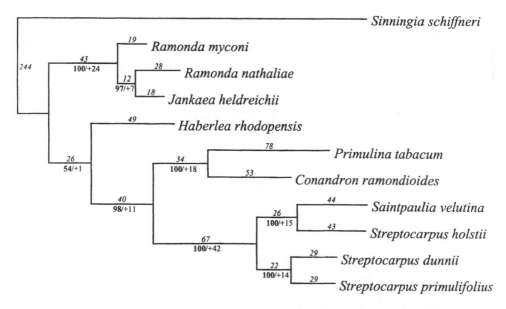

Figure 17.5 Phylogram of the most parsimonious tree for the 'genus' data set of 904 steps length, based on parsimony analysis of the combined *trnL+F*, ITS, and *Gcyc* nucleotides data sets (CI = 0.829; RI = 0.733; RC = 0.607 inclusive uninformative sites). Numbers along branches are as in Figure 17.4.

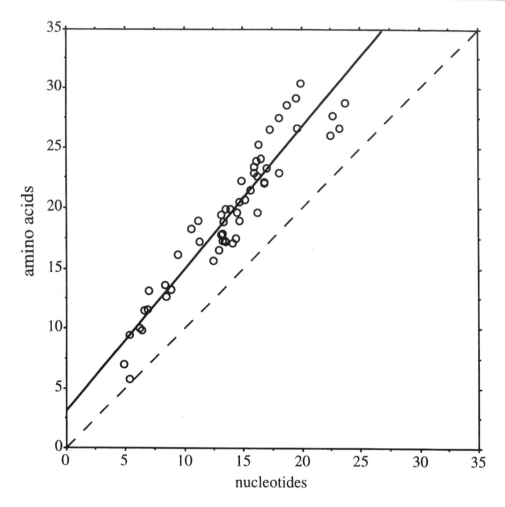

Figure 17.6 Relation between *Gcyc* nucleotides and *Gcyc* amino acid divergence (%) in pairwise sequence comparisons of the 'genus' data set (gaps excluded in pairwise comparisons) (r = 0.946).

synonymous and non-synonymous substitutions revealed a significant correlation (r = 0.860; $P > 0.001$) (Fig. 17.7); the intercept of the regression curve at 4.8% accords with the fact that at shorter taxonomic distances more synonymous substitutions occur than non-synonymous changes. However, with increasing divergence between the taxa, proportionally more non-synonymous substitutions have taken place, indicated by a slope of less than 45° in the regression curve, resulting in proportionally higher values of amino acid changes, thus divergence, compared to nucleotide changes.

Plotted against sequence divergence figures, ITS values are consistently higher (r = 0.876; $P > 0.001$), and *trnL+F* consistently lower (r = 0.86; $P > 0.001$) than *Gcyc* figures (Fig. 17.8). In pairwise sequence comparisons, divergence ranged from 0.2% (*Ramonda myconi* – *Jankaea heldreichii*) to 7.0% (*Sinningia schiffneri* – *Primulina*

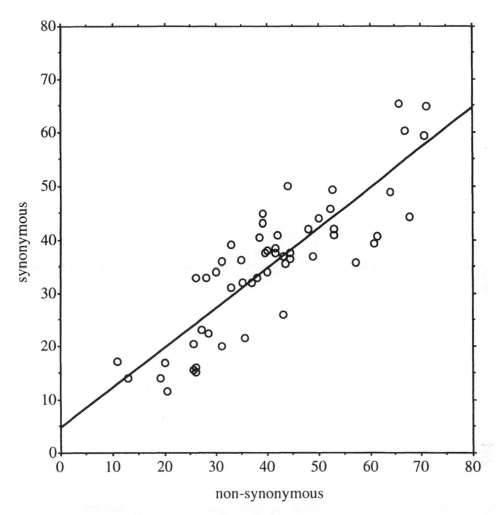

Figure 17.7 Relation between synonymous and non-synonymous changes across the *Gcyc* nucleotide data matrix of the 'genus' data set in pairwise comparisons (gaps excluded in pairwise comparisons) (r = 0.86).

tabacum) in cpDNA, 2% (*Jankaea heldreichii* – *Ramonda myconi*/*Ramonda nathaliae*) to 34.5% (*Sinningia schiffneri* – *Streptocarpus dunnii*) in ITS, and 4.9% (*Ramonda nathaliae* – *Jankaea heldreichii*) to 23.7% (*Sinningia schiffneri* – *Streptocarpus dunnii*) in *Gcyc* (Table 17.3).

17.3.2 The 'species' data set

17.3.2.1 Sequence comparison between ITS and Gcyc

Unlike the 'genus' data set the alignment of ITS sequences of the 'species' analysed required the insertion of 17 gaps only, of 1–4 bp length. Not unexpectedly, a high

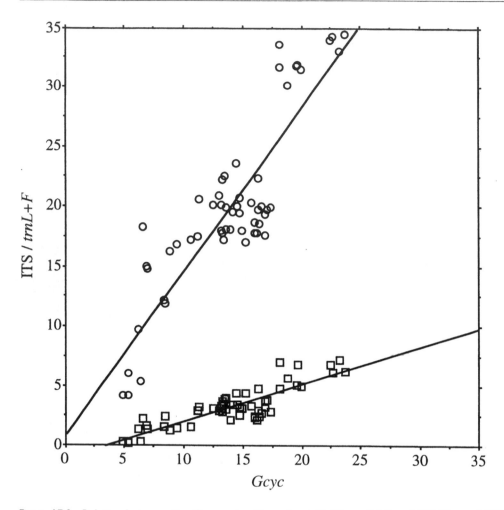

Figure 17.8 Relation between the *Gcyc* nucleotide and *trnL*+F (□, r = 0.86) and ITS (○, r = 0.876) sequence divergences (%) in pairwise comparisons of the 'genus' data set (gaps excluded in pairwise comparisons).

proportion of the unambiguously aligned sites (77.8%) were constant, only 8.7% were potentially informative phylogenetically, and 13.5% were autapomorphies (Table 17.4).

The *Gcyc* gene fragment used for the species comparison was 372 bp long, and required no alignment gaps. Similar to ITS sequences, 82.5% of the sites were constant, but only half as many, 4.9%, were potentially informative (Table 17.4). Translation of the nucleotide matrix to amino acids resulted in 124 amino acids. Compared to the nucleotides, the percentage of constant sites decreased to 70.2%, while the percentages of informative sites and of autapomorphies increased to 8.9% and 21.0%, respectively. *Saintpaulia difficilis*, *S. magungensis*, *S. teitensis*, and *S. grotei* and *S. magungensis* var. *occidentalis* had identical amino acid compositions (Table 17.4).

Table 17.4 Sequence characteristic of ITS 1 and ITS 2, and *Gcyc* regions of *Streptocarpus holstii* and 20 *Saintpaulia* taxa: 'species' data set.

Parameter	ITS 1 and ITS 2	Gcyc nucleotides	Gcyc amino acids
Length range, bp	462–471	–	–
Length mean, bp	466.7	372	124
Aligned length, bp	483	372	124
G+C content mean, %	53.1	42.0	–
Sequence divergence (in–outgroup)	46–58	26–33	14–19
Sequence divergence, (in–outgroup) %	10.0–12.5	7.0–8.9	11.3–15.3
Sequence divergence (ingroup)	0–73	0–19	0–10
Sequence divergence, (ingroup) %	0–15.8	0–5.1	0–8.1
Number of indels, (informative)	17 (7)	0	0
Size of indels, bp	1–4	–	–
Number of constant sites (%)	376 (77.8)	307 (82.5)	87 (70.2)
Number of variable sites (%)	107 (22.2)	65 (17.5)	37 (29.8)
Number of autapomorphic sites (%)	65 (13.5)	47 (12.6)	26 (21.0)
Number of informative sites (%)	42 (8.7)	18 (4.9)	11 (8.9)
Transitions (min)	70	37	–
Transversions (min)	55	30	–
Transitions/transversions	1.27	1.23	–
Average number of steps per character	0.254	0.185	0.331

17.3.2.2 Phylogenetic reconstruction: ITS vs Gcyc

Parsimony analysis of both sets of 'species' data resulted generally in numerous most parsimonious trees. Analysis of aligned ITS1 and ITS2 sequences resulted in four most parsimonious trees, of 123 steps length (CI = 0.919; RI = 0.851; RC = 0.782). The strict consensus of those trees resolved biogeographic relations amongst *Saintpaulia* taxa (Fig. 17.9). *Saintpaulia goetzeana* (BS = 98%, DI = +12) and *Saintpaulia teitensis* (BS = 100%, DI = +10) are separated from the rest, which form a polytomy of three species (*S. intermedia, S. pendula, S. rupicola*) and two groups, one consisting of *S. magungensis* var. *occidentalis, S. brevipilosa* and *S. nitida,* and the second group consisting of the *S. ionantha*-complex (Möller and Cronk, 1997b) and including the rest of the taxa (BS = 64%, DI = +1).

For the *Gcyc* nucleotide matrix 2210 most parsimonious trees of 69 steps length (CI = 0.829; RI = 0.733; RC = 0.608) were obtained. The strict consensus tree was less resolved than the rDNA tree and the clades only weakly supported (Fig 17.10). *Saintpaulia goetzeana* was separated from the rest of the taxa (BS = 95%, DI = +5), but *S. teitensis* nested within the main '*ionantha*'-complex. The rDNA based '*ionantha*'-complex was not completely reflected in the *Gcyc* topology (compare Figs. 17.9 and 17.10): *S.* cf. *ionantha, S.* Sigi Falls and *S. diplotricha* formed a group outside the complex (DI = +1); *S. magungensis* was part of a polytomy involving the latter group and the Nguru Mt taxa (*S. brevipilosa, S. nitida*) and *S. rupicola* (DI = +1); and *S. pendula* and *S. intermedia* were included in the '*ionantha*'-complex.

Parsimony analysis on the *Gcyc* amino acid data set gave 3890 most parsimonious trees of 41 steps length (CI = 0.829; RI = 0.708; RC = 0.587). As expected, the resolution of the strict consensus tree was very low (data not shown) and only *S. goetzeana* separated from the rest of the taxa which were arranged in a polytomy.

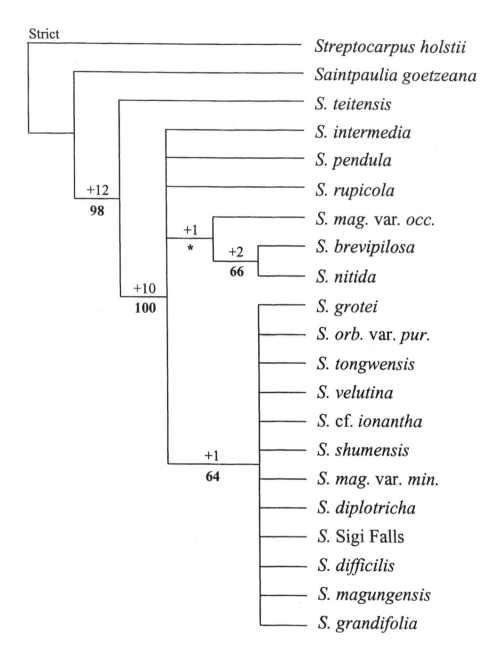

Figure 17.9 Strict consensus tree of 4 most parsimonious trees for 20 *Saintpaulia* taxa of 123 steps length based on ITS 1 and ITS 2 sequence data sets (CI = 0.919; RI = 0.851; RC = 0.782; inclusive uninformative sites). Numbers above branches indicate decay indices. Numbers below branches (**bold** print) indicate bootstrap values of 1000 replicates. * indicates branch that collapses in bootstrap analysis.

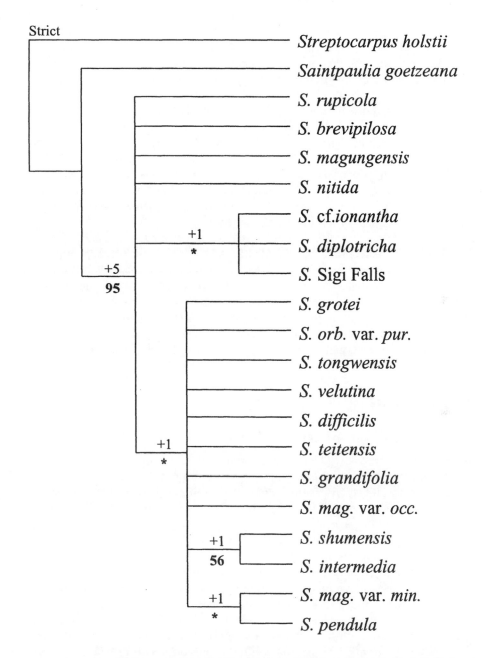

Figure 17.10 Strict consensus tree of 2210 most parsimonious trees for 20 *Saintpaulia* taxa of 69 steps length based on *Gcyc* nucleotide data (CI = 0.829; RI = 0.733; RC = 0.608; inclusive uninformative sites). Numbers above branches indicate decay indices. Numbers below branches (**bold** print) indicate bootstrap values of 200 replicates. * indicates branches that collapse in bootstrap analysis.

17.3.2.3 Comparative pairwise sequence divergence

Within the *Saintpaulia* ingroup accessions, sequence divergence of ITS ranged from 0–15.8% (*S. goetzeana* – *S. nitida*), and from 10.0–12.5% between ingroups and the outgroup. No sequence variation was observed (excluding gap positions) between nine *Saintpaulia* accessions: *S. difficilis*, *S. diplotricha*, *S. grandifolia*, *S. grotei*, *S. magungensis*, *S. magungensis* var. *minima*, *S.* Sigi Falls, *S. tongwensis*, and *S. velutina*.

Sequence divergence of the *Gcyc* nucleotide matrix within the ingroup was 0–5.1% (*S. goetzeana* – *S. diplotricha*), similar to ITS. However, sequence divergence between in- and outgroup was 7–8.9%, lower than the ITS matrix. *S. grotei* and *S. magungensis* var. *occidentalis* had identical sequences, and those and *S. difficilis*, *S. grotei*, *S. magungensis* var. *minima*, *S. orbicularis* var. *purpurea*, *S. magungensis* var. *occidentalis* and *S. tongwensis*, differed by two sites at most in pairwise sequence divergence comparisons. Maximum sequence divergence was observed between *S. goetzeana* and *S. diplotricha* (Table 17.4).

As in the 'genus' data set, translation of *Gcyc* nucleotides into amino acids resulted in higher divergence percentage values (Table 17.4). Amongst the ingroup taxa sequence divergence was between 0 (*S. grotei* – *S. magungensis* var. *occidentalis*) and 8.1% (*S. goetzeana* – *S. pendula*), and 11.3–15.3% between in- and outgroup.

A graph of the sequence divergence values of pairwise comparisons for rDNA and *Gcyc* indicates a bipartite pattern for the 'species' data set ($r = 0.627$; $P > 0.001$) (Fig. 17.11). At low divergence *Gcyc* was more variable than ITS, up to values of around 2% sequence divergence. Above this divergence ITS showed a greater sequence divergence.

17.4 DISCUSSION

17.4.1 Comparative rates of gene evolution

In common with other chloroplast regions, the *trnL+F* intron/spacer is relatively slowly evolving (although the spacer appears to be noticeably more variable than the intron, probably due to functional constraints on the latter) (Fig. 17.2). The two nuclear genes (*Gcyc* and ITS) both evolve more rapidly than the cpDNA, even though *Gcyc* is a coding region: this is concordant with the general phenomenon that chloroplast genome evolution is conservative. What is more interesting, however, is the rate comparison between *Gcyc* and ITS. At high levels of sequence divergence (inter-genus comparisons) ITS appears to be more divergent (Fig. 17.11). However, at low level of sequence divergence (inter-species comparisons) *Gcyc* appears to be evolving more quickly. How can this be? Our explanation is that this is an artefact of the direct sequencing approach of PCR products employed. Because PCR-based sequencing generated a consensus sequence in a repeat family like ITS (Hershkovitz *et al.*, 1999 – this volume), mutations will not be seen unless they are near complete fixation. At lower levels they will either be scored as polymorphic or not recognised above background noise. The intercept of the line in Fig. 17.11 is therefore a result of the time taken for fixation, both in the genome and in the population. *Gcyc* is a single copy gene and mutations can be fixed more rapidly, so the intercept is at or near zero. At low levels of divergence ITS

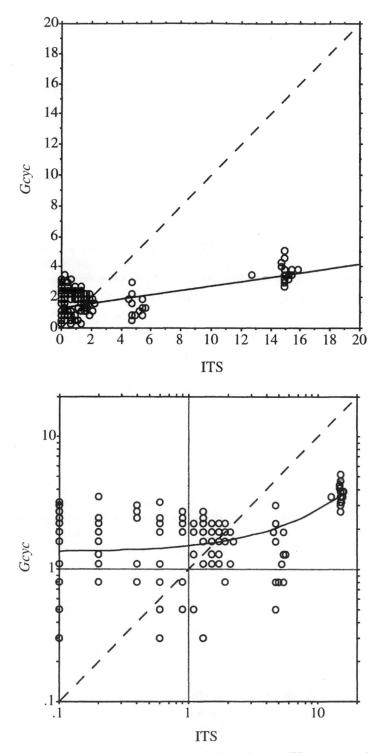

Figure 17.11 Relation between *Gcyc* nucleotide and ITS sequence divergences (%) in pairwise comparisons of the 'species' data set (gaps excluded in pairwise comparisons) (r = 627). Top: linear scale; bottom: log scale.

mutation will not be seen, whereas *Gcyc* mutations will be detected. At higher divergence levels however, ITS can continue to accumulate changes rapidly due to the low functional constraint on ITS sequences (as reflected by the slope of the graph). Although post-transcriptional processes require conservation in ITS at short putative recognition sites (Liu and Schardl, 1994), there is otherwise little functional constraint as preservation of secondary structure does not necessarily require sequence conservation (Mai and Coleman, 1997). On the other hand, *Gcyc* is more functionally constrained. This would be expected to place considerable limits on divergence rates.

17.4.1.1 Rate constancy and molecular clocks

DNA molecules with no functional constraints are believed to evolve linearly over time; they accumulate mutational changes (substitutions, insertion/deletion events) independently from selective pressures, making them useful as 'molecular clocks' (Clegg *et al.*, 1994; Sang *et al.*, 1995). Chloroplast spacers are thought to evolve in a clock-like way (Clegg *et al.*, 1994) as they are not subject to molecular drive and are not considered to be under any strong functional constraint. However, when comparing divergence rates between cpDNA and *Gcyc* it appears that there exists a strong linear relationship, and the divergence rate seems to be constant. The same is true for ITS and *Gcyc* divergence (Fig. 17.8). This indicates that all three sequences evolve in a clock-like manner, and could be used for estimating divergence times if the date of the origin of Gesneriaceae were known in order to provide a calibration point. Unfortunately the relationship between the Gesneriaceae and its closest allied families, e.g. Scrophulariaceae, Bignoniaceae, Lentibulariaceae, or Acanthaceae is not very well established (Chase *et al.*, 1993).

17.4.2 Utility of cyc and ITS for phylogenetic analysis

One consequence of molecular drive is that at very low levels of divergence ITS is not suitable for resolving patterns of hierarchical relationships (Möller and Cronk, 1997b). At this level the rate of synonymous substitutions allows single-copy nuclear genes, such as *Gcyc*, to be phylogenetically more useful (Doyle and Doyle, 1999 – this volume). At higher levels, however, it is a useful conservative source of phylogenetic information, in the same way as *trnL+F*. Its variation pattern brackets that of ITS, being both slower and faster. The similarity of phylogenetic trees derived from *Gcyc*, ITS and *trnL+F* (Fig. 17.4) indicates that all these gene regions have phylogenetic utility. However, any allelic variation in scn genes may cause problems for phylogenetic reconstruction.

17.4.3 Gcyc and the evolution of floral symmetry patterns

Cycloidea is a functional gene involved in establishing floral zygomorphy (Luo *et al.*, 1996), and it is interesting to consider how its evolution varies over morphological transitions to actinomorphy in the Gesneriaceae. If actinomorphy is caused by a loss-of-function mutation at the *Gcyc* locus, we would expect that, relieved from selection, the *Gcyc* gene would evolve rapidly and quickly become frameshifted or would contain stop codons. We checked the sequences of the actinomorphic taxa, *Conandron ramondioides*, *Ramonda myconi* and *Ramonda nathaliae*, for frame shifts or stop codons and found none. Furthermore, indels in actinomorphic species are all in multiples of three nucleotides, thus preserving the reading frame. This is a strong indication that the gene in those taxa is still intact and potentially functional. Additionally, there is a reversal to zygomorphy in *Jankaea heldreichii* from an actinomorphic *Ramonda*, further indicating a conservation of functional *Gcyc*.

To investigate the possibility of a differential substitution rate of *Gcyc* for zygomorphic and actinomorphic taxa, two approaches were investigated: (1) the branch lengths of the *Gcyc* tree (Fig. 17.4c) were plotted, for actinomorphic and zygomorphic taxa separately, against the branch lengths of the cpDNA (Fig. 17.4a) and ITS (Fig. 17.4b) trees, respectively (Fig. 17.12). For this analysis it was necessary to alter the tree topology of the *Gcyc* tree to match the tree based on combined data of all three genes (Fig. 17.5); and (2) for a *Gcyc* internal analysis of different rates of evolution between actinomorphic and zygomorphic taxa, separate trees were produced using either the 1st and 2nd or the 3rd codon position only. Trees that matched the topology of the combined data tree were chosen, and the branch lengths of trees derived from 1st and 2nd codons were plotted against those from the 3rd codon position.

In the first case, *Gcyc* vs. ITS or *Gcyc* vs. cpDNA, the branch lengths for actinomorphic taxa were not significantly different from zygomorphic taxa, indicating no differences in the rate of evolution (Fig. 17.12). In the second case, plotting branch lengths for the different codon positions did not indicate a different rate of evolution for actinomorphic taxa either (Fig. 17.13). If the gene is not functional, those taxa would have similar rates at all codon positions, and thus would follow a 45° diagonal line, whereas taxa with coding *Gcyc* genes would have significantly higher rates at the 3rd codon position, as indicated in Fig. 17.13. There is no notable shift in substitution rates over changes to actinomorphy, and no obvious difference in the substitution rate between codon positions (Fig. 17.13) – strongly implying further that the gene is functional and that it is still under selectional constraint.

Naturally occurring reversion to actinomorphy is therefore very different from actinomorphy caused by loss-of-function 'peloric' mutants such as those in *Antirrhinum majus* (Luo *et al.*, 1996). We suggest therefore that *Gcyc* is still expressed, probably at an early stage of development. Early expression of *Gcyc* may be involved in controlling the number of primordia and establishing them in register with the axis of symmetry of the flower (Luo *et al.*, 1996). We suggest that expression has been reduced in the later stages of floral development, which is the stage at which differential growth of the primordia establishes asymmetry (zygomorphy). Thus evolution of actinomorphy from zygomorphy may have come about gradually by selection on modifying genes which progressively alter the expression of *Gcyc*. This idea could be tested by examining the expression pattern of *Gcyc* in secondarily actinomorphic flowers using RNA *in situ* hybridisation. It could also be tested by genetic analysis of the putative modifiers by means of crosses. In the zygomorphic–actinomorphic hybrid *Jankaea heldreichii* × *Ramonda myconi* (× *Jankaemonda vandedemii*) the flowers are always actinomorphic, implying that this trait is dominant. However, the segregation in the F2 has never

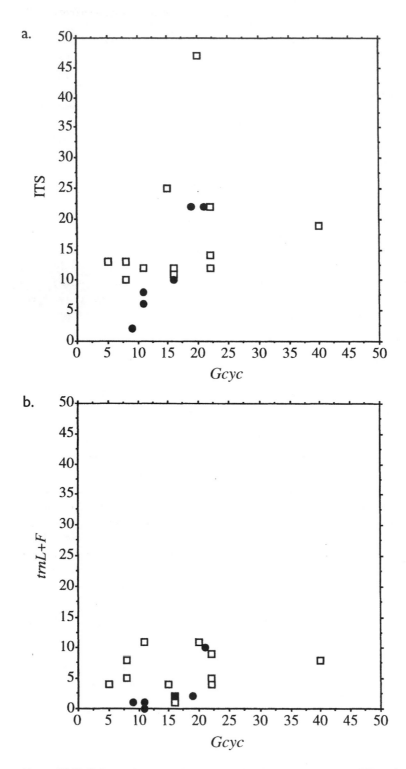

Figure 17.12 Relation between branch lengths for actinomorphic (●) and zygomorphic (□) taxa of the 'genus' data set for trees of *Gcyc* nucleotides and a) ITS or b) *trnL+F*.

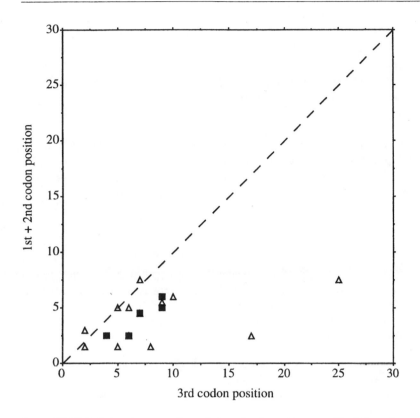

Figure 17.13 Relation between branch lengths for actinomorphic (■) and zygomorphic (△) taxa of trees based on the 1st and 2nd or the 3rd codon position of the *Gcyc* nucleotide 'genus' data set.

been studied because the plants are difficult to grow and the F1 shows a high degree of sterility. In induced peloric mutants of *Antirrhinum* actinomorphy is recessive. Alternative explanations of the intact ORF in the actinomorphic species are: (1) *Gcyc* expression has been reduced in later stages of flower development by mutations in its own regulatory region that still preserve the early expression; (2) the *Gcyc* gene studied is not the orthologue of the *Antirrhinum cyc* gene and the real orthologue is mutated in actinomorphic species; and (3) in Gesneriaceae, the floral asymmetry is generated independently of the *cyc*-like genes. We believe these are less likely hypotheses than the modifier hypothesis presented above. Nevertheless they could be tested by performing *in situ* RNA hybridisations using *Ramonda*. This study reveals the caution which has to be exercised in extrapolating from loss-of-function mutations induced in developmental genes to real evolutionary processes.

ACKNOWLEDGEMENTS

The authors thank Prof. E. Coen, John Innes Centre, Norwich, for kindly supplying materials relating to the *cyc* gene and for much help and advice. We also thank

Prof. D. Schweizer and Dr. A. Bachmeir, Dept. for Cytology and Genetics, Institute of Botany and Botanical Garden, University of Vienna, Austria, for suggestions on data analysis; H.-E. Wichert, Herrenhäuser Gärten, Hannover, Germany, and M.-J. Evans, American Gloxinian and Gesneriad Society, Randolph, NJ, USA, for kind donation of plant material of *Saintpaulia*, and Prof. D. Vokou for sending flower material of *Jankaea*; H. Sluiman, J. Preston and C. Guihal for their technical support; U. Gregory, D. Mitchell, and J. Main for the maintenance and expansion of the Gesneriaceae collection; the Regius Keeper and staff at the Royal Botanic Garden Edinburgh for the use of their excellent research facilities. We thank the Institute of Cell and Molecular Biology, University of Edinburgh, for access to sequencing facilities and N. Preston for assistance. The receipt of a Leverhulme Trust Award, No. F/771/B, is gratefully acknowledged.

REFERENCES

Bakker, F. T., Culham, A. and Gibby, M. (1999) Phylogenetics and diversification in *Perlargonium*, in *Molecular systematics and plant evolution*, (eds P. M. Hollingsworth, R. M. Bateman and R. J. Gornall), Taylor & Francis, London, pp. 353–374.

Baldwin, B. G. (1992) Phylogenetic utility of the internal transcribed spacers of nuclear ribosomal DNA in plants, an example from the Compositae. *Molecular Phylogenetics and Evolution*, **1**, 3–16.

Bremer, K. (1988) The limits of amino acid sequence data in angiosperm phylogenetic reconstruction. *Evolution*, **42**, 795–803.

Burtt, B. L. and Wiehler, H. (1996) Classification of the family Gesneriaceae. *Gesneriana*, **1(1)**, 1–4.

Chase, M. W., Soltis, D. E., Olmstead, R. G., Morgan, D., Les, D. H., Mishler, B. D. *et al.* (1993) Phylogenetics of seed plants: an analysis of nucleotide sequences from the plastid gene *rbc*L. *Annals of the Missouri Botanical Garden*, **80**, 528–580.

Clegg, M. T., Gaut, B. S., Learn, G. H. and Morton, B. R. (1994) Rates and patterns of chloroplast DNA evolution. *Proceedings of the National Academy of Sciences of the USA*, **91**, 6795–6801.

Coen, E. S. and Nugent, J. M. (1994) Evolution of flowers and inflorescences. *Development* (Suppl.), 107–116.

Cubas, P., Lauter, N., Doebley, J. and Coen, E. (1999) The TCP domain: a motif found in proteins regulating plant growth and development. *The Plant Journal*, **18**, 215–222.

Doebley, J., Stec, A. and Hubbard, L. (1997) The evolution of apical dominance in maize. *Nature*, **386**, 485–488.

Donoghue, M. J., Olmstead, R. G., Smith, J. F. and Palmer, J. D. (1992) Phylogenetic relationships of Dipsacales based on *rbc*L sequences. *Annals of the Missouri Botanic Garden*, **79**, 333–345.

Dover, G. A. (1986) Molecular drive in multigene families: how biological novelties arise, spread and are assimilated. *Trends in Genetics*, **2**, 159–165.

Downie, S. R. and Katz-Downie, D. S. (1996) A molecular phylogeny of Apiaceae subfamily Apioideae: evidence from nuclear ribosomal DNA internal spacer sequences. *American Journal of Botany*, **83**, 234–251.

Doyle, J. J. and Doyle, J. L. (1987) A rapid DNA isolation procedure for small quantities of fresh leaf tissue. *Phytochemistry Bulletin*, **19**, 11–15.

Doyle, J. J. and Doyle, J. L. (1999) Nuclear protein-coding genes in phylogeny reconstruction and homology assessment: some examples from Leguminosae, in *Molecular systematics and plant evolution*, (eds P. M. Hollingsworth, R. M. Bateman and R. J. Gornall), Taylor & Francis, London, pp. 229–254.

Elder, J. F. and Turner, B. J. (1995) Concerted evolution of repetitive DNA sequences in eukaryotes. *The Quarterly Review of Biology*, **70**, 297–320.

Ennos, R. A., Sinclair, W. T., Hu, X-S. and Langdon. A. (1999) Using organelle markers to elucidate the history, ecology and evolution of plant populations, in *Molecular systematics and plant evolution*, (eds P. M. Hollingsworth, R. M. Bateman and R. J. Gornall), Taylor & Francis, London, pp. 1–19.

Farris, J. S. (1989) The retention index and homoplasy excess. *Systematic Zoology*, **38**, 406–407.

Felsenstein, J. (1985) Confidence limits on phylogenies: an approach using the bootstrap. *Evolution*, **39**, 783–791.

Hagemann, R. and Schröder, M.-B. (1989) The cytological basis of chloroplast inheritance in angiosperms. *Protoplasma*, **152**, 57–64.

Hershkovitz, M. A., Zimmer, E. A. and Hahn, W. J. (1999) Ribosomal DNA sequences and angiosperm systematics, in *Molecular systematics and plant evolution*, (eds P. M. Hollingsworth, R. M. Bateman and R. J. Gornall), Taylor & Francis, London, pp. 268–326.

Hillis, D. M., Moritz, C., Porter, C. A. and Baker, R. J. (1991) Evidence for biased gene conversion in concerted evolution of ribosomal DNA. *Science*, **251**, 308–310.

Kluge, A. G. and Farris, J. S. (1969) Quantitative phyletics and the evolution of anurans. *Systematic Zoology*, **18**, 1–32.

Kumar, S., Tamura, K. and Nei, M. (1993) *MEGA: molecular evolutionary genetics analysis, version 1.01*. The Pennsylvania State University, University Park, PA.

Liu, J.-S. and C. L. Schardl (1994) A conserved sequence in internal transcribed spacer 1 of plant nuclear rRNA genes. *Plant Molecular Biology*, **26**, 775–778.

Luo, D., Carpenter, R., Vincent, C., Copsey, L. and Coen, E. (1996) Origin of floral asymmetry in *Antirrhinum*. *Nature*, **383**, 794–799.

Maddison, W. P. and Maddison, D. R. (1992) *MacClade, version 3.01*. Sinauer, Sunderland, MA.

Mai, J. C. and Coleman, W. (1997) The internal transcribed spacer 2 exhibits a common secondary structure in green algae and flowering plants. *Journal of Molecular Evolution*, **44**, 258–271.

Mogensen, H. L. (1996) The hows and whys of cytoplasmic inheritance in seed plants. *American Journal of Botany*, **83**, 383–404.

Möller, M. and Cronk, Q. C. B. (1997a) Origin and relationships of *Saintpaulia* H.Wendl. (Gesneriaceae) based on ribosomal DNA internal transcribed spacer (ITS) sequences. *American Journal of Botany*, **84**, 956–965.

Möller, M. and Cronk, Q. C. B. (1997b) Phylogeny and disjunct distribution: evolution of *Saintpaulia* (Gesneriaceae). *Proceedings of the Royal Society of London, Series B.* **264**, 1827–1836.

Reboud, X. and Zeyl, C. (1994) Organelle inheritance in plants. *Heredity*, **72**, 132–140.

Sang, T., Crawford, D. J., Stuessy, T. F. and O, M. S. (1995) ITS sequences and the phylogeny of the genus *Robinsonia* (Asteraceae). *Systematic Botany*, **20**, 55–64.

Schuler, G. (1995) *MACAW: Multiple alignment construction & analysis workbench*. National Center for Biotechnology Information (NCBI), Bethesda, Maryland.

Smith, J. F. (1996) Tribal relationships within Gesneriaceae: a cladistic analysis of morphological data. *Systematic Botany*, **21**, 497–513.

Smith, J. F., Wolfram, J. C., Brown, K. D., Carroll, C. L. and Denton, S. (1997) Tribal relationships within Gesneriaceae: evidence from DNA sequences of the chloroplast gene *ndh*F. *Annals of the Missouri Botanical Garden*, **80**, 50–66.

Soltis, D. E. and Kuzoff, R. K. (1995) Discordance between nuclear and chloroplast phylogenies in the *Heuchera* group (Saxifragaceae). *Evolution*, **49**, 727–742.

Susanna, A., Jacas, N. G., Soltis, D. E. and Soltis, P. S. (1995) Phylogenetic relationships in tribe Cardueae (Asteraceae) based on ITS sequences. *American Journal of Botany*, **82**, 1056–1068.

Swofford, D. L. (1993) *PAUP: phylogenetic analysis using parsimony, version 3.1.1.* Illinois Natural History Survey, Champaign, IL.

Taberlet, P., Gielly, L., Pautou, G. and Bouvet, J. (1991) Universal primers for amplification of three non-coding regions of chloroplast DNA. *Plant Molecular Biology*, **17**, 1103–1109.

Tilney-Bassett, R. A. E. (1994) Nuclear controls of chloroplast inheritance in higher plants. *Journal of Heredity*, **85**, 347–354.

White, T. J., Bruns, T., Lee, S. and Taylor, J. (1990) Amplification and direct sequencing of fungal ribosomal RNA genes for phylogenetics, in *PCR protocols*, (eds M. A. Innis, D. H. Gelfand, J. J. Sninsky and T. J. White), Academic Press, London, pp. 315–322.

Wojciechowski, M. F., Sanderson, M. J., Baldwin, B. G. and Donoghue, M. J. (1993) Monophyly of aneuploid *Astragalus* (Fabaceae): evidence from nuclear ribosomal DNA internal transcribed spacer sequences. *American Journal of Botany*, **80**, 711–722.

Chapter 18

Inferior ovaries and angiosperm diversification

M. H. G. Gustafsson and V. A. Albert

ABSTRACT

The nature of the inferior ovary has been a controversial issue in angiosperm morphology. Recent phylogenetic and molecular developmental findings allow an improved understanding of the evolution of inferior ovaries, the phenomenon known as epigyny. Character optimisation experiments using phylogenetic models based on DNA sequence data show that epigyny has evolved from hypogyny many times independently. Perigyny and hemi-epigyny are not necessary intermediate stages in this evolutionary process. Reversals are relatively infrequent and in many large groups the condition is fixed, perhaps as a result of functional constraints. Hypotheses about the evolutionary significance of epigyny include increased ovary protection and potential for fruit diversification. The inferior ovary is formed ontogenetically through intercalary growth in regions below the floral organ primordia. The outer layers of inferior ovaries often show some degree of structural similarity to floral appendages and/or floral axes. In many cases, however, features unique to the inferior ovary wall predominate, giving the ovary its own morphological character. Studies of gene expression patterns in inferior ovaries of *Gerbera* and *Malus* indicate that MADS box genes play a role in their development, but that the nature of this role is not entirely congruent with the ABC model for appendicular organ determination in flowers.

18.1 Introduction

Epigynous flowers have an inferior ovary, which means that the perianth and stamens are inserted above the ovary, rather than below. In such flowers, it appears that the basal parts of the perianth are fused with the ovary or that the ovary has sunk into the floral base. Epigyny occurs in numerous families in various parts of the angiosperm system, in a few basal and ancient groups (e.g. Chloranthaceae) as well as several derived, more recent ones (e.g. Asteraceae).

Epigyny has prompted different morphological interpretations, resulting in contrasting conclusions about its evolutionary history. For example, the wall of the inferior ovary has been considered to be either receptacular (part of the floral axis), appendicular (resulting from fusion of floral organs with the ovary), or a combination of both precursor tissues. Whereas some taxa seem, based largely on anatomical evidence, to be clearly receptacular or appendicular, there are many controversial cases (Douglas, 1944, 1957; Kaplan, 1967).

In *Molecular systematics and plant evolution* (1999) (eds P. M. Hollingsworth, R. M. Bateman and R. J. Gornall), Taylor & Francis, London, pp. 403–431.

Recently, the rapidly expanding field of molecular developmental genetics has provided a fresh perspective on the homology of floral organs. Morphology and ontogeny aside, organs and tissues can be said to be homologous in the genetic sense (process orthologous in the terminology of Albert *et al.* (1998)); that is, they express to some extent the same genes during development. It is possible that genes normally responsible for floral organ determination are also expressed in the developing wall of the inferior ovary, which would explain the sepal- and/or petal-like features of some inferior ovaries, such as the petaloid ovary of *Begonia*.

Looking at the distribution of epigyny, it becomes clear that it is present in several large, diverse and indisputably monophyletic families, including Asteraceae, Orchidaceae, Rubiaceae and Araliaceae-Apiaceae. There are very few cases of hypogyny in these families, and from these observations arises the question of whether epigyny in general can be traced to the origin of species-rich clades. It is also interesting to investigate which other characters may have evolved in concert with epigyny. So far, speculations about the functional implications of epigyny have mostly concerned the improved (reinforced) protection that the condition provides for ovules, but other aspects meriting evaluation include its possible implications for fruit diversification and dispersal mode switches.

The present study considers the evolution of epigyny in the light of recent phylogenetic findings resulting from DNA sequence analysis. Widely sampled analyses, such as that of Chase *et al.* (1993), provide for the first time an encompassing framework for the analysis of character evolution among angiosperms. Complementary DNA-based information on the determination of organ identity holds great promise for understanding angiosperm ontogeny and morphology. This paper attempts to synthesise the many anatomical and ontogenetic investigations on epigyny from these perspectives.

18.2 The evolution of epigyny – evidence from phylogenetic trees

18.2.1 An optimisation framework

In order to investigate the evolutionary history of epigyny, variously defined characters and character states describing ovary position were mapped onto the most taxonomically detailed cladogram of angiosperms yet published (i.e. that of Chase *et al.*, 1993; reanalysed by Rice *et al.*, 1997). These results are presented here descriptively, and are used throughout the text where examples of the evolution of epigyny are described. For uniformity, optimisations on the trees resulting from the 'Search II' *rbc*L matrix of Chase *et al.* (1993) were performed (cf. Albert *et al.*, 1998). The tree of Rice *et al.* (1997) was based on the same matrix as Search II of Chase *et al.* (1993), but was the subject of a more computer-intensive search for shortest trees; although some topological differences were found, none of these were strongly supported by parsimony jackknife analysis (J. S. Farris, M. Källersjö and V. A. Albert, unpublished). The trees used here from each analysis are single most-parsimonious trees, retained from the original Chase *et al.* analysis (V. A. Albert, unpublished) or from the Rice *et al.* analysis, downloaded from the 'Treezilla' web site (http://www.herbaria.harvard.edu/~rice/treezilla/). The number of terminals included is 499, representing all major extant groups of seed plants. Optimisations were performed using the computer program MacClade 3.0 (Maddison and Maddison, 1992).

Information about floral morphology was extracted from Cronquist (1981), Dahlgren *et al.* (1985), Heywood (1985) and Takhtajan (1997). Codings were based on family generalisations; each placeholding terminal was given all states occurring in its family. The family classifications of Cronquist (1981), Takhtajan (1997) and, for monocots, Dahlgren *et al.* (1985) were used in separate optimization experiments. Systematic error is expected with these standardisations, especially in the Cronquist-based codings, which use more inclusive terminals. In a first set of optimisations, 'complete' epigyny was treated as a separate state, and the varying degrees of hemi-epigyny were also treated as a single character state. Hypogyny and perigyny were treated as separate states. The character states are illustrated in Fig. 18.1. In the second optimisation experiment, all degrees of epigyny were treated as a single character state. This was also the case in the third optimisation, where perigyny was also treated as homologous to (i.e. merged with) hypogyny.

In addition to the analysis of the entire angiosperm system, a large epigynous clade, the Asterales (*sensu* Gustafsson, 1996), was investigated in greater detail in order to study patterns of reversal from epigyny. The phylogenetic tree of Bremer and Gustafsson (1997) was used, and character information was compiled from the aforementioned general textbooks and from Engler and Prantl (1887–1915). The family classification of Gustafsson (1996) was followed.

One family of Asterales, the Goodeniaceae, is variable in terms of ovary position and shows various interesting intermediate stages. This family was chosen for yet another set of optimisations. In this analysis the states of individual species were used, based on Krause (1912) and the unpublished data and tree of M. H. G. Gustafsson.

18.2.2 Distribution of epigyny among angiosperms

121 of the 385 families in Cronquist's (1981) system include epi- or hemi-epigynous taxa, but only 54 families are uniformly epigynous. In numerous cases there is variation within families; for example, epigyny, hemi-epigyny, hypogyny and perigyny are all represented in the Rosaceae. When the system of Takhtajan

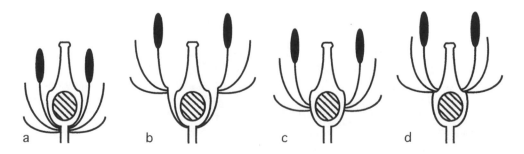

Figure 18.1   Semidiagrammatic representation of the relationship between ovary cavity position (hatched) and perianth insertion in a) hypogynous, b) perigynous, c) hemi-epigynous and d) epigynous flowers.

(1997) is used, the number of polymorphic terminals decreases substantially as compared to Cronquist's system, partly as a result of the breakup of polymorphic and non-monophyletic terminals such as the Grossulariaceae and Cornaceae *sensu* Cronquist. In the following, reference will be made to the set of optimisations made using Takhtajan's (1997) system.

The optimisations show that hypogyny is clearly the plesiomorphic condition in angiosperms. Considering unambiguous changes only, epigyny *s.l.* (i.e. including hemi-epigyny) has evolved at least 64 times from this condition and occurs in most major clades of angiosperms, though it is rare in the Magnoliales–Laurales, and absent from the Ranunculales and large parts of the Rosidae. However, a few major clades optimise as basally epigynous: Zingiberales, Orchidales–Asparagales, Asterales–Dipsacales–Araliales, Cornales, Myrtales, and Begoniales–Fagales. When the tree of Rice *et al.* (1997) was used, a clade including the Hamamelidaceae and several saxifragalean families also optimises as plesiomorphically epigynous.

Epigyny is apparently derived from hypogyny much more often than the converse. With two states (hypogyny including perigyny, and epigyny including hemi-epigyny) there are 64 unambiguous changes from hypogyny to epigyny but only 24 in the opposite direction.

No major clades optimise as perigynous, but perigyny occurs in the Calycanthaceae-Idiospermaceae clade and consistently in a few small families (Bretschneideriaceae, Cephalotaceae, Chrysobalanaceae, Crossosomataceae, Heteropyxidaceae, Lythraceae, Moringaceae, Tropaeolaceae). In addition, perigyny occurs as one of two or more states within many families. When perigyny was considered a separate state together with epigyny and hypogyny, there were 30 unambiguous changes from hypogyny to perigyny and four changes from epigyny to perigyny, but not a single change from perigyny to any other character state.

When, in addition to perigyny, hemi-epigyny was considered a separate state, the optimisation remained the same on major branches. Only a few small families are consistently hemi-epigynous (Altingiaceae, Barclayaceae, Iteaceae, Tecophilaeaceae, Trapaceae) and no clades above the family level show this state unambiguously. Hemi-epigyny does occur as one of two or more states within many families, however. There are 12 unambiguous changes from hypogyny to hemi-epigyny, and eight from epigyny to hemi-epigyny, but none from perigyny to hemi-epigyny. There were no unambiguous reversals from hemi-epigyny to any other state.

18.2.3 Epigyny in the Asterales

Focusing on one of the larger, mostly epigynous clades, the Asterales are included in a larger clade with the Apiales, Dipsacales and families such as Escalloniaceae and Bruniaceae. Epigyny is plesiomorphic in the Asterales, but several genera are secondarily hemi-epigynous or hypogynous (Fig. 18.2). Hypogyny is found in the Carpodetaceae (*Abrophyllum, Cuttsia*), Menyanthaceae (*Menyanthes, Nymphoides*), Phellinaceae (*Phelline*), Campanulaceae (e.g. *Cyananthus* [actually perigynous] (Schönland, 1889) and *Unigenes* (Wimmer, 1943)), and Goodeniaceae (*Brunonia*). Hemi-epigyny is found in the Menyanthaceae (*Villarsia* and *Nephrophyllidium*), Argophyllaceae (*Argophyllum*), and several members of Goodeniaceae and Campanulaceae.

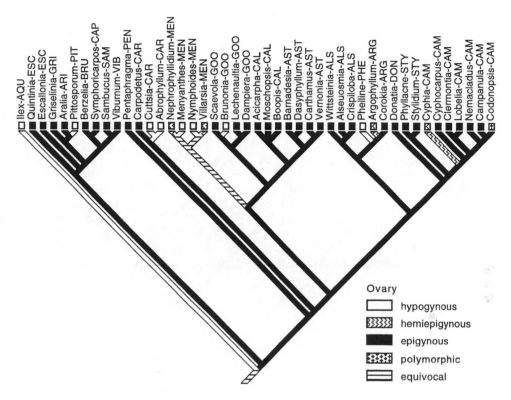

Figure 18.2   Optimisation analysis of ovary position in the order Asterales. Character states were optimised onto one of the three trees resulting from the analysis of Bremer and Gustafsson (1997). The character states were assigned for each individual genus.

18.2.4 Epigyny in the Goodeniaceae

The Goodeniaceae, a family embedded within the Asterales, optimise as plesiomorphically epigynous. *Brunonia* is the only completely hypogynous genus, but species of *Coopernookia*, *Goodenia* and *Velleia* are hemi-epigynous. The most parsimonious interpretation is four independent reversals to hypogyny/hemi-epigyny in *Brunonia*, *Coopernookia strophiolata*, *Goodenia scapigera*, and the clade comprising *Goodenia macroplectra*, *G. mimuloides* and *Velleia* (Fig. 18.3). Whereas only a fraction of all Goodeniaceae species have been sampled (23 out of c. 400), this pattern is unlikely to become less complex as more species are added. It is therefore justifiable to say that the Goodeniaceae (or at least the *Goodenia* group *sensu* Gustafsson *et al.,* 1996) are unusually labile in terms of ovary position.

18.2.5 Fixation and instability of epigyny

There are relatively few reversals from epigyny in angiosperms, as discussed above. Moreover, the reversals that do occur are unevenly distributed among the epigynous groups. It appears that some groups were fixed for epigyny early in their evolution, whereas others remained labile.

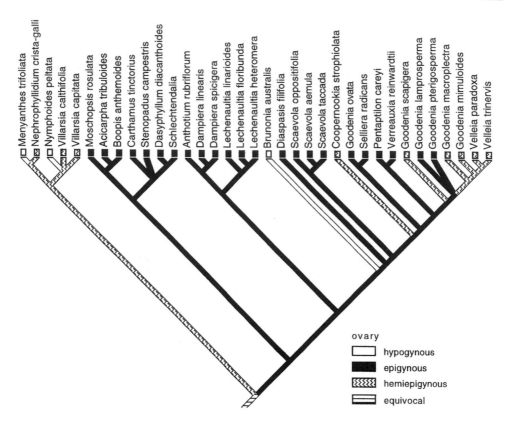

Figure 18.3 Optimisation analysis of ovary position in the Goodeniaceae, with outgroups from the Asteraceae (*Carthamus* through *Schlechtendalia*), Calyceraceae (*Moschopsis* through *Boopis*) and Menyanthaceae (*Menyanthes* through *Villarsia*). All genera of Goodeniaceae are represented. According to optimisation experiments of the entire angiosperm system (described in text) and the order Asterales (described in the text and shown in Fig. 18.2), the plesiomorphic condition in the group is epigyny. Character states were optimised onto the strict consensus tree of Gustafsson (manuscript). This tree resulted from a cladistic analysis of *rbc*L sequences and 98 morphological characters, applying the successive approximations weighting procedure (Farris, 1969).

The Orchidaceae, one of the largest angiosperm families, are strictly epigynous with no known reversals. That is true for the entire Orchidales–Asparagales with the exception of one clade nested within the group that comprises several small families (Anthericaceae, Asphodelaceae, Dasypogonaceae, Dracaenaceae, Hemerocallidaceae, Hyacinthaceae, Nolinaceae, Ruscaceae, Xanthorrhoeaceae: Fig. 18.4). These families appear to be completely hypogynous, lacking any 'intermediate' stages. In this case there seems to have been a single return to complete hypogyny, with no intermediates or secondary reversals.

The 'Hamamelidales–Saxifragales' clade does map out as plesiomorphically epigynous, but there is a reversal in the Crassulaceae–Paeoniaceae clade, as well as in the Tetracarpaeaceae (Fig. 18.5). There is some intrafamilial variation (e.g. within Saxifragaceae and Hamamelidaceae) and hemi-epigyny occurs in families such as

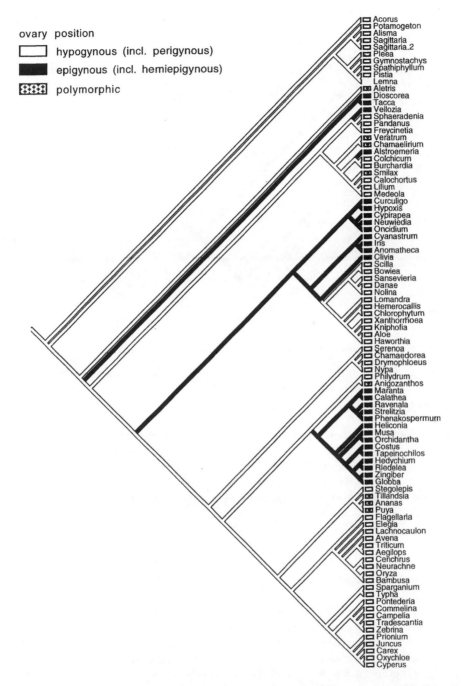

ovary position
☐ hypogynous (incl. perigynous)
■ epigynous (incl. hemiepigynous)
▦ polymorphic

Figure 18.4 Optimisation analysis of ovary position in angiosperms, exemplified by monocots. Character states were optimised (using MacClade; Maddison and Maddison, 1992) onto part of the single shortest tree of Rice *et al.* (1997) available on the Web (http://www.herbaria.harvard.edu/~rice/treezilla/). Character states are assigned based on whole familial circumscriptions, following the classification of Takhtajan (1997).

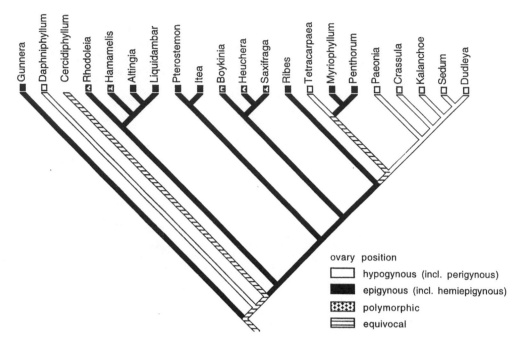

Figure 18.5   Optimisation analysis of ovary position in angiosperms, exemplified by the Saxifragales clade. Character states were optimised onto the same tree used for Fig. 18.4. Character states are assigned based on whole familial circumscriptions, following the classification of Takhtajan (1997).

the Saxifragaceae and Penthoraceae. Interestingly, many of the hypogynous members of this clade are apocarpous (Tetracarpaeaceae, Crassulaceae, Paeoniaceae). The entire clade appears to be unstable so far as gynoecial fusions are concerned.

Within the Asteridae, there is a relatively well supported clade (the Asterid I clade of Chase *et al.*, 1993) comprising the orders Scrophulariales, Lamiales, Gentianales, Solanales and Boraginales. This group of orders is plesiomorphically hypogynous, one of the very few exceptions being the large family Rubiaceae. Of the *ca.* 650 genera of Rubiaceae, only *Gaertnera* and *Pagamea* are reported to be hypogynous (Takhtajan, 1997), but these genera are likely nested well within the family and show an atypical hypogyny that is only manifest late in ontogeny (Igersheim *et al.*, 1994).

The other major clade of Asteridae (Asterid II of Chase *et al.*, 1993) is, as already mentioned, basally epigynous. Some families and orders show reversals (i.e. Goodeniaceae, Menyanthaceae), whereas others do not (i.e. Asteraceae, Dipsacales). Within the Apiaceae–Araliaceae a single fully hypogynous species is known, *Tetraplasandra gymnocarpa* (Eyde and Tseng, 1969), which is nested well within the family (Plunkett *et al.*, 1997). Interestingly, *T. gymnocarpa* is endemic to Hawaii, and this highly isolated island environment has apparently produced another secondarily hypogynous species, *Hillebrandia sandwicensis* in the Begoniaceae (Charpentier *et al.*, 1989).

Within the Asterales, most families are epigynous without exceptions (Fig. 18.2). A few, such as the Campanulaceae are variable although this is not evident with the sparse sampling in the tree from Bremer and Gustafsson (1997). The Goodeniaceae also appear unstable, with several reversals, but most of these are to hemi-epigyny (as is clear from the more detailed optimisation of this family shown in Fig. 18.3). The sister group of the Goodeniaceae is the Calyceraceae plus Asteraceae. Both families lack reversals to hypogyny. Note that the fruit type in both of these families is constant: a one-seeded, dry-walled, indehiscent fruit crowned by a modified calyx. The inflorescence in both families is a very dense capitulum or more complex form (as in many Calyceraceae), but in both cases the flowers are densely packed. The fixation of epigyny in these two families may be related to their fixed fruit and inflorescence morphology.

The fact that there are several instances of both 'unstable' epigyny (with a prevalence of reversals, as in the Goodeniaceae and the Saxifragaceae–Hamamelidaceae clade), as well as complete and immediate switches to epigyny with few or no reversals (as in the Rubiaceae and the Orchidales–Asparagales), may indicate that different morphogenetic and/or evolutionary mechanisms have been active in these phylogenetically independent instances of epigyny, resulting in different degrees of structural fixation of the gynoecium.

18.3 Covariation with other kinds of organ fusion/synorganisation

A trend towards increased organ fusion occurs in several groups of angiosperms, both within and between organ categories, connation and adnation, respectively. Synsepaly, sympetaly, stamen–corolla adnation and stamen connation are all widespread phenomena. Epigyny can be seen as yet another, extreme form of organ fusion, involving all four floral whorls. Interestingly, epigyny often co-occurs with other forms of floral fusion, most frequently with sympetaly (notably in the Asterales, Dipsacales, Rubiaceae, Cucurbitaceae and Ericaceae–Vaccinioideae). Also common are various forms of stamen connation (e.g. in many Campanulaceae and Asteraceae), and there are also a few cases of gynostemium formation (Orchidaceae, Stylidiaceae). In the Calyceraceae, there is not only a stamen–corolla tube, but this is basally fused to the style, forming a structure comprising three whorls (Erbar, 1993). Apart from anther connation, these fusions are all congenital, and are produced by intercalary growth that occurs most frequently at a very late stage in ontogeny, as in the case of stamen–corolla adnation (Erbar, 1991).

Erbar (1991) distinguished between two ontogenetically distinct types of sympetaly: early and late. In early sympetaly, the formation of a ring-shaped primordium is the first stage in corolla development, whereas in late sympetaly free petal primordia form first and later join as a result of interprimordial growth. Erbar (1991) summarised data on the Asteridae *sensu stricto*, and looking at this group it becomes clear that there is a strong correlation between epigyny and early sympetaly. All epigynous groups studied have early sympetaly, and Acanthaceae (including Thunbergiaceae), Oleaceae and some Gentianaceae are the only hypogynous groups with early sympetaly. In addition, the epigynous families Apiaceae and Araliaceae, although choripetalous, have an ontogeny in its early stages closely similar to early sympetaly (Erbar and Leins, 1988). In epigynous taxa, sympetalous or not, the entire floral apex frequently becomes concave and ring-shaped early in development. This most likely affects the morphology of the region where petals are initiated, resulting in a ring primordium. Early sympetaly could thus be viewed as an effect of epigyny. To understand better the relationship between early sympetaly and epigyny, additional study of the few hypogynous taxa with early sympetaly is highly desirable.

A trait with some features in common with epigyny is syncarpy. Angiosperms are plesiomorphically apocarpous, and syncarpy has evolved several times from this condition (as in Nymphaeaceae and derived clades of monocotyledons and eudicots). Syncarpy is now present in *ca*. 80% of angiosperms (Endress and Stumpf, 1990). The relationship between syncarpy and epigyny is problematic, as it is difficult to determine whether an inferior ovary is syncarpous in the normal sense or whether the carpels are connivent merely as a result of being surrounded by other tissues. It is clear from the present optimisations, however, that most epigynous groups at or above family level have syncarpous sister groups, and can be hypothesised to be syncarpous-derived. Exceptions are the unicarpellate families Hernandiaceae and Chloranthaceae in the Magnoliales. Nymphaeaceae, Nupharaceae and Barclayaceae form a clade with a hypogynous and apocarpous sister-group, the Cabombaceae (Chase *et al.*, 1993). Nupharaceae are hypogynous (ambiguously interpreted in our optimisation) and syncarpous. In the case of the Saxifragales (Fig. 18.5), their close relatives are syncarpous (Daphniphyllaceae, Gunneraceae) or unicarpellate. However, the groups showing reversal from epigyny (Tetracarpaeaceae, Paeoniaceae, Crassulaceae) are apocarpous. A group intermediate in both senses is Penthoraceae, which have hemi-epigynous ovaries with apically free carpels. In many Saxifragaceae, likewise, the carpels are fused only at the very base. It is clear that epigyny and syncarpy share some phylogenetic correlation, and it would be interesting to investigate whether these phenomena are also functionally, developmentally or genetically linked.

A large proportion of epigynous angiosperms have valvate corolla aestivation (including virtually all Santalales, Asterales and Apiales, many Cornales and Rubiaceae). Valvation as such has been mentioned as a condition associated with (or as a precondition facilitating) fusion and synorganisation in general – fusion and complex interactions are more easily achieved between organs that are arranged precisely edge-to-edge, as with valvate petals (Endress, 1994). It is therefore possible that valvate aestivation of floral parts was a condition favouring fusion events leading to epigyny.

18.4 Origins and evolutionary (adaptational) significance of epigyny

18.4.1 Origins and spread of epigyny

Epigyny is not a recent phenomenon. Friis *et al.* (1994) described several epigynous flowers from the early Cretaceous of Portugal. Some of these are among the oldest unambiguous angiosperm fossils known. The taxonomic affinities of many of these fossil flowers are ambiguous, but some resemble extant members of the Hernandiaceae and Chloranthaceae, which are among the few epigynous, extant magnoliids.

The two largest angiosperm families are the Asteraceae and Orchidaceae, each with well over 20,000 species; both are without exception epigynous. Additional large families completely or predominantly epigynous are the Rubiaceae, with *ca*. 11,000 species (Takhtajan, 1997), Melastomataceae (4500, a few hypogynous), Myrtaceae (4000) and Apiaceae (3200). Given that most of the largest angiosperm families are epigynous, it is tempting to hypothesise a relationship between epigyny and the radiation of these groups. To be sure, there are many species-rich hypogynous groups as well, including Fabaceae (18,000), Poaceae (11,000), and Cyperaceae (5300, species numbers again from Takhtajan, 1997), and although epigyny has existed for at least 130 million years, it still has far from outcompeted hypogyny in phylogenetic terms. The fact that epigyny is at least locally common among angiosperms does, however, prompt speculation about its possible adaptational value or status as a formidable structural constraint.

18.4.2 Ovary protection

The evolution of seed plants reveals a progression toward increased ovule protection (Albert *et al.*, 1998). The second (outer) integument in angiosperms (and possibly gnetophytes) is one such character. The angiosperm carpel is another, and one that has been considered to be one of the keys to the success of angiosperms. Syncarpy has also sometimes been mentioned in this context (e.g. Grant, 1950), but it is debatable to what extent syncarpy offers better protection of ovules *per se*. Epigyny could then be another step toward increased protection. An efficient protection of the ovules resulting from epigyny can be seen in taxa such as the Cactaceae, where the fruit is surrounded by a pre-existing protective layer bearing spiny areoles.

Grant (1950) considered epigyny to be largely a condition developed in bird-pollinated plants, as an adaptive response to this pollination mode. Epigyny results in the protection of ovules through their spatial separation from the part of the flower with which pollinators interact. According to the results of a survey by Grant (1950), there is a correlation between the occurrence of bird pollination syndrome and epigyny. A higher percentage of bird-flower genera (genera with representatives that are bird pollinated) are epigynous than, for example, long-tongued lepidoptera-pollinated genera (37.8% compared with 25.8%). Rather than an adaptation, however, epigyny could be seen as a pre-adaptation for interaction with large, destructive pollinators like birds, which might otherwise incidentally or actively damage the ovary/ovules. Whatever the causal mechanism, it seems clear that there is a correlation between epigyny and bird pollination, and that ovules in some senses are better protected in an inferior ovary. This apparent correlation should be investigated further using well-corroborated cladograms to establish whether bird pollination is indeed associated with the origin (the basal nodes) of epigynous groups.

Eyde and Tseng (1969) suggested two causes for the evolution of secondary hypogyny in the Hawaiian araliad *Tetraplasandra gymnocarpa*. In agreement with the view of epigyny as a character increasing ovule protection, relaxed seed-predator pressure in the isolated island environment of the Hawaiian islands could enhance the survival of hypogynous mutants. Eyde and Tseng also mentioned the possibility that hypogyny had evolved within *Tetraplasandra* in order to ensure outcrossing, a change of breeding system said to characterise plants of oceanic islands. This would be achieved via the instantly altered positions of floral parts in the case of hypogyny, with stamens positioned below the stigmata, rendering self-fertilisation difficult. (Change of relative lengths of style or stigma could of course have had the same effect in other plants, without involving the ovary.) It could be added that fixation of mutations in open (low-competition) environments such as those of young oceanic islands is frequently rapid, as a result of founder effects and inbreeding (Albert *et al.*, 1998) and this could account for the fixation of both morphological novelties and reversals like that of *Tetraplasandra gymnocarpa* (and, perhaps, *Hillebrandia*).

18.4.3 Morphological consequences for fruit morphology and diversity

One aspect of epigyny that is of potential evolutionary importance is calyx position. The calyx very frequently persists on fruits, and in many cases it continues to differentiate during fruit development into a dispersal aid, as in the case of the pappus of Asteraceae. The distal position of the calyx in epigynous plants is probably advantageous in cases of wind and animal (epizoochorous) dispersal, but whether this advantage is strong enough to act as an evolutionary force is uncertain. It should be noted that there are also wind-dispersed taxa with basally attached dispersal devices (e.g. *Platanus*: Platanaceae).

Epigyny apparently constrains fruit dehiscence: the commonly occurring septicidal or loculicidal dehiscence through zones of weakness in demarcation lines between fused carpels or along their midveins, respectively, cannot be easily achieved, as new layers of tissue with different properties have been added onto the ovary. In the Campanulaceae *s.l.*, hemi-epigynous genera usually dehisce septicidally in the protruding, superior part of the fruit, whereas completely epigynous genera have 'reinvented' dehiscence in mechanically novel ways, as seen in *Campanula* and *Musschia*. There is a high proportion of indehiscent fruit-types among epigynous families, which may also reflect a loss of the 'normal' lines of dehiscence. Admittedly, there are also several epigynous groups that do have septicidally or loculicidally dehiscing capsules with dehiscence also in the epigynous zones, such as in the Goodeniaceae, Stylidiaceae, and many monocots.

Another feature that may be correlated with epigyny is fruit fleshiness. (Groups where fleshiness is common include the Santalales, Cornales, Myrtales as well as Caprifoliaceae, Chloranthaceae, Grossulariaceae *s.s.*, Cucurbitaceae, Ericaceae and Rubiaceae.) Epigyny may constitute a pre-adaptation for fruit fleshiness, but it should be noted that numerous hypogynous groups also have fleshy fruits. In general, it is fair to say that the increased fruit complexity that is a consequence of epigyny allows for the evolution of new attributes. One example that illustrates this is *Scaevola* (Goodeniaceae), which has an indehiscent, usually 2-seeded, completely epigynous fruit. Carolin (1966) studied the anatomy of the fruit of this and other genera in the family. The outermost layer (epicarp) was considered to be derived from the outer floral whorls, whereas the inner layers (mesocarp and endocarp) were thought to correspond to the carpels. This interpretation was presumably based on vascular characters, as the epicarp is generally traversed by vascular bundles leading to perianth and stamens. The three tissue layers often have quite different histologies. The epicarp may be thin and dry or fleshy, the meso-carp (if distinct from endocarp) fleshy, corky (with suberised cell walls), or hard (with sclerified cell walls), while the endocarp is always hard. *Scaevola taccada* is a species with three distinct tissue types in the mature fruit: a fleshy endocarp and a corky mesocarp gradually passing into a hard endocarp. This fruit and that of a few other species are distributed by oceanic currents, the corky layer presumably giving them the necessary buoyancy for this dispersal mode (Krause, 1912). The proliferation of tissues resulting from epigyny apparently has contributed to the considerable fruit diversity in *Scaevola*.

18.5 Morphological evidence for the evolutionary history of epigyny

18.5.1 Earlier theories

In her extensive reviews of the earlier work on the nature of the inferior ovary, Douglas (1944) described the various theories regarding the derivation of epigyny, and the following is a summary of her treatment. These theories were considered by some of their proponents to apply to angiosperm ovaries in general, although other authors recognised that more than one theory may be valid, and that different models may be applicable to different plant groups.

According to what has been referred to as the appendicular theory, the interior of the inferior ovary corresponds to the carpels, and the outer layer of the ovary wall comprises the fused bases of the other floral appendages (i.e. sepals, petals and stamens). A variant of this theory considers the most central part of the ovary (i.e. the placentae) to be an extension of the floral axis. This applies to ovaries with axile placentation only, ignoring those with parietal placentae.

In contrast, the entire ovary has sometimes been viewed as receptacular (i.e. non-carpellate), with the possible exception of parts of the ovary roof. In a slight modification of this theory, the ovary is seen as receptacular except for the ovary roof and central placentae, which are considered carpellary. In another basically receptacular model, only the inner part of the wall and the ovary roof are consid-ered carpellary. In yet another model, the ovary is considered receptacular except for the central part, including the inner part of the wall, the placentae and the roof, which are all considered carpellary.

Some authors, especially those focusing on ontogenetic evidence, did not attempt to homologise the inferior ovary with other structures/organs, but viewed it as a structure *sui generis*; others simply described it as resulting from intercalary growth, evading any attempt to infer historical homology whatsoever.

18.5.2 Conditions transitional to epigyny

Many flowers are neither hypogynous nor typically epigynous but have an ovary–perianth relationship that is, in one way or another, intermediate between these two conditions. Semi-inferior ovaries are distally free from the perianth and stamens, and the free parts of sepals, petals and stamens variously depart from the same level or, more rarely, at successively higher levels (as in some Campanulaceae; Moeliono and Tuyn, 1960). The sepals of *Goodenia scapigera* (Goodeniaceae) proximally pass into ridges that are attached only along a median, adaxial bridge of tissue, leaving the lateral parts free. These conditions are, each in their own distinctive way, intermediate between hypo- and epigyny. The same could be said of perigynous flowers, in which the ovary is free but surrounded by a variously shaped hypanthial cup, bearing the floral appendages along the rim. Examples of this can be found in several families, and a particularly well developed free hypanthium is found in many Rosaceae, Chrysobalanaceae and Lythraceae. Free hypanthia (also termed floral cups or floral tubes) can occur in combination with complete or partial epigyny, as in the Rosaceae–Maloideae (Steeves *et al.*, 1991).

Half-inferior ovaries and superior ovaries surrounded by a hypanthial cup have been perceived as stages 'transitional' between hypo- and epigyny, and have been discussed in connection with the origin and evolution of epigyny (e.g. Puri, 1952). Perigyny is compatible with the hypothesised receptacular origin of epigyny, in which the floral axis is thought to have expanded gradually into a cup-like structure, and then fused with the ovary. For example, some species of *Pereskia* are perigynous and the genus probably has a basal position in the otherwise epigynous Cactaceae, which seems to support such a scenario for that family (Boke, 1966). There is of course also the possibility that the bases of the floral appendages have become adnate to each other to form a free hypanthium, which could then in turn become adnate to the ovary (Puri, 1952). Those semi-inferior ovaries with floral parts diverging at various levels (like those of many Goodeniaceae) can, on the other hand, be viewed as supporting the appendicular theory, showing that each of the floral whorls has gradually become adnate to the ovary. The distribution of these forms has sometimes been used to establish the (non-)homology of epigyny; Carolin (1977) considered the epigyny of Campanulaceae to be the result of the fusion of a free hypanthium like that of *Cyananthus* (a presumed basally divergent member of the family) to the ovary, whereas that of the closely related Goodeniaceae was seen as the result of successive fusion of the floral whorls to the ovary, ostensibly indicated by the progressive series found in *Velleia*. This should then indicate separate origins of epigyny in the two groups. However, the presence in Campanulaceae of hemi-epigynous taxa, e.g. species of *Campanumoea* and *Codonopsis*, genera palynologically similar to *Cyananthus* (Dunbar 1975a, b) allows alternative scenarios. Optimisation of characters describing ovary position onto the phylogeny of the Asterales shows epigyny to be plesiomorphic in the entire order; all occurrences of hypo- and perigyny are therefore interpreted as derived.

The very high relative incidence of transitions directly from hypogyny to epigyny on the Chase *et al.* tree suggests that hemi-epigyny may not be a required transitional stage, but rather that a switch directly from hypogyny to epigyny can occur. Remarkably, there are no unambiguous state changes at all from hemi-epigyny to either hypogyny or epigyny. It should be noted, however, that hemi-epigyny rarely occurs as the single state within terminals, but usually as a polymorphism in combination with epigyny or hypogyny.

The present results suggest also that perigyny is usually not a transitional stage to epigyny, but has evolved several times independently. There is in fact not a single unambiguous reversal from perigyny. Peri- and epigyny thus appear evolutionarily dissociated. However, the same reservation expressed regarding hemi-epigyny can also be stated for perigyny, namely that considerable within-terminal variation is not considered when only unambiguous state changes are counted, as in the present study.

18.5.3 Evidence from floral vasculature

The idea of the evolutionary conservation of the vascular system of plants is an old one, possibly influenced by a perceived analogy with vertebrate bones (Carlquist, 1970). Features of the vasculature have thus been considered particularly valuable when assessing homologies, and have figured prominently in the discussion of the origin of the inferior ovary (for reviews see Douglas, 1944, 1957; Puri, 1952). Vestigial vascular traces often remain when an organ has been lost evolutionarily (e.g., Eames and MacDaniels, 1947, p. 356). Moreover, vascular systems often remain distinct even when organs are congenitally fused. One obvious example is sympetalous corollas, which are usually the result of congenital fusion (Endress, 1994) and contain the same number of major veins and constituent petals (indicated by lobation or by number of primordia). Even in calyces the number of constituent sepals is often evident from the vasculature (Puri, 1951).

Numerous exceptions to this rule of conservation exist, for example where rudimentary or reduced organs or organs simply reduced in size can lack vasculature secondarily, showing that a vascular strand may be lost even before an organ is. For example, the stamens of extremely small-flowered species of *Senecio* lack vasculature (Endress and Stumpf, 1990). In the case of fusion, vasculature may evolve novel traits such as the extensive reticulation in the vascular systems of the corolla tube of some Campanulaceae, which even obscures the borders between the vasculature of individual petals (Gustafsson, 1996).

In many inferior ovaries, vascular strands diverge acropetally to the various floral organs, just as they do in a typical hypogynous flower. This type of vasculature has been taken to support the appendicular theory for the origin of the inferior ovary. Particularly convincing in this respect is the reduction/fusion series documented by Eames (1931) for Ericaceae, which shows the progressive fusion of vasculature following the evolution of epigyny in this family. In some genera, such as *Gaylussacia*, the vasculature is still similar to the hypogynous genera *Pyrola* and *Andromeda* because the vascular traces are still largely free from each other. In the genus *Vaccinium*, on the other hand, the traces are fused to varying degrees, in some species to an extent that a hypothesised history of organ fusion is obscured. Assuming that this example is correctly interpreted phylogenetically, free/separate traces can indeed indicate congenital fusion. An isolated case of fused traces in fused structures, however, is difficult to interpret. The notion that free traces leading through the hypanthium to the floral appendages supports the appendicular theory was criticised by Puri (1952), who argued that separate traces in the hypanthium behave much like foliar traces traversing the cortex of the stem, which would instead support the view of the wall of the inferior ovary as an axial extension.

Although there has been general agreement among anatomists on the appendicular nature of most inferior ovaries, a few cases have been put forward where on vascular evidence the ovary is considered to have congenitally sunk into the floral axis (Douglas, 1957; Kaplan, 1967). Two examples are the Cactaceae and the Santalaceae, both of which, according to the character optimisations, represent independent derivations of epigyny from hypogynous ancestors (in the case of Santalaceae this is true of the whole order Santalales). In the Cactaceae, rather than ascending to the ovary from the base, the ovary traces run from the top of the inferior ovary, where they depart from the vascular cylinder and then descend to the placentae. As expected, they are 'inverted', after turning downwards, with the phloem abaxial. A progression of ovary forms within the Cactaceae shows increased 'sinking' of the ovary locules (Tiagi, 1955). The vasculature of the floral axes of the strongly perigynous genera *Rosa* and *Calycanthus* are similar, and these appear to represent morphologically intermediate stages in the formation of the axial type of inferior ovary. Phylogenetic support for perigyny as an intermediate evolutionary step is meagre, and according to the present results, this condition is typically evolutionarily derived. The aforementioned case of perigynous *Pereskia* species, however, actually seems to reveal such a transitional step (Boke, 1966).

An interesting feature of the vasculature of some inferior ovaries is that it sometimes links certain tissue layers with particular floral appendages. Such a connection can be made when the vascular traces leading to floral organs traverse the ovary wall with little or no branching or fusion. Exemplary are the petal and sepal traces of Goodeniaceae, which in some species traverse sectors of the inferior ovary that are also externally petal- or sepal-like to some extent (see also Carolin, 1959). The 'carpellary dorsals' that traverse the central parts of the pome have been used as an argument for the carpellate identity of the more or less well delimited tissue surrounding these vascular bundles (MacDaniels, 1940; Steeves, 1991).

In conclusion, vascular characters can change independently of external structure, and are not always more conservative than external morphology (Puri, 1952). Clearly, vasculature alone does not tell the full evolutionary story; it should simply be regarded as a character complex among others, often providing information useful for phylogenetic analysis and determination of homology of structures, including determining the history of fused organs.

18.5.4 Ontogeny and organogenesis

Endress (1994) distinguished three phases in the ontogeny of floral organs: initiation, attainment of shape by differential growth, and tissue differentiation. As pointed out by Endress, the view of the nature of the inferior ovary is likely to be affected by which growth phase is emphasised. Most ontogeneticists have been concerned primarily with the first two stages. Considering these developmental phases, the entire range of arrangements of the floral appendages in relation to the ovary – the different degrees of epigyny combined with variously developed free hypanthia – can be explained by variation in the exact location, direction and extent of intercalary growth in the zone below the appendicular primordia after these are initiated (Leins, 1972). Numerous ontogenetic studies of various epigynous groups have largely confirmed this observation, for example in Agavaceae (Mogensen, 1968), Apiaceae (Leins and Erbar, 1985), Asteraceae (Harris, 1995; Leins and Erbar, 1987), Araliaceae (Erbar and Leins, 1988), Begoniaceae (Charpentier *et al.*, 1989),

Calyceraceae (Erbar, 1993), Campanulaceae (Kaplan, 1967), Goodeniaceae (Leins and Erbar, 1989), maloid Rosaceae (Steeves *et al.*, 1991) and Stylidiaceae (Erbar, 1992).

A survey of this ontogenetic work shows that the floral axis of epigynous flowers generally passes a bowl-shaped, concave stage (e.g. Rubiaceae, Asteraceae), a crucial difference in early floral ontogeny compared with taxa with superior ovaries that possess a rounded apex without a depression. Postgenital fusion of floral organs never seems to be responsible for the formation of inferior ovaries, in contrast to cases of connation of petals and anthers, where fusion occasionally occurs (Endress, 1994). Floral parts are thus congenitally fused to form the inferior ovary. There is accordingly no recapitulation in ontogeny of evolutionary fusion events. It should be noted, however, that ontogeny is also subject to evolutionary change, and short-cuts (shortening of pathways) can evolve in ontogeny, such that it no longer reflects phylogeny (see for example Steeves *et al.*, 1991).

Based on a study of the floral ontogeny of *Begonia* and *Hillebrandia* (Begoniaceae), Charpentier *et al.* (1989) concluded that the ovary of this family cannot be interpreted either as axial or appendicular, but should be considered an intermediate structure. It is produced by intercalary growth in a zone considered transitional between the bases of the floral appendages and the floral axis. The placentae have a complex ontogeny. The basal, axile placentae develop at the floral apex, which is situated in a deep pit at time of initiation. This zone of the ovary is greatly elongated via intercalary growth. Distal to the axile placentae, parietal placentae develop as outgrowths from the walls of the upper part of the pit.

Comparisons and phylogenetic ideas not supported by or 'soluble' by ontogeny have sometimes been considered futile by morphologists (e.g. Sattler, 1974); the concept of congenital fusion, which incorporates a hypothesis of evolutionary history, was described as 'metaphysical' by Leins and Erbar (1985). Sattler (1974) did not even accept the use of the term carpel in describing many syncarpous (let alone epigynous) ovaries, where ovule-bearing structures do not arise as separate organs. It seems, however, that considering only what is directly observable during ontogeny (which the above authors require) means applying an unnecessary constraint on the study of morphology. Comparative morphology analysed in conjunction with well corroborated phylogenies can significantly enhance an evolutionary understanding of morphological phenomena, including various kinds of congenital fusion. In the case of the inferior ovary, it may in many cases be justified to refer to parts of it as carpellary, staminal, petalar, sepalar or axile even though these parts are never free and distinct during ontogeny; the evolutionary and morphological/positional connection (historical and positional orthology *sensu* Albert *et al.* 1998) to the corresponding free structures is often quite obvious.

18.5.5 Histology and tissue identity

As noted by Endress (1994), it is often not until the late stages of differentiation that the histological identity of floral organs becomes evident. As many ontogeneticists study primarily organogenesis and early organ development, later-stage tissue differentiation and its implications for homologising structures is not always considered. In the case of the inferior ovary, there can be recognisable tissue differentiation with more or less distinct layers or sectors of tissue that may be similar to sepals, petals, stamens and pistils (the protruding parts of the gynoecium). This is, however, by no means always the case; the ovary wall may incorporate unique tissue types. In addition, there is of course an obvious difference in the ultimate fate of the tissues in the ovary wall. Most petals and stamens wilt and/or dehisce after anthesis, whereas the ovary wall continues to differentiate, sometimes becoming fleshy or woody, or developing special dehiscence structures. Sepals, on the other hand, often remain attached to the fruit, and can continue to differentiate along with it.

In some Goodeniaceae, parts of the inferior ovary wall are similar histologically to sepals and to some extent also to stamens, petals and carpels. Ribs occur along the ovary of *Coopernookia strophiolata* and many species of *Goodenia* (e.g. *G. mimuloides*, *G. scapigera*). These ribs are only loosely attached to the ovary wall and are apparently proximal continuations of the free sepals, which they resemble in internal and external structure. The tissue sectors alternating with these ribs also resemble continuations of the petals, with no apparent distinction in external structure or vasculature between the free parts and those fused to the ovary. Many species have nectariferous pits in the ovary wall, delimited by the continuations of the petals. The nectar-secreting tissue is on the inside (i.e. on what appears to be the actual carpellary tissue), retaining in this 'closed-in' position the nectar producing activity it otherwise performs on exposed parts, for example, as an apical nectariferous disc in other Goodeniaceae like *Scaevola* and the related family Asteraceae, or as a zone surrounding the ovary base as in *Velleia* (Goodeniaceae) and the related family Menyanthaceae. The nectar pits of *Goodenia* and *Coopernookia* are very similar to those of *Pentaphragma* (Pentaphragmataceae; see Poulsen, 1903), the basal-most genus of the Asterales (Gustafsson, 1996).

The morphological nature of the pome (the fruit of Rosaceae–Maloideae) has been much debated (Steeves *et al.*, 1991, and references therein). The delimitation of the carpels is one of the controversial issues. A line of demarcation or a difference in structure or colour delimits the central part of the ovary and fruit in many apples (*Malus domestica*). This central region has been considered homologous to the carpels. The vasculature is generally compatible with an appendicular interpretation (MacDaniels, 1940). In *Amelanchier alnifolia*, another member of the Maloideae, the styles and the central part of the ovary are the only parts susceptible to frost damage, suggesting at least a physiological similarity between these tissues (Steeves *et al.*, 1991). Ontogenetically, however, only a small part of the ovary apart from the styles can be said to be derived from carpel primordia or the zone immediately below. In this case, anatomy and histology reveal characteristics of parts of the ovary that are similar to those of the floral whorls, whereas ontogenetic history does not.

The ovaries of cacti have a vasculature that indicates an axial origin, as discussed above, and even casual observation of the external structure of the cactus ovary and fruit appears to confirm this. Areoles highly similar to those of the stem of cacti may cover the exterior of the fruit, giving the impression that the ovary has been depressed into a stem segment. The 'bracteoles' on the ovary of *Pereskia* and *Opuntia* sometimes even bear flowers in their axils (Tiagi, 1955). The outer layer of the ovary in this case seems to be highly similar to an axis, both in structure and in function.

In summary, the tissues of the inferior ovary are often structurally similar to those of floral appendages, indicating a likely process orthology (*sensu* Albert *et al.*, 1998). Likewise, similarity to an axis, as in cacti, indicates that the outer layer of the ovary is process orthologous to vegetative stems. In many cases, however, the inferior ovary wall has its own character, possessing several unique features that are present in neither floral appendages nor axes.

18.6 Loss of organ identity following fusion

Organ identity indistinction has followed organ fusion in many angiosperms. In the Santalales, which happen also to be epigynous, the ovules usually lack an integument and are enclosed in (and hardly distinguishable from) the placental tissue (Takhtajan, 1997). Within this order, the Eremolepidaceae represent an extreme condition, as they have an ovule virtually reduced to just a female gametophyte. In cases like the Santalales, with their indehiscent fruits, the seed is never exposed to the environment but remains well protected until germination. It is not surprising that the integument, which normally produces the outer protective layers of the seed, has become reduced under these circumstances. Similarly, the carpels of the inferior ovary are largely enclosed in other tissues that have taken over much of the protective functions, with reduction and loss of distinction of the carpels resulting.

In syncarpous ovaries the number of constituent carpels is often quite obvious. If the demarcation lines are not evident, the vasculature and stigma lobation clearly shows the carpel number in most cases. The Goodeniaceae and Asteraceae have distinctly bicarpellate ovaries, as revealed by ovary partition and vasculature (in the case of Goodeniaceae) and stigma lobation (in the Asteraceae). In the closely related Calyceraceae there is no partition of the ovary cavity nor any lobation of the style, and the only feature indicating a bicarpellate origin is the presence of two equally strong vascular strands in the style. The presence of five ledges in the ovary locule led Erbar (1993) to hypothesise a pentacarpellary origin. The Calyceraceae is clearly an ambiguous case where carpel identity has nearly been lost.

In the genus *Clusia* (Clusiaceae) fusion of stamens is common, and synandria of fused stamens or staminodes result. The fusion is sometimes complete, as in synandria of sections *Androstylium*, *Phloianthera* and *Retinostemon*, in which anthers are scattered on the surface of massive organs. In sections *Polythecandra* and *Gomphanthera* it is even difficult to delineate anthers and thecae (Engler, 1925; Bittrich and Amaral, 1996; M. H. G. Gustafsson, unpublished). In the staminodial rings of female flowers of several sections there is often no trace of individual parts, either in lobation or external structure although there frequently is in vasculature.

In *Stenopadus*, one of the most basal genera of Asteraceae (Karis *et al.*, 1992), there are ten vascular traces in the outer layer of the inferior ovary, as well as an inner circle of gynoecial bundles (Carlquist, 1961). The outer (peripheral) bundles are proximal continuations of the strands of stamens and corolla. Most other Asteraceae have lost the midvein in the petals, and the number of ovary wall bundles is accordingly reduced to five. (There is also commonly a reduction in the number of gynoecial bundles.) Several derived Asteraceae show a further reduction of vascular trace number in the ovary wall to four or two, but this is not accompanied by reduction in the corolla vasculature (Carlquist, 1961). The vasculature of the inferior ovary thus seems to have evolved to a certain extent in concert with that of the floral appendages, but subsequently it has undergone additional, independent reductions.

18.7 Genetic homology and process orthology

18.7.1 MADS box genes[1]

The identities of flower organs are controlled by a limited set of genes. Many of these so-called 'homeotic genes' are members of the same eukaryotic gene superfamily, known as the MADS box genes (Theißen et al., 1996). Many MADS box genes have a determination function analogous to the homeobox pattern formation genes of animals (e.g., Shubin et al., 1997). Both gene superfamilies encode transcription factors, proteins that regulate the expression of other genes. MADS box genes are known from animals and fungi, but unlike those of plants, most of these genes are involved in regulating cellular differentiation. Likewise, many homeobox genes of plants are members of signal transduction pathways (Mattsson et al., 1992; Carabelli et al., 1993; Sessa et al., 1993; Söderman et al., 1994), although there are exceptions (e.g. Vollbrecht et al., 1991; Reiser et al., 1995). MADS and homeobox genes do interact in yeast (Herschbach et al., 1994; Mead et al., 1996), suggesting that they may form regulatory relationships during plant development.

MADS box genes encode proteins with at least two distinctive domains; the MADS domain (Schwarz-Sommer et al., 1990), conserved across eukaryotic life, and the K domain (Ma et al., 1991), which is known only from plants. The MADS domain may be a site of DNA binding (Nurrish and Treisman, 1995) and both the MADS and K domains may function in protein dimerisation (Davies and Schwarz-Sommer, 1994; Shore and Sharrocks, 1995).

Plant MADS box genes fall into different functional classes that strongly correlate with membership in particular gene lineages (Doyle, 1994; Purugganan et al., 1995; Tandre et al., 1995). These lineages, or orthologues, appear to be descended from a common ancestral gene through several duplication events. An ancient derivation for several of these orthologues is supported by including both angiosperm and Norway spruce MADS box genes in phylogenetic analysis (Tandre et al., 1995). The spruce sequences intercalate within previously-recognized angiosperm gene lineages rather than forming a group of their own. Therefore, several MADS box orthologues are older than the divergence of the ancestors of conifers and angiosperms, which occurred over 300 million years ago.

Gene duplication is a common theme in the diversification of gene function (e.g., Helariutta et al., 1996). MADS box genes have various known and putative activities, ranging from root, embryo and vegetative development, pre-patterning of the floral meristem, and determination of the four flower whorls, to ovule development

---

[1] Sections 18.7.1 and 18.7.2 are modified from Albert et al. (1998).

(Rounsley *et al.*, 1995). The ancestral function of plant MADS box genes remains unknown. MADS box genes being isolated from Lycopodiopsida (M. Svensson, pers. comm., 1997; Svensson and Engström, 1997), the sister group of all other vascular plants (Raubeson and Jansen, 1992; Kenrick and Crane, 1997), may help resolve this issue.

18.7.2 The ABC model

Genetic studies of *Arabidopsis* and *Antirrhinum* mutants have led to a model for flower organ determination that included the combinatorial action of three functions (Coen and Meyerowitz, 1991). Two of these functions are now known to be encoded solely by MADS box genes, the third by MADS box plus other, unrelated genes in *Arabidopsis*.

The A function was seen to be expressed in sepals and petals. By itself, expression of A would lead to sepals. Combined with B, A would lead to petals. The *Arabidopsis* MADS box gene *APETALA1* (*AP1*: Mandel *et al.*, 1992) and possibly also its orthologues (e.g. *SQUAMOSA* of *Antirrhinum*: Huijser *et al.*, 1992) are A-function genes. The other principal A-function gene is *APETALA2* (*AP2*: Jofuku *et al.*, 1994), a putative transcription factor that is part of a unique gene and protein family. It would later be found that only *AP1* expression is truly restricted to the first and second floral whorls (Mandel *et al.*, 1992); *AP2* is expressed in carpels and the outer integuments of ovules (Jofuku *et al.*, 1994; Modrusan *et al.*, 1994), although expression patterns differ among different *AP2* alleles; C. S. Gasser, pers. comm., 1997). *AP1* orthologues reside in a complex, larger orthologue that includes genes with diverse vegetative, infloral, and floral expression patterns (Purugganan *et al.*, 1995; Theißen *et al.*, 1996). Thus far, however, firm evidence for an *AP1*- or *AP2*-like A function is missing for *Antirrhinum*, which may suggest that the A function is mechanistically non-uniform among angiosperms and therefore likely derived relative to the B and C functions, as sepals are thought to be relative to petals (Albert *et al.*, 1998).

The B function is expressed principally in petals and stamens, but also in the outer integuments of ovules (Jack *et al.*, 1994). All B-function genes identified so far are MADS box genes. B-function genes form a single lineage with at least two orthologues, one corresponding to the *Arabidopsis* gene *APETALA3* (*AP3*: Jack *et al.*, 1992), the other to *PISTILLATA* (*PI*: Kramer *et al.*, 1998). *Antirrhinum* orthologues of these genes are *DEFICIENS* (*DEF*: Schwarz-Sommer *et al.*, 1990) and *GLOBOSA* (*GLO*: Tröbner *et al.*, 1992), respectively. Protein products of these genes may form AP3/PI and DEF/GLO heterodimers (Tröbner *et al.*, 1992; Goto and Meyerowitz, 1994) that regulate further expression of both gene pairs (B orthologue dynamics differ in some species, including *Petunia* (van der Krol *et al.*, 1993); see below). The prevention of B-function gene expression in the sepal whorl remains poorly characterised.

The C function is expressed in the reproductive organs. C expression alone leads to carpels. Combined with B, C generates stamens. All known C-function genes are MADS box genes related to the *Arabidopsis* gene *AGAMOUS* (*AG*: Yanofsky *et al.*, 1990). The *Antirrhinum* orthologue is *PLENA* (*PLE*: Bradley *et al.*, 1993). Some plants (e.g. *Petunia*: Tsuchimoto *et al.*, 1993) possess two *AG* orthologues.

Some phylogenetic analyses suggest that the C- and B-function orthologues may be sister lineages, descended from a single gene duplication event (Doyle, 1994; Tandre *et al.*, 1995).

To summarize, in its simplest form, the ABC model specifies that A = sepals, A + B = petals, B + C = stamens, and C = carpels. Recently, ovules have been recognised as a fifth organ category, and ovule-specific D-function MADS box genes have been identified (Colombo *et al.*, 1995). Further complexity beyond the simple ABC model is indicated by 'ectopic' expression of A- and B-function genes in developing ovules (e.g. Jack *et al.*, 1994; Modrusan *et al.*, 1994).

An additional complication is that organ identity may be conferred in part by a concentration effect. For example, co-suppression transgenics for the *Petunia* MADS box genes *PMADS1* and *FBP2* show different degrees of organ identity transformation depending upon the number of gene copies inserted (van der Krol *et al.*, 1993; Angenent *et al.*, 1994). Transcription factors such as MADS domain proteins act as on/off switches in the regulation of other genes, but this activity is modulated by the intracellular concentrations of these factors. Thus, greater or lesser transcription and/or post-translational modification (e.g. phosphorylation, dimerisation, or nuclear localisation) of A- or B-function genes could lead to aB or Ab petals (lower case = weak function, upper case = strong function) with different phenotypes, the latter possibly sepaloid. Ectopic expression of C in the petal whorl or A in the staminal whorl could likewise give rise to petal-stamen intermediates. Even though the C and A functions may share a negative regulatory relationship (in *Arabidopsis*, AP2 is a negative regulator of *AG* in the outer floral whorls, and AG is a negative regulator of *AP1* in the inner whorls; see Theißen and Saedler (1995), p. 633), PLE (a C-function protein) and SQUA (a potential A-function protein) are nevertheless known to interact within some cells of *Antirrhinum* (Davies *et al.*, 1996).

18.7.3 Process orthology in the inferior ovary

Several examples have been mentioned in which the parts of the inferior ovary are histologically similar to perianth parts. In these cases, the possibility exists that this similarity is based upon common expression of the same organ identity-determining genes (process orthology *sensu* Albert *et al.*, 1998). One particularly interesting example in this context, and one for which the genetic background is increasingly understood, is the Hawaiian lobelioid genus *Clermontia* (Campanulaceae) (Albert *et al.*, 1998). In a large group of species in this epigynous genus, *ca.* 65% of those recognised by Lammers (1991), there is no typical calyx, but instead a whorl that is identical, or nearly so, to the corolla in every morphological and anatomical aspect. These species thus have a 'double corolla' phenotype. This feature appears to have evolved once (from preliminary phylogenetic studies of 5S-NTS sequences; Di Laurenzio *et al.*, 1997), and apparently represents a one-for-one homeotic switch. Genetically, this is apparently caused by an extension of the zone of B-function MADS box gene expression to the outermost floral whorl, which then receives the typical petal expression complement of A + B (data are available for B function expression only). Interestingly, the double corolla mutation also affects the morphology of the ovary. In species with 'normal' calyx structure, the exterior of the developing fruit is much like the exterior of the calyx (e.g. *Clermontia arborescens, C. tuberculata).* The outer layer of the mature fruits is tougher and thicker, and does not appear to dehisce as readily as that of the double-corolla species *C. montis-loa* and *C. parviflora* (V. A. Albert, pers. obs.).

Molecular developmental genetic studies of *Gerbera* (Asteraceae) by Yu *et al.* (1999) and M. Kotilainen (pers. comm., 1998) have shown that MADS box genes of classes A, B and C are variously involved in the formation of the inferior ovary of this genus, the first epigynous plant to be studied in this respect. Overexpression of a native B-function MADS box gene, *GGLO1*, was shown to affect the shape of the inferior ovary. In ectopically expressing transgenics the ovary is twice as long as that of wild-type plants, whereas the ovule is unaffected. The style was not altered morphologically but it did wilt prematurely. In *in situ* hybridisation experiments, however, no late expression of B-function MADS box genes was found in the ovary wall (i.e. to correlate potentially with ovary elongation), but an *AP1/SQUAMOSA* historical orthologue, *GSQUA1*, showed strong expression in developing vascular bundles of the ovary wall. Two C-function genes, *GAGA1* and *GAGA2*, are expressed in the pistil and stamen primordia, in tissue below these organs, in the ovule, and in the placenta. At least in those developmental stages studied, there is no expression in the ovary wall, apart from the ovary roof. A continuous zone of expression surrounding the ovary would have been expected if the inner part of the inferior ovary is homologous with the superior ovary of hypogynous plants such as *Arabidopsis* and *Antirrhinum*. A partial process orthology with such free ovaries does exist, as expression occurs in some (albeit restricted) sectors of the ovary in *Gerbera*. From the data presently available, there is not much molecular marker support for an appendicular derivation of the *Gerbera* ovary given the absence of B and C function gene expression in the ovary wall. The situation is far more complex than a simple addition of expression zones of floral appendages, which would be a central zone of C-type gene expression surrounded by, successively, a B + C expression zone, an A + B zone and a zone where A alone may be expressed. Such a pattern may have been present in ancestors of *Gerbera*, however, and the one presently found may be the result of subsequent spatial and temporal modifications in gene expression, in analogy with the evolution of floral vasculature in the family. Gene expression patterns from several basal and derived Asteraceae should be studied in order to evaluate this hypothesis further.

In apple (*Malus domestica*) an *AP1*-like MADS-box gene, *MdMADS5*, is expressed in the outer fleshy layer and skin (Yao *et al.*, 1999). No expression of B-class genes was detected in the fruit. This could be taken to indicate a process orthology between sepals and the outer layers of the ovary, but it has not been established that *MdMADS5* actually is expressed in sepals of apple. An A-function as described from *Arabidopsis* still remains to be documented from other systems and may in fact be a derived function within this group of genes. An important role of *AP1* and at least some of its orthologues, such as *SQUAMOSA* of *Antirrhinum*, is in the control of meristem identity, and this may thus be a more general and perhaps the plesiomorphic function of these genes. Modulation of floral meristem activity is what ontogenetically makes the ovary inferior, and it is a distinct possibility that the *AP1* orthologue in apple has temporally extended its expression beyond floral meristem initiation. The expression of no less than seven different MADS-box genes in the ovary of apple (Yao *et al.*, 1999), indicates a complex situation not at all conforming to a simple 'addition of whorls' model, as discussed above. The finding that a B-function MADS box gene does express in the superior ovary wall of *Solanum* (Garcia-Maroto *et al.*, 1993) further adds to an emerging picture of diversity and complexity in the developmental genetics of angiosperm ovaries, as does the *GGLO1* over-expression phenotype described for *Gerbera* previously.

One case of organ fusion that has become partly understood genetically is the stamen–corolla tube *sensu* Erbar (1991), that is, the basal part of a sympetalous corolla to which the stamen filaments are attached, as observed in many Asteridae. The stamen–corolla tube of *Petunia* is lost when expression of a petal-specific B-function MADS box gene, *PMADS1*, is eliminated (van der Krol *et al.*, 1993; Albert *et al.*, 1998). The intercalary growth that normally produces the stamen–corolla tube does not occur. This constitutes a very interesting similarity to the *Gerbera* case, where B-function genes have an effect on a synorganised zone through modulation of intercalary growth (M. Kotilainen, pers. comm., 1998), and points to the possibility that B-function genes may be generally involved in floral organ fusion/synorganisation (cf. Albert *et al.*, 1998). If so, elongation of the *Gerbera* ovary may also be a response to early expression of B-function MADS box genes.

18.8 The limits of process orthology assertions

To conclude, it is evident that there is no single answer to what epigyny is in terms of homology: it must be characterised as a phenomenon with very diverse manifestations and multiple origins. Establishment of process orthology with other organs of the plant is sometimes possible, directly, when developmental genetic evidence is available, or indirectly, when morphological features strongly indicate a relationship to such organs. However, process orthology can be a matter of degree – gene expression patterns can differ quantitatively, positionally and temporally. Such partial process orthology likely underlies the different degrees of morphological similarities between the inferior ovary on the one hand and floral appendages and floral axes on the other. Evolution may alter gene expression patterns and new features may be acquired, eventually to an extent completely obscuring a history of process orthology. A phylogenetic perspective could illuminate such cases, if ancestral (plesiomorphic) conditions in a group include manifestations of process orthology. A synthesis of phylogenetic, positional, morphological and developmental genetic data is necessary for a full understanding of complex homology problems such as the nature of the inferior ovary.

ACKNOWLEDGEMENTS

This research was supported by the Lewis B. and Dorothy Cullman Foundation. MHGG was supported by a postdoctoral fellowship from the Swedish Natural Science Research Council (NFR). Thanks are due to Richard Bateman, Anders Kvarnheden, Mika Kotilainen and the Teeri Lab (Institute of Biotechnology, Helsinki) for their helpful contributions.

REFERENCES

Albert, V. A., Gustafsson, M. H. G. and DiLaurenzio, L. (1998) Ontogenetic systematics, molecular developmental genetics, and the angiosperm petal, in *Molecular systematics of plants II: DNA sequencing*, (eds D. E. Soltis, P. S. Soltis and J. J. Doyle), pp. 349–374. Kluwer Academic Publishers, Boston.

Angenent, G. C., Franken, J., Busscher, M., Weiss, D and van Tunen, A. J. (1994) Co-suppression of the petunia homeotic gene *fbp2* affects the identity of the generative meristem. *The Plant Journal*, **5**, 33–44.

Bittrich, V. and Amaral, M. C. E. (1996) Flower morphology and pollination biology of *Clusia* species from the Gran Sabana (Venezuela). *Kew Bulletin*, **51**, 681–694.

Boke, N. H. (1966) Ontogeny and structure of the flower and fruit of *Pereskia aculeata*. *American Journal of Botany*, **53**, 534–542.

Bradley, D., Carpenter, R., Sommer, H., Hartley, N. and Coen, E. S. (1993) Complementary floral homeotic phenotypes result from opposite orientations of a transposon at the *PLENA* locus of *Antirrhinum*. *Cell*, **72**, 85–95.

Bremer, K. and Gustafsson, M. H. G. (1997) East Gondwana ancestry of the sunflower alliance of families. *Proceedings of the National Academy of Sciences USA*, **94**, 9188–9190.

Carabelli, M., Sessa, G., Baima, S., Morelli, G. and Ruberti, I. (1993) The *Arabidopsis Athb-2* and *-4* genes are strongly induced by far-red-rich light. *The Plant Journal*, **4**, 469–479.

Carlquist, S. (1961) *Comparative plant anatomy*. Holt, Rinehart and Winston, New York.

Carlquist, S. (1970) Towards acceptable evolutionary interpretations of floral anatomy. *Phytomorphology*, **19**, 332–362.

Carolin, R. C. (1959) Floral structure and anatomy in the family Goodeniaceae. *Proceedings of the Linnean Society of New South Wales*, **84**, 242–255.

Carolin, R. C. (1966) Seeds and fruit of the Goodeniaceae. *Proceedings of the Linnean Society of New South Wales*, **91**, 58–83.

Carolin, R. C. (1977) The systematic relationships of *Brunonia*. *Brunonia*, **1**, 9–29.

Charpentier, A., Brouillet, L. and Barabé, D. (1989) Organogénèse de la fleur pistillée du *Begonia dregei* et de l'*Hillebrandia sandwicensis* (Begoniaceae*). *Canadian Journal of Botany*, **67**, 3625–3639.

Chase, M. W., Soltis, D. E., Olmstead, R. G., Morgan, D., Les, D. H., Mishler, B.D. *et al.* (1993) Phylogenetics of seed plants: an analysis of nucleotide sequences from the plastid gene *rbcL*. *Annals of the Missouri Botanical Garden*, **80**, 528–580.

Coen, E. S. and Meyerowitz, E. M. (1991) The war of the whorls: genetic interactions controlling flower development. *Nature*, **353**, 31–37.

Colombo, L., Franken, J., Koetje, E., van Went, J., Dons, H. J. M., Angenent, C. G. *et al.* (1995) The petunia MADS box gene *FBP11* determines ovule identity. *The Plant Cell*, **7**, 1859–1868.

Cronquist, A. (1981) *An integrated system of classification of flowering plants*. Columbia University Press, New York.

Dahlgren, R., Clifford, H. T and Yeo, P. (1985) *The families of the monocotyledons*. Springer, Berlin.

Davies, B. and Schwarz-Sommer, Z. (1994) Control of floral organ identity by homeotic MADS box transcription factors, in *Results and problems in cell differentiation*, (ed. L. Nover), pp. 235–258, Springer, Berlin.

Davies, B., Egea-Cortines, M., de Andrade Silva, E., Saedler, H. and Sommer, H. (1996) Multiple interactions amongst floral homeotic MADS box proteins. *The EMBO Journal*, **15**, 4330–4343.

DiLaurenzio, L., Struwe, L., Pepper, A. S., Kizirian, D. and Albert, V. A. (1997) Gene expression analysis of sepal identity in *Clermontia* (Lobelioideae: Campanulaceae): homeosis and floral diversification in the Hawaiian archipelago, in *Evolution of plant development*, Keystone symposium B1, Taos, New Mexico, p. 24.

Douglas, G. E. (1944) The inferior ovary. *The Botanical Review*, **10**, 125–186.

Douglas, G. E. (1957) The inferior ovary. II. *The Botanical Review*, **23**, 1–46.

Doyle, J. J. (1994) Evolution of a plant homeotic multigene family: toward connecting molecular systematics and molecular developmental genetics. *Systematic Biology*, **43**, 307–328.

Dunbar, A. (1975a) On pollen of Campanulaceae and related families with special reference to the surface ultrastructure. I. Campanulaceae subfam. Campanuloidae. *Botaniska Notiser*, **128**, 73–101.

Dunbar, A. (1975b) On pollen of Campanulaceae and related families with special reference to the surface ultrastructure. II. Campanulaceae subfam. Cyphioidae and subfam. Lobelioidae; Goodeniaceae; Sphenocleaceae. *Botaniska Notiser*, **128**, 102–118.

Eames, A. J. (1931) The vascular anatomy of the flower with refutation of the theory of carpel polymorphism. *American Journal of Botany*, **18**, 147–188.

Eames, A. J. and MacDaniels, L. H. (1947) *An introduction to plant anatomy*. McGraw-Hill, New York.

Endress, P. K. (1994) *Diversity and evolutionary biology of tropical flowers*. Cambridge University Press, Cambridge.

Endress, P. K. and Stumpf, S. (1990) Non-tetrasporangiate stamens in the angiosperms: structure, systematic distribution and evolutionary aspects. *Botanische Jahrbücher für Systematik, Pflanzengeschichte und Pflanzengeographie*, **112**, 193–240.

Engler, A. (1925) Guttiferae, in *Die natürlichen pflanzenfamilien*, ed. 2, vol. 21, (eds. A. Engler and K. Prantl), pp. 154–237, Engelmann, Leipzig.

Engler, A. and Prantl, K. (1887–1915) *Die natürlichen pflanzenfamilien*. Engelmann, Leipzig.

Erbar, C. (1991) Sympetaly – a systematic character? *Botanische Jahrbücher für Systematik, Pflanzengeschichte und Pflanzengeographie*, **112**, 417–451.

Erbar, C. (1992) Floral development of two species of *Stylidium* (Stylidiaceae) and some remarks on the systematic position of the family Stylidiaceae. *Canadian Journal of Botany*, **70**, 258–271.

Erbar, C. (1993) Studies on the floral development and pollen presentation *in Acicarpha tribuloides* with a discussion of the systematic position of the family Calyceraceae. *Botanische Jahrbücher für Systematik, Pflanzengeschichte und Pflanzengeographie*, **115**, 325–350.

Erbar, C. and Leins, P. (1988) Blütenentwicklungsgeschichtliche Studien an *Aralia* und *Hedera* (Araliaceae). *Flora*, **180**, 391–406.

Eyde, R. H. and Tseng, C. C. (1969) Flower of *Tetraplasandra gymnocarpa*. Hypogyny with epignous ancestry. *Science*, **166**, 506–508.

Farris, J. S. (1969) A successive approximations approach to character weighting. *Systematic Zoology*, **18**, 374–385.

Friis, E. M., Pedersen, K. R. and Crane, P. R. (1994) Angiosperm floral structures from the Early Cretaceous of Portugal. *Plant Systematics and Evolution, Supplement*, **8**, 31–49.

Garcia-Maroto, F., Salamini, F. and Rohde, W. (1993) Molecular cloning and expression patterns of three alleles of the *DEFICIENS*-homologous gene *St-DEFICIENS* from *Solanum tuberosum*. *Plant Journal*, **4**, 771–780.

Goto, K. and E. M. Meyerowitz. (1994) Function and regulation of the *Arabidopsis* floral homeotic gene *PISTILLATA*. *Genes and Development*, **8**, 1548–1560.

Grant, V. (1950) The protection of the ovules in flowering plants. *Evolution*, **4**, 179–201.

Gustafsson, M. H. G. (1996) Phylogenetic studies in the Asterales. *Comprehensive summaries of Uppsala dissertations from the Faculty of Science and Technology*, **218**, 36 pp. Acta Universitatis Upsaliensis, Uppsala.

Gustafsson, M. H. G., Backlund, A. and Bremer, B. (1996) Phylogeny of the Asterales sensu lato based on *rbcL* sequences with particular reference to the Goodeniaceae. *Plant Systematics and Evolution*, **199**, 217–242.

Harris, E. M. (1995) Inflorescence and floral ontogeny in Asteraceae: a synthesis of historical and current concepts. *The Botanical Review*, **61**, 93–278.

Helariutta, Y., Kotilainen, M, Elomaa, P., Kalkkinen, N., Bremer, K. Teeri, T. H. and Albert, V. A. (1996) Duplication and functional divergence in the chalcone synthase gene family of Asteraceae: evolution with substrate change and catalytic simplification. *Proceedings of the National Academy of Sciences USA*, **93**, 9033–9038.

Herschbach, B. M., Arnaud, M. B. and Johnson, A. D. (1994) Transcriptional repression directed by the yeast alpha 2 protein in vitro. *Nature*, **370**, 309–311.

Heywood, V.H. (ed.) (1985) *Flowering plants of the world*. B.T. Batsford, London.

Huijser, P., Klein, J., Lönnig, W. -E., Meijer, H., Saedler, H. and Sommer, H. (1992) Bracteomania, an inflorescence anomaly, is caused by the loss of function of the MADS box gene *SQUAMOSA* in *Antirrhinum majus*. *The EMBO Journal*, **11**, 1239–1249.

Igersheim, A., Puff, C., Leins, P. and Erbar, C. (1994) Gynoecial development of *Gaertnera* Lam. and of presumably allied taxa of the Psychotrieae (Rubiaceae): secondarily 'superior' vs. inferior ovaries. *Botanische Jahrbücher für Systematik, Pflanzengeschichte und Pflanzengeographie*, **116**, 401–414.

Jack, T., Brockman, L. L. and Meyerowitz, E. M. (1992) The homeotic gene *APETALA3* of *Arabidopsis thaliana* encodes a MADS-box and is expressed in petals and stamens. *Cell*, **68**, 683–687.

Jack, T., Fox, G. L. and Meyerowitz, E. M. (1994) *Arabidopsis* homeotic gene *APETALA3* ectopic expression: transcriptional and posttranscriptional regulation determine floral organ identity. *Cell*, **76**, 703–716.

Jofuku, K. D., den Boer, B. G., van Montagu, M. and Okamuro, J. K.(1994) Control of *Arabidopsis* flower and seed development by the homeotic gene *APETALA2*. *The Plant Cell*, **6**, 1211–1225.

Kaplan, D. R. (1967) Floral morphology, organogenesis and interpretation of the inferior ovary in *Downingia bacigalupii*. *American Journal of Botany*, **54**, 1274–1290.

Karis, P. O., Källersjö, M. and Bremer, K. (1992) Phylogenetic analysis of the Cichorioideae (Asteraceae), with emphasis on the Mutisieae. *Annals of the Missouri Botanical Garden*, **79**, 416–427.

Kenrick, P. and Crane, P. R. (1997) *The origin and early diversification of land plants: a cladistic study*. Smithsonian Institution Press, Washington.

Kramer, E. M., Dorit, R. L. and Irish, V. F. (1998) Molecular evolution of genes controlling petal and stamen development: duplication and divergence within the *APETALA3* and *PISTILLATA* MADS-box gene lineages. *Genetics*, **149**, 765–783.

Krause, K. (1912) Goodeniaceae und Brunoniaceae, in *Das Pflanzenreich*, (ed. A. Engler), vol. IV. 277 and 277b, Wilhelm Engelmann, Leipzig.

Lammers, T. G. (1991) Systematics of *Clermontia* (Campanulaceae, Lobelioideae). *Systematic Botany Monographs*, **32**, 1–97.

Leins, P. (1972) Das Karpell im ober- und unterständigen Gynoeceum. *Berichte der Deutschen Botanischen Gesellshaft*, **85**, 291–294.

Leins, P. and Erbar, C. (1985) Zur frühen Entwicklungsgeschichte des Apiaceen-Gynoeceums. *Botanische Jahrbücher für Systematik, Pflanzengeschichte und Pflanzengeographie*, **106**, 53–60.

Leins, P. and Erbar, C. (1987) Studien zur Blütenentwicklung an Compositen. *Botanische Jahrbücher für Systematik, Pflanzengeschichte und Pflanzengeographie*, **108**, 381–401.

Leins, P. and Erbar, C. (1989) Zur Blütenentwicklung und sekundären Pollenpräsentation bei *Selliera radicans* Cav. (Goodeniaceae). *Flora*, **182**, 43–56.

Ma, H., Yanofsky, M. F. and Meyerowitz, E. M. (1991) *AGL1-AGL6*, an *Arabidopsis* gene family with similarity to floral homeotic and transcription factor genes. *Genes and Development*, **5**, 484–495.

MacDaniels, L. H. 1940. The morphology of the apple and other pome fruits. *Memoir Cornell University Agricultural Experiment Station*, **230**, 1–32.

Maddison, W. P. and Maddison, D. R. (1992) *MacClade: Interactive analysis of phylogeny and character evolution, Version 3.0*. Sinauer Associates, Sunderland, Massachusetts.

Mandel, M. A., Gustafson-Brown, C., Savidge, B. and Yanofsky, M. F. (1992) Molecular characterization of the *Arabidopsis* floral homeotic gene *APETALA1*. *Nature*, **360**, 273–277.

Mattsson, J., Söderman, E., Svenson, M., Borkird, C. and Engström, P. (1992) A new homeobox-leucine zipper gene from *Arabidopsis thaliana*. *Plant Molecular Biology*, **18**, 1019–1022.

Mead, J., Zhong, H., Acton, T. B. and Vershon, A. K. (1996) The yeast alpha2 and Mcm1 proteins interact through a region similar to a motif found in homeodomain proteins of higher eukaryotes. *Molecular and Cell Biology*, **16**, 2135–2143.

Modrusan, Z., Reiser, L., Feldmann, K. A., Fischer, R. L. and Haughn, G. W. (1994) Homeotic transformation of ovules into carpel-like structures in *Arabidopsis*. *The Plant Cell*, **6**, 333–349.

Moeliono, B. and Tuyn, P. (1960) Campanulaceae, in *Flora Malesiana*, (ed. C. G. G. J. van Steenis), vol. 6(1), pp. 107–141, Noordhoff-Kolff, Djakarta.

Mogensen, H. L. (1968) Floral ontogeny and interpretation of the inferior ovary in *Agave parryi*. *Canadian Journal of Botany*, **47**, 23–26.

Nurrish, S. J. and Treisman, R. (1995) DNA binding specificity determinants in MADS-box transcription factors. *Molecular and Cell Biology*, **15**, 4076–4085.

Plunkett, G. M., Soltis, D. E. and Soltis, P. S. (1997) Clarification of the relationship between Apiaceae and Araliaceae based on *mat*K and *rbc*L sequence data. *American Journal of Botany*, **84**, 565–580.

Poulsen, V. A. (1903) *Pentaphragma ellipticum* sp. nov. Et Bidrag til Kundskab om Slægten *Pentaphragma* Wall. *Videnskabelige Meddelelser fra Dansk Naturhistorisk Forening i København*, pp. 319–330.

Puri, V. (1951) The rôle of floral anatomy in the solution of morphological problems. *The Botanical Review*, **17**, 471–553.

Puri, V. (1952) Floral anatomy and inferior ovary. *Phytomorphology*, **2**, 122–129

Purugganan, M. D., Rounsley, S. D., Schmidt, R. J. and Yanofsky, M. F. (1995) Molecular evolution of flower development: diversification of the plant MADS-box regulatory gene family. *Genetics*, **140**, 345–356.

Raubeson, L. A. and Jansen, R. K. (1992) Chloroplast DNA evidence on the ancient evolutionary split in vascular land plants. *Science*, **255**, 1697–1699.

Reiser, L., Modrusan, Z., Margossian, L., Samach, A., Ohad, N., Haughn, G. W. and Fischer, R. L. (1995) The *BELL1* gene encodes a homeodomain protein involved in pattern formation in the *Arabidopsis* ovule primordium. *Cell*, **83**, 735–742.

Rice, K. A., Donoghue, M. J. and Olmstead, R. G. (1997) Analyzing large data sets: *rbc*L 500 revisited. *Systematic Biology*, **46**, 554–563.

Rounsley, S. D., Ditta, G. S. and Yanofsky, M. F. (1995) Diverse roles for MADS box genes in *Arabidopsis* development. *The Plant Cell*, **7**, 1259–1269.

Sattler, R. (1974) A new approach to gynoecial morphology. *Phytomorphology*, **24**, 22–34.

Schönland, S. (1889) Campanulaceae, in *Die natürlichen Pflanzenfamilien*, (eds. A. Engler and K. Prantl), vol. 4(5), pp. 40–70, Engelmann, Leipzig.

Schwarz-Sommer, Z., Huijser, P., Nacken, W., Saedler, H. and Sommer, H. (1990) Genetic control of flower development: homeotic genes in *Antirrhinum majus*. *Science*, **250**, 931–936.

Sessa, G., Morelli, G. and Ruberti, I. (1993) The Athb-1 and -2 HD-Zip Domains homo-dimerize forming complexes of different DNA binding specificities. *The EMBO Journal*, **12**, 3507–3517.

Shore, P. and Sharrocks, A. D. (1995) The MADS-box family of transcription factors. *European Journal of Biochemistry*, **229**, 1–13.

Shubin, N., Tabin, C. and Carroll, S. (1997). Fossils, genes and the evolution of animal limbs. *Nature*, **388**, 639–648.

Söderman, E., Mattsson, J., Svensson, M., Borkird, C. and Engström, P. (1994) Expression pattern of novel genes encoding homeodomain leucine-zipper proteins in *Arabidopsis thaliana*. *Plant Molecular Biology*, **26**, 145–154.

Steeves, T. A., Steeves, M. W. and Olson, A. R. (1991) Flower development in *Amelanchier alnifolia* (Maloideae). *Canadian Journal of Botany*, **69**, 844–857.

Svensson, M. and Engström, P. (1997) The *L421* gene isolated from *Lycopodium annotinum* L. strobilus cDNA library is a distant relative to seed plant MADS-box genes. *Evolution of Plant Development*, Keystone Symposium B1, Taos, New Mexico, p. 27 [abstract].

Takhtajan, A. (1997) *Diversity and classification of flowering plants*. Columbia University Press, New York.

Tandre, K., Albert, V. A., Sundås, A. and Engström, P. (1995) Conifer homologues to genes that control floral development in angiosperms. *Plant Molecular Biology*, **27**, 69–78.

Theißen, G., Kim, J. T. and Saedler, H. (1996) Classification and phylogeny of the MADS-box multigene family suggest defined roles of MADS-box gene subfamilies in the morphological evolution of eukaryotes. *Journal of Molecular Evolution*, **43**, 484–516

Theißen, G. and Saedler, H. (1995) MADS-box genes in plant ontogeny and phylogeny: Haeckel's 'biogenetic law' revisited. *Current Opinion in Genetics and Development*, **5**, 628–639.

Tiagi, Y. D. (1955) Studies in floral morphology. II. Vascular anatomy of the flower of certain species of the Cactaceae. *The Journal of the Indian Botanical Society*, **34**, 408–428.

Tröbner, W., Ramirez, L., Motte, P., Hue, I., Huijser, P., Lönnig, W. *et al.* (1992) *GLOBOSA*: a homeotic gene which interacts with *DEFICIENS* in the control of *Antirrhinum* floral organogenesis. *The EMBO Journal*, **11**, 4693–4704.

Tsuchimoto, S., van der Krol, A. R. and Chua, N. -H. (1993) Ectopic expression of *pMADS3* in transgenic petunia phenocopies the petunia blind mutant. *Plant Cell*, **5**, 843–853.

van der Krol, A. R., Brunelle, A., Tsuchimoto, S. and Chua, N. -H. (1993) Functional analysis of petunia floral homeotic MADS box gene *pMADS1*. *Genes and Development*, **7**, 1214–1228.

Vollbrecht, E., Veit, B., Sinha, N. and Hake, S. (1991) The developmental gene *KNOTTED-1* is a member of a maize homeobox gene family. *Nature*, **350**, 241–243.

Wimmer, F. E. (1943) Campanulaceae–Lobelioideae, I. Teil, in *Das Pflanzenreich*, (ed. R. Mansfeld), vol 4(276)b, pp. i–vi + 1–260, Engelmann, Leipzig.

Yanofsky, M. F., Ma, H., Bowman, J. L., Drew, G. N., Feldmann, K. A. and. Meyerowitz, E. M. (1990) The protein encoded by the *Arabidopsis* homeotic gene *AGAMOUS* resembles transcription factors. *Nature*, **346**, 35–39.

Yao, J., Dong, Y., Kvarnheden, A. and Morris, B. (1999) Seven apple MADS-box genes are expressed in different parts of the fruit. *Journal of the American Society for Horticultural Science*, **124**, 8–13.

Yu, D., Kotilainen, M., Pöllänen, E., Mehto, M., Elomaa, P., Helariutta, Y. *et al.* (1999) Organ identity genes and modified patterns of flower development in *Gerbera hybrida*. *Plant Journal*, **17**, 51–62.

Chapter 19

Integrating molecular and morphological evidence of evolutionary radiations

R. M. Bateman

ABSTRACT

Effective studies of evolutionary radiations require broad-minded practitioners who can: (a) critically assess *a priori*, process-based rhetoric (especially when pertaining to adaptation); (b) effectively synthesise both phenotypic and genotypic information in order to obtain credible answers to profound evolutionary questions; and (c) transcend the current paucity of interest in relative rates of evolution in morphology (the direct expression, and mediator, of evolution) versus the segments of DNA currently in fashion for phylogenetic analysis (in most cases without morphological expression and thus mere evolutionary fingerprints – observers rather than players in the great evolutionary game). Genomes are at least vaguely clock-like, however unreliable the time that they keep. Morphology is held in equilibrium by stabilising selection for much of evolutionary time, punctuated by relatively rapid speciation events. Thus, rapid radiations are from first principles better tracked by morphology than molecules, and the seductively simple solution of mapping across molecular phylogenies a few morphological characters selected *a priori* (a practice currently much in vogue) gives priority to the wrong type of phylogenetic data. The wide range of evolutionary processes potentially underlying any specific radiation can usefully be narrowed by comparing the topologies and relative branch lengths of phylogenies generated from molecular and morphological data for the same range of coded taxa (a practice here termed 'cross-matrix disparity'). Compared with molecular trees, misnamed 'total evidence' trees, generated via the fusion of acceptably congruent matrices of molecular and discrete morphological data, probably offer a better framework for the over-rated procedure of mapping continuous morphological characters and extrinsic (ecological and environmental) factors often implicated in radiations. Evolutionary phylogenetics will only transcend these self-imposed barriers to optimal interpretations by taking a genuinely integrative but also genuinely empirical approach to the interpretation of evolutionary radiations – an approach that synthesises phylogenetic pattern, evolutionary process and ecological context.

General Abercrombie struggled to his feet, a sheaf of notes in his hand.
Your Excellency, my lords, Admiral Bertie, and gentlemen. We are met here together,' two bars of silence, 'on this happy occasion,' two more bars, 'to celebrate what I may perhaps be permitted to call an unparalleled feat of *combined*

In *Molecular systematics and plant evolution* (1999) (eds P. M. Hollingsworth, R. M. Bateman and R. J. Gornall), Taylor & Francis, London, pp. 432–471.

operations, of *combination*, valour, organisation and, I may say, of indomitable will.' Pause. 'I take no credit to myself.' Cries of No, no; and cheers. 'No. It is all due,' pause, 'to a young lady in Madras.'

'Sir, sir,' hissed his aide-de-camp, 'you have turned over two pages. You have come to the joke.'

(Patrick O'Brian, 1977, *The Mauritius Command*: 346–7)

19.1 Introduction

All topics are of course equal in evolutionary systematics, but some are more equal than others. Among the popular review topics of the past century, evolutionary radiations (almost always described as *adaptive* radiations) have drawn especially large quantities of ink. However, in recent times this largely morphologically-based literature has been swamped by a deluge of molecular biological studies. In the ensuing scientific and sociological tension, some hard-line morphologists have routinely criticised molecular biologists for diverting much of the ever-dwindling resources available to systematics into suboptimal evolutionary studies driven more by technology than biology. For their part, some hard-line molecular biologists have viewed morphologists as archaic, irrelevant and unnecessarily obstructive to a remarkable technologically-driven paradigm shift. As usual, the truth is likely to reside somewhere between the two polarised extremes.

This concluding paper to this volume attempts to link the preceding phylogenetic contributions within the broadly defined topic of radiations. Although I draw my few empirical examples exclusively from the plant kingdom, in accord with the remit of this volume, I believe that most of my assertions are broadly applicable to the remaining kingdoms of the Earth's biota. Admittedly, the paper is also an unashamedly biased personal manifesto, intended to convince the reader that:

(1) Most evolutionary biologists use the term 'radiation' with insufficient care, having failed to pay explicit attention to the many potential defining criteria.

(2) Similarly, most authors firmly manacle the qualifier 'adaptive' to the term 'radiation' throughout their texts without seriously considering the (in my view, highly questionable) general primacy of adaptation in evolutionary biology.

(3) Persistent ideological conflict between morphologists and molecular biologists is profoundly misconceived; rather, the identification and interpretation of evolutionary radiations cannot be satisfactorily achieved without constructing morphological *and* molecular matrices for the same range of analysed species and comparing the resulting phylogenies.

19.2 The ambiguous nature of the evolutionary radiation

19.2.1 Conceptual framework for evolutionary radiations

Although many scientific terms are subject to a wide range of implicit or explicit definitions and careless applications, the evolutionary radiation may take pride of place as the most routinely ill-defined among the first rank of biological concepts.

In practice, the many ambiguities boil down to a single question: just when *is* evolution sufficiently profound to constitute a radiation?

Review of the extensive literature shows a strong dichotomy in emphasis between palaeobiological and neobiological attempts to answer this question. Neobiologists tend to focus on case-studies at low taxonomic levels (Sanderson, 1998), and have the opportunity to include direct and detailed observations on molecular processes, genotypic expression and ecological interactions. In contrast, palaeobiologists have a stronger grasp of the significance of time in evolution, and are increasingly becoming the last bastions of morphology, since it is of necessity their primary comparative tool. Radiations tend to be discussed in terms of higher taxonomic ranks (Bateman *et al.*, 1998), invoking a sweeping (if often somewhat unconstrained) conceptual grandeur.

One of the more interesting palaeobiological syntheses was offered by Erwin (1992), who recognised six types of radiation defined by four criteria: species productivity (broadly, speciation, and its relationship with diversity), degree of morphological divergence (disparity, and its relationship with complexity), rate of morphological divergence, and the phylogenetic breadth of the taxa actively involved in the radiation (a scheme summarised here as Table 19.1). In Erwin's classification, novelty radiations generate higher taxa through rapid and profound morphological divergence, often occupying previously vacant ecospace. Diversification radiations are driven by new extrinsic ecological opportunities offered to several clades; they are characterised by greater speciation rates but less profound divergence. Economic radiations are similarly extrinsic but involve fewer lineages and a more restricted environmental range. All the aforementioned classes of radiation can involve adaptation, but true adaptive radiations should be *prompted* by adaptations; they typically evolve in a single clade and promote rapid speciation (Erwin recognised two subcategories, according to the degree of morphological divergence exhibited). Lastly, divergence radiations reflect 'steady state' evolution, being progressive increases in species number within a clade by relatively slow and morphologically superficial speciation.

Even if Erwin's classification *per se* is provisional and intended primarily to stimulate debate, the criteria that he selected to define his classification undoubtedly

Table 19.1 Classification of radiation types suggested by Erwin (1992, figs. 3, 4). H = high, M = moderate, L = low.

Radiation type	Species productivity	Degree of morphologic divergence	Rate of morphologic divergence	Phylogenetic breadth
Novelty	L	H	H	(L–)H
Extrinsic I (Diversification)	H	M–H	L–H	H
Extrinsic II (Economic)	L–H	L–M	L–M	L(–M)
Adaptive I	H	M–H	M–H	L
Adaptive II	H	L–M	M–H	L
Divergence	M–H	L	L	L

form a suitable basis for further discussion. In particular, he used intrinsic properties of the organisms to define the radiations but was often obliged to resort to extrinsic properties of their environments to explain radiations.

In the following overview of potential criteria for defining and categorising an evolutionary radiation, I set aside the additional complicating (albeit important) issue of whether the evolutionary events driving the radiation generate novel adaptations that confer increased fitness, and may even qualify for yet another much-debated status, the 'key innovation' (cf. Nitecki, 1990; Erwin, 1992; Bateman, 1995, 1996; Givnish, 1997; Bateman *et al.*, 1998; Sanderson, 1998). It is, however, worth noting at the outset that much of the often heated debate regarding the best definition of a radiation has been caused by imprecise formulation of concepts, ambiguous use of terminology, and especially by incongruence between those who view radiations as adaptive by definition and those who take a broader and less judgmental view.

19.2.2 Intrinsic criteria (i.e. reflecting phenotype and/or genotype)

Criterion 1. Taxonomic diversity. Linnean classifications are by far the most commonly used basis for organismal diversity estimates, usually focusing on a single taxonomic rank. In my opinion, species constitute the most attractive basic analytical units, as they are potentially self-defining, in contrast with less readily defined higher taxonomic categories. Having said this, in practice most plant species remain the product of alpha-taxonomy alone; the implicit hypothesis that they possess the genetic cohesion required of *bona fide* biological species has not been tested, so that their circumscription arguably remains as ambiguous as that of higher taxa (higher taxa are especially popular as fundamental analytical units in palaeontological studies of radiations, as species circumscription among fossils is even more troublesome, being biologically untestable).

Criterion 2. Character diversity. Any phylogenetic analysis is based on one or more matrices of descriptive characters, each character being represented by at least two readily differentiable states (e.g. the now seriously clichéd red versus blue flowers in morphology, and A versus G for a specified base-pair in DNA sequences). It is the character states rather than the characters that are the basic unit of phylogenetic analysis, since states can be assigned to characters in many different ways, essentially at the whim of the analyst (e.g. Scotland and Pennington, 1999). Obviously, the number of *bona fide* character states included in a matrix is positively correlated with the degree of effort devoted to investigating their 'host' organisms, but it is also constrained by a theoretical upper limit that reflects the complexity of the organism (this assertion applies equally to phenotypic and genotypic data sources).

The net rate of species increase within a specified clade may be an inadequate measure of a radiation. For example, niche partitioning could saturate an ecosystem with species without generating new morphological innovations, simply by increasingly partitioning among lineages the pre-existing repertoire of features. Alternatively, rapid divergence of lineages could increase the pool of features within the clade even if its species-level diversity declined (e.g. Barrett and Graham, 1997).

Thus, the best measure of rate of change in phenotypic diversity may be the product of the rate of increase in species number and the rate of increase in the overall pool of character-states.

However, at present our knowledge of the full sequences of plants is confined to a few model organisms of particular economic importance, such as *Arabidopsis* and *Zea/Oryza*, and even then this information does not truly define the overall genetic diversity as we have poor knowledge of intraspecific sequence variation. Even most morphological phylogenetic matrices fail to describe at least some major elements of phenotype, and comparison among morphological matrices is further handicapped by profound differences in criteria applied when delimiting and selecting character states (e.g. Bateman and Simpson, 1998; Bateman, 1999; Scotland and Pennington, 1999). Thus, in practice, species richness remains more easily and accurately estimated than character-state richness.

Codicil 1. We have yet to set any boundaries on the biological entity whose diversity we are attempting to assess; these can be geographic, ecological and/or taxonomic. Geographic and ecological factors are by definition extrinsic to the organism and thus discussed below, but taxonomy is dominantly intrinsic. Many studies of radiations have knowingly focused on sets of species that shared specific habitats but were not taxonomically related, and in many other cases single higher taxa were extracted directly from non-phylogenetic classifications, thereby incurring a serious risk that the suite of species under scrutiny was not monophyletic. Hopefully, few systematists would now disagree with the premise that radiations should be assessed within well-defined monophyletic groups, given their single evolutionary origin and inclusive nature (e.g. Smith and Patterson, 1988).

Codicil 2. There is strong phylogenetic evidence that the present-day vascular land-plant flora, currently estimated at 270,000 species (Walter and Gillett, 1998), began to diverge from a single ancestor about 420 My ago (Kenrick and Crane, 1997; Bateman *et al.*, 1998). Thus, we would have some justification in regarding the last 420 My of land-plant speciation as a remarkable evolutionary radiation – provided, of course, that we made this assertion in comparison with the previous *ca.* 3,000 My (88% duration) of life on Earth. Radiations by definition have a temporal context, but often this is not explicitly stated. If an evolutionary radiation is essentially a phase of exceptional biotic diversification, then it must be defined as a period when the rate of diversification is considerably higher than the known background rate for that clade. In other words, we need acceptable estimates of the diversity of that clade throughout its evolutionary history before we can confidently identify radiations within the clade.

Codicil 3. Although many previous reviewers of this topic chose to define radiations using rates of diversification, they often employed taxonomic measures of diversity, most commonly species richness. However, an increased rate of origination of species (or, if we employ the more ambitious Criterion 2, an increased rate of origination of character states) will not increase the overall diversity of the clade if there is an increase of similar magnitude in the rate of extinction of species (or, for Criterion 2, of loss of character states). Thus, it is clear that a radiation should be defined as a differential between origination rate and extinction rate; *both taxonomic and character-state diversity are best assessed as rates of origination over extinction per unit time within a single well defined clade.*

Codicil 4. As well as delimiting monophyletic groups, a phylogenetic context is of more general value in the study of radiations. A cladogram (a) tests by congruence initial assertions of character-state homology (e.g. Patterson, 1988), (b) identifies homoplasy and thus minimises confusion of phenotypically similar but genotypically dissimilar character-state transitions, and (c) simplifies the study of character change by emphasising apomorphies. The explicit sequence of acquisition (or loss) of sets of character states on each cladistic branch provides biologically meaningful phylogenetic correlations of dynamic character-state *transitions*, rather than biologically meaningless statistical correlations of static character states. Among other benefits, this offers the possibility of identifying a key innovation underlying a radiation. However, note that this is likely to be first recognised as the phenotypic expression of an actively evolving (typically developmental) gene, rather than the non-phenotypically (or, to be more accurate, non-morphologically) expressed genes currently favoured for phylogenetic fingerprinting.

Moreover, when the species under comparison are placed in the context of a well supported cladogram, at least three additional measures of radiations can in theory be applied:

Criterion 3. Topological asymmetry. Differences in the relative frequency of cladogenesis across a higher-level clade during the specified time interval has the potential to generate strong topological asymmetry (imbalance), and the points of origin of such asymmetry can be viewed as the initiation points of radiations. Literature on this approach has expanded rapidly in the 1990s and of necessity entails a strong element of mathematical modelling (cf. Nee *et al.*, 1992; Guyer and Slowinski, 1993; Sanderson and Donoghue, 1994, 1996a; Fusco and Cronk, 1995; Heard, 1996). The complexity of the modelling process increases greatly with the size of the phylogenetic matrix, leading to much controversy regarding the most appropriate parameters to select from a wide range of options (e.g. Sanderson and Donoghue, 1996a, table 1).

Criterion 4. Single-matrix phylogenetic disparity. At its most basic, phylogenetic disparity is the genealogical dissimilarity between a pair of coded taxa, as determined by counting the number of character-state transitions from the base of the cladogram to their respective branch tips (or, alternatively, from one coded taxon to another by the shortest possible route across the cladogram; that is, via their most recent shared hypothetical ancestor). The approach has found particular favour among palaeobiologists for assessing degrees of morphological divergence within events such as the Cambrian invertebrate radiation (Briggs *et al.*, 1992) and later marine radiations (Foote, 1996).

Criterion 5. Cross-matrix phylogenetic disparity. Another increasingly popular activity is to compare molecular phylogenies generated for the same range of coded taxa from more than one gene. However, these discussions generally focus on topological similarities and the strength of internal nodes, asking which gene is likely to give the more robust and more accurate phylogeny for the group under scrutiny (e.g. Soltis *et al.*, 1998a). I believe that an opportunity has been largely missed to use cross-matrix disparity to unravel evolutionary patterns in general and radiations in particular, not by comparing different molecular trees but by comparing the congruence of molecular and *morphological* trees. Also, it is necessary to compare topologies *and* branch lengths, within a specified clade during a specified time interval (Bateman, 1996, 1999).

19.2.3 Extrinsic criteria (ecology s.l.)

A second tranche of potential criteria for defining radiations necessitate consideration of more than mere characters intrinsic to a single clade; they require in addition extrinsic information reflecting the interactions of species with each other and with the environment.

Criterion 6. The number of clades showing simultaneous rapid increases in intrinsic criteria (i.e. taxonomic and/or character diversity). Even if quantitative assessments of radiations are best restricted to a narrowly-delimited clade, there may be compelling reasons for comparing diversification patterns among clades. For example, the palaeontological literature frequently documents events where a profound environmental change appears to have prompted synchronous accelerated evolution in two or more clades that shared a specific habitat or lifestyle (a concept also increasingly employed in studies of recent ecological refugia). Moreover, some such events result in the acquisition by unrelated clades of similar novel characteristics during speciation, implying parallel adaptive responses following the inferred environmental perturbation. One problem with this approach is selecting appropriate clades and determining whether they are truly comparable entities (e.g. if they possess similar levels of species richness or similar times of origination).

Criterion 7. The sum total of niches occupied by the clade during the specified time interval. If this criterion is to differ from species richness it must make the questionable assumption that niches can be defined without reference to particular species – in other words, that plants can be viewed as ecomorphs playing specific roles within predefined ecosystem frameworks (e.g. Behrensmeyer *et al.*, 1992; Erwin, 1992).

Most authors agree that the simplest scenario is to radiate into empty ecospace. However, it is important to note that two very different types of empty ecospace exist. Primary niche vacancies occur in landscapes or habitats that have never been occupied (for example, the entire terrestrial realm prior to 420 My ago: Bateman *et al.*, 1998). Successful occupants of such niches are unlikely to exhibit extensive pre-adaptation, as the lineage will not previously have encountered the environmental challenges posed by the vacant niche. As most habitats had been invaded (if not wholly conquered) by land-plants within 120 My of terrestrialisation (Behrensmeyer *et al.*, 1992), primary vacancies are rare or absent from the modern flora. In contrast, secondary niche vacancies were previously occupied but have been temporarily vacated, typically due to environmental perturbations that proved sufficiently profound to suspend incumbent advantage. A potentially critical factor, incumbent advantage is effectively a 'top-slice' of fitness, conferred not by the intrinsic properties of the organism but by the mere fact of its extensive occupancy of the niche, which constitutes a significant deterrent to any potential competitors (Gilinsky and Bambach, 1987; Pimm, 1991; Rosenzweig and McCord, 1991; DiMichele and Bateman, 1996). Any successful colonists of these temporarily vacant niches are likely to be characterised by extensive pre-adaptation, being well fitted for the niche but having been obliged to wait patiently for environmental happenstance to locally eliminate (strictly, extirpate) the previous incumbent.

Criterion 8. The sum total of ecospace occupied by the clade during the specified time interval. This may be thought to equate with Criterion 7, but the two concepts are not fully congruent if evolution leads to subdivision of previous niches within a specified ecospace. For example, the ability to accommodate many species within limited ecospace appears to be the great strength of angiosperms (e.g. Crane, 1989), relative to coarser niche division among other highly ecologically successful plant groups that evolved earlier in the fossil record such as the rhyniophytes in the Late Silurian–Early Devonian, archaeopterid progymnosperms in the Late Devonian, rhizomorphic lycopsids in the Late Carboniferous, and conifers in the Jurassic (e.g. Behrensmeyer et al., 1992). Searches for elusive 'key innovations' have been questionably successful for primary radiations of both the angiosperms (Sanderson and Donoghue, 1994) and land-plants (Bateman et al., 1998).

Criterion 9. The proportion of global biomass contributed by the clade during the specified time interval. Biomass is the measure of biotic success most beloved of ecologists. However, it need not reflect either taxonomic diversity or character-state diversity; for example, in the present cold-temperate zone of the Northern Hemisphere huge amounts of biomass are sequestered in remarkably few species of broadly phenotypically similar coniferous trees. Also, the palaeobotanical record suggests that most of the species that achieved exceptional ecological success were relatively short-lived, perhaps due to over-specialisation to narrow environmental parameters and consequent extinction when those parameters changed too rapidly for successful adaptive responses.

Erwin (1992) omitted extrinsic criteria from his classification of radiations, though his text discussed them in detail as essential causal factors. Other authors have explicitly added ecological diversification to phenotypic diversification (both essential criteria for assessing adaptation) and lineage diversification as a defining framework for radiations (cf. Givnish, 1997; Jackman et al., 1997; Sanderson, 1998). Pragmatically, however, it is difficult to see how extrinsic factors could be satisfactorily built into phylogenies; any phylogenetic approach to the study of extrinsically-driven radiations must in practice rely on character 'mapping' (see below).

19.2.4 Temporal context of radiations

Most if not all of the intrinsic and extrinsic criteria are potentially relevant to identifying and interpreting evolutionary radiations. Nonetheless, the most practical definition, given present levels of data, is that a radiation is a large surplus of species birth over species death (natality over mortality in the terminology of ecological survivorship), each calculated as rates averaged over specified time intervals. These calculations of species turnover should be constrained physiographically – to a specific geographical area and/or range of habitats – and systematically – preferably to monophyletic clades rather than the more conventional higher taxa prescribed by pre-cladistic Linnean hierarchies. Admittedly, some authors downplay the importance of temporally defined rates in their preferred definitions of a radiation (cf. Sanderson and Donoghue, 1996b; Barrett and Graham, 1997; Givnish, 1997).

Obviously, if we choose to focus on rates, calculations of rates of species mortality and natality require an effective measure of time. Setting aside laboratory-based breeding experiments, which involve overly simple organisms, overly simple ecosystems, and observations over too few generations, there are two viable methods of assessing biodiversity levels through time:

Method 1. We can document first and last appearances of species in the fossil record. Although a valuable exercise, this is fraught with difficulties. Both the fossil record and the fossil plants themselves are highly fragmentary. Fragmentation of the plants means that only the most skilfully reconstructed whole-plant species can be compared confidently with their modern descendants in a morphological phylogenetic analysis. The temporally and geographically sporadic nature of the terrestrial fossil record means that the observed origin and extinction dates of a species are liable to incur statistical errors measured in many millions of years (e.g. Fortey and Jefferies, 1982). This realisation led some authors to argue that the timing of speciation events may be more accurately predicted indirectly, via cladograms, than observed directly in the fossil record (e.g. Norrell, 1992). Such temporal resolution will be wholly inadequate if (as is argued below) most major evolutionary radiations are extremely rapid.

Method 2. More recently, the molecular phylogenetic revolution has provided the molecular clock hypothesis. We now routinely obtain nucleic acid sequences from a range of species for one or more specified genes to generate a molecular phylogeny (or, more precisely, a gene tree *sensu* Doyle, 1997) – typically for extant rather than extinct species and putatively physiologically rather than morphologically expressed genes. This approach underpins many of the contributions to this volume. In theory, once the preferred most-parsimonious tree based on nucleic acid sequences has been identified, we can determine the number of base substitutions separating each speciation event as a measure of molecular disparity. If in addition we can date at least one branch point (most commonly using the first appearance in the fossil record), it can be related to the present-day time-line. We can then interpolate dates of the remaining branch points in the phylogeny by assuming a constant rate of base substitution for the targeted fragment of genome throughout the evolutionary history of the clade under scrutiny.

Because morphological branch lengths are not proportional to time elapsed since speciation, morphological phylogenies lack the horizontal alignment of coded taxa along the present time-plane that is predicted by the 'molecular clock' in molecular phylogenies. However, the burgeoning data-base of molecular phylogenies has made it abundantly clear that the molecular clock is also a rash assumption, subject to many modifying factors (e.g. Avise, 1994; Sanderson, 1998b). Earlier authors tended to excuse profound mutation rate variations among species as tending to characterise distant relatives, as though mutation rate itself was a phylogenetic character (it may well be under considerable phylogenetic constraint, but hardly conforms to the requirement for discontinuity between alternate states of a character). For example, an 18S rRNA neighbour-joining tree of 500 species representing all eukaryotic kingdoms (Van De Peer and De Wachter, 1997) reveals apparently accelerated rates of mutation in several distinct clades, such as the *Cladophora*, *Acetabularia* and *Chlorochion* groups (green algae), *Plasmodium* group (Apicomplexa), red algae and higher animals.

However, at a smaller scale, a stepwise arrangement of internal branches implies a gradual increase in mutation rate during the evolution of some clades (e.g. dinoflagellates, *Penaeus* and *Dicyema* groups of Animalia). Several more recent studies at lower taxonomic levels have revealed profound mutation rate contrasts among closely related taxa – they can even occur among recently diverged sister-species. In some groups, there is a clear biological explanation for the rate change.

For example, the relatively species-poor subtribe Neottieae of the family Orchidaceae – dominantly temperate and terrestrial – demonstrates an interesting contrast in relative branch lengths between nuclear ITS and plastid *trnL* phylogenies (Hollingsworth *et al.*, 1998, unpublished; Molvray *et al.*, 1998). Specifically, the clade exhibits several independent origins of obligate saprophytism, and the saprophytic species possess unusually long branches in the *trnL* phylogeny but not in the ITS phylogeny. The obvious interpretation is that mutation rate increases greatly in the functionless plastids, thereby undermining any credibility for applying a molecular clock across the clade as a whole. Not surprisingly, facultatively saprophytic individuals that occasionally appear within populations of otherwise characteristically photosynthetic species of Neottieae possess plastid mutation rates similar to those of conspecific photosynthetic individuals (Hollingsworth *et al.*, unpublished).

Comparison of the plastid genomes of the holoparasitic *Epifagus* and fully photosynthetic *Nicotiana* (both Scrophulariaceae: Wolfe *et al.*, 1992; Nickrent *et al.*, 1998) revealed extensive deletions that left intact only 42 genes, plus the inverted repeat, in the former, but still allowed near-colinearity with the latter. More precisely, most photosynthetic genes have been deleted, whereas the gene expression machinery related to ribosomal activity has been left largely functional.

However, not all mutation rate transitions within clades are so readily explained as in saprophytes and parasites. As data accrue, it will be interesting to judge the performance of more recently advocated protein clocks, which combine many enzyme sequences (e.g. Doolittle *et al.*, 1996; see also Stoebe *et al.*, 1999 – this volume).

To return to our original question, we have finally arrived at a working definition of a radiation: *a large surplus in the rate of natality over the rate of mortality for species and/or character states within a specified clade over a specified time interval.* We have also accepted that the practical applicability of this definition depends on our ability to at least approximately date evolutionary events, preferably using degrees of sequence divergence calibrated via the fossil record. But most importantly, note that evolutionary radiation as defined here is purely a historical pattern; we have assumed that the cladogram reflects evolution but we have as yet made *no* assumptions with regard to underlying biological processes.

19.3 Radiation is a pattern whereas adaptation is a process

19.3.1 Both adaptive and non-adaptive processes can underlie punctuational patterns

As I have previously noted (Bateman, 1999: 13), although 'radiation' is almost always firmly manacled to the term 'adaptive', the adaptive nature of evolutionary radiations is treated as reliably axiomatic – its veracity is rarely considered, let alone tested. In several recent papers (e.g. Bateman and DiMichele, 1994; Bateman, 1996, 1999; DiMichele and Bateman, 1996; Bateman *et al.*, 1998) Bill DiMichele and I have argued that the assumption that adaptation is the dominant driving force of radiations requires far more empirical support than is currently available, and that in many circumstances non-adaptive scenarios possess greater credibility. My preferred prompt for radiations – at least, for strongly divergent 'novelty' radiations – is saltational evolution over a single generation via germ-cell mutation or allopolyploidy. Moreover, several parasaltational mechanisms that allow radical phenotypic change over a few generations also merit greater attention than they currently receive (for a more detailed discussion, including precise definitions of these terms, see Bateman and DiMichele, 1994). Adaptive radiations that are couched in terms of classic Hardy–Weinberg equilibria within infinite panmictic populations constitute the reddest of herrings if selection pressure is low, adaptation of secondary importance and speciation occurs via mutant 'hopeful monsters' (Bateman, 1995; see also Patterson, 1999).

I have further argued that models of adaptive evolutionary change are best summarised in Wright's (1968) evocative concept of an adaptive landscape (or, in my process-neutral parlance, a *fitness* landscape), with each genetically cohesive population gradually climbing the slope of an adaptive peak towards its goal of a local fitness optimum (Fig. 19.1). Major problems highlighted by this caricature of neoDarwinism are how to move from peak to peak across the intervening fitness valleys (particularly if such low levels of fitness are in practice lethal), and how to colonise peaks already occupied by other populations. The first problem is readily solved by generating a great number and diversity of radically altered mutants that could potentially leap across the valleys in a single generation, rather than attempting to traverse the valleys in many generations (the *generative phase*) – a process often accompanied by niche-switching. The second problem is solved if we are dealing not with a slowly changing and recalcitrant fitness *landscape* but with a constantly changing and more fluid fitness *seascape* (Bateman, 1995, 1999; Fig. 19.1). Under this model, the fitness peaks wax and wane too rapidly to be optimised (and thus occupied) by relatively slow adaptive mechanisms, leaving them reliably vacant for any extremely fortunate mutant (termed a 'prospecies') that possesses near-optimal characteristics for that peak at that moment in time (the *establishment phase*). Fitness landscapes are most likely to become less stable seascapes immediately following major environmental perturbations.

Having outlined my somewhat unconventional neoGoldschmidtian perspective on evolution, I will not attempt further 'supraDarwinian' justifications here, beyond noting that empirical results consistent with these hypotheses are summarised below. Rather, I will step back from the brink of heterodoxy and instead merely predicate my subsequent discussion with a single less radical assumption – that the punctuational *pattern* of evolution perceived and widely popularised by S. J. Gould and others (Eldredge and Gould, 1972; Gould and Lewontin, 1979; Eldredge, 1992; Gould and Eldredge, 1993) is real, rather than an artefact of the absence from the fossil and living records of a myriad of phylogenetically intermediate species ('ghost' species that would reflect non-preservation due to rapid extinction, persistent rarity and/or relatively poor preservation potential).

The popular punctuationist–gradualist controversy has survived two decades and is likely to continue to entertain palaeobiologists and neobiologists for many years to come, echoing (and to some degree reviving) the far older catastrophist–gradualist debate that still overarches the earth sciences (e.g. Hallam, 1983). However, in my view the best-documented case-studies of evolution in the fossil record

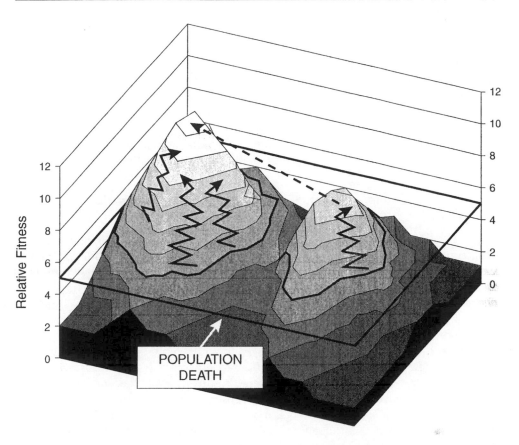

Figure 19.1 A hypothetical fitness 'seascape', indicating how a fortuitous macromutation can allow a lineage to saltationally cross a lethal fitness depression in a single generation (solid arrow) and subsequently drive towards a fitness optimum via traditional neoDarwinian selection (dashed arrow). Repeated environmental fluctuations mean that in practice the fitness optimum shifts too frequently to be attained.

(e.g. Jackson and Cheetham, 1994, 1999; Wray, 1995) give great credibility to the reality of the rectangular (i.e. phenogram-like) pattern inherent in punctuated equilibrium, and recent breakthroughs in our understanding of the developmental–genetic underpinnings of macroevolution (e.g. Slack *et al.*, 1993; Davies and Schwarz-Sommer, 1994; Thiessen *et al.*, 1996; Reeves and Olmstead, 1998; Gustafsson and Albert, 1999 – this volume; Möller *et al.*, 1999 – this volume) offer seductive causal processes to explain punctuational patterns – at least, those evident in morphological phylogenies.

Review of the recent literature suggests that the reality of long periods of evolutionary stasis in specific lineages (i.e. 'equilibrium') is increasingly accepted by both neoDarwinian and supraDarwinian researchers. The most obvious explanation for stasis is that, in most ecological circumstances, neoDarwinian natural selection enhances genetic stability. In other words, the background mode of natural selection is stabilising selection, which inhibits evolutionarily meaningful morphological change. The relatively rapid intervening periods of evolution and speciation (i.e. 'punctuation') may occur under the relatively high directional or disruptive selection pressures that underpin neoDarwinian microevolution, or they may occur even more rapidly under the relatively low selection pressures that are more characteristic of the various supraDarwinian macroevolutionary scenarios.

In either case, the assumed reality of the pattern of long periods of stasis separated by much shorter periods of evolutionary change has profound implications for our interpretation of both molecular and morphological phylogenies – implications that have been largely neglected in the evolutionary literature.

19.3.2 Punctuational patterns in morphological and molecular phylogenies

In order to set the stage for subsequent discussion of three genuine case-studies, I have concocted a simple hypothetical phylogeny of eight species, four extant and four extinct, all ultimately derived from a single ancestor (Fig. 19.2). Patterns of divergence are contrasted for morphological data (left) and sequence data (right) under two scenarios (top versus bottom). Both scenarios share geologically instantaneous (punctuational) morphological divergence but constant, clock-like sequence divergence. Extinction events are evenly spaced and occur at the same points in time in both scenarios. In contrast, speciation events are evenly spaced in the first scenario, but highly basally concentrated in the second scenario. Moreover, the degree of morphological divergence shows no overall trends in the first scenario, but in the second the basal divergences are far more profound (i.e. more disparate) than those associated with later speciation events. In other words, in the second scenario, both the rate and morphological magnitude of lineage divergence decrease through time.

Both scenarios could in theory be viewed as modest radiations, since each begins with one extant species and ends with four. However, if we impose our additional criterion – requiring a species natality/mortality ratio well above the background rate for the clade – the true radiation occurs only in the second scenario, where it is confined to the brief stippled period. Moreover, the fractal pattern of diminishing speciation rate and degree of morphological disparity through time is consistent with Erwin's (1992) 'novelty radiation' – rapid and profound morphological divergence into free ecospace – and hence with the saltational evolutionary processes outlined by Bateman and DiMichele (1994). In other words, we can use phylogenetic pattern to (a) identify radiations and (b) constrain (or, more accurately, to falsify some elements of) the potential range of underlying evolutionary processes.

But this interpretation is based primarily on the morphological phylogenies; what do the corresponding sequence phylogenies tell us? Well, their clock-like properties accurately date the relative divergence times of the extant lineages. However, we learn nothing of the relationships of the extinct species, nor can we meaningfully separate the most recently divergent of the four extant species from its sister-species (in this example there has been insufficient time for the novel species to acquire molecular autapomorphies). Moreover, in the radiation scenario, the much earlier divergences of the remaining three extant lineages are too closely spaced in time to satisfactorily resolve their relationships. In other words, first

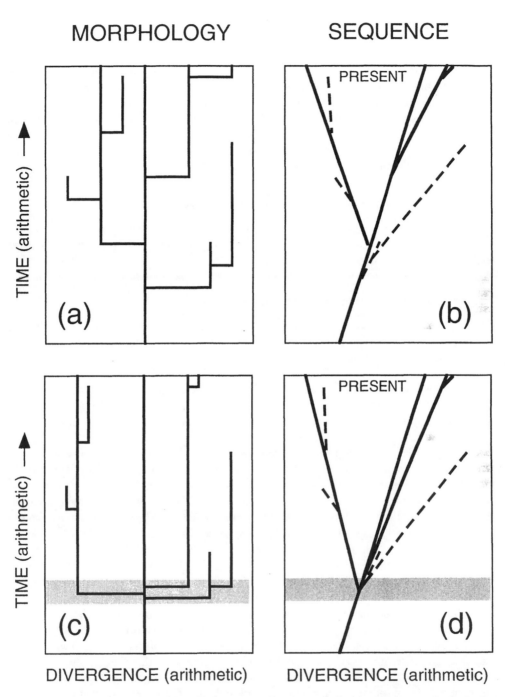

MORPHOLOGY   SEQUENCE

TIME (arithmetic) →

(a)   PRESENT

                              (b)

TIME (arithmetic) →

(c)   PRESENT

                              (d)

DIVERGENCE (arithmetic)   DIVERGENCE (arithmetic)

Figure 19.2 Comparison of the patterns of degree of divergence through time for morphological data (left) and sequence data (right) for a small hypothetical clade of eight species (four extinct). In the first scenario (a–b), speciation rate and degree of morphological divergence are random through time, whereas in the second scenario (c–d) early divergences constitute a radiation (stippled), being far more temporally concentrated and more profound than later divergences.

principles suggest that sequence data are poor for delimiting recently evolved species (cf. Davis, 1999 – this volume) and poor for resolving rapid radiations within phylogenies (Bateman, 1996, 1999).

To emphasise a key point of this essay: If morphological evolution follows a punctuational pattern (dictated by long periods of stabilising selection that are only occasionally broken by temporary release from selection and consequent speciation) and thus there is no morphological clock, but if in contrast genomic mutation is broadly clock-like, then in phylogenetic terms *the vast majority of morphological character-state transitions occur **during** speciation events and the vast majority of molecular character-state transitions occur **between** them* (Fig. 19.2).

19.4 Three molecular phylogenetic case-studies

To further investigate the implications of these theoretical observations, I now turn to three selected phylogenetic case-studies that compare, either implicitly or explicitly, molecular and morphological data.

19.4.1 Northern hemisphere Orchidinae

The first study concerns a moderately large (127 northern hemisphere ingroup species, plus two southern hemisphere outgroups) molecular matrix based on the full ITS assembly; spanning the largely temperate and terrestrial subtribe Orchidinae of the family Orchidaceae, it is primarily the product of a joint research project between the Royal Botanic Gardens of Kew and Edinburgh (Bateman *et al.*, 1997, unpublished a, b; Pridgeon *et al.*, 1997). Plastid sequences (*trnL*) are also now available for the majority of the species, though regrettably I have not yet completed the corresponding morphological matrix.

The *ca.* 3400 most-parsimonious trees (e.g. Fig. 19.3) collectively reveal a need for generic recircumscriptions, involving reassignment of 10% of the species in the subtribe, in order to achieve a monophyletic classification. This most profoundly affected the type genus of this massive family, *Orchis s.l.*, which became phylogenetically fragmented during the last 150 years as taxonomists periodically 'cherry picked' morphologically apomorphic monophyletic genera, thereby leaving a polyphyletic 'plesiomorphic rump' *sensu* Bateman *et al.* (1999). This taxonomic revision usefully reduced the number of putative intergeneric hybrid combinations reported for the subtribe, since most were revealed to be infrageneric. More importantly from an evolutionary perspective, the phylogeny provided a framework for the mapping of non-sequence characters. This revealed a strong phylogenetic signal in chromosome numbers, which identified one tentatively delimited clade as prone to polyploidy and another clade as delimited by chromosomal fusions. Tuber morphology also proved to be phylogenetically conserved, whereas several other morphological characters are all highly homoplastic; examples include the presence of a hood-like galea, presence and size of a labellar spur, number and shape of bursicles, retinacles and viscidia on the gynostemium, discrete anthocyanin markings on vegetative organs, and dominant floral pigments (e.g. Strack *et al.*, 1989).

Much of this evolutionary lability is likely to reflect adaptive co-evolution with pollinators (e.g. Dafni, 1992). This conclusion was also drawn from a more detailed

Figure 19.3 Randomly selected example of *ca.* 3400 most-parsimonious cladograms for ITS1–5.8S–ITS2 assemblies of 129 species of the Orchidinae (Orchidaceae). This provisional tree is part of ongoing work; it therefore lacks confidence figures and should be interpreted with caution (see also Pridgeon *et al.*, 1997).

ITS investigation of the orchidiniid genus *Platanthera* by Hapeman and Inoue (1997) though, interestingly, frequent reduction and/or loss of the spur in the Orchidinae challenges Hodges' (1997) conclusion that floral spurs tend to be key innovations in angiosperms. In addition, combinations of dominantly plesiomorphic morphological states characterise even some of the most derived orchidiniid clades, partly explaining previous taxonomic ambiguities (Bateman *et al.*, 1997; Pridgeon *et al.*, 1997).

Even more interesting patterns emerge from the ITS tree in Fig. 19.3 when branch lengths are taken into account. Setting aside the perennially troublesome iterative allotetraploids within *Dactylorhiza* (cf. Averyanov, 1990; Hedrén, 1996), near-polytomies characterise those genera that are also most controversial with regard to morphological species delimitation and relationships, namely *Ophrys*, *Serapias*, *Himantoglossum* and *Gymnadenia–Nigritella* (cf. Sundermann, 1980; Delforge, 1995). In the case of the first three genera, morphological divergence appears negligible in phylogenetically useful characters; the differences among supposed species are largely quantitative and believed to reflect very recent fine-tuned co-evolution with pollinating insects (e.g. Dafni, 1992; Devillers and Devillers-Terschuren, 1994). It is, of course, equally logical to assume that, given negligible molecular *and* morphological divergence, at least some of the putative species are in fact spurious (it will be interesting to see the results of several population-level molecular studies currently being conducted on these genera).

Passing on to the *Gymnadenia–Nigritella* clade (*Gymnadenia s.l.* as delimited by Bateman *et al.*, 1997), the surprisingly long ITS branches detected among supposed varieties of *Gymnadenia s.s.* suggest the presence of several morphologically cryptic species (Bateman *et al.*, 1997, unpublished), whereas sequence divergence is negligible among the several putative species within *Nigritella* (Bateman *et al.*, unpublished). More interestingly, there is also little sequence divergence between the paraphyletic *Gymnadenia s.s.* and monophyletic *Nigritella*, despite several profound differences in floral morphology that separate the two supposed genera. Here we have identified a molecular short branch that, in contrast, corresponds with a presumed morphological long branch (strictly, this interpretation still requires rigorous testing via a corresponding morphological matrix). One evolutionary hypothesis consistent with this phylogenetic pattern is a recent saltation event; a mutation in a key floral developmental gene occurred in an individual of *Gymnadenia s.s.*, simultaneously and profoundly affecting several floral features and thereby instantly generating the first individual of *Nigritella*. This prospecies successfully established itself, with polyploidy and autogamy (Teppner and Klein, 1985; Gerbaud, 1998) playing key roles in the subsequent putative species-level radiation.

19.4.2 Macaronesian Poteriae

Ever since Darwin's love–hate relationship with the Galapagos Islands, it has been clear that oceanic volcanic islands would play a major role in prompting advances in evolutionary theory. They constitute a relatively simple and discrete biological system that is nonetheless natural – or to be more precise *was* natural until historical times, when most such islands were overrun by man-transported invasives (Cronk and Fuller, 1995; Quammen, 1996). Islands are physiographically isolated, encompass a limited range of habitats (e.g. MacArthur and Wilson, 1967; Wagner and Funk, 1995; Clark and Grant, 1996; Thornton, 1996; Givnish and Sytsma, 1997a; Baldwin *et al.*, 1998; Steussy and Ono, 1998), and are rarely niche-satu-rated. When first formed they offer primarily vacant niches in a sterile landscape. Subsequent frequent and localised disturbances – especially common phenomena on volcanic islands – maintain fitness as a mobile 'seascape', and periodically clear specific habitats to create generally more transient, secondarily vacant niches. Both primary and secondary vacancies are occupied by a limited range of immigrant species and their descendants. As I noted previously (Bateman, 1999: 16), 'in theory, islands offer a relatively simple evolutionary system that might allow the "unholy trinity" of origination, radiation and migration to be untangled *a posteriori*.'

Increasing interest is being shown by molecular phylogeneticists in oceanic islands, with a clear preference for chains that are remote but possess floras that are relatively well documented morphologically. To date, the classic example is the near-ideal evolutionary microcosm of the Hawaiian chain (Wagner and Funk, 1995; Givnish and Sytsma, 1997a), but more recent studies have probed other regions of the globe. Among these is the Macaronesian group (including the Canaries, Azores and Madeira), located offshore from subtropical West Africa and formed within the last 21 My by tectonic activity (Bramwell, 1976). Pioneering molecular studies in Macaronesia have considered the Asteraceae (Kim *et al.*, 1996; Francisco-Ortega *et al.*, 1997) and the Rosaceae (Welsh, 1997), both of which show genus-level endemism.

Welsh's (1997) study of the subtribe Poteriae of the Rosaceae involved a 50-character morphological cladistic analysis of eight endemic ingroup species currently assigned to three genera. These were compared with a morphologically heterogeneous ensemble of five outgroup species from four genera – genera that, despite collectively spanning both hemispheres, have been accused in various non-phylogenetic studies of being potential sister-groups to the Macaronesian endemics. The resulting morphological cladogram (Fig. 19.4a) reveals robust, putatively monophyletic groups in both the ingroup and outgroup, and broadly supports the current generic delimitation of the endemic genera of the ingroup (albeit requiring the transfer of '*Marcetella*' *maderensis* to *Bencomia*: Welsh, 1997). The topology is consistent with the classic pattern of a single immigration event into Macaronesia followed by radiation among the islands. However, at least some species of the geographically widespread (and potentially paraphyletic) ecolog-ical generalist *Sanguisorba* appear to be sister to the ingroup, thus reducing the chances of identifying the most likely embarkation point for the migrating ancestor of the ingroup.

Despite the notorious recalcitrance of the Rosaceae to the extraction of sequence-able DNA, ITS1 sequences were eventually obtained from the majority of the outgroup and ingroup species (Welsh, 1997). The resulting single most-parsimo-nious sequence cladogram (Fig. 19.4b) confirms the distinctness of *Hagenia* and *Agrimonia* in the outgroup relative to *Sanguisorba*, which is again sister to an ingroup that, on present evidence, is unequivocally monophyletic. However, there is no meaningful resolution of the ingroup, where molecular branches are short or non-existent, in stark contrast to the long molecular branches subtending the more phylogenetically divergent outgroup members.

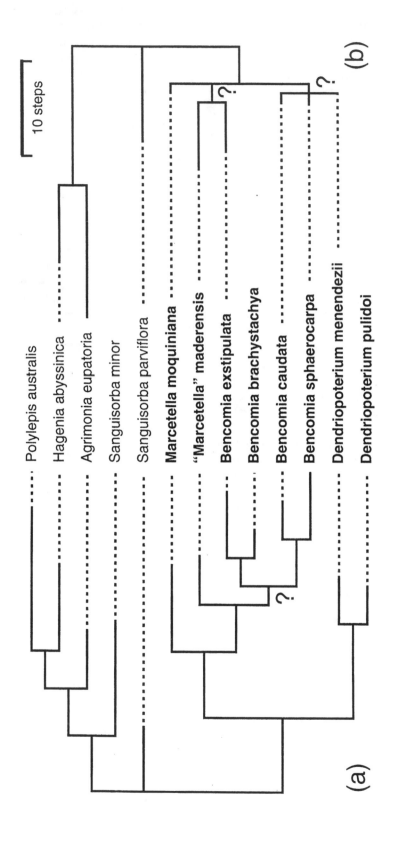

Figure 19.4 One of two most-parsimonious morphological cladograms under Deltran optimisation (a) and the single most-parsimonious ITS1 sequence cladogram (b) for endemic Poterieae (Rosaceae) of the Macaronesian Islands; modified after Welsh (1997, figs. 4.8b and 3.1 respectively). Nodes bearing question marks have less than 50% bootstrap support.

If we attempt to interpret Fig. 19.4, the topologies of both the morphological and molecular trees are consistent with derivation of the entire ingroup from a single recent immigrant to Macaronesia. *Sanguisorba parviflora* provides the closest credible model for a potential ancestor among the outgroups; not only is it sister to the ingroup but it is subtended by far shorter branches than the other outgroup members for both morphology and ITS, suggesting relative plesiomorphy. The short branches evident within the ingroup for ITS – one of the fastest evolving phylogenetically utilitarian segments of the plant genome (Baldwin *et al.*, 1995; Hershkovitz *et al.*, 1999 – this volume) – indicate a recent radiation within the Macaronesian Islands. However, in morphological terms the ingroup members are highly disparate and well differentiated; they possess many diagnostic character states distributed among several organs, and their strong reproductive specialisations are especially noteworthy (Bramwell, 1978; Welsh, 1997).

Overall, exceptionally rapid morphological divergence is indicated. This could be viewed as the result of a genuine adaptive radiation driven by directional or disruptive selection, or of non-adaptive processes such as drift/shifting balance within different isolated populations (cf. Wright, 1968; Whitlock *et al.*, 1995; Coyne *et al.*, 1997) or saltation within single populations (Bateman and DiMichele, 1994). In either case, it is likely to have been prompted by the *removal* of stabilising selection from the ingroup ancestor upon its arrival in the relatively liberal selective regime of Macaronesia.

19.4.3 Hawaiian Asteraceae

According to Lowrey (1995), the genus *Tetramolopium* (Asteraceae) evolved in New Guinea and radiated within the Hawaiian Islands from a New Guinean immigrant that arrived in East Maui within the last 2 My (the maximum age of the island). The extant New Guinean species were therefore aggregated into a single outgroup (admittedly thereby weakening any subsequent interpretations of migration patterns) in order to phylogenetically polarise the 11 cladistically coded species of *Tetramolopium* from Hawaii. The matrix consisted of 15 bistate and seven multistate characters, including considerable architectural variation (Lowrey, 1995; Bateman, 1999).

The resulting morphological cladogram (Fig. 19.5a) shows that the most primitive Hawaiian species of *Tetramolopium*, *T. humile*, is similar to the outgroup. Both are separated by a long phylogenetic branch from two major clades comprising the remaining *Tetramolopium* species; each of the two clades is also well supported by long branches reflecting several features. However, few morphological characters change within these two major clades, and few of the coded species are delimited by morphological autapomorphies; in phylogenetic terms at least, most of these species are relatively indistinct.

In 1996 I wrote:

> This pattern of basally concentrated morphological character-state transitions is consistent with two possible evolutionary interpretations. The early evolution of the clade on the Hawaiian archipelago could have been characterised by radical morphological transitions into ecological vacuums, whereas finer niche

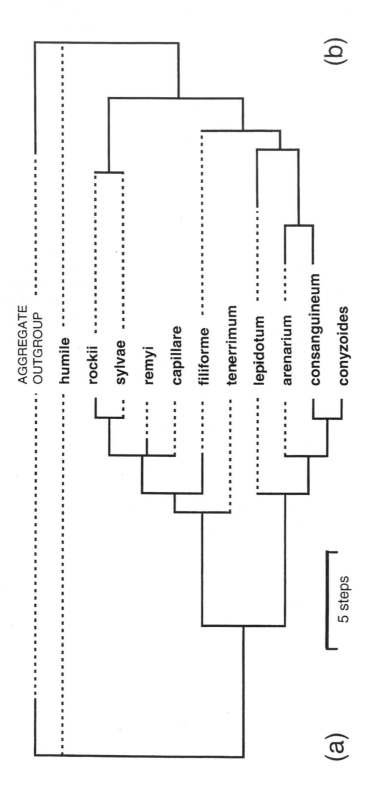

Figure 19.5 One of ten most-parsimonious morphological cladograms under Acctran optimisation (a) and the single most-parsimonious RFLP fragment cladogram (b) for *Tetramolopium* (Asteraceae) of the Hawaiian Islands; modified after Lowrey (1995, fig. 11.5) and Okada *et al.* (1997, fig. 6) respectively. All nodes apparently have greater than 50% bootstrap support.

partitioning was necessary among the more recently evolved species as ecospace filled, suppressing more radical evolutionary steps (including major changes in gross architecture) of the kind that characterised earlier speciation events [i.e. the hypothesis advanced in section 19.3.2 to explain the patterns of disparity evident in the hypothetical phylogeny of Fig. 19.2c]. Alternatively, the basal long branches of the phylogeny are artefacts of preferential extinction of early-formed species of *Tetramolopium*, rather than accurately reflecting patterns of character acquisition during speciation [cf. Bateman *et al.*, 1998: 283].

One potential arbiter of these competing scenarios would be the fossil record – specifically, whether any added fossil species could interpolate onto the near-basal long branches, thereby shortening them (Bateman and DiMichele, 1994; Bateman, 1996). A second approach would be to construct a molecular phylogeny for the same range of species, . . . [though] even the rapidly mutating gene ITS (Baldwin *et al.*, 1995) reportedly shows less than 1% sequence divergence within the genus (Chan *et al.*, 1995).

The subsequent low copy number nuclear RFLP cladogram generated by Okada *et al.* (1997, fig. 6), including only seven of the original 11 ingroup members and rooted on a single extensive population from the Cook Islands, is reproduced as Fig. 19.5b. Topologically, the molecular tree contradicts the morphological tree only in the placement of *T. filiforme* (cf. Figs 19.5a, b). The RFLP branches are longer and nodes more robust than in the ITS1 tree described above for Macaronesian Poteriae, though still shorter than the corresponding morphological branches near the base of the trees.

The relatively uniform molecular branch lengths suggest a less sudden radiation (it is also tempting to claim that the *Tetramolopium* radiation began earlier than the Poteriae radiation, but of course the RFLP and ITS sequence data are very different in nature). In particular, the hypothesis of preferential extinction of the earliest-formed species on the archipelago receives little support from the molecular phylogeny, suggesting that a fractal morphological radiation into empty ecospace from a *T. humile*-like ancestor is the more credible explanation of the pattern observed in Fig. 19.5a.

19.5 Why evolutionary radiations are best interpreted by comparing molecular and morphological phylogenies

19.5.1 Preferred protocol for comparing molecular and morphological trees

Thus far, I have failed to generate a corresponding morphological tree for the ITS tree generated for the Orchidinae (Fig. 19.3), and our consequent pragmatic decision to map morphological characters across the preferred molecular tree considerably weakens the ensuing interpretations (see below). But even in the case of the Poteriae (Fig. 19.4) and *Tetramolopium* (Fig. 19.5) studies, comparison of the morphological and molecular trees within each study, and of the results of these two studies of island radiations, is compromised by several readily remedied deficiencies.

Firstly, due to sampling or extraction problems, in both island studies the species yielding molecular data are only a subset of the species yielding morphological data. This inevitably complicates topological comparison. Secondly, information was incomplete on the treatment of characters, particularly the choice of optimisation algorithm and whether autapomorphic states were included in branch-length calculations (unlikely, as autapomorphies are now routinely excluded from calculations of ensemble homoplasy indices). Thirdly, the studies varied in the nature of the tree-building algorithm, in whether the trees were rooted or unrooted, and if rooted whether the outgroup was treated as monophyletic or paraphyletic. All of these decisions influence the perceived lengths of individual branches, which I regard as of great interpretative value (see below). And lastly, information on the robustness of the trees was inadequate; as noted by Sanderson (1998), phylogenetic consumers have learned to expect nodes festooned with bootstrap, jack-knife and/or decay index values.

The first rule should obviously be that, ideally, the molecular and morphological matrices should be analysed under the same protocols. Furthermore, in order to allow comparison not only of topologies but also the equally important branch lengths, trees should be rooted, and autapomorphies included in branch-length calculations (they are, after all, *bona fide* evolutionary steps: Yeates, 1992; Bateman and Simpson, 1998). Finally, both of the two extreme optimisation options (Acctran and Deltran) should be applied to all matrices and the resulting maximum ranges of length of each branch should then be compared.

19.5.2 Inferring evolutionary process from phylogenetic pattern: Cross-matrix disparity as an under-used interpretative tool

Having outlined the presentational weaknesses of the three case-studies, can they nonetheless offer valuable overviews on phylogenetic interpretation in general and the use of cross-matrix disparity in particular?

The two studies of island radiations are more informative, as they were based on both molecular and morphological phylogenies. There is already a vast literature recently published on comparing topologies of phylogenies based on different types of data but the same range of species, yet if phylogenetic studies are genuinely as robust as their practitioners routinely claim we should in most cases expect broadly congruent topologies (cf. Givnish and Sytsma, 1997b; Sanderson, 1998; though see Comes and Abbott, 1999 – this volume). And that anticipated robustness is certainly evident in the studies of island radiations.

If we can take as read the broad accuracy of the topology, we can access a more potentially fruitful level in the interpretative hierarchy. Specifically, we can focus on contrasts in branch lengths within and between phylogenies based on different data types for the same range of coded taxa. Thus, in the case-studies of section 19.4, it is the highly contrasting branch lengths *between* the morphological and molecular trees, and the contrasting branch lengths between outgroup and ingroup *within* each tree, that usually constrain the range of most probable evolutionary scenarios underpinning each specific radiation.

Our interpretation of the Poteriae as a recent radiation from a single immigrant to Macaronesia is based on the contrast in the ingroup between long morphological branches and short molecular branches (Fig. 19.4). This morphological versus mole-cular contrast is less marked in *Tetramolopium*, but there is an interesting pattern of variation within the morphological tree; the longer basal branches suggest a niche-filling radiation (Fig. 19.5). In the Orchidinae tree (Fig. 19.3), supposed species of genera such as *Ophrys*, *Serapias* and *'Nigritella'* are subtended by demonstrably short molecular and putatively short morphological branches, indicating that they are not species at all or alternatively are the product of very recent radiations. In contrast, the transition from *Gymnadenia s.s.* to *'Nigritella'* is depicted as a mole-cular short branch but a presumed morphological long branch – the classic pattern for a recent saltational evolutionary event, facilitating the subsequent species-level radiation in *'Nigritella'*.

Other similar empirical studies would undoubtedly reveal other distinct patterns of contrast between molecular and morphological phylogenies, requiring additional causal explanations and much more careful consideration of the issues than I have offered in this hasty, broad-brush essay. Sadly, the surprising dearth of phyloge-netic literature directly comparing morphological and sequence trees (see below) limits our ability to test the method further at present.

However, even these few studies clearly demonstrate that in recent radiations (adaptive or otherwise) morphological evolution reliably outpaces even the most rapidly mutating segments of the plant genome that are of phylogenetic utility. Admittedly, loci with high mutation rates such as microsatellites (Jarne and Lagoda, 1996) may change sufficiently fast to resolve at least some rapid evolutionary events, but only if the events in question are very recent. A deeper radiation would be difficult to resolve with microsatellites (or other analogous regions), as post-radiation mutations soon obscure any phylogenetic signal.

Additional examples of radiations on ecological islands from the Asteraceae, including East African *Dendrosenecio* (Knox and Palmer, 1995) and Hawaiian silverswords–tarweeds (Baldwin and Robichaux, 1995; Baldwin, 1997), reveal similar phylogenetic patterns. It should be noted that although recent radiations are more readily studied on oceanic islands, the more evolutionarily profound radi-ations typically occur on continental land-masses where the resulting increased diversity is less geographically constrained. Nonetheless, this generalisation of patterns of evolution on islands supports (albeit equivocally) a punctuational view of evolutionary radiations.

19.5.3 The perils of punctuation and implications for 'total evidence' trees

I now return to my phylogenetic mantra for this essay: If morphological evolution describes a punctuational pattern but molecular evolution describes a far more gradual pattern, then the vast majority of morphological character-state transitions occur *during* evolutionary events and the vast majority of molecular character-state transitions occur *between* them (Fig. 19.2).

I have never seen this principle stated elsewhere, yet I believe that few currently active phylogeneticists would, on reflection, disagree. Little remains of the formerly strong support for morphological anagenesis (also known as phyletic gradualism – the slow, stately accumulation of modified features by a non-branching lineage; e.g.

Fortey and Jefferies, 1982), having fallen victim to paucity of credible evidence. And the increasing body of developmental–genetic evidence for macromutational processes is irrefutable; major evolutionary transitions reflect a few key developmental genes – genes that control the morphological changes that allow us as biologists to recognise speciation events. We now know that such genes can be profoundly modified by a single base-pair mutation or the insertion of a single transposon. However, thus far developmental genes have rarely been used as molecular phylogenetic tools (though see contributions to this volume by Doyle and Doyle, 1999; Gustaffson and Albert, 1999; Möller et al., 1999); in the vast majority of cases, the physiological genes and presumed non-coding regions that are the bread and butter of molecular phylogenetics record a mere historical fingerprint of evolution rather than evolution *per se.*

Does this matter? Well, in my opinion it all depends on the patterns of evolution within the group under scrutiny. Let us take for example the recent Poteriae radiation documented in Fig. 19.4. Here, the morphological data have yielded a far more robust and credible representation of ingroup relationships than has ITS1. What pattern would emerge if no further speciation and no extinction occurred in the clade and an enthusiastic phylogeneticist reanalysed the eight species long (say 5 My) into the future? Morphology would show little if any change, but ITS would have accumulated many more mutations and thereby greatly lengthened the terminal branches. However, the non-terminal branches that determine our perception of relationships would not have acquired additional synapomorphies; indeed, they would be vulnerable to further mutation of the very few currently informative sites, while the ever-lengthening terminal branches would become increasingly prone to long-branch attraction and the resulting false topologies (e.g. Penny et al., 1994). The only hope of improving the apparent robustness in the molecular phylogeny would be extinctions; these would increase the perceived resolution of the near-polytomy by concentrating the few apparent synapomorphies onto fewer internal branches, which would appear longer and hence better survive the now standard (if often thoughtlessly applied) rigours of bootstrap and decay index calculations.

However, unambiguous and widespread extinctions have not been sufficient to rescue molecular attempts to reconstruct the phylogeny of the fungal, bryophyte and pteridophyte lineages that together constituted the earliest terrestrial plant communities; the most credible phylogenies remain morphological (e.g. Kenrick and Crane, 1997; Bateman et al., 1998), partly because resolution is better and partly because coverage of major groups is rendered far more complete by the inclusion of fossils lacking close living descendants. In short, it is asking a great deal of molecular data to resolve a radiation that spanned as little as 35 My but ended at least 350 My ago (Bateman et al., 1998). Moreover, the few extant species morphologically similar to those early land-plants belie any attempt to correlate the age of a clade with the age of its extant species (Bateman, 1996; Bateman et al., 1998). For example, the homosporous lycopsid genus *Huperzia* has close morphological analogues among the earliest vascular land-plants (Kenrick and Crane, 1997), yet its many extant representatives among tropical epiphytes yield near-identical *rbcL* sequences (Wikström and Kenrick, 1997).

One obvious solution (perhaps in retrospect *too* obvious) to both the Poteriae and early land-plant conundra is to sequence additional genes, yet of course the relative branch lengths in trees generated from those extra genes are likely to be similar to those that already exist; we will still add many character-state transitions to the terminal branches and few to the crucial internal branches that presumably reflect a genuinely rapid radiation immediately following the immigration of the ancestor (i.e. we have detected a 'hard' polytomy). Indeed, it is striking how data for additional genes tend to reinforce those portions of a clade best resolved by the original gene, leaving the targeted weaker nodes still frustratingly ambiguous; for example, this occurred when *trnL* data were added to the ITS phylogeny of the Orchidinae shown in Fig. 19.3 (Bateman *et al.*, unpublished).

Rather, the best chance of resolving *different* time-depths of a phylogeny is surely to add morphological data to molecular data, since there is no reason to expect congruence of branch lengths between molecular and morphological phylogenies. The few empirical attempts to combine molecular and morphological data in botany (e.g. Pennington, 1996) have demonstrated the potential power of this approach to generating maximally resolved trees.

Mixing two or more matrices of contrasting data types for the same range of coded taxa in a 'total evidence' parsimony analysis has become increasingly popular over the last decade, despite engendering much controversy (cf. Hillis, 1987; Swofford, 1991; Donoghue and Sanderson, 1992; Bull *et al.*, 1993; Eernisse and Kluge, 1993; Patterson *et al.*, 1993; de Queiroz *et al.*, 1995; Miyamoto and Fitch, 1995; Huelsenbeck *et al.*, 1996; Nixon and Carpenter, 1996; Kluge, 1998). The term is inappropriate – 'maximum available evidence', or Nixon and Carpenter's 'simultaneous analysis', would perhaps be a more accurate description of the technique. Nonetheless, the underlying argument that the approach maximises the range of data available to the parsimony algorithm, and thereby maximises the effectiveness of the character congruence test of homology, is irrefutable.

Several authors have argued that to combine the multiple matrices without first analysing them separately is disadvantageous, as on conceptual grounds only matrices yielding broadly congruent trees should be combined. I suggest that there is an additional, more pragmatic reason for not immediately combining matrices, for it squanders the opportunity to compare potentially evolutionarily informative contrasts in both topologies and branch lengths. Once this comparison has been made, an acceptably congruent 'total evidence' tree can be generated – a tree that is likely to form the best framework for 'mapping' additional characters deliberately excluded from tree-building (see below and Fig. 19.6).

19.6 Why molecular phylogenies should supplement rather than supplant morphological phylogenies

19.6.1 Supposed advantages of molecular data

In the past few years, molecular technological advances have prompted a profound change in the approach of comparative biologists to phylogeny reconstruction. Most early phylogenetic studies focused on morphological characters, so that as recently as 1989 Sanderson and Donoghue experienced considerable difficulty when trying to amass sufficient molecular phylogenetic matrices for comparison with morphological matrices in their cross-matrix survey of homoplasy. In contrast, 7–8 years

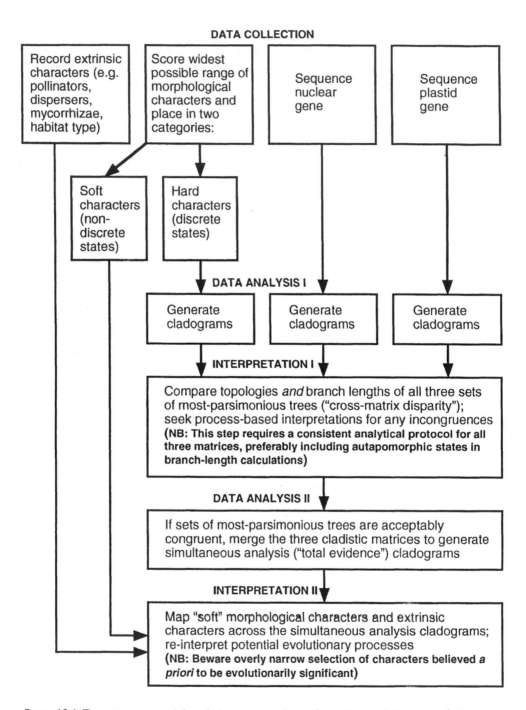

Figure 19.6 Tentative protocol for phylogenetic analysis of putative evolutionary radiations.

later Sanderson and Donoghue (1996b) and Givnish and Sytsma (1997b) were able to assess homoplasy in both types of data for a far larger sample of matrices of both categories, the latter authors deciding (not entirely convincingly, as noted by Sanderson, 1998) that molecular data were on average less homoplastic than morphological data and thus seen as a better basis for phylogeny reconstruction.

Relatively low homoplasy is one of several theoretical advantages attributed to sequence data. Most existing plant molecular phylogenies are derived from a surprisingly small portion of the total organismal genome. Restriction-site matrices, especially for cpDNA, are still widely used in botanical phylogenetic studies and are reputed to carry on average the lowest homoplasy levels (e.g. Givnish and Sytsma, 1997a, b). Nonetheless, my impression is that they are increasingly employed as adjuncts to sequence data. Initially, uniparentally inherited plastid genes found much favour and still underpin the largest sequence data matrices (e.g. *rbcL*, *trnL-F*, *matK*, *ndhF*, *atpB*: Chase *et al.*, 1993; Albert *et al.*, 1994; Olmstead and Palmer, 1994; Chase and Albert, 1998; Soltis *et al.*, 1998b; Stoebe *et al.*, 1999 – this volume), though note recent increases in the popularity of biparentally inherited nuclear rDNA (notably 18S, 26S and especially the ITS1–5.8S–ITS2 assembly: Baldwin *et al.*, 1995; Hershkovitz *et al.*, 1999 – this volume). Yet more recently, we have seen an expansion in single-clade studies that document sequences for two or more genes. Selecting at least one gene from the plastid genome and at least one gene from the nucleus is attractive, as it allows comparison of phylogenetic patterns that are potentially incongruent, given the contrasting modes of inheritance and expression of the organelle and the 'host' cell (Comes and Abbott, 1999 – this volume.)

In contrast with nucleic acid sequences, the pool of genuinely discrete morphological character states is decidedly finite. As an analyst adds successive species to a matrix, the number of new morphological states encountered rapidly diminishes, such that further species are diagnosed mainly by different combinations of character states already coded into the matrix (Sanderson, 1998; Bateman, 1999). Even when supplemented with anatomical and ontogenetic characters, the phenotype remains a limited source of viable states. Thus, in an era that increasingly admires the machismo of 'large' (strictly, species-rich) matrices, morphology struggles to generate resolved trees (a matrix needs a considerable excess of apomorphic character states over analysed species to generate fully resolved trees). In contrast, despite the fact that the 'language of the genes' is written in only four letters, the number of bases present in even the most depauperate plant genome is enormous; molecular phylogenies are limited more by the patience and consumables budget of the analyst than by the intrinsic constraints of the organism. Full resolution of large sequence matrices is guaranteed (albeit usually yielding many most-parsimonious trees and thus many contrasting fully resolved 'truths', as in the Orchidinae casestudy), and as a result the qualifier 'robust' is now thoughtlessly attached to the phrase 'molecular phylogeny' almost as often as the equally unjustified qualifier 'adaptive' is found in lockstep with 'radiation'.

Moving from the practical to the conceptual, the near-neutrality with regard to evolution of physiological genes and intergenic spacers is generally seen as advantageous – they are viewed as less likely than morphology to be under selective pressures, and thus less likely to show convergent homoplasy (cf. Givnish, 1997;

Sanderson, 1998). Given the observations made above regarding the asymptotic diminution of new character states in increasingly species-rich matrices, species-rich higher taxa are likely to be especially prone to phenotypic homoplasy. Also, the genotype is not subject to the vagaries of ecophenotypic variability, and any sexual dimorphism evident in the phenotype is confined to small, discrete regions in the nuclear and mitochondrial genomes that are rarely used for phylogeny reconstruction. To summarise, the tendency of the phenotype to evolve is increasingly viewed as a regrettable disadvantage for phylogeny reconstruction, at least where process-based interpretation is the primary goal – a somewhat bizarre predicament.

19.6.2 Uses and abuses of character 'mapping'

Conceptual arguments invoking increased resolving power and evolutionary neutrality have undoubtedly aided the molecular revolution, but more pragmatic technological advances – notably increased automation and decreased costs, especially in time commitments to projects – probably represent greater incentives. Consequently, it has rapidly become fashionable to either ignore morphology as a quantifiable phenomenon, or more commonly to score cladistically a few characters supposed *a priori* to be of particular evolutionary interest and then 'map' them across a sequence tree *a posteriori*. Such analytical apartheid is undeniably logical, but is it biologically rational?

Common sense suggests that good morphological characters offering discrete states generate good, discrete phylogenies that, in many evolutionary circumstances, are more likely to yield accurate topologies than are corresponding molecular data. What are *not* welcome in morphological phylogenetic matrices are 'indiscrete' intrinsic characters – those that are in practice continuous. Yet if my view of adaptation is correct, and its primary role is in most cases the subtle fine-tuning of evolutionary lineages whose pre-existing disparity is due to other, non-adaptive processes, it is typically the continuous phenotypic characters that encapsulate the adaptive process (for example, changes in spur length in response to the proboscis length of co-evolving pollinators: Hapeman and Inoue, 1997; Hodges, 1997). These characters, most amenable to selection, *are* undoubtedly best mapped across pre-existing phylogenies.

However, the mapping process abrogates the extraordinary exploratory power of cladistics. Rarely do morphological characters supposed *a priori* to be diagnostic – characters inherited from traditional taxonomy – survive a parsimony analysis with their honour intact. Many 'diagnostic' characters are in practice downgraded *a posteriori*, to be replaced by more reliable synapomorphies whose value (or, in the case of many characters such as SEM-based micromorphology, whose very existence) had escaped the attention of all previous taxonomists. In contrast, mapping does not require time-consuming investigation of the study organisms; rather, scoring is confined to a few characters of 'obvious' evolutionary utility, such as those directly involved in pollinator specificity in angiosperms, often cribbed from the existing literature. Unfortunately, the one clear rule learned thus far about evolutionary processes is that they do not feel obliged to be obvious – they can be so subtle, or indeed so blatant, that they are wholly overlooked. In practice, the mapping approach to interpreting morphology is far less objective than comparing the molecular tree(s) with a well-founded morphological tree – a tree that allows the characters themselves to find their most appropriate level in the phylogenetic hierarchy (Bateman and Simpson, 1998) and provides a genuine test of *a priori* assertions of character-state homology (Patterson, 1988).

The use of phylogenies for inferring evolutionary process from patterns of character change has increased in tandem with molecular phylogenetics. Phylogenetic systematists were inclined to interpret phenotypic change directly from matrices including phenotypic characters, seeking patterns of character correlation consistent with adaptation (including putative key innovations: cf. Larson and Losos, 1996; Lauder, 1996; Sanderson and Hufford, 1996) or with saltation (Bateman and DiMichele, 1994; Bateman, 1999).

Ecologists who adopted phylogenetics promptly re-christened the discipline 'the comparative method' (e.g. Harvey and Pagel, 1991; Silvertown *et al.*, 1997). The concept of phylogenetic constraint – that changes in the ecological properties of species in general and adaptations in particular needed to be distinguished from features inherited from common ancestors – led to a comparative approach termed 'phylogenetically independent contrasts' (PICS). PICS was used to berate proponents of traditional comparative methods that treat species as independent units (TIPS) and thereby wholly ignore phylogeny (a debate well summarised by Ricklefs, 1996). Despite criticisms of underlying models (e.g. the brief critique of Brownian motion by Bateman, 1998), PICS is now a well-established technique (or, more accurately, suite of techniques) for mapping and interpreting characters on phylogenies.

Most importantly, we need an effective means of phylogenetically testing the heterogeneous wealth of *extrinsic* characters relevant to radiations – the co-evolutionary, ecological, geographical and temporal criteria that we outlined in section 19.2.3 but that we could not readily fit into testable intrinsic definitions of the radiations summarised in section 19.2.2 and Table 19.1. Here is where mapping becomes of greatest interpretative value – though I note with some irony that extrinsic characters may be as effectively mapped across morphological trees as across their molecular equivalents (Fig. 19.6).

19.6.3 Morphology is dead – long live morphology!

Although I have in the past (e.g. Bateman, 1999) waxed lyrical about the insights provided by several excellent recent phylogenetic studies of island radiations, it is salutary to realise that of seven studies of Hawaiian plant groups in Wagner and Funk (1995), of eight studies of putative adaptive radiations of plants in Givnish and Sytsma (1997a), and of ten evolutionary case-studies of (mostly) Pacific island plants in Steussy and Ono (1998), not one study diagrammatically compared morphological and sequence phylogenies. Where phylogenies *were* compared, they either reflected sequences for two or more different genes, sequences for one gene and a (generally older) restriction site data matrix, or in two cases a restriction site matrix versus morphology (Sakai *et al.*, 1997; Setoguchi *et al.*, 1998), the former study lacking details of the all-important branch lengths. And the supplanting of morphological phylogenies by molecular phylogenies over the past decade is by no means confined to island studies.

The recent exponential increase in sequencing activity (at least, in the better funded developed world) reflects in part the success of such studies in unravelling plant phylogeny, but also exploits tremendous advances in the speed, accuracy and safety of generating nucleic acid sequences, and of computationally generating phylogenies from the resulting vast bodies of sequence data. Consequently, well-founded characterisation of the morphology of a range of species now requires considerably more time and biological knowledge than does an equivalent sequencing study. Acknowledging recent, rapid and seductive molecular advances, many funding bodies have in practice created a positive feedback loop that strongly encourages the development of further similar studies, contributing to the increasingly familiar phenomenon of 'research homogenisation'.

To echo the fictional quote inserted at the beginning of this paper, the evolutionary biology community has in its haste turned over two pages of the book of phylogeny at once, thereby overlooking what I view as a crucial intermediate paragraph in evolutionary phylogenetics. I agree with Sanderson's (1998: 1654) defence of morphology over molecules that 'it seems reckless (and potentially wasteful) to use the *class* of data as a surrogate for the reliability of any *particular* data set or any *particular* clade', but I would go still further. As documented in this volume, recent advances in molecular systematics, phylogenetic data analysis and developmental genetics have given the systematic biology community its first genuine opportunity to test macroevolutionary patterns and processes. Yet, thus far we have only half grasped the nettle, due to:

(1) Fundamental inconsistencies among workers in the approaches taken to the analysis and presentation of phylogenetic work, thus minimising comparability of trees both within and among clades;
(2) The increasing inclination to by-pass *bona fide* morphological phylogenetics, arguing that morphological characters should be protected from the potential embarrassment of direct parsimony analysis and instead be mapped onto molecular frameworks, and;
(3) A regrettable tendency to 'know' the answer before asking the question; for example, by assuming that all radiations are adaptive, and testing this assumption by mapping only those phenotypic characters suspected *a priori* of being under direct selection at the time speciation occurred.

As these remaining constraints are largely self-imposed, their removal is in theory easily accomplished – it is certainly desirable, as it will allow plant molecular biology to make even greater contributions to evolutionary phylogenetics in the future.

19.7 Conclusions

(1) Evolutionary radiations have long been a popular review topic. Consequently, many different definitions of a radiation are available and they are based on many criteria, both intrinsic properties of the organisms (e.g. taxonomic diversity, character diversity) and extrinsic properties of their environments and ecological interactions (e.g. number of clades simultaneously radiating, ranges of niches and/or ecospace occupied or amount of biomass sequestered). The preferred definition in

this study is wholly intrinsic and time-constrained: Within a specified clade and a specified period of time, a radiation is best defined as an exceptionally high surplus of species natality (speciation) over species mortality (extinction) and/or of character-state acquisition over character-state loss within the overall pool of species. The primary difficulty with this definition is obtaining a sufficiently precise temporal framework for the rate estimates. Note that this definition describes only pattern – it makes no assumptions regarding underlying processes.

(2) The still widely held neobiological notion than radiations are reliably and wholly adaptive is rejected, in favour of a more informed palaeobiological perspective on the multiplicity of patterns and processes encompassed by the term radiation. A further model involves saltational transitions across fitness 'seascapes', the only means of transgressing lethal fitness valleys. Rather than supporting prior assumptions of process, phylogenies are better used more objectively to falsify at least some of the many potential mechanisms underlying radiations. Despite increasing evidence of perturbations in base substitution rates at various hierarchical levels, molecular phylogenies do record broadly clock-like behaviour. In contrast, first principles suggest that there is no corresponding morphological clock. Rather, morphological evolution follows a pattern of punctuated equilibria – long periods of stasis (generally reflecting stabilising selection) separate brief periods of rapid phenotypic change (generally reflecting either directional/disruptive selection or release from selection). During radiations, considerable phenotypic divergence occurs with proportionately negligible divergence in the metabolic genes currently favoured for molecular phylogenetic studies. Consequently, the vast majority of morphological character-state transitions occur *during* speciation events and the vast majority of molecular character-state transitions occur *between* them.

(3) This conclusion means that molecular data alone are inferior to morphological data alone for resolving radiations; there is insufficient time between speciation events to accumulate sufficient molecular synapomorphies, and perceived resolution can only be ostensibly improved by the extinction of phylogenetic intermediates. Radiations are best interpreted by *comparing* molecular and morphological phylogenies for the same range of coded taxa, focusing not only on topological congruence but also on contrasts in branch lengths – an approach here termed 'cross-matrix disparity' (Fig. 19.6). Limited practical application of this approach suggests that the circumstances under which relative branch lengths and underlying topologies will differ between morphological and molecular matrices are both predictable and potentially highly informative. However, comparison of phylogenies is currently undermined by the lack of a consistent protocol for data collection and analysis: key variables include tree-building algorithms, rooting methods, exclusion (or, more accurately, inclusion) of autapomorphies, and choice of optimisation algorithms.

(4) Despite the advantages of cross-matrix comparison, evolutionary phylogeneticists increasingly emphasise molecular trees at the expense of morphological trees. This preference is usually justified on conceptual grounds; morphological characters supposedly incur higher levels of homoplasy, offer a limited pool of potential character states and so cannot fully resolve species-rich matrices, respond directly to evolution and hence are vulnerable to convergence, and are prone to non-genetic (ecophenotypic) variation. However, practical advantages may in truth be more compelling; recent technological advances in molecular data-generation and analysis

have rendered molecular phylogenies more easily constructed (and more fundable) than well-founded morphological phylogenies.

(5) Sadly, very few studies compare morphological and molecular phylogenies of the same clade (and most of these reflect recent island radiations); it has become more commonplace to compare either trees resulting from contrasting molecular analyses or to practice analytical apartheid – mapping a limited number of morphological characters across pre-existing molecular phylogenies to protect them from the rigours of parsimony analysis. This effective downgrading of morphological phylogenetics squanders the opportunity to discover novel phenotypic characters within the clade, and reduces the objectivity of any mechanistic interpretation by restricting the analysis to characters that are presupposed to be directly involved in the evolutionary events targeted by the study.

(6) A better analytical protocol is to generate separate molecular and morphological phylogenies and compare their relative signals. If sufficiently topologically congruent, they may then be combined in a so-called 'total evidence' phylogeny (Fig. 19.6). This approach not only maximises the congruence test of homology but also takes advantage of the probability that the two categories of data will maximise resolution at different hierarchical levels. The 'total evidence' tree resulting from the combination of molecular and genuinely discrete morphological characters then forms the best framework for mapping other weaker intrinsic properties (e.g. the continuous morphological characters most amenable to selection) and all the extrinsic properties of ecology and environment that facilitate selection – not to mention the many other evolutionary processes capable of driving radiations. This methodology would fit comfortably into a truly synthetic approach to the definition, identification and interpretation of evolutionary radiations.

ACKNOWLEDGEMENTS

I am grateful to Josie Welsh, Alec Pridgeon, Mark Chase, Pete Hollingsworth and Jill Preston for generously contributing the (mostly unpublished) Poteriae and Orchidinae case-studies outlined in this paper. I also thank Pete Hollingsworth, Toby Pennington and Paula Rudall for critically reading the manuscript.

REFERENCES

Albert, V. A., Backlund, A., Bremer, K., Chase, M. W., Manhart, J. R., Mishler, B. D. *et al.* (1994) Functional constraints and *rbc*L evidence for land plant phylogeny. *Annals of the Missouri Botanical Garden*, **81**, 534–67.

Albert, V. A., Gustafsson, M. H. G. and Di Laurenzio, L. (1998) Ontogenetic systematics, molecular developmental genetics, and the angiosperm petal, in *Molecular systematics of plants II: DNA sequencing*, (eds D. E. Soltis, P. S. Soltis and J. J. Doyle), Chapman & Hall, London, pp. 349–74.

Averyanov, L. V. (1990) A review of the genus *Dactylorhiza*, in *Orchid biology: reviews and perspectives, V.* (ed. J. Arditti), Timber Press, Portland, Oregon, pp. 159–206.

Avise, J. C. (1994) *Molecular markers, natural history, and evolution*, Chapman and Hall, London.

Baldwin, B. G. (1997) Adaptive radiation of the Hawaiian silversword alliance: congruence and conflict of phylogenetic evidence from molecular and non-molecular investigations, in *Molecular evolution and adaptive radiation*, (eds T. J. Givnish and K. J. Sytsma), Cambridge University Press, Cambridge, pp. 103–28.

Baldwin, B. G. and Robichaux, R. H. (1995) Historical biology and ecology of the Hawaiian silversword alliance (Asteraceae): new molecular phylogenetic perspectives, in *Hawaiian biogeography: evolution on a hot spot archipelago*, (eds W. L. Wagner and V. A. Funk), Smithsonian Institution, Washington, DC, pp. 259–87.

Baldwin, B. G., Crawford D. J., Francisco-Ortega, J., Kim, S.-C., Sang, T. and Steussy, T. F. (1998) Molecular phylogenetic insights on the origin and evolution of oceanic island plants, in *Molecular systematics of plants II: DNA sequencing* (eds D. E. Soltis, P. S. Soltis and J. J. Doyle), Kluwer, New York, pp. 410–41.

Baldwin, B. G., Sanderson, M. J., Porter, J. M., Wojchiechowski, M. F., Campbell, C. S. and Donoghue, M. J. (1995) The ITS region of nuclear ribosomal DNA: a valuable source of evidence on angiosperm phylogeny. *Annals of the Missouri Botanical Garden*, **82**, 247–77.

Barrett, S. C. H. and Graham, S. W. (1997) Adaptive radiation in the plant family Pontederiaceae: insights from phylogenetic analysis, in *Molecular evolution and adaptive radiation*, (eds T. J. Givnish and K. J. Sytsma), Cambridge University Press, Cambridge, pp. 225–58.

Bateman, R. M. (1995) Explaining higher plant radiations: do radical developmental mutations drive phylogenetic change across fitness seascapes?, in *Evolutionary radiations and development*, (symposium), Abstracts of the Fifth Congress of the European Society for Evolutionary Biology (Edinburgh), p. 23.

Bateman, R. M. (1996) Non-floral homoplasy and evolutionary scenarios in living and fossil land-plants, in *Homoplasy and the evolutionary process*, (eds M. J. Sanderson and L. Hufford), Academic Press, London, pp. 91–130.

Bateman, R. M. (1998) Review of: *Plant life histories: ecology, phylogeny and evolution* (eds J. Silvertown, M. Franco and J. L. Harper, 1997). *Botanical Journal of the Linnean Society*, **127**, 176–7.

Bateman, R. M. (1999). Architectural radiations cannot be optimally interpreted without morphological and molecular phylogenies, in *The evolution of plant architecture*, (eds M. H. Kurmann and A. R. Hemsley), Royal Botanic Gardens, Kew, in press.

Bateman, R. M. and DiMichele, W. A. (1994) Saltational evolution of form in vascular plants: a neoGoldschmidtian synthesis, in *Shape and form in plants and fungi*, (eds D. S. Ingram and A. Hudson), Academic Press, London, pp. 63–102.

Bateman, R. M. and Simpson, N. J. (1998). Comparing phylogenetic signals from reproductive and vegetative organs, *Reproductive biology in systematics, conservation and economic botany*, (eds S. J. Owens and P. Rudall), Royal Botanic Gardens, Kew, pp. 231–53.

Bateman, R. M., Pridgeon, A. M. and Chase, M. W. (1997) Phylogenetics of subtribe Orchidinae (Orchidoideae, Orchidaceae) based on nuclear ITS sequences. 2. Infrageneric relationships and taxonomic revision to achieve monophyly of *Orchis sensu stricto*. *Lindleyana*, **12**, 113–41.

Bateman, R. M., Crane, P. R., DiMichele, W. A., Kenrick, P., Rowe, N. P., Speck, T. *et al.* (1998) Early evolution of land plants: phylogeny, physiology, and ecology of the primary terrestrial radiation. *Annual Review of Ecology and Systematics*, **29**, 263–92.

Bateman, R. M., Chase, M. W., Hollingsworth, P. M., Preston, J., Pridgeon, A. M., Williams, N. H. *et al.* (1999) Rapidly expanding genera of European Orchidaceae: biological realities or taxonomic artefacts? in *Systematics in South Africa*, Abstracts of the First Conference of the South African Society for Systematic Biology (Stellenbosch), pp. 16–17.

Behrensmeyer, A. K., Damuth, J. D., DiMichele, W. A., Potts, R., Sues, H. -D. and Wing, S. L. (1992) *Terrestrial ecosystems through time*, Chicago University Press, Chicago.

Björklund, M. (1994) The independent contrast method in comparative biology. *Cladistics*, **10**, 425–33.

Bramwell, D. (1976) The endemic flora of the Canary Islands: distribution, relationships and phytogeography, in *Biogeography and ecology of the Canary Islands*, (ed. G. Kunkel), The Hague, Netherlands, pp. 207–40.

Bramwell, D. (1978) The endemic genera of Rosaceae (Poteriae). *Botanica Macaronesia*, **6**, 67–73.

Briggs, D. E. G., Fortey, R. A. and Wills, M. A. (1992) Morphological disparity in the Cambrian. *Science*, **256**, 1670–3.

Bull, J. J., Huelsenbeck, J. P., Cunningham, C. W., Swofford, D. L. and Waddell, P. J. (1993) Partitioning and combining data in phylogenetic analysis. *Systematic Biology*, **42**, 384–97.

Chan, R., Lowrey, T., Natvig, D. and Whitkus, R. (1995) Phylogenetic analysis of internal transcribed spacer (ITS) sequences from nuclear ribosomal DNA of Hawaiian, Cook Island and New Guinea *Tetramolopium* (Compositae; Asterae) [abstract]. *American Journal of Botany* (suppl.), **82**, 118–9.

Chase, M. W. and Albert, V. A. (1998) A perspective on the contribution of plastid *rbc*L sequences to angiosperm phylogenetics, in *Molecular systematics of plants II: DNA sequencing*, (eds D. E. Soltis, P. S. Soltis and J. J. Doyle), Kluwer, New York, pp. 488–507.

Chase, M. W., Soltis, D. E., Olmstead, R. G., Morgan, D., Les, D. H., Mishler, B. D. *et al.* (1993) Phylogenetics of seed plants: an analysis of nucleotide sequences from the plastid gene *rbc*L. *Annals of the Missouri Botanical Garden*, **80**, 528–80.

Clark, B. and Grant, P. R., eds (1996) *Evolution on islands.* Philosophical Transactions of the Royal Society of London, **B1341**.

Comes, H. P. and Abbott, R. J. (1999) Reticulate evolution in the Mediterranean species complex of *Senecio* sect. *Senecio*: uniting phylogenetic and population-level approaches, in *Molecular systematics and plant evolution*, (eds P. M. Hollingsworth, R. M. Bateman and R. J. Gornall), Taylor & Francis, London, pp. 171–98.

Coyne, J. A., Barton, N. H. and Turelli, M. (1997) A critique of Sewall Wright's Shifting Balance theory of evolution. *Evolution*, **51**, 643–71.

Crane, P. R. (1989) Patterns of evolution and extinction in vascular plants, in *Evolution and the fossil record*, (eds K. C. Allen and D. E. G. Briggs), Belhaven, Chichester, pp. 153–87.

Cronk, Q. C. B. and Fuller, J. L. (1995) *Invasive plants: the threat to natural ecosystems worldwide*, WWF Plant Conservation Handbook 2, Chapman & Hall, London.

Dafni, A. (1992) *Pollination ecology: a practical approach*, Oxford University Press, Oxford.

Davies, B. and Schwarz-Sommer, Z. (1994) Control of floral organ identity by homeotic MADS-box transcription factors. *Results and Problems in Cell Differentiation*, **20**, 235–58.

Davis, J. I. (1999) Monophyly, populations, and species, in *Molecular systematics and plant evolution*, (eds P. M. Hollingsworth, R. M. Bateman and R. J. Gornall), Taylor & Francis, London, pp. 139–70.

de Queiroz, A., Donoghue, M. J. and Kim, J. (1995). Separate versus combined analysis of phylogenetic evidence. *Annual Review of Ecology and Systematics*, **26**, 657–81.

Delforge, P. (1995) *Orchids of Britain & Europe*, HarperCollins, London.

Devillers, P. and Devillers-Terschuren, J. (1994) Essai d'analyse systématique du genre *Ophrys. Naturaliste Belges (Orchidaceae 7 suppl.)*, **75**, 273–400.

DiMichele, W. A. and Bateman, R. M. (1996) Plant paleoecology and evolutionary inference: two examples from the Paleozoic. *Review of Palaeobotany and Palynology*, **90**, 223–47.

Donoghue, M. J. and Sanderson, M. J. (1992) The suitability of molecular and morphological evidence in reconstructing plant phylogeny, in *Molecular systematics of plants* (eds P. S. Soltis, D. E. Soltis and J. J. Doyle), Chapman and Hall, London, pp. 340–68.

Doolittle, R. F., Feng, D.-F., Tsang, S., Cho, G. and Little, E. (1996) Determining divergence times of the major kingdoms of living organisms with a protein clock. *Science*, **271**, 470–7.

Doyle, J. J. (1994) Evolution of a plant homeotic multigene family: toward connecting molecular systematics and molecular developmental genetics. *Systematic Biology*, **43**, 307–28.

Doyle, J. J. (1997) Trees within trees: genes and species, molecules and morphology. *Systematic Biology*, **46**, 537–53.

Doyle, J. J. and Doyle, J. L. (1999) Nuclear protein-coding genes in phylogeny reconstruction and homology assessment: some examples from Leguminosae, in *Molecular systematics and plant evolution*, (eds P. M. Hollingsworth, R. M. Bateman and R. J. Gornall), Taylor & Francis, London, pp. 229–54.

Eernisse, D. J. and Kluge, A. G. (1993) Taxonomic congruence versus total evidence, and amniote phylogeny inferred from fossils, molecules, and morphology. *Molecular Biology and Evolution*, **10**, 1170–95.

Eldredge, N., ed. (1992) *Systematics, ecology, and the biodiversity crisis*, Columbia University Press, New York.

Eldredge, N. and Gould, S. J. (1972) Punctuated equilibria: an alternative to phyletic gradualism, in *Models in paleobiology*, (ed. T. J. M. Schopf), Freeman, San Francisco, pp. 82–115.

Erwin, D. H. (1992) A preliminary classification of evolutionary radiations. *Historical Biology*, **6**, 133–47.

Foote, M. (1996) Models of morphological diversification, in *Evolutionary paleobiology*, (eds D. Jablonski, D. H. Erwin and J. H. Lipps), Chicago University Press, Chicago, pp. 62–86.

Fortey, R. A. and Jefferies, R. P. S. (1982) Fossils and phylogeny – a compromise approach, in *Problems of phylogenetic reconstruction* (eds K. A. Joysey and A. E. Friday), Academic Press, London, pp. 197–234.

Francisco-Ortega, J., Crawford, D. J., Santos-Guerra, A. and Jansen, R. K. (1997) Origin and evolution of *Argyranthemum* (Asteraceae: Anthemideae) in Macaronesia, in *Molecular evolution and adaptive radiation*, (eds T. J. Givnish and K. J. Sytsma), Cambridge University Press, Cambridge, pp. 407–31.

Fusco, G. and Cronk, Q. C. B. (1995) A new method for evaluating the shape of large phylogenies. *Journal of Theoretical Biology*, **175**, 235–43.

Gerbaud, O. (1998) *Gymnadenia* R. Br. et *Nigritella* Rich.: un seul et même genre? *Cahors de la Societé Française Orchidée*, **4**, 80–93.

Gilinsky, N. L. and Bambach, R. K. (1987) Asymmetrical patterns of origination and extinction in higher taxa. *Paleobiology*, **13**, 427–45.

Givnish, T. J. (1997) Adaptive radiation and molecular systematics: issues and approaches, in *Molecular evolution and adaptive radiation*, (eds T. J. Givnish and K. J. Sytsma), Cambridge University Press, Cambridge, pp. 1–54.

Givnish, T. J. and Sytsma, K. J., eds (1997a) *Molecular evolution and adaptive radiation*, Cambridge University Press, Cambridge.

Givnish, T. J. and Sytsma, K. J. (1997b) Homoplasy in molecular vs. morphological data: the likelihood of correct phylogenetic inference, in *Molecular evolution and adaptive radiation*, (eds T. J. Givnish and K. J. Sytsma), Cambridge University Press, Cambridge, pp. 55–101.

Gould, S. J. and Eldredge, N. (1993) Punctuated equilibrium comes of age. *Nature*, **366**, 223–7.

Gould, S. J. and Lewontin, R. C. (1979) The spandrels of San Marco and the Panglossian paradigm: a critique of the adaptationist program, *Proceedings of the Royal Society of London*, **B205**, 581–98.

Gustafsson, M. H. G. and Albert, V. A. (1999) Inferior ovaries and angiosperm diversification, in *Molecular systematics and plant evolution*, (eds P. M. Hollingsworth, R. M. Bateman and R. J. Gornall), Taylor & Francis, London, pp. 403–31.

Guyer, C. and Slowinski, J. B. (1993) Adaptive radiation and the topology of large phylogenies. *Evolution*, **47**, 253–63.

Hallam, A. (1983) *Great geological controversies.* Oxford University Press, Oxford.

Hapeman, J. R. and Inoue, K. (1997) Plant–pollinator interactions and floral radiation in *Platanthera* (Orchidaceae), in *Molecular evolution and adaptive radiation*, (eds T. J. Givnish and K. J. Sytsma), Cambridge University Press, Cambridge, pp. 433–54.

Harvey, P. H. and Pagel, M. D. (1991) *The comparative method in evolutionary biology*, Oxford University Press, Oxford.

Heard, S. B. (1996) Patterns in phylogenetic tree balance with variable and evolving speciation rates. *Evolution*, **50**, 2141–48.

Hedrén, M. (1996) Genetic differentiation, polyploidization and hybridization in northern European *Dactylorhiza* (Orchidaceae): evidence from allozyme markers. *Plant Systematics and Evolution*, **201**, 31–55.

Hershkovitz, M. A., Zimmer, E. A. and Hahn, W. J. (1999) Ribosomal DNA sequences and angiosperm systematics, in *Molecular systematics and plant evolution*, (eds P. M. Hollingsworth, R. M. Bateman and R. J. Gornall), Taylor & Francis, London, pp. 268–326.

Hillis, D. M. (1987) Molecular versus morphological approaches to systematics. *Annual Review of Ecology and Systematics*, **18**, 23–42.

Hodges, S. A. (1997) Rapid radiation due to a key innovation in columbines (Ranunculaceae: *Aquilegia*), in *Molecular evolution and adaptive radiation*, (eds T. J. Givnish and K. J. Sytsma), Cambridge University Press, Cambridge, pp. 391–405.

Hollingsworth. P. M., Bateman, R. M., Tebbitt, M., Squirrell, J., Cameron, H. M. and Dickson, J. H. (1998) Population genetics and phylogenetic context of European *Epipactis* (Orchidaceae: Neottieae). *Abstracts of the Second International Conference on the Comparative Biology of the Monocotyledons*, University of New South Wales, Sydney, p. 29.

Huelsenbeck, J. P., Bull, J. J. and Cunningham, C. W. (1996) Combining data in phylogenetic analysis. *Trends in Ecology and Evolution*, **11**, 152–7.

Jackman, T., Losos, J. B., Larson, A. and de Queiroz, K. (1997) Phylogenetic studies of convergent adaptive radiations in Caribbean *Anolis* lizards, in *Molecular evolution and adaptive radiation*, (eds T. J. Givnish and K. J. Sytsma), Cambridge University Press, Cambridge, pp. 535–57.

Jackson, J. B. C. and Cheetham, A. H. (1994) Phylogeny reconstruction and the tempo of speciation in cheilostome bryozoa. *Paleobiology*, **20**, 407–23.

Jackson, J. B. C. and Cheetham, A. H. (1999) Tempo and mode of speciation in the sea. *Trends in Ecology and Evolution*, **14**, 72–7.

Jarne, P. and Lagoda, P. J. L. (1996) Microsatellites, from molecules to populations and back. *Trends in Ecology and Evolution*, **11**, 424–9.

Kenrick, P. and Crane, P. R. (1997) *The origin and early diversification of land plants: a cladistic study.* Smithsonian Series in Comparative Evolutionary Biology, Smithsonian Institution Press, Washington DC.

Kim, S. C., Crawford, D. J., Francisco-Ortega, J. and Santos-Guerra, A. (1996) A common origin for woody *Sonchus* and five related genera in the Macaronesian Islands: molecular evidence for extensive radiation. *Proceedings of the National Academy of Sciences USA*, **93**, 7743–8.

Kluge, A. G. (1998) Total evidence or taxonomic congruence: cladistics or consensus classi-fication. *Cladistics*, **14**, 151–8.

Knox, E. B. and Palmer, J. D. (1995) Chloroplast DNA variation and the recent radiation of the giant senecios (Asteraceae) on the tall mountains of eastern Africa. *Proceedings of the National Academy of Sciences USA*, **92**, 10349–53.

Larson, A. and Losos, J. B. (1996) Phylogenetic systematics of adaptation, in *Adaptation* (eds M. R. Rose and G. V. Lauder), Academic Press, London, pp. 187–220.

Lauder, G. V. (1990) Functional morphology and systematics: studying functional patterns in a historical context. *Annual Review of Ecology and Systematics*, **21**, 317–40.

Lauder, G. V. (1996) The argument from design, in *Adaptation* (eds M. R. Rose and G. V. Lauder), Academic Press, New York, pp. 55–91.

Lowrey, T. K. (1995) Phylogeny, adaptive radiation, and biogeography of Hawaiian *Tetramolopium* (Asteraceae, Astereae), in *Hawaiian biogeography: evolution on a hot spot archipelago*, (eds W. L. Wagner and V. A. Funk), Smithsonian Institution, Washington, DC, pp. 195–220.

MacArthur, R. H. and Wilson, E. O. (1967) *The theory of island biogeography*, Monographs in Population Biology 1. Princeton University Press, Princeton, NJ.

Miyamoto, M. M. and Fitch, W. M. (1995) Testing species phylogenies and phylogenetic methods with congruence. *Systematic Biology*, **44**, 64–76.

Möller, M., Clokie, M., Cubas, P. and Cronk, Q. C. B. (1999) Integrating molecular phylo-genies and developmental genetics: a Gesneriaceae case study, in *Molecular systematics and plant evolution*, (eds P. M. Hollingsworth, R. M. Bateman and R. J. Gornall), Taylor & Francis, London, pp. 375–402.

Molvray, M., Kores, P. J. and Chase, M. W. (1998) Saprophytism in Orchidaceae. *Abstracts of the Second International Conference on the Comparative Biology of the Monocotyledons*, University of New South Wales, Sydney, p. 39.

Nee, S., Mooers, A. Ø. and Harvey, P. H. (1992) Tempo and mode of evolution revealed from molecular phylogenies. *Proceedings of the National Academy of Sciences USA*, **89**, 8322–6.

Nickrent, D. L., Duff, R. J., Colwell, A. E., Wolfe, A. D., Young, N. D., Steiner, K. E. *et al.* (1998) Molecular phylogenetic and evolutionary studies of parasitic plants, in *Molecular systematics of plants II: DNA sequencing*, (eds D. E. Soltis, P. S. Soltis and J. J. Doyle), Kluwer, New York, pp. 211–41.

Nitecki, M., ed. (1990) *Evolutionary innovations*, Chicago University Press, Chicago.

Nixon, K. C. and Carpenter, J. M. (1996) On simultaneous analysis. *Cladistics*, **12**, 221–41.

Norrell, M. A. (1992) Taxic origin and temporal diversity: the effect of phylogeny, in *Extinction and phylogeny*, (eds M. J. Novacek and Q. D. Wheeler), Columbia University Press, New York, pp. 89–118.

Okada, M., Whitkus, R. and Lowrey, T. K. (1997) Genetics of adaptive radiation in Hawaiian and Cook Island species of *Tetramolopium* (Asteraceae; Astereae). I. Nuclear RFLP marker diversity. *American Journal of Botany*, **84**, 1236–46.

Olmstead, R. G. and Palmer, J. D. (1994) Chloroplast DNA systematics: a review of methods and data analysis. *American Journal of Botany*, **81**, 1205–24.

Patterson, C. (1988) Homology in classical and molecular biology. *Molecular Biology and Evolution*, **5**, 603–25.

Patterson, C. (1999) *Evolution*, 2nd edn., Natural History Museum, London.

Patterson, C., Williams, D. M. and Humphries, C. J. (1993) Congruence between molecular and morphological phylogenies. *Annual Review of Ecology and Systematics*, **24**, 153–88.

Pennington, R. T. (1996) Molecular and morphological data provide resolution at different levels in *Andira*. *Systematic Biology*, **45**, 496–515.

Penny, D., Lockhart, P. J., Steel, M. A. and Hendy, M. D. (1994) The role of models in reconstructing evolutionary trees, in *Models in phylogeny reconstruction*, (eds R. W. Scotland, D. M. Williams and D. J. Siebert), Oxford University Press, Oxford, pp. 211–30.

Pimm, S. L. (1991) *The balance of nature?* Chicago University Press, Chicago.

Pridgeon, A. M., Bateman, R. M., Cox, A. V., Hapeman, J. R. and Chase, M. W. (1997) Phylogenetics of subtribe Orchidinae (Orchidoideae, Orchidaceae) based on nuclear ITS sequences. 1. Intergeneric relationships and polyphyly of *Orchis sensu lato*. *Lindleyana*, **12**, 89–109.

Quammen, D. (1996) *The song of the dodo: island biogeography in an age of extinctions*, Pimlico, London.

Reeves, P. A. and Olmstead, R. G. (1998) Evolution of novel morphological traits in a clade containing *Antirrhinum majus* (Scrophulariaceae). *American Journal of Botany*, **85**, 1047–56.

Ricklefs, R. E. (1996) Phylogeny and ecology. *Trends in Ecology and Evolution*, **11**, 229–30.

Rose, M. R. and Lauder, G. V., eds (1996) *Adaptation*, Academic Press, London.

Rosenzweig, M. L. and McCord, R. D. (1991) Incumbent replacement: evidence for long-term evolutionary progress. *Paleobiology*, **17**, 202–13.

Sakai, A. K., Weller, S. G., Wagner, W. L., Soltis, P. S. and Soltis, D. E. (1997). Phylogenetic perspectives on the evolution of dioecy: adaptive radiation in the endemic Hawaiian genera *Scheidea* and *Alsinodendron* (Caryophyllaceae: Alsinoideae), in *Molecular evolution and adaptive radiation*, (eds T. J. Givnish and K. J. Sytsma), Cambridge University Press, Cambridge, pp. 455–73.

Sanderson, M. J. (1998a). Reappraising adaptive radiation. *American Journal of Botany*, **85**, 1650–5.

Sanderson, M. J. (1998b). Estimating rate and time in molecular phylogenies: beyond the molecular clock, in *Molecular systematics of plants II: DNA sequencing* (eds D. E. Soltis, P. S. Soltis and J. J. Doyle), Kluwer, New York, pp. 242–64.

Sanderson, M. J. and Donoghue, M. J. (1989). Patterns of variation in levels of homoplasy. *Evolution*, **43**, 1781–95.

Sanderson, M. J. and Donoghue, M. J. (1994) Shifts in diversification rate with the origin of the angiosperms. *Science*, **264**, 1590–3.

Sanderson, M. J. and Donoghue, M. J. (1996a). Reconstructing shifts in diversification rates on phylogenetic trees. *Trends in Ecology and Evolution*, **11**, 15–20.

Sanderson, M. J. and Donoghue, M. J. (1996b). The relationship between homoplasy and confidence in a phylogenetic tree, in *Homoplasy: the recurrence of similarity in evolution*, (eds M. J. Sanderson and L. Hufford), Academic Press, London, pp. 67–89.

Sanderson, M. J. and Hufford, L., eds (1996) *Homoplasy: the recurrence of similarity in evolution*, Academic Press, London.

Scotland, R. W. and Pennington, R. T., eds (1999) *Homology in systematics*, Taylor & Francis, London.

Setoguchi, H., Ohba, H. and Tobe, H. (1998) Evolution in *Crossostylis* (Rhizophoraceae) on the South Pacific Islands, in *Evolution and speciation of island plants*, (eds T. F. Steussy and M. Ono), Cambridge University Press, Cambridge, pp. 203–29.

Silvertown, J., Franco, M. and Harper, J. L., eds (1997) *Plant life histories: ecology, phylogeny and evolution*, Cambridge University Press, Cambridge.

Slack, J. M. W., Holland, P. W. H. and Graham, C. F. (1993) The zootype and the phylo-typic stage. *Nature*, **361**, 490–2.

Smith, A. B. and Patterson, C. (1988) The influence of taxonomy on the perception of patterns of evolution. *Evolutionary Biology*, **23**, 127–216.

Soltis, D. E., Soltis, P. S., Mort, M. E., Chase, M. W., Savolainen, V., Hoot, S. B. and Morton, C. M. (1998a) Inferring complex phylogenies using parsimony: an empirical approach using three large DNA data sets for angiosperms. *Systematic Biology*, **47**, 32–42.

Soltis, D. E., Soltis, P. S. and Doyle, J. J. (eds) (1998b). *Molecular systematics of plants II: DNA sequencing*, Kluwer, New York.

Steussy, T. F. and Ono, M., eds (1998) *Evolution and speciation of island plants*, Cambridge University Press, Cambridge.

Stoebe, B., Hansmann, S., Goremykin, V., Kowallik, K. V. and Martin, W. (1999) Proteins encoded in sequenced chloroplast genomes: an overview of gene content, phylogenetic information and endosymbiotic transfer to the nucleus, in *Molecular systematics and plant evolution*, (eds P. M. Hollingsworth, R. M. Bateman and R. J. Gornall), Taylor & Francis, London, pp. 327–52.

Strack, D., Busch, E. and Klein, E. (1989) Anthocyanin patterns in European orchids and their taxonomic and phylogenetic relevance. *Phytochemistry*, **28**, 2127–39.

Sundermann, H. (1980) *Europäische und Mediterrane Orchideen – ein Bestimmungsflora* (3rd edn). Schmersow, Hildesheim.

Swofford, D. L. (1991) When are phylogeny estimates from molecular and morphological data incongruent?, in *Phylogenetic analysis of DNA sequences*, (eds M. Miyamoto and J. Cracraft), Oxford University Press, Oxford, pp. 295–333.

Teppner, H. and Klein, E. (1985) Karyologie und Fortpflanzungsmodus von *Nigritella* (Orchidaceae, Orchideae). *Phyton*, **25**, 147–76.

Thiessen, G., Kim, J. T. and Saedler, H. (1996) Classification and phylogeny of the MADS-box gene subfamilies in the morphological evolution of eukaryotes. *Journal of Molecular Evolution*, **10**, 484–516.

Thornton, I. (1996) *Krakatau: the destruction and reassembly of an island ecosystem*, Harvard University Press, Boston, Mass.

Van De Peer, Y. and De Wachter, R. (1997) Evolutionary relationships among the eukaryotic crown taxa, taking into account site-to-site rate variation in 18S rRNA. *Journal of Molecular Evolution*, **45**, 619–30.

Wagner, W. L. and Funk, V. A., eds (1995) *Hawaiian biogeography: evolution on a hot spot archipelago*, Smithsonian Institution, Washington DC.

Walter, K. S. and Gillett, H. J., eds (1998) *1997 IUCN red list of threatened plants*. WCMC, Cambridge, UK and IUCN, Gland, Swizerland.

Welsh, J. (1997) *Morphological and molecular cladistic analysis of the* Bencomia *alliance (Rosaceae: Poteriae): adaptive radiation on the Macaronesian Islands*. Unpublished MSc thesis, Royal Botanic Garden Edinburgh/University of Edinburgh, UK.

Whitlock, M. C., Phillips, P. C., Moore, F. B.-G. and Tonsor, S. J. (1995) Multiple fitness peaks and epistasis. *Annual Review of Ecology and Systematics*, **26**, 601–29.

Wikström, N. and Kenrick, P. (1997) Phylogeny of Lycopodiaceae (Lycopsida) and the relationships of *Phylloglossum drummondii* Kunze based on *rbc*L sequences. *International Journal of Plant Sciences*, **158**, 862–71.

Wolfe, K. H., Morden, C. W. and Palmer, J. D. (1992) Function and evolution of a minimal plastid genome from a nonphotosynthetic parastitic plant. *Proceedings of the National Academy of Sciences USA*, **89**, 10648–52.

Wray, G. A. (1995) Punctuated evolution of embryos. *Science*, **267**, 1115–6.

Wright, S. (1968) *Evolution and the genetics of populations*, Chicago University Press, Chicago.

Yeates, D. (1992) Why remove autapomorphies? *Cladistics*, **8**, 387–9.

Taxon Index

Subject Index

Systematics Association Publications

1. Bibliography of key works for the identification of the British fauna and flora, 3rd edition (1967)[†]
Edited by G.J. Kerrich, R.D. Meikle and N. Tebble
2. Function and taxonomic importance (1959)[†]
Edited by A.J. Cain
3. The species concept in palaeontology (1956)[†]
Edited by P.C. Sylvester-Bradley
4. Taxonomy and geography (1962)[†]
Edited by D. Nichols
5. Speciation in the sea (1963)[†]
Edited by J.P. Harding and N. Tebble
6. Phenetic and phylogenetic classification (1964)[†]
Edited by V.H. Heywood and J. McNeill
7. Aspects of Tethyan biogeography (1967)[†]
Edited by C.G. Adams and D.V. Ager
8. The soil ecosystem (1969)[†]
Edited by H. Sheals
9. Organisms and continents through time (1973)[†]
Edited by N.F. Hughes
10. Cladistics: a practical course in systematics (1992)[*]
P.L. Forey, C.J. Humphries, I.J. Kitching, R.W. Scotland, D.J. Siebert and D.M. Williams
11. Cladistics: the theory and practice of parsimony analysis (2nd edition) (1998)[*]
I.J. Kitching, P.L. Forey, C.J. Humphries and D.M. Williams

[*]Published by Oxford University Press for the Systematics Association
[†]Published by the Association (out of print)

Systematics Association Special Volumes

1. The new systematics (1940)
Edited by J.S. Huxley (reprinted 1971)
2. Chemotaxonomy and serotaxonomy (1968)[*]
Edited by J.G. Hawkes
3. Data processing in biology and geology (1971)[*]
Edited by J.L. Cutbill
4. Scanning electron microscopy (1971)[*]
Edited by V.H. Heywood
5. Taxonomy and ecology (1973)[*]
Edited by V.H. Heywood
6. The changing flora and fauna of Britain (1974)[*]
Edited by D.L. Hawksworth

7. Biological identification with computers (1975)*
Edited by R.J. Pankhurst
8. Lichenology: progress and problems (1976)*
Edited by D.H. Brown, D.L. Hawksworth and R.H. Bailey
9. Key works to the fauna and flora of the British Isles and northwestern Europe, 4th edition (1978)*
Edited by G.J. Kerrich, D.L. Hawksworth and R.W. Sims
10. Modern approaches to the taxonomy of red and brown algae (1978)
Edited by D.E.G. Irvine and J.H. Price
11. Biology and systematics of colonial organisms (1979)*
Edited by G. Larwood and B.R. Rosen
12. The origin of major invertebrate groups (1979)*
Edited by M.R. House
13. Advances in bryozoology (1979)*
Edited by G.P. Larwood and M.B. Abbott
14. Bryophyte systematics (1979)*
Edited by G.C.S. Clarke and J.G. Duckett
15. The terrestrial environment and the origin of land vertebrates (1980)
Edited by A.L. Pachen
16. Chemosystematics: principles and practice (1980)*
Edited by F.A. Bisby, J.G. Vaughan and C.A. Wright
17. The shore environment: methods and ecosystems (2 volumes) (1980)*
Edited by J.H. Price, D.E.G. Irvine and W.F. Farnham
18. The Ammonoidea (1981)*
Edited by M.R. House and J.R. Senior
19. Biosystematics of social insects (1981)*
Edited by P.E. House and J.-L. Clement
20. Genome evolution (1982)*
Edited by G.A. Dover and R.B. Flavell
21. Problems of phylogenetic reconstruction (1982)*
Edited by K.A. Joysey and A.E. Friday
22. Concepts in nematode systematics (1983)*
Edited by A.R. Stone, H.M. Platt and L.F. Khalil
23. Evolution, time and space: the emergence of the biosphere (1983)*
Edited by R.W. Sims, J.H. Price and P.E.S. Whalley
24. Protein polymorphism: adaptive and taxonomic significance (1983)*
Edited by G.S. Oxford and D. Rollinson
25. Current concepts in plant taxonomy (1983)*
Edited by V.H. Heywood and D.M. Moore
26. Databases in systematics (1984)*
Edited by R. Allkin and F.A. Bisby
27. Systematics of the green algae (1984)*
Edited by D.E.G. Irvine and D.M. John
28. The origins and relationships of lower invertebrates (1985)[‡]
Edited by S. Conway Morris, J.D. George, R. Gibson and H.M. Platt
29. Infraspecific classification of wild and cultivated plants (1986)[‡]
Edited by B.T. Styles

30. Biomineralization in lower plants and animals (1986)[‡]
Edited by B.S.C. Leadbeater and R. Riding
31. Systematic and taxonomic approaches in palaeobotany (1986)[‡]
Edited by R.A. Spicer and B.A. Thomas
32. Coevolution and systematics (1986)[‡]
Edited by A.R. Stone and D.L. Hawksworth
33. Key works to the fauna and flora of the British Isles and northwestern Europe, 5th edition (1988)[‡]
Edited by R.W. Sims, P. Freeman and D.L. Hawksworth
34. Extinction and survival in the fossil record (1988)[‡]
Edited by G.P. Larwood
35. The phylogeny and classification of the tetrapods (2 volumes) (1988)[‡]
Edited by M.J. Benton
36. Prospects in systematics (1988)[‡]
Edited by D.L. Hawksworth
37. Biosystematics of haematophagous insects (1988)[‡]
Edited by M.W. Service
38. The chromophyte algae: problems and perspective (1989)[‡]
Edited by J.C. Green, B.S.C. Leadbeater and W.L. Diver
39. Electrophoretic studies on agricultural pests (1989)[‡]
Edited by H. D. Loxdale and J. den Hollander
40. Evolution, systematics, and fossil history of the Hamamelidae (2 volumes) (1989)[‡]
Edited by P. R. Crane and S. Blackmore
41. Scanning electron microscopy in taxonomy and functional morphology (1990)[‡]
Edited by D. Claugher
42. Major evolutionary radiations (1990)[‡]
Edited by P.D. Taylor and G.P. Larwood
43. Tropical lichens: their systematics, conservation and ecology (1991)[‡]
Edited by G.J. Galloway
44. Pollen and spores: patterns of diversification (1991)[‡]
Edited by S. Blackmore and S.H. Barnes
45. The biology of free-living heterotrophic flagellates (1991)[‡]
Edited by D.J. Patterson and J. Larsen
46. Plant-animal interactions in the marine benthos (1992)[‡]
Edited by D.M. John, S.J. Hawkins and J.H. Price
47. The Ammonoidea: environment, ecology and evolutionary change (1993)[‡]
Edited by M.R. House
48. Designs for a global plant species information system (1993)[‡]
Edited by F.A. Bisby, G.F. Russell and R.J. Pankhurst
49. Plant galls: organisms, interactions, populations (1994)[‡]
Edited by M.A.J. Williams
50. Systematics and conservation evaluation (1994)[‡]
Edited by P.L. Forey, C.J. Humphries and R.I. Vane-Wright
51. The Haptophyte algae (1994)[‡]
Edited by J.C. Green and B.S.C. Leadbeater

[*] Published by Academic Press for the Systematics Association
[†] Published by the Palaeontological Association in conjunction with Systematics Association
[‡] Published by the Oxford University Press for the Systematics Association
[**] Published by Chapman & Hall for the Systematics Association

Printed and bound by CPI Group (UK) Ltd, Croydon, CR0 4YY

04/11/2024

01783523-0001